High Performance Computing in Remote Sensing

CHAPMAN & HALL/CRC COMPUTER and INFORMATION SCIENCE SERIES

Series Editor: Sartaj Sahni

PUBLISHED TITLES

ADVERSARIAL REASONING: COMPUTATIONAL APPROACHES TO READING THE OPPONENT'S MIND
Alexander Kott and William M. McEneaney

DISTRIBUTED SENSOR NETWORKS
S. Sitharama Iyengar and Richard R. Brooks

DISTRIBUTED SYSTEMS: AN ALGORITHMIC APPROACH
Sukumar Ghosh

FUNDEMENTALS OF NATURAL COMPUTING: BASIC CONCEPTS, ALGORITHMS, AND APPLICATIONS
Leandro Nunes de Castro

HANDBOOK OF ALGORITHMS FOR WIRELESS NETWORKING AND MOBILE COMPUTING
Azzedine Boukerche

HANDBOOK OF APPROXIMATION ALGORITHMS AND METAHEURISTICS
Teofilo F. Gonzalez

HANDBOOK OF BIOINSPIRED ALGORITHMS AND APPLICATIONS
Stephan Olariu and Albert Y. Zomaya

HANDBOOK OF COMPUTATIONAL MOLECULAR BIOLOGY
Srinivas Aluru

HANDBOOK OF DATA STRUCTURES AND APPLICATIONS
Dinesh P. Mehta and Sartaj Sahni

HANDBOOK OF DYNAMIC SYSTEM MODELING
Paul A. Fishwick

HANDBOOK OF REAL-TIME AND EMBEDDED SYSTEMS
Insup Lee, Joseph Y-T. Leung, and Sang H. Son

HANDBOOK OF SCHEDULING: ALGORITHMS, MODELS, AND PERFORMANCE ANALYSIS
Joseph Y.-T. Leung

HIGH PERFORMANCE COMPUTING IN REMOTE SENSING
Antonio J. Plaza and Chein-I Chang

THE PRACTICAL HANDBOOK OF INTERNET COMPUTING
Munindar P. Singh

SCALABLE AND SECURE INTERNET SERVICES AND ARCHITECTURE
Cheng-Zhong Xu

SPECULATIVE EXECUTION IN HIGH PERFORMANCE COMPUTER ARCHITECTURES
David Kaeli and Pen-Chung Yew

High Performance Computing in Remote Sensing

Edited by

Antonio J. Plaza

University of Extremadura
Caceres, Spain

Chein-I Chang

University of Maryland, Baltimore County
Baltimore, MD, U.S.A.

National Chung Hsing University
Taichung, Taiwan, Republic of China

CRC Press is an imprint of the
Taylor & Francis Group, an **informa** business
A CHAPMAN & HALL BOOK

CRC Press
Taylor & Francis Group
6000 Broken Sound Parkway NW, Suite 300
Boca Raton, FL 33487-2742

First issued in paperback 2019

ISBN-13: 978-1-58488-662-4 (hbk)
ISBN-13: 978-0-367-38847-8 (pbk)

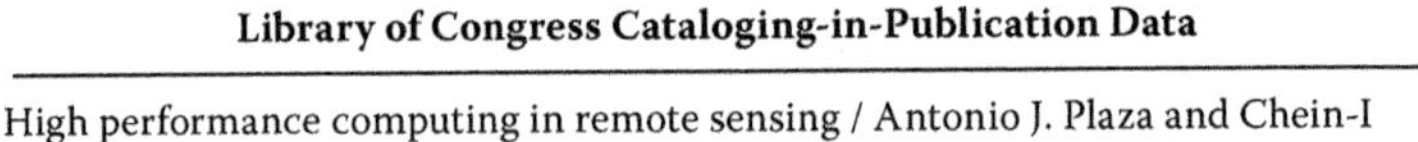

Library of Congress Cataloging-in-Publication Data

High performance computing in remote sensing / Antonio J. Plaza and Chein-I Chang, editors.
p. cm. -- (Chapman & Hall/CRC computer & information science series)
Includes bibliographical references and index.
ISBN 978-1-58488-662-4 (alk. paper)
1. High performance computing. 2. Remote sensing. I. Plaza, Antonio J. II. Chang, Chein-I. III. Title. IV. Series.

QA76.88.H5277 2007
621.36'78028543--dc22 2007020736

Visit the Taylor & Francis Web site at
http://www.taylorandfrancis.com

and the CRC Press Web site at
http://www.crcpress.com

Contents

List of Tables

List of Figures

Acknowledgments

The editors would like to thank all the contributors for all their help and support during the production of this book, and for sharing their vast knowledge with readers. In particular, Profs. Javier Plaza and David Valencia are gratefully acknowledged for their help in the preparation of some of the chapters of this text. Last but not least, the editors gratefully thank their families for their support on this project.

About the Editors

Antonio Plaza received the M.S. degree and the Ph.D. degree in computer engineering from the University of Extemadura, Spain, where he was awarded the outstanding Ph.D. dissertation award in 2002. Dr. Plaza is an associate professor with the Department of Technology of Computers and Communications at University of Extremadura. He has authored or co-authored more than 140 scientific publications including journal papers, book chapters, and peer-reviewed conference proceedings. His main research interests comprise remote sensing, image and signal processing, and efficient implementations of large-scale scientific problems on high-performance computing architectures, including commodity Beowulf clusters, heterogeneous networks of workstations, grid computing facilities, and hardware-based computer architectures such as field-programmable gate arrays (FPGAs) and graphics processing units (GPUs).

He has held visiting researcher positions at several institutions, including the Computational and Information Sciences and Technology Office (CISTO) at NASA/Goddard Space Flight Center, Greenbelt, Maryland; the Remote Sensing, Signal and Image Processing Laboratory (RSSIPL) at the Department of Computer Science and Electrical Engineering, University of Maryland, Baltimore County; the Microsystems Laboratory at the Department of Electrical & Computer Engineering, University of Maryland, College Park; and the AVIRIS group at NASA/Jet Propulsion Laboratory, Pasadena, California.

Dr. Plaza is a senior member of the IEEE. He is active in the IEEE Computer Society and the IEEE Geoscience and Remote Sensing Society, and has served as proposal evaluator for the European Commission, the European Space Agency, and the Spanish Ministry of Science and Education. He is also a frequent manuscript reviewer for more than 15 highly-cited journals (including several IEEE *Transactions*) in the areas of computer architecture, parallel/distributed systems, remote sensing, neural networks, image/signal processing, aerospace and engineering systems, and pattern analysis. He is also a member of the program committee of several international conferences, such as the European Conference on Parallel and Distributed Computing; the International Workshop on Algorithms, Models and Tools for Parallel Computing on Heterogeneous Networks; the Euromicro Workshop on Parallel and Distributed Image Processing, Video Processing, and Multimedia; the Workshop on Grid Computing Applications Development; the IEEE GRSS/ASPRS Joint Workshop on Remote Sensing and Data Fusion over Urban Areas; and the IEEE International Geoscience and Remote Sensing Symposium.

Dr. Plaza is the project coordinator of HYPER-I-NET (Hyperspectral Imaging Network), a four-year Marie Curie Research Training Network (see http://www.hyperinet.eu) designed to build an interdisciplinary European research community focused on remotely sensed hyperspectral imaging. He is guest editor (with Prof. Chein-I Chang) of a special issue on high performance computing for hyperspectral imaging for the the *International Journal of High Performance Computing Applications*. He is associate editor for the *IEEE Transactions on Geoscience and Remote Sensing* journal in the areas of *Hyperspectral Image Analysis and Signal Processing*. Additional information is available at http://www.umbc.edu/rssipl/people/aplaza.

Chein-I Chang received his B.S. degree from Soochow University, Taipei, Taiwan; the M.S. degree from the Institute of Mathematics at National Tsing Hua University, Hsinchu, Taiwan; and the M.A. degree from the State University of New York at Stony Brook, all in mathematics. He also received his M.S., and M.S.E.E. degrees from the University of Illinois at Urbana-Champaign and the Ph.D. degree in electrical engineering from the University of Maryland, College Park.

Dr. Chang has been with the University of Maryland, Baltimore County (UMBC) since 1987 and is currently professor in the Department of Computer Science and Electrical Engineering. He was a visiting research specialist in the Institute of Information Engineering at the National Cheng Kung University, Tainan, Taiwan, from 1994 to 1995. He received an NRC (National Research Council) senior research associateship award from 2002 to 2003 sponsored by the U.S. Army Soldier and Biological Chemical Command, Edgewood Chemical and Biological Center, Aberdeen Proving Ground, Maryland. Additionally, Dr. Chang was a distinguished lecturer chair at the National Chung Hsing University sponsored by the Ministry of Education in Taiwan from 2005 to 2006 and is currently holding a chair professorship of diaster reduction technology from 2006 to 2009 with the Environmental Restoration and Disaster Reduction Research Center, National Chung Hsing University, Taichung, Taiwan, ROC.

He has three patents and several pending on hyperspectral image processing. He is on the editorial board of the *Journal of High Speed Networks* and was an associate editor in the area of hyperspectral signal processing for *IEEE Transactions on Geoscience and Remote Sensing*. He was the guest editor of a special issue of the *Journal of High Speed Networks* on telemedicine and applications and co-guest edits three special issues on Broadband Multimedia Sensor Networks in Healthcare Applications for the *Journal of High Speed Networks*, 2007 and on high-performance computing for hyperspectral imaging for the *International Journal of High Performance Computing Applications*.

Dr. Chang is the author of *Hyperspectral Imaging: Techniques for Spectral Detection and Classification* published by Kluwer Academic Publishers in 2003 and the editor of two books, *Recent Advances in Hyperspectral Signal and Image Processing*, Trivandrum, Kerala: Research Signpost, Trasworld Research Network, India, 2006, and *Hyperspectral Data Exploitation: Theory and Applications*, John Wiley & Sons, 2007. Dr. Chang is currently working on his second book, *Hyperspectral Imaging: Algorithm Design and Analysis*, John Wiley & Sons due 2007. He is a Fellow of the SPIE and a member of Phi Kappa Phi and Eta Kappa Nu. Additional information is available at http://www.umbc.edu/rssipl.

Contributors

Giovanni Aloisio, *Euromediterranean Center for Climate Change & University of Salento, Italy*
Gregory P. Asner, *Carnegie Institution of Washington, Stanford, California*
José I. Benavides, *University of Córdoba, Spain*
Jeffrey H. Bowles, *Naval Research Laboratory, Washington, DC*
Massimo Cafaro, *Euromediterranean Center for Climate Change & University of Salento, Italy*
Chein-I Chang, *University of Maryland Baltimore County, Baltimore, Maryland*
Roberto Cossu, *European Space Agency, ESA-Esrin, Italy*
Qian Du, *Missisipi State University, Mississippi*
Esam El-Araby, *George Washington University, Washington, DC*
Tarek El-Ghazawi, *George Washington University, Washington, DC*
Italo Epicoco, *Euromediterranean Center for Climate Change & University of Salento, Italy*
Sandro Fiore, *Euromediterranean Center for Climate Change & University of Salento, Italy*
Luigi Fusco, *European Space Agency, ESA-Esrin, Italy*
Samuel D. Gasster, *The Aerospace Corporation, El Segundo, California*
David Gillis, *Naval Research Laboratory, Washington, DC*
José González-Mora, *University of Málaga, Spain*
Robert O. Green, *Jet Propulsion Laboratory & California Institute of Technology, California*
Nicolás Guil, *University of Málaga, Spain*
Robert S. Haxo, *Carnegie Institution of Washington, Stanford, California*
Luis O. Jiménez-Rodríguez, *University of Puerto Rico at Mayaguez*
David E. Knapp, *Carnegie Institution of Washington, Stanford, California*
Craig A. Lee, *The Aerospace Corporation, El Segundo, California*
Jacqueline Le Moigne, *NASA's Goddard Space Flight Center, Greenbelt, Maryland*
Pablo Martínez, *University of Extremadura, Cáceres, Spain*
James W. Palko, *The Aerospace Corporation, El Segundo, California*
Rosa Pérez, *University of Extremadura, Cáceres, Spain*
Antonio Plaza, *University of Extremadura, Cáceres, Spain*
Javier Plaza, *University of Extremadura, Cáceres, Spain*
Manuel Prieto, *Complutense University of Madrid, Spain*
Gianvito Quarta, *Institute of Atmospheric Sciences and Climate, CNR, Bologna, Italy*
Christian Retscher, *European Space Agency, ESA-Esrin, Italy*
Wilson Rivera-Gallego, *University of Puerto Rico at Mayaguez, Puerto Rico*
Edmundo Sáez, *University of Córdoba, Spain*
Javier Setoain, *Complutense University of Madrid, Spain*
Mohamed Taher, *George Washington University, Washington, DC*
Christian Tenllado, *Complutense University of Madrid, Spain*
James C. Tilton, *NASA Goddard Space Flight Center, Greenbelt, Maryland*

Francisco Tirado, *Complutense University of Madrid, Spain*
David Valencia, *University of Extremadura, Cáceres, Spain*
Miguel Vélez-Reyes, *University of Puerto Rico at Mayaguez, Puerto Rico*
Jianwei Wang, *University of Maryland Baltimore County, Baltimore, Maryland*
Emilio L. Zapata, *University of Málaga, Spain*

Chapter 1

Introduction

Antonio Plaza
University of Extremadura, Spain

Chein-I Chang
University of Maryland, Baltimore County

Contents

1.1 Preface

Advances in sensor technology are revolutionizing the way remotely sensed data are collected, managed, and analyzed. The incorporation of latest-generation sensors to airborne and satellite platforms is currently producing a nearly continual stream of high-dimensional data, and this explosion in the amount of collected information has rapidly created new processing challenges. In particular, many current and future applications of remote sensing in Earth science, space science, and soon in exploration science require real- or near-real-time processing capabilities. Relevant examples include environmental studies, military applications, tracking and monitoring of hazards such as wild land and forest fires, oil spills, and other types of chemical/biological contamination.

To address the computational requirements introduced by many time-critical applications, several research efforts have been recently directed towards the incorporation of high-performance computing (HPC) models in remote sensing missions. HPC is an integrated computing environment for solving large-scale computational demanding problems such as those involved in many remote sensing studies. With the aim of providing a cross-disciplinary forum that will foster collaboration and development in those areas, this book has been designed to serve as one of the first available references specifically focused on describing recent advances in the field of HPC applied to remote sensing problems. As a result, the content of the book has been organized to appeal to both remote sensing scientists and computer engineers alike. On the one hand, remote sensing scientists will benefit by becoming aware of the extremely high computational requirements introduced by most application areas in Earth and space observation. On the other hand, computer engineers will benefit from the wide range of parallel processing strategies discussed in the book. However, the material presented in this book will also be of great interest to researchers and practitioners working in many other scientific and engineering applications, in particular, those related with the development of systems and techniques for collecting, storing, and analyzing extremely high-dimensional collections of data.

1.2 Contents

The contents of this book have been organized as follows. First, an introductory part addressing some key concepts in the field of computing applied to remote sensing, along with an extensive review of available and future developments in this area, is provided. This part also covers other application areas not necessarily related to remote sensing, such as multimedia and video processing, chemical/biological standoff detection, and medical imaging. Then, three main application-oriented parts follow, each of which illustrates a specific parallel computing paradigm. In particular, the HPC-based techniques comprised in these parts include multiprocessor (cluster-based) systems, large-scale and heterogeneous networks of computers, and specialized hardware architectures for remotely sensed data analysis and interpretation. Combined, the four parts deliver an excellent snapshot of the state-of-the-art in those areas, and offer a thoughtful perspective of the potential and emerging challenges of applying HPC paradigms to remote sensing problems:

- *Part I: General.* This part, comprising Chapters 2 and 3, develops basic concepts about HPC in remote sensing and provides a detailed review of existing and planned HPC systems in this area. Other areas that share common aspects with remote sensing data processing are also covered, including multimedia and video processing.
- *Part II: Multiprocessor systems.* This part, comprising Chapters 4–8, includes a compendium of algorithms and techniques for HPC-based remote sensing data analysis using multiprocessor systems such as clusters and networks of computers, including massively parallel facilities.
- *Part III: Large-scale and heterogeneous distributed computing.* The focus of this part, which comprises Chapters 9–13, is on parallel techniques for remote sensing data analysis using large-scale distributed platforms, with special emphasis on grid computing environments and fully heterogeneous networks of workstations.

- *Part IV: Specialized architectures.* The last part of this book comprises Chapters 14–18 and is devoted to systems and architectures for at-sensor and real-time collection and analysis of remote sensing data using specialized hardware and embedded systems. The part also includes specific aspects about current trends in remote sensing sensor design and operation.

1.2.1 Organization of Chapters in This Volume

The first part of the book (*General*) consists of two chapters that include basic concepts that will appeal to both students and practitioners who have not had a formal education in remote sensing and/or computer engineering. This part will also be of interest to remote sensing and general-purpose HPC specialists, who can greatly benefit from the exhaustive review of techniques and discussion on future data processing perspectives in this area. Also, general-purpose specialists will become aware of other application areas of HPC (e.g., multimedia and video processing) in which the design of techniques and parallel processing strategies to deal with extremely large computational requirements follows a similar pattern as that used to deal with remotely sensed data sets. On the other hand, the three application-oriented parts that follow (*Multiprocessor systems*, *Large-scale and heterogeneous distributed computing*, and *Specialized architectures*) are each composed of five selected chapters that will appeal to the vast scientific community devoted to designing and developing efficient techniques for remote sensing data analysis. This includes commercial companies working on intelligence and defense applications, Earth and space administrations such as NASA or the European Space Agency (ESA) – both of them represented in the book via several contributions – and universities with programs in remote sensing, Earth and space sciences, computer architecture, and computer engineering. Also, the growing interest in some emerging areas of remote sensing such as hyperspectral imaging (which will receive special attention in this volume) should make this book a timely reference.

1.2.2 Brief Description of Chapters in This Volume

We provide below a description of the chapters contributed by different authors. It should be noted that all the techniques and methods presented in those chapters are well consolidated and cover almost entirely the spectrum of current and future data processing techniques in remote sensing applications. We specifically avoided repetition of topics in order to complete a timely compilation of realistic and successful efforts in the field. Each chapter was contributed by a reputed expert or a group of experts in the designed specialty areas. A brief outline of each contribution follows:

- **Chapter 1. Introduction.** The present chapter provides an introduction to the book and describes the main innovative contributions covered by this volume and its individual chapters.

- **Chapter 2. High-Performance Computer Architectures for Remote Sensing Data Analysis: Overview and Case Study.** This chapter provides a review of the state-of-the-art in the design of HPC systems for remote sensing. The chapter also includes an application case study in which the pixel purity index (PPI), a well-known remote sensing data processing algorithm included in Kodak's Research Systems ENVI (a very popular remote sensing-oriented commercial software package), is implemented using different types of HPC platforms such as a massively parallel multiprocessor, a heterogeneous network of distributed computers, and a specialized hardware architecture.
- **Chapter 3. Computer Architectures for Multimedia and Video Analysis.** This chapter focuses on multimedia processing as another example application with a high demanding computational power and similar aspects as those involved in many remote sensing problems. In particular, the chapter discusses new computer architectures such as graphic processing units (GPUs) and multimedia extensions in the context of real applications.
- **Chapter 4. Parallel Implementation of the ORASIS Algorithm for Remote Sensing Data Analysis.** This chapter presents a parallel version of ORASIS (the Optical Real-Time Adaptive Spectral Identification System) that was recently developed as part of a U.S. Department of Defense program. The ORASIS system comprises a series of algorithms developed at the Naval Research Laboratory for the analysis of remotely sensed hyperspectral image data.
- **Chapter 5. Parallel Implementation of the Recursive Approximation of an Unsupervised Hierarchical Segmentation Algorithm.** This chapter describes a parallel implementation of a recursive approximation of the hierarchical image segmentation algorithm developed at NASA. The chapter also demonstrates the computational efficiency of the algorithm using remotely sensed data collected by the Landsat Thematic Mapper (a multispectral instrument).
- **Chapter 6. Computing for Analysis and Modeling of Hyperspectral Imagery.** In this chapter, several analytical methods employed in vegetation and ecosystem studies using remote sensing instruments are developed. The chapter also summarizes the most common HPC-based approaches used to meet these analytical demands, and provides examples with computing clusters. Finally, the chapter discusses the emerging use of other HPC-based techniques for the above purpose, including data processing onboard aircraft and spacecraft platforms, and distributed Internet computing.
- **Chapter 7. Parallel Implementation of Morphological Neural Networks for Hyperspectral Image Analysis.** This chapter explores in detail the utilization of parallel neural network architectures for solving remote sensing problems. The chapter further develops a new morphological/neural parallel algorithm for the analysis of remotely sensed data, which is implemented using both massively parallel (homogeneous) clusters and fully heterogeneous networks of distributed workstations.

- **Chapter 8. Parallel Wildland Fire Monitoring and Tracking Using Remotely Sensed Data.** This chapter focuses on the use of HPC-based remote sensing techniques to address natural disasters, emphasizing the (near) real-time computational requirements introduced by time-critical applications. The chapter also develops several innovative algorithms, including morphological and target detection approaches, to monitor and track one particular type of hazard, wildland fires, using remotely sensed data.
- **Chapter 9. An Introduction to Grids for Remote Sensing Applications.** This chapter introduces grid computing technology in preparation for the chapters to follow. The chapter first reviews previous approaches to distributed computing and then introduces current Web and grid service standards, along with some end-user tools for building grid applications. This is followed by a survey of current grid infrastructure and science projects relevant to remote sensing.
- **Chapter 10. Remote Sensing Grids: Architecture and Implementation.** This chapter applies the grid computing paradigm to the domain of Earth remote sensing systems by combining the concepts of remote sensing or sensor Web systems with those of grid computing. In order to provide a specific example and context for discussing remote sensing grids, the design of a weather forecasting and climate science grid is presented and discussed.
- **Chapter 11. Open Grid Services for Envisat and Earth Observation Applications.** This chapter first provides an overview of some ESA Earth Observation missions, and of the software tools that ESA currently provides for facilitating data handling and analysis. Then, the chapter describes a dedicated Earth-science grid infrastructure, developed by the European Space Research Institute (ESRIN) at ESA in the context of DATAGRID, the first large European Commission-funded grid project. Different examples of remote sensing applications integrated in this system are also given.
- **Chapter 12. Design and Implementation of a Grid Computing Environment for Remote Sensing.** This chapter develops a new dynamic Earth Observation system specifically tuned to manage huge quantities of data coming from space missions. The system combines recent grid computing technologies, concepts related to problem solving environments, and other HPC-based technologies. A comparison of the system to other classic approaches is also provided.
- **Chapter 13. A Solutionware for Hyperspectral Image Processing and Analysis.** This chapter describes the concept of an integrated process for hyperspectral image analysis, based on a *solutionware* (i.e., a set of catalogued tools that allow for the rapid construction of data processing algorithms and applications). Parallel processing implementations of some of the tools in the Itanium architecture are presented, and a prototype version of a hyperspectral image processing toolbox over the grid, called Grid-HSI, is also described.
- **Chapter 14. AVIRIS and Related 21st Century Imaging Spectrometers for Earth and Space Science.** This chapter uses the NASA Jet Propulsion

Laboratory's Airborne Visible/Infrared Imaging Spectrometer (AVIRIS), one of the most advanced hyperspectral remote sensing instrument currently available, to review the critical characteristics of an imaging spectrometer instrument and the corresponding characteristics of the measured spectra. The wide range of scientific research as well as application objectives pursued with AVIRIS are briefly presented. Roles for the application of high-performance computing methods to AVIRIS data set are discussed.

- **Chapter 15. Remote Sensing and High-Performance Reconfigurable Computing Systems.** This chapter discusses the role of reconfigurable computing using field programmable gate arrays (FPGAs) for onboard processing of remotely sensed data. The chapter also describes several case studies of remote sensing applications in which reconfigurable computing has played an important role, including cloud detection and dimensionality reduction of hyperspectral imagery.
- **Chapter 16. FPGA Design for Real-Time Implementation of Constrained Energy Minimization for Hyperspectral Target Detection.** This chapter describes an FPGA implementation of the constrained energy minimization (CEM) algorithm, which has been widely used for hyperspectral detection and classification. The main feature of the FPGA design provided in this chapter is the use of the Coordinate Rotation DIgital Computer (CORDIC) algorithm to convert a Givens rotation of a vector to a set of shift-add operations, which allows for efficient implementation in specialized hardware architectures.
- **Chapter 17. Real-Time Online Processing of Hyperspectral Imagery for Target Detection and Discrimination.** This chapter describes a real-time online processing technique for fast and accurate exploitation of hyperspectral imagery. The system has been specifically developed to satisfy the extremely high computational requirements of many practical remote sensing applications, such as target detection and discrimination, in which an immediate data analysis result is required for (near) real-time decision-making.
- **Chapter 18. Real-Time Onboard Hyperspectral Image Processing Using Programmable Graphics Hardware.** Finally, this chapter addresses the emerging use of graphic processing units (GPUs) for onboard remote sensing data processing. Driven by the ever-growing demands of the video-game industry, GPUs have evolved from expensive application-specific units into highly parallel programmable systems. In this chapter, GPU-based implementations of remote sensing data processing algorithms are presented and discussed.

1.3 Distinguishing Features of the Book

Before concluding this introduction, the editors would like to stress several distinguishing features of this book. First and foremost, this book is the first volume that is entirely devoted to providing a perspective on the state-of-the-art of HPC techniques in the context of remote sensing problems. In order to address the need for a consolidated reference in this area, the editors have made significant efforts to invite highly recognized experts in academia, institutions, and commercial companies to write relevant chapters focused on their vast expertise in this area, and share their knowledge with the community. Second, this book provides a compilation of several well-established techniques covering most aspects of the current spectrum of processing techniques in remote sensing, including supervised and unsupervised techniques for data acquisition, calibration, correction, classification, segmentation, model inversion and visualization. Further, many of the application areas addressed in this book are of great social relevance and impact, including chemical/biological standoff detection, forest fire monitoring and tracking, etc. Finally, the variety and heterogeneity of parallel computing techniques and architectures discussed in the book are not to be found in any other similar textbook.

1.4 Summary

The wide range of computer architectures (including homogeneous and heterogeneous clusters and groups of clusters, large-scale distributed platforms and grid computing environments, specialized architectures based on reconfigurable computing, and commodity graphic hardware) and data processing techniques covered by this book exemplifies a subject area that has drawn together an eclectic collection of participants, but increasingly this is the nature of many endeavors at the cutting edge of science and technology.

In this regard, one of the main purposes of this book is to reflect the increasing sophistication of a field that is rapidly maturing at the intersection of many different disciplines, including not only remote sensing or computer architecture/engineering, but also signal and image processing, optics, electronics, and aerospace engineering. The ultimate goal of this book is to provide readers with a peek at the cutting-edge research in the use of HPC-based techniques and practices in the context of remote sensing applications. The editors hope that this volume will serve as a useful reference for practitioners and engineers working in the above and related areas. Last but not least, the editors gratefully thank all the contributors for sharing their vast expertise with the readers. Without their outstanding contributions, this book could not have been completed.

Chapter 2

High-Performance Computer Architectures for Remote Sensing Data Analysis: Overview and Case Study

Antonio Plaza,
University of Extremadura, Spain

Chein-I Chang,
University of Maryland, Baltimore

Contents

Advances in sensor technology are revolutionizing the way remotely sensed data are collected, managed, and analyzed. In particular, many current and future applications of remote sensing in earth science, space science, and soon in exploration science require real- or near-real-time processing capabilities. In recent years, several efforts have been directed towards the incorporation of high-performance computing (HPC) models to remote sensing missions. In this chapter, an overview of recent efforts in the design of HPC systems for remote sensing is provided. The chapter also includes an application case study in which the pixel purity index (PPI), a well-known remote sensing data processing algorithm, is implemented in different types of HPC platforms such as a massively parallel multiprocessor, a heterogeneous network of distributed computers, and a specialized field programmable gate array (FPGA) hardware architecture. Analytical and experimental results are presented in the context of a real application, using hyperspectral data collected by NASA's Jet Propulsion Laboratory over the World Trade Center area in New York City, right after the terrorist attacks of September 11th. Combined, these parts deliver an excellent snapshot of the state-of-the-art of HPC in remote sensing, and offer a thoughtful perspective of the potential and emerging challenges of adapting HPC paradigms to remote sensing problems.

2.1 Introduction

The development of computationally efficient techniques for transforming the massive amount of remote sensing data into scientific understanding is critical for space-based earth science and planetary exploration [1]. The wealth of information provided by latest-generation remote sensing instruments has opened groundbreaking perspectives in many applications, including environmental modeling and assessment for Earth-based and atmospheric studies, risk/hazard prevention and response including wild land fire tracking, biological threat detection, monitoring of oil spills and other types of chemical contamination, target detection for military and defense/security purposes, urban planning and management studies, etc. [2]. Most of the above-mentioned applications require analysis algorithms able to provide a response in real- or near-real-time. This is quite an ambitious goal in most current remote sensing missions, mainly because the price paid for the rich information available from latest-generation sensors is the enormous amounts of data that they generate [3, 4, 5].

A relevant example of a remote sensing application in which the use of HPC technologies such as parallel and distributed computing are highly desirable is hyperspectral imaging [6], in which an image spectrometer collects hundreds or even thousands of measurements (at multiple wavelength channels) for the same area on the surface of the Earth (see Figure 2.1). The scenes provided by such sensors are often called "data cubes," to denote the extremely high dimensionality of the data. For instance, the NASA Jet Propulsion Laboratory's Airborne Visible Infra-Red Imaging Spectrometer (AVIRIS) [7] is now able to record the visible and near-infrared spectrum (wavelength region from 0.4 to 2.5 micrometers) of the reflected light of an area 2 to 12 kilometers wide and several kilometers long using 224 spectral bands (see Figure 3.8). The resulting cube is a stack of images in which each pixel (vector) has an associated spectral signature or 'fingerprint' that uniquely characterizes the underlying objects, and the resulting data volume typically comprises several GBs per flight. Although hyperspectral imaging

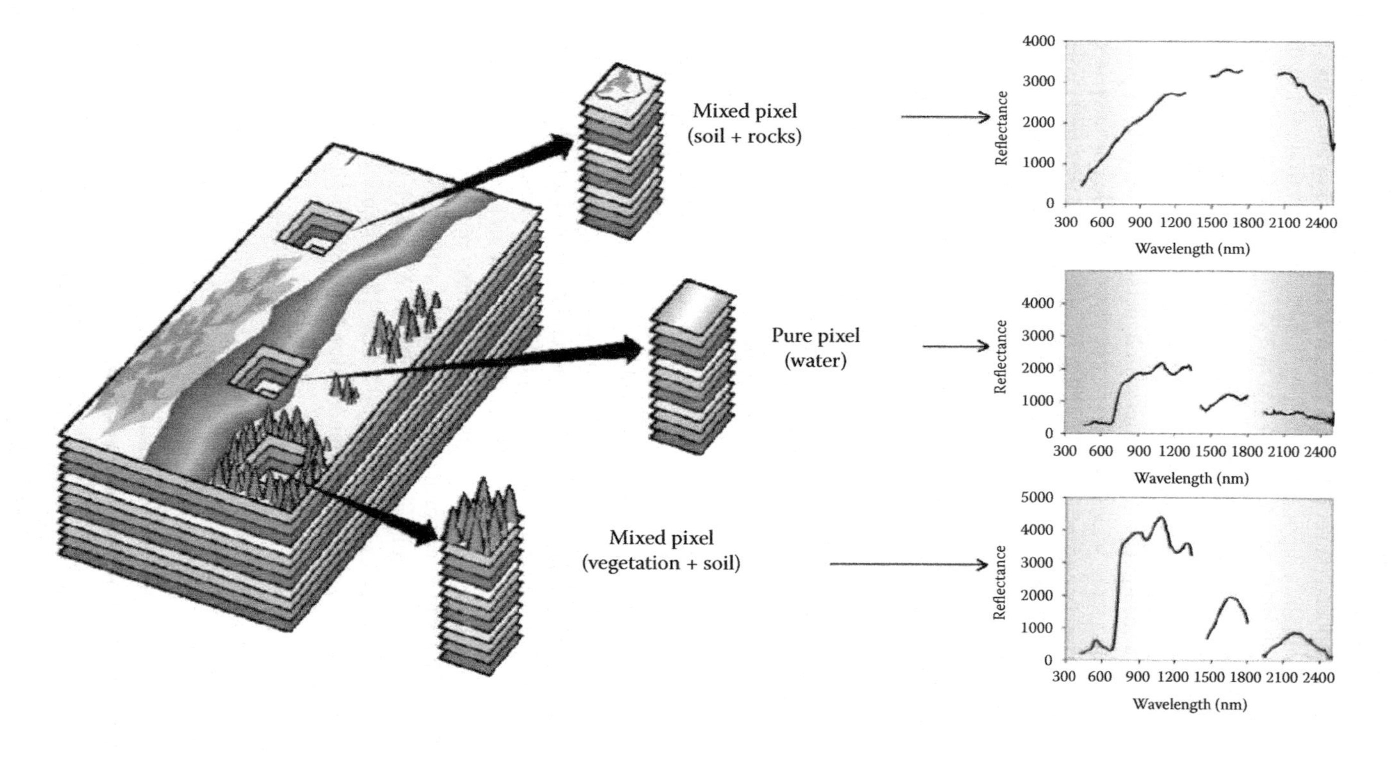

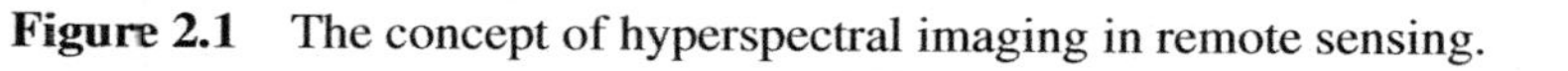

Figure 2.1 The concept of hyperspectral imaging in remote sensing.

is a good example of the computational requirements introduced by remote sensing applications, there are many other remote sensing areas in which high-dimensional data sets are also produced (several of them are covered in detail in this book). However, the extremely high computational requirements already introduced by hyperspectral imaging applications (and the fact that these systems will continue increasing their spatial and spectral resolutions in the near future) make them an excellent case study to illustrate the need for HPC systems in remote sensing and will be used in this chapter for demonstration purposes.

Specifically, the utilization of HPC systems in hyperspectral imaging applications has become more and more widespread in recent years. The idea developed by the computer science community of using COTS (commercial off-the-shelf) computer equipment, clustered together to work as a computational "team," is a very attractive solution [8]. This strategy is often referred to as Beowulf-class cluster computing [9] and has already offered access to greatly increased computational power, but at a low cost (commensurate with falling commercial PC costs) in a number of remote sensing applications [10, 11, 12, 13, 14, 15]. In theory, the combination of commercial forces driving down cost and positive hardware trends (e.g., CPU peak power doubling every 18–24 months, storage capacity doubling every 12–18 months, and networking bandwidth doubling every 9–12 months) offers supercomputing performance that can now be applied a much wider range of remote sensing problems.

Although most parallel techniques and systems for image information processing employed by NASA and other institutions during the last decade have chiefly been homogeneous in nature (i.e., they are made up of identical processing units, thus simplifying the design of parallel solutions adapted to those systems), a recent trend in the design of HPC systems for data-intensive problems is to utilize highly heterogeneous computing resources [16]. This heterogeneity is seldom planned, arising mainly as a result of technology evolution over time and computer market sales and trends. In this regard, networks of heterogeneous COTS resources can realize a very high level of aggregate performance in remote sensing applications [17], and the pervasive availability of these resources has resulted in the current notion of *grid computing* [18], which endeavors to make such distributed computing platforms easy to utilize in different application domains, much like the World Wide Web has made it easy to distribute Web content. It is expected that grid-based HPC systems will soon represent the tool of choice for the scientific community devoted to very high-dimensional data analysis in remote sensing and other fields.

Finally, although remote sensing data processing algorithms generally map quite nicely to parallel systems made up of commodity CPUs, these systems are generally expensive and difficult to adapt to onboard remote sensing data processing scenarios, in which low-weight and low-power integrated components are essential to reduce mission payload and obtain analysis results in real time, i.e., at the same time as the data are collected by the sensor. In this regard, an exciting new development in the field of commodity computing is the emergence of programmable hardware devices such as field programmable gate arrays (FPGAs) [19, 20, 21] and graphic processing units (GPUs) [22], which can bridge the gap towards onboard and real-time analysis of remote sensing data. FPGAs are now fully reconfigurable, which allows one to adaptively select a data processing algorithm (out of a pool of available ones) to be applied onboard the sensor from a control station on Earth.

On the other hand, the emergence of GPUs (driven by the ever-growing demands of the video-game industry) has allowed these systems to evolve from expensive application-specific units into highly parallel and programmable commodity components. Current GPUs can deliver a peak performance in the order of 360 Gigaflops (Gflops), more than seven times the performance of the fastest $\times$86 dual-core processor (around 50 Gflops). The ever-growing computational demands of remote sensing applications can fully benefit from compact hardware components and take advantage of the small size and relatively low cost of these units as compared to clusters or networks of computers.

The main purpose of this chapter is to provide an overview of different HPC paradigms in the context of remote sensing applications. The chapter is organized as follows:

- Section 2.2 describes relevant previous efforts in the field, such as the evolution of cluster computing in remote sensing applications, the emergence of distributed networks of computers as a cost-effective means to solve remote sensing problems, and the exploitation of specialized hardware architectures in remote sensing missions.
- Section 2.3 provides an application case study: the well-known Pixel Purity Index (PPI) algorithm [23], which has been widely used to analyze hyperspectral images and is available in commercial software. The algorithm is first briefly described and several issues encountered in its implementation are discussed. Then, we provide HPC implementations of the algorithm, including a cluster-based parallel version, a variation of this version specifically tuned for heterogeneous computing environments, and an FPGA-based implementation.
- Section 2.4 also provides an experimental comparison of the proposed implementations of PPI using several high-performance computing architectures. Specifically, we use Thunderhead, a massively parallel Beowulf cluster at NASA's Goddard Space Flight Center, a heterogeneous network of distributed workstations, and a Xilinx Virtex-II FPGA device. The considered application is based on the analysis of hyperspectral data collected by the AVIRIS instrument over the World Trade Center area in New York City right after the terrorist attacks of September 11th.
- Finally, Section 2.5 concludes with some remarks and plausible future research lines.

2.2 Related Work

This section first provides an overview of the evolution of cluster computing architectures in the context of remote sensing applications, from the initial developments in Beowulf systems at NASA centers to the current systems being employed for remote sensing data processing. Then, an overview of recent advances in heterogeneous computing systems is given. These systems can be applied for the sake of distributed processing of remotely sensed data sets. The section concludes with an overview of hardware-based implementations for onboard processing of remote sensing data sets.

2.2.1 Evolution of Cluster Computing in Remote Sensing

Beowulf clusters were originally developed with the purpose of creating a cost-effective parallel computing system able to satisfy specific computational requirements in the earth and space sciences communities. Initially, the need for large amounts of computation was identified for processing multispectral imagery with only a few bands [24]. As sensor instruments incorporated hyperspectral capabilities, it was soon recognized that computer mainframes and mini-computers could not provide sufficient power for processing these kinds of data. The Linux operating system introduced the potential of being quite reliable due to the large number of developers and users. Later it became apparent that large numbers of developers could also be a disadvantage as well as an advantage.

In 1994, a team was put together at NASA's Goddard Space Flight Center (GSFC) to build a cluster consisting only of commodity hardware (PCs) running Linux, which resulted in the first Beowulf cluster [25]. It consisted of 16 100Mhz 486DX4-based PCs connected with two hub-based Ethernet networks tied together with channel bonding software so that the two networks acted like one network running at twice the speed. The next year Beowulf-II, a 16-PC cluster based on 100Mhz Pentium PCs, was built and performed about 3 times faster, but also demonstrated a much higher reliability. In 1996, a Pentium-Pro cluster at Caltech demonstrated a sustained Gigaflop on a remote sensing-based application. This was the first time a commodity cluster had shown high-performance potential.

Up until 1997, Beowulf clusters were in essence engineering prototypes, that is, they were built by those who were going to use them. However, in 1997, a project was started at GSFC to build a commodity cluster that was intended to be used by those who had not built it, the HIVE (highly parallel virtual environment) project. The idea was to have workstations distributed among different locations and a large number of compute nodes (the compute core) concentrated in one area. The workstations would share the computer core as though it was apart of each. Although the original HIVE only had one workstation, many users were able to access it from their own workstations over the Internet. The HIVE was also the first commodity cluster to exceed a sustained 10 Gigaflop on a remote sensing algorithm.

Currently, an evolution of the HIVE is being used at GSFC for remote sensing data processing calculations. The system, called Thunderhead (see Figure 2.2), is a 512-processor homogeneous Beowulf cluster composed of 256 dual 2.4 GHz Intel Xeon nodes, each with 1 GB of memory and 80 GB of main memory. The total peak performance of the system is 2457.6 GFlops. Along with the 512-processor computer core, Thunderhead has several nodes attached to the core with a 2 Ghz optical fibre Myrinet.

NASA is currently supporting additional massively parallel clusters for remote sensing applications, such as the Columbia supercomputer at NASA Ames Research

Figure 2.2 Thunderhead Beowulf cluster (512 processors) at NASA's Goddard Space Flight Center in Maryland.

Center, a 10,240-CPU SGI Altix supercluster, with Intel Itanium 2 processors, 20 terabytes total memory, and heterogeneous interconnects including InfiniBand network and a 10 GB Ethernet. This system is listed as #8 in the November 2006 version of the Top500 list of supercomputer sites available online at http://www.top500.org.

Among many other examples of HPC systems included in the list that are currently being exploited for remote sensing and earth science-based applications, we cite three relevant systems for illustrative purposes. The first one is MareNostrum, an IBM cluster with 10,240 processors, 2.3 GHz Myrinet connectivity, and 20,480 GB of main memory available at Barcelona Supercomputing Center (#5 in Top500). Another example is Jaws, a Dell PowerEdge cluster with 3 GHz Infiniband connectivity, 5,200 GB of main memory, and 5,200 processors available at Maui High-Performance Computing Center (MHPCC) in Hawaii (#11 in Top500). A final example is NEC's Earth Simulator Center, a 5,120-processor system developed by Japan's Aerospace Exploration Agency and the Agency for Marine-Earth Science and Technology (#14 in Top500). It is highly anticipated that many new supercomputer systems will be specifically developed in forthcoming years to support remote sensing applications.

2.2.2 Heterogeneous Computing in Remote Sensing

In the previous subsection, we discussed the use of cluster technologies based on multiprocessor systems as a high-performance and economically viable tool for efficient processing of remotely sensed data sets. With the commercial availability of networking hardware, it soon became obvious that networked groups of machines distributed among different locations could be used together by one single parallel remote sensing code as a distributed-memory machine [26]. Of course, such networks were originally designed and built to connect heterogeneous sets of machines. As a result, heterogeneous networks of workstations (NOWs) soon became a very popular tool for distributed computing with essentially unbounded sets of machines, in which the number and locations of machines may not be explicitly known [16], as opposed to cluster computing, in which the number and locations of nodes are known and relatively fixed.

An evolution of the concept of distributed computing described above resulted in the current notion of grid computing [18], in which the number and locations of nodes are relatively dynamic and have to be discovered at run-time. It should be noted that this section specifically focuses on distributed computing environments without meta-computing or grid computing, which aims at providing users access to services distributed over wide-area networks. Several chapters of this volume provide detailed analyses of the use of grids for remote sensing applications, and this issue is not further discussed here.

There are currently several ongoing research efforts aimed at efficient distributed processing of remote sensing data. Perhaps the most simple example is the use of heterogeneous versions of data processing algorithms developed for Beowulf clusters, for instance, by resorting to heterogeneous-aware variations of homogeneous algorithms, able to capture the inherent heterogeneity of a NOW and to load-balance the computation among the available resources [27]. This framework allows one to easily port an existing parallel code developed for a homogeneous system to a fully heterogeneous environment, as will be shown in the following subsection.

Another example is the Common Component Architecture (CCA) [28], which has been used as a plug-and-play environment for the construction of climate, weather, and ocean applications through a set of software components that conform to standardized interfaces. Such components encapsulate much of the complexity of the data processing algorithms inside a black box and expose only well-defined interfaces to other components. Among several other available efforts, another distributed application framework specifically developed for earth science data processing is the Java Distributed Application Framework (JDAF) [29]. Although the two main goals of JDAF are flexibility and performance, we believe that the Java programming language is not mature enough for high-performance computing of large amounts of data.

2.2.3 Specialized Hardware for Onboard Data Processing

Over the last few years, several research efforts have been directed towards the incorporation of specialized hardware for accelerating remote sensing-related calculations aboard airborne and satellite sensor platforms. Enabling onboard data processing introduces many advantages, such as the possibility to reduce the data down-link bandwidth requirements at the sensor by both preprocessing data and selecting data to be transmitted based upon predetermined content-based criteria [19, 20]. Onboard processing also reduces the cost and the complexity of ground processing systems so that they can be affordable to a larger community. Other remote sensing applications that will soon greatly benefit from onboard processing are future web sensor missions as well as future Mars and planetary exploration missions, for which onboard processing would enable autonomous decisions to be made onboard.

Despite the appealing perspectives introduced by specialized data processing components, current hardware architectures including FPGAs (on-the-fly reconfigurability) and GPUs (very high performance at low cost) still present some limitations that need to be carefully analyzed when considering their incorporation to remote sensing missions [30]. In particular, the very fine granularity of FPGAs is still not efficient, with extreme situations in which only about 1% of the chip is available for logic while 99% is used for interconnect and configuration. This usually results in a penalty in terms of speed and power. On the other hand, both FPGAs and GPUs are still difficult to radiation-harden (currently-available radiation-tolerant FPGA devices have two orders of magnitude fewer equivalent gates than commercial FPGAs).

2.3 Case Study: Pixel Purity Index (PPI) Algorithm

This section provides an application case study that is used in this chapter to illustrate different approaches for efficient implementation of remote sensing data processing algorithms. The algorithm selected as a case study is the PPI [23], one of the most widely used algorithms in the remote sensing community. First, the serial version of the algorithm available in commercial software is described. Then, several parallel implementations are given.

2.3.1 Algorithm Description

The PPI algorithm was originally developed by Boardman et al. [23] and was soon incorporated into Kodak's Research Systems ENVI, one of the most widely used commercial software packages by remote sensing scientists around the world. The underlying assumption under the PPI algorithm is that the spectral signature associated to each pixel vector measures the response of multiple underlying materials at each site. For instance, it is very likely that the pixel vectors shown in Figure 3.8 would actually contain a mixture of different substances (e.g., different minerals, different types of soils, etc.). This situation, often referred to as the "mixture problem" in hyperspectral analysis terminology [31], is one of the most crucial and distinguishing properties of spectroscopic analysis.

Mixed pixels exist for one of two reasons [32]. Firstly, if the spatial resolution of the sensor is not fine enough to separate different materials, these can jointly occupy a single pixel, and the resulting spectral measurement will be a composite of the individual spectra. Secondly, mixed pixels can also result when distinct materials are combined into a homogeneous mixture. This circumstance occurs independent of

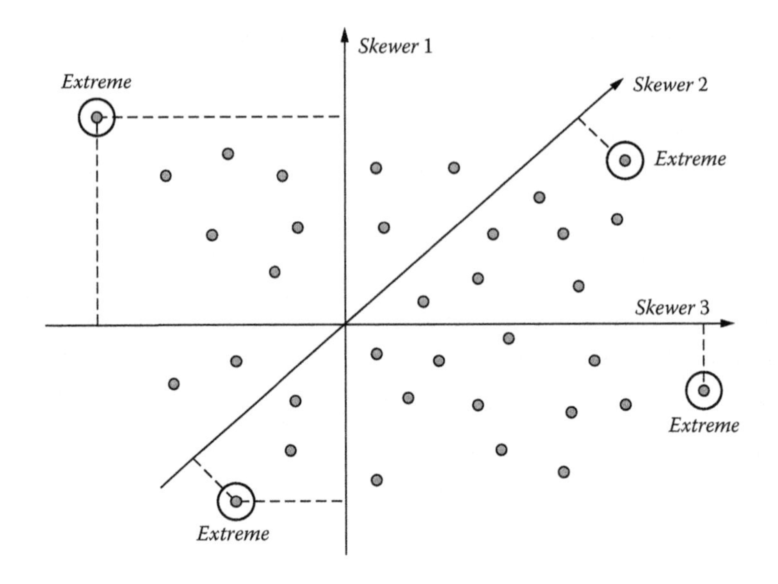

Figure 2.3 Toy example illustrating the performance of the PPI algorithm in a 2-dimensional space.

the spatial resolution of the sensor. A hyperspectral image is often a combination of the two situations, where a few sites in a scene are pure materials, but many others are mixtures of materials.

To deal with the mixture problem in hyperspectral imaging, spectral unmixing techniques have been proposed as an inversion technique in which the measured spectrum of a mixed pixel is decomposed into a collection of spectrally pure constituent spectra, called *endmembers* in the literature, and a set of correspondent fractions, or *abundances*, that indicate the proportion of each endmember present in the mixed pixel [6].

The PPI algorithm is a tool to automatically search for endmembers that are assumed to be the vertices of a convex hull [23]. The algorithm proceeds by generating a large number of random, N-dimensional unit vectors called "skewers" through the data set. Every data point is projected onto each skewer, and the data points that correspond to extrema in the direction of a skewer are identified and placed on a list (see Figure 2.3). As more skewers are generated, the list grows, and the number of times a given pixel is placed on this list is also tallied. The pixels with the highest tallies are considered the final endmembers.

The inputs to the algorithm are a hyperspectral data cube $\mathbf{F}$ with N dimensions; a maximum number of endmembers to be extracted, E; the number of random skewers to be generated during the process, k; a cut-off threshold value, t_v, used to select as final endmembers only those pixels that have been selected as extreme pixels at least t_v times throughout the PPI process; and a threshold angle, t_a, used to discard redundant endmembers during the process. The output of the algorithm is a set of E final endmembers $\{\mathbf{e}_e\}_{e=1}^{E}$. The algorithm can be summarized by the following steps:

1. *Skewer generation.* Produce a set of k randomly generated unit vectors $\{\mathbf{skewer}_j\}_{j=1}^k$.
2. *Extreme projections.* For each $\mathbf{skewer}_j$, all sample pixel vectors $\mathbf{f}_i$ in the original data set $\mathbf{F}$ are projected onto $\mathbf{skewer}_j$ via dot products of $|\mathbf{f}_i \cdot \mathbf{skewer}_j|$ to find sample vectors at its extreme (maximum and minimum) projections, thus forming an extrema set for $\mathbf{skewer}_j$ that is denoted by $S_{extrema}(\mathbf{skewer}_j)$. Despite the fact that a different $\mathbf{skewer}_j$ would generate a different extrema set $S_{extrema}(\mathbf{skewer}_j)$, it is very likely that some sample vectors may appear in more than one extrema set. In order to deal with this situation, we define an indicator function of a set S, denoted by $I_S(\mathbf{x})$, to denote membership of an element $\mathbf{x}$ to that particular set as follows:

$$I_S(\mathbf{f}_i) = \left\{ \begin{matrix} 1 \text{ if } \mathbf{x} \in S \\ 0 \text{ if } \mathbf{x} \notin S \end{matrix} \right\} \tag{2.1}$$

3. *Calculation of PPI scores.* Using the indicator function above, we calculate the PPI score associated to the sample pixel vector $\mathbf{f}_i$ (i.e., the number of times that given pixel has been selected as extreme in step 2) using the following equation:

$$N_{PPI}(\mathbf{f}_i) = \sum_{j=1}^{k} I_{S_{extrema}(\mathbf{skewer}_j)}(\mathbf{f}_i) \tag{2.2}$$

4. *Endmember selection.* Find the pixel vectors with scores of $N_{PPI}(\mathbf{f}_i)$ that are above t_v and form a unique set of endmembers $\{\mathbf{e}_e\}_{e=1}^E$ by calculating the spectral angle distance (SAD) for all possible vector pairs and discarding those pixels that result in an angle value below t_a. It should be noted that the SAD between a pixel vector $\mathbf{f}_i$ and a different pixel vector $\mathbf{f}_j$ is a standard similarity metric for remote sensing operations, mainly because it is invariant in the multiplication of the input vectors by constants and, consequently, is invariant to unknown multiplicative scalings that may arise due to differences in illumination and sensor observation angle:

$$\begin{aligned} \text{SAD}(\mathbf{f}_i, \mathbf{f}_j) &= \cos^{-1}(\mathbf{f}_i \cdot \mathbf{f}_j / \|\mathbf{f}_i\| \cdot \|\mathbf{f}_j\|) \\ &= \cos^{-1}\left(\frac{\sum_{l=1}^{N} f_{il} f_{jl}}{\sqrt{\sum_{l=1}^{N} f_{il}^2}\sqrt{\sum_{l=1}^{N} f_{jl}^2}} \right) \end{aligned} \tag{2.3}$$

From the algorithm description above, it is clear that the PPI is not an iterative algorithm [33]. In order to set parameter values for the PPI, the authors recommend using as many random skewers as possible in order to obtain optimal results. As a result, the PPI can only guarantee to produce optimal results asymptotically and its computational complexity is very high. According to our experiments using standard AVIRIS hyperspectral data sets (typically, 614×512 pixels per frame and 224 spectral bands), the PPI generally requires a very high number of skewers (in the order of $k = 10^4$ or $k = 10^5$) to produce an accurate final set of endmembers [32] and results in processing times above one hour when the algorithm is run on a latest-generation desktop PC. Such response time is unacceptable in most remote sensing applications. In the following section, we provide an overview of HPC paradigms applied to speed up computational performance of the PPI using different kinds of parallel and distributed computing architectures.

2.3.2 Parallel Implementations

This section first develops a parallel implementation of the PPI algorithm that has been specifically developed to be run on massively parallel, homogeneous Beowulf clusters. Then, the parallel version is transformed into a heterogeneity-aware implementation by introducing an adaptive data partitioning algorithm specifically developed to capture in real time the specificities of a heterogeneous network of distributed workstations. Finally, an FPGA implementation aimed at onboard PPI-based processing is provided.

2.3.2.1 Cluster-Based Implementation of the PPI Algorithm

In this subsection, we describe a master-slave parallel version of the PPI algorithm. To reduce code redundancy and enhance reusability, our goal was to reuse much of the code for the sequential algorithm in the parallel implementation. For that purpose, we adopted a spatial-domain decomposition approach [34, 35] that subdivides the image cube into multiple blocks made up of entire pixel vectors, and assigns one or more blocks to each processing element (see Figure 2.4).

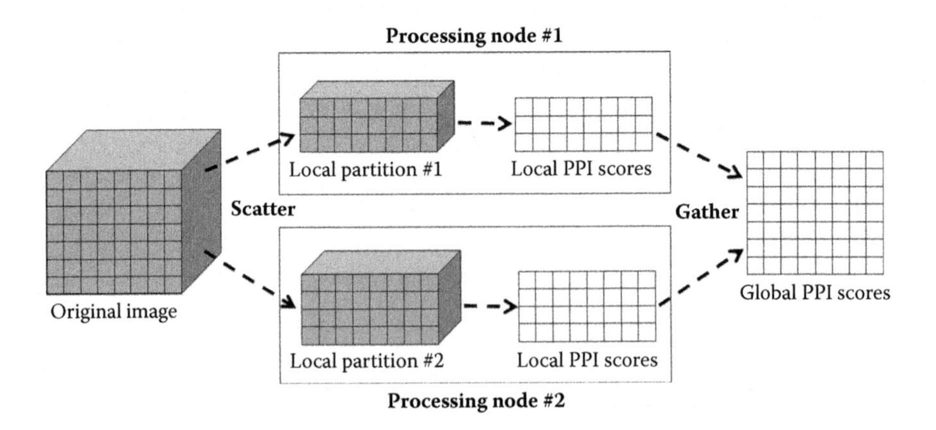

Figure 2.4 Domain decomposition adopted in the parallel implementation of the PPI algorithm.

It should be noted that the PPI algorithm is mainly based on projecting pixel vectors that are always treated as a whole. This is a result of the convex geometry process implemented by the PPI, which is based on the spectral "purity" or "convexity" of the entire spectral signature associated to each pixel. Therefore, a spectral-domain partitioning scheme (which subdivides the whole multi-band data into blocks made up of contiguous spectral bands or sub-volumes, and assigns one or more sub-volumes to each processing element) is not appropriate in our application [8]. This is because the latter approach breaks the spectral identity of the data because each pixel vector is split amongst several processing elements.

A further reason that justifies the above decision is that, in spectral-domain partitioning, the calculations made for each hyperspectral pixel need to originate from several processing elements, and thus require intensive inter-processor communication. Therefore, in our proposed implementation, a master-worker spatial domain-based decomposition paradigm is adopted, where the master processor sends partial data to the workers and coordinates their actions. Then, the master gathers the partial results provided by the workers and produces a final result.

As it was the case with the serial version, the inputs to our cluster-based implementation of the PPI algorithm are a hyperspectral data cube $\mathbf{F}$ with N dimensions; a maximum number of endmembers to be extracted, p; the number of random skewers to be generated during the process, k; a cut-off threshold value, t_v; and a threshold angle, t_a. The output of the algorithm is a set of E endmembers $\{\mathbf{e}_e\}_{e=1}^{E}$. The parallel algorithm is given by the following steps:

1. *Data partitioning*. Produce a set of L spatial-domain homogeneous partitions of $\mathbf{F}$ and scatter all partitions by indicating all partial data structure elements that are to be accessed and sent to each of the workers.
2. *Skewer generation*. Generate k random unit vectors $\{\mathbf{skewer}_j\}_{j=1}^{k}$ in parallel and broadcast the entire set of skewers to all the workers.
3. *Extreme projections*. For each $\mathbf{skewer}_j$, project all the sample pixel vectors at each local partition l onto $\mathbf{skewer}_j$ to find sample vectors at its extreme projections, and form an extrema set for $\mathbf{skewer}_j$ that is denoted by $S^{(l)}_{extrema}(\mathbf{skewer}_j)$. Now calculate the number of times each pixel vector $\mathbf{f}_i^{(l)}$ in the local partition is selected as extreme using the following expression:

$$N_{PPI}^{(l)}\left(\mathbf{f}_i^{(l)}\right) = \sum_{j=1}^{k} I_{S^{(l)}_{extrema}(\mathbf{skewer}_j)}\left(\mathbf{f}_i^{(l)}\right) \tag{2.4}$$

4. *Candidate selection*. Select those pixels with $N_{PPI}^{(l)}\left(\mathbf{f}_i^{(l)}\right) > t_v$ and send them to the master node.
5. *Endmember selection*. The master gathers all the individual endmember sets provided by the workers and forms a unique set $\{\mathbf{e}_e\}_{e=1}^{E}$ by calculating the SAD for all possible pixel vector pairs in parallel and discarding those pixels that result in angle values below t_a.

It should be noted that the proposed parallel algorithm has been implemented in the C++ programming language, using calls to message passing interface (MPI) [36]. We emphasize that, in order to implement step 1 of the parallel algorithm, we resorted to MPI-derived data types to directly scatter hyperspectral data structures, which may be stored non-contiguously in memory, in a single communication step. As a result, we avoid creating all partial data structures on the root node (thus making better use of memory resources and compute power).

2.3.2.2 Heterogeneous Implementation of the PPI Algorithm

In this subsection, we provide a simple application case study in which the standard MPI-based implementation of the PPI is adapted to a heterogeneous environment by reutilizing most of the code available for the cluster-based system [27]. This approach is generally preferred due to the relatively large amount of data processing algorithms and parallel software developed for homogeneous systems. Before introducing our implementation of the PPI algorithm for heterogeneous NOWs, we must first formulate a general optimization problem in the context of fully heterogeneous systems (composed of different-speed processors that communicate through links at different capacities) [16]. Such a computing platform can be modeled as a complete graph $G = (P, E)$, where each node models a computing resource p_i weighted by its relative cycle-time w_i. Each edge in the graph models a communication link weighted by its relative capacity, where c_{ij} denotes the maximum capacity of the slowest link in the path of physical communication links from p_i to p_j (we assume that the system has symmetric costs, i.e., $c_{ij} = c_{ji}$.

With the above assumptions in mind, processor p_i should accomplish a share of $\alpha_i \cdot W$ of the total workload, denoted by W, to be performed by a certain algorithm, with $\alpha_i \geq 0$ for $1 \leq i \leq P$ and $\sum_{i=1}^{P} \alpha_i = 1$. With the above assumptions in mind, an abstract view of our problem can be simply stated in the form of a master-worker architecture, much like the commodity cluster-based homogeneous implementation described in the previous section. However, in order for such parallel algorithms to be also effective in fully heterogeneous systems, the master program must be modified to produce a set of L spatial-domain heterogeneous partitions of $\mathbf{F}$ in step 1.

In order to balance the load of the processors in the heterogeneous environment, each processor should execute an amount of work that is proportional to its speed. Therefore, two major goals of our partitioning algorithm should be: (i) to obtain an appropriate set of workload fractions $\{\alpha_i\}_{i=1}^{P}$ that best fit the heterogeneous environment; and (ii) to translate the chosen set of values into a suitable decomposition of the input data, taking into account the properties of the heterogeneous system.

To accomplish the above goals, we use a workload estimation algorithm (WEA) that assumes that the workload of each processor p_i must be directly proportional to its local memory and inversely proportional to its cycle-time w_i. Below, we provide a description of a WEA algorithm, which replaces step 1 in the implementation of PPI provided in our previous section. Steps 2–5 of the parallel algorithm in the previous section would be executed immediately after WEA and remain the same as those outlined in the algorithmic description provided in the previous section (thus greatly enhancing code reutilization). The input to WEA is **F**, an N-dimensional data cube, and the output is a set of L spatial-domain heterogeneous partitions of **F**:

1. Obtain necessary information about the heterogeneous system, including the number of available processors P, each processor's identification number $\{p_i\}_{i=1}^{P}$, and processor cycle-times $\{\alpha_i\}_{i=1}^{P}$.
2. Set $\alpha_i = \lfloor \frac{(P/w_i)}{\sum_{i=1}^{P}(1/w_i)} \rfloor$ for all $i \in \{1, \cdots, P\}$. In other words, this step first approximates the $\{\alpha_i\}_{i=1}^{P}$ so that the amount of work assigned to each processor is proportional to its speed and $\alpha_i \cdot w_i \approx const$ for all processors.
3. Iteratively increment some α_i until the set of $\{\alpha_i\}_{i=1}^{P}$ best approximates the total workload to be completed, W, i.e., for $m = \sum_{i=1}^{P} \alpha_i$ to W, find $k \in \{1, \cdots, P\}$ so that $w_k \cdot (\alpha_k + 1) = min\{w_i \cdot (\alpha_i + 1)\}_{i=1}^{P}$, and then set $\alpha_k = \alpha_k + 1$.
4. Once the set $\{\alpha_i\}_{i=1}^{P}$ has been obtained, a further objective is to produce P partitions of the input hyperspectral data set. To do so, we proceed as follows:
   - Obtain a first partitioning of the hyperspectral data set so that the number of rows in each partition is proportional to the values of $\{\alpha_i\}_{i=1}^{P}$.
   - Refine the initial partitioning taking into account the local memory associated to each processor.

The parallel algorithm described above has been implemented using two approaches. The first one is based on the C++ programming language with calls to standard MPI functions. A second implementation was developed using HeteroMPI [37], a heterogeneous version of MPI that automatically optimizes the workload assigned to each heterogeneous processor (i.e., this implementation automatically determines the load distribution accomplished by our proposed WEA algorithm). Experimentally, we tested that both implementations resulted in very similar results, and, hence, the experimental validation provided in the following section will be based on the performance analysis achieved by the first implementation (i.e., using our proposed WEA algorithm to estimate the workloads).

2.3.2.3 FPGA-Based Implementation of the PPI Algorithm

In this subsection, we describe a hardware-based parallel strategy for implementation of the PPI algorithm that is aimed at enhancing replicability and reusability of slices in FPGA devices through the utilization of systolic array design [38]. One of the main advantages of systolic array-based implementations is that they are able to provide a systematic procedure for system design that allows for the derivation of a well-defined processing element-based structure and an interconnection pattern that can then be easily ported to real hardware configurations. Using this procedure, we can also calculate the data dependencies prior to the design, and in a very straightforward manner. Our proposed design intends to maximize computational power of the hardware and minimize the cost of communications. These goals are particularly relevant in our specific application, where hundreds of data values will be handled for each intermediate result, a fact that may introduce problems related with limited resource availability and inefficiencies in hardware replication and reusability.

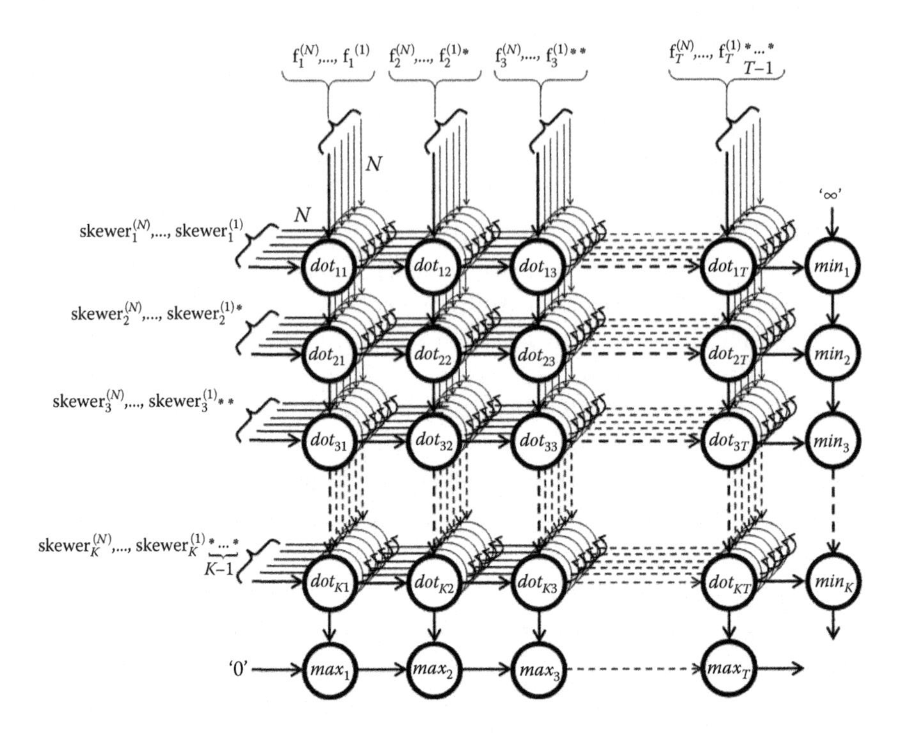

Figure 2.5 Systolic array design for the proposed FPGA implementation of the PPI algorithm.

After several empirical experiments using real data sets, we opted for a configuration in which local results remain static at each processing element, while pixel vectors are input to the systolic array from top to bottom and skewer vectors are fed to the systolic array from left to right. Figure 2.5 illustrates the above principle, in which local results remain static at each processing element, while pixel vectors are input to the systolic array from top to bottom and skewer vectors are fed to the systolic array from left to right. In Figure 2.5, asterisks represent delays while $\mathbf{skewer}_j^{(n)}$ denotes the value of the n-th band of the j-th skewer, with $j \in \{1, \cdots, K\}$ and $n \in \{1, \cdots, N\}$, where N is the number of bands of the input hyperspectral scene. Similarly, $\mathbf{f}_i^{(n)}$ denotes the reflectance value of the n-th band of the i-th pixel, with $i \in \{1, \cdots, T\}$, where T is the total number of pixels in the input image. The processing nodes labeled as *dot* in Figure 2.5 perform the individual products for the skewer projections. On the other hand, the nodes labeled as *max* and *min* respectively compute the maxima and minima projections after the dot product calculations have been completed. In fact, the *max* and *min* nodes can be respectively seen as part of a 1-dimensional systolic array that avoids broadcasting the pixel while simplifying the collection of the results.

Basically, a systolic cycle in the architecture described in Figure 2.5 consists in computing a single dot product between a pixel and a skewer. A full vector dot-product calculation requires N multiplications and $N - 1$ additions, where N is the number of spectral bands. It has been shown in previous work that the skewer values can be limited to a very small set of integers when N is large, as in the case of hyperspectral images. A particular and interesting set is $\{1, -1\}$, since it avoids the multiplication [39]. The dot product is thus reduced to an accumulation of positive and negative values (the self-connections in the *dot* nodes of Figure 2.5 represent the accumulation of intermediate results in those nodes). With the above assumptions in mind, the *dot* nodes only need to accumulate the positive or negative values of the pixel input according to the skewer input. These units are thus only composed of a single 16-bit addition/subtraction operator. If we suppose that an addition or a subtraction is executed every clock cycle, then the calculation of a full dot product requires N clock cycles. During the first systolic cycle, dot_{11} starts processing the first band of the first pixel vector, $\mathbf{f}_1$. During the second systolic cycle, the node dot_{12} starts processing the first band of pixel $\mathbf{f}_2$, while the node dot_{11} processes the second band of pixel $\mathbf{f}_1$, and so on.

The main advantage of the systolic array described above is its scalability. Depending on the resources available on the reconfigurable board, the number of processors can be adjusted without modifying the control of the array. In order to reduce the number of passes, we decided to allocate the maximum number of processors in the available FPGA components. In other words, although in Figure 2.5 we represent an ideal systolic array in which T pixels can be processed, this is not the usual situation, and the number of pixels usually has to be divided by P, the number of available processors. In this scenario, after T/P systolic cycles, all the nodes are working. When all the pixels have been flushed through the systolic array, T/P additional systolic cycles are thus required to collect the results for the considered set of P pixels and a new set of P different pixels would be flushed until processing all T pixels in the original image.

Finally, to obtain the vector of endmember abundances $\{a_{i1}, a_{i2}, \cdots, a_{iE}\}$ for each pixel $\mathbf{f}_i$, we multiply each $\mathbf{f}_i$ by $(\mathbf{M}^T\mathbf{M})^{-1}\mathbf{M}^T$, where $\mathbf{M} = \{\mathbf{e}_e\}_{e=1}^{E}$ and the superscript T denotes the matrix transpose operation. As recently described [40], this operation can be done using a so-called parallel block algorithm, which has been adopted in this work to carry out the final spectral unmixing step added to our description of PPI algorithm using part of the systolic array design outlined above.

Based on the design described above, we have developed a high-level implementation of PPI using Handel-C [41], a design and prototyping language that allows using a pseudo-C programming style. The final decision on implementing our design using Handel-C instead of other well-known hardware description languages such as VHDL or Verilog was taken on the account that a high-level language may allow users to generate hardware versions of available hyperspectral analysis algorithms in relatively short time. For illustrative purposes, the source code in Handel-C corresponding to the *extreme projections* step of our FPGA implementation of the PPI algorithm is shown below. The skewer initialization and spectral unmixing-related portions of the code are not shown for simplicity. For a detailed understanding of the piece of code shown in the listing below, we point to reference material and documentation on Handel-C and Xilinx [41, 42].

Listing 1 Source code of the Handel-C (high level) FPGA implementation of the PPI algorithm.

```
void main(void) {

  unsigned int 16 max[E]; //E is the number of endmembers
  unsigned int 16 end[E];
  unsigned int 16 i;
  unsigned int 10000 k; //k denotes the number of skewers
  unsigned int 224 N; //N denotes the number of bands

  par(i = 0;i < E;i++) max[i]=0;
  par(k = 0;k < E;k++) {
    par(k = 0;k < E;k++) {
      par(j = 0;j < N;j++) {
        Proc_Element[i][k](pixels[i][j],skewers[k][j],0@i,0@k);
      }
    }
  }

  for(i = 0;i < E;i++) {
    max[i]=Proc_Element[i][k](0@max[i], 0, 0@i, 0@k);
  }

 phase_1_finished=1
  while(!phase_2) { //Waiting to enter phase 2 }
  for(i = 0;i < E;i++) end[i]=0;
  for(i = 0;i < E;i++) {
    par(k = 0;k < E;k++) {
      par(j = 0;j < N;j++) {
        end[i]=end[i]&&Proc_Element[i][k](pixels[i][j],skewers[k][j],0,0);
      }
    }
  }

  phase_2_finished=1
  global_finished=0
  for(i = 0;i < E;i++) global_finished=global_finished&&end[i];
```

The implementation above was compiled and transformed into an EDIF specification automatically by using the DK3.1 software package [43]. We also used other tools such as Xilinx ISE 6.1i [42] to carry out automatic place and route, and to adapt the final steps of the hardware implementation to the Virtex-II FPGA used in the experiments.

2.4 Experimental Results

This section provides an assessment of the effectiveness of the parallel versions of PPI described throughout this chapter. Before describing our study on performance analysis, we first describe the HPC computing architectures used in this work. These include Thunderhead, a massively parallel Beowulf cluster made up of homogeneous commodity components and available at NASA's GSFC; four different networks of heterogeneous workstations distributed among different locations; and a Xilinx Virtex-II XC2V6000-6 FPGA. Next, we describe the hyperspectral data sets used for evaluation purposes. A detailed survey on algorithm performance in a real application is then provided, along with a discussion on the advantages and disadvantages of each particular approach. The section concludes with a discussion of the results obtained for the PPI implemented using different HPC architectures.

2.4.1 High-Performance Computer Architectures

This subsection provides an overview of the HPC platforms used in this study for demonstration purposes. The first considered system is Thunderhead, a 512-processor homogeneous Beowulf cluster that can be seen as an evolution of the HIVE project, started in 1997 to build a homogeneous commodity cluster to be exploited in remote sensing applications. It is composed of 256 dual 2.4 GHz Intel Xeon nodes, each with 1 GB of memory and 80 GB of main memory. The total peak performance of the system is 2457.6 GFlops. Along with the 512-processor computer core, Thunderhead has several nodes attached to the core with a 2 Ghz optical fibre Myrinet. The proposed cluster-based parallel version of the PPI algorithm proposed in this chapter was run from one of such nodes, called *thunder1*. The operating system used in the experiments was Linux Fedora Core, and MPICH [44] was the message-passing library used.

To explore the performance of the heterogeneity-aware implementation of PPI developed in this chapter, we have considered four different NOWs. All of them were custom-designed in order to approximate a recently proposed framework for evaluating heterogeneous parallel algorithms [45], which relies on the assumption that a heterogeneous algorithm cannot be executed on a heterogeneous network faster than its homogeneous version on an equivalent homogeneous network. In this study, a homogeneous computing environment was considered equivalent to the heterogeneous one based when the three requirements listed below were satisfied:

1. Both environments should have exactly the same number of processors.
2. The speed of each processor in the homogeneous environment should be equal to the average speed of the processors in the heterogeneous environment.
3. The aggregate communication characteristics of the homogeneous environment should be the same as those of the heterogeneous environment.

With the above three principles in mind, a heterogeneous algorithm may be considered optimal if its efficiency on a heterogeneous network is the same as that evidenced by

TABLE 2.1 Specifications of Heterogeneous Computing Nodes in a Fully Heterogeneous Network of Distributed Workstations

Processor Number	Architecture Overview	Cycle-Time (Seconds/Mflop)	Memory (MB)	Cache (KB)
p_1	Intel Pentium 4	0.0058	2048	1024
p_2, p_5, p_8	Intel Xeon	0.0102	1024	512
p_3	AMD Athlon	0.0026	7748	512
p_4, p_6, p_7, p_9	Intel Xeon	0.0072	1024	1024
p_{10}	UltraSparc-5	0.0451	512	2048
$p_{11} - p_{16}$	AMD Athlon	0.0131	2048	1024

its homogeneous version on the equivalent homogeneous network. This allows using the parallel performance achieved by the homogeneous version as a benchmark for assessing the parallel efficiency of the heterogeneous algorithm. The four networks are considered approximately equivalent under the above framework. Their descriptions follow:

- *Fully heterogeneous network.* Consists of 16 different workstations and four communication segments. Table 2.1 shows the properties of the 16 heterogeneous workstations, where processors $\{p_i\}_{i=1}^{4}$ are attached to communication segment s_1, processors $\{p_i\}_{i=5}^{8}$ communicate through s_2, processors $\{p_i\}_{i=9}^{10}$ are interconnected via s_3, and processors $\{p_i\}_{i=11}^{16}$ share the communication segment s_4. The communication links between the different segments $\{s_j\}_{j=1}^{4}$ only support serial communication. For illustrative purposes, Table 2.2 also shows the capacity of all point-to-point communications in the heterogeneous network, expressed as the time in milliseconds to transfer a 1-MB message between each processor pair (p_i, p_j) in the heterogeneous system. As noted, the communication network of the fully heterogeneous network consists of four relatively fast homogeneous communication segments, interconnected by three slower communication links with capacities $c^{(1,2)} = 29.05$, $c^{(2,3)} = 48.31$, $c^{(3,4)} = 58.14$ in milliseconds, respectively. Although this is a simple architecture, it is also a quite typical and realistic one as well.
- *Fully homogeneous network.* Consists of 16 identical Linux workstations with processor cycle-time of $w = 0.0131$ seconds per Mflop, interconnected via

TABLE 2.2 Capacity of Communication Links (Time in Milliseconds to Transfer a 1-MB Message) in a Fully Heterogeneous Network

Processor	$p_1 - p_4$	$p_5 - p_8$	$p_9 - p_{10}$	$p_{11} - p_{16}$
$p_1 - p_4$	19.26	48.31	96.62	154.76
$p_5 - p_8$	48.31	17.65	48.31	106.45
$p_9 - p_{10}$	96.62	48.31	16.38	58.14
$p_{11} - p_{16}$	154.76	106.45	58.14	14.05

a homogeneous communication network where the capacity of links is $c = 26.64$ ms.

- *Partially heterogeneous network.* Formed by the set of 16 heterogeneous workstations in Table 2.1 but interconnected using the same homogeneous communication network with capacity $c = 26.64$ ms.
- *Partially homogeneous network.* Formed by 16 identical Linux workstations with cycle-time of $w = 0.0131$ seconds per Mflop, interconnected using the communication network in Table 2.2.

Finally, in order to test the proposed systolic array design in hardware-based computing architectures, our parallel design was implemented on a Virtex-II XC2V6000-6 FPGA of the Celoxica's ADMXRC2 board. It contains 33,792 slices, 144 Select RAM Blocks, and 144 multipliers (of 18-bit $\times$ 18-bit). Concerning the timing performances, we decided to pack the input/output registers of our implementation into the input/output blocks in order to try to reach the maximum achievable performance.

2.4.2 Hyperspectral Data

The image scene used for experiments in this work was collected by the AVIRIS instrument, which was flown by NASA's Jet Propulsion Laboratory over the World Trade Center (WTC) area in New York City on September 16, 2001, just 5 days after the terrorist attacks that collapsed the two main towers and other buildings in the WTC complex. The data set selected for the experiments was geometrically and atmospherically corrected prior to data processing, and consists of 614×512 pixels, 224 spectral bands, and a total size of 140 MB. The spatial resolution is 1.7 meters per pixel. Figure 2.6(left) shows a false color composite of the data set selected for the experiments using the 1682, 1107, and 655 nm channels, displayed. A detail of the WTC area is shown in a rectangle.

At the same time of data collection, a small U.S. Geological Survey (USGS) field crew visited lower Manhattan to collect spectral samples of dust and airfall debris deposits from several outdoor locations around the WTC area. These spectral samples were then mapped into the AVIRIS data using reflectance spectroscopy and chemical analyses in specialized USGS laboratories. For illustrative purposes, Figure 2.6(right) shows a thermal map centered at the region where the buildings collapsed. The map shows the target locations of the thermal hot spots.

An experiment-based cross-examination of endmember extraction accuracy was first conducted to assess the SAD-based spectral similarity scores obtained after comparing the ground-truth USGS reference signatures with the corresponding five endmembers extracted by the three parallel implementations of the PPI algorithm. This experiment revealed that the three considered parallel implementations did not produce exactly the same results as those obtained by the original PPI algorithm implemented in Kodak's Research Systems ENVI 4.0, although the spectral similarity scores with regards to the reference USGS signatures were very satisfactory in all cases.

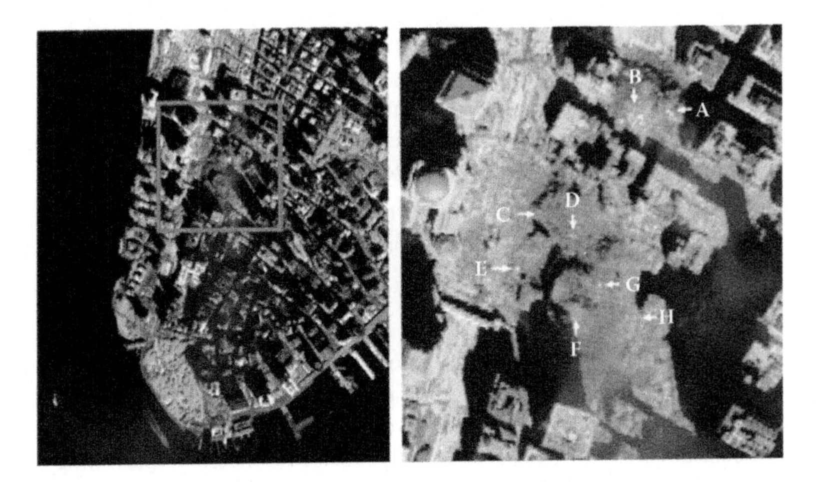

Figure 2.6 AVIRIS hyperspectral image collected by NASA's Jet Propulsion Laboratory over lower Manhattan on Sept. 16, 2001 (left), and location of thermal hot spots in the fires observed in the World Trade Center area (right).

Table 2.3 shows the spectral angle distance (SAD) between the most similar target pixels detected by the original ENVI implementation and our three proposed parallel implementations with regards to the USGS signatures. In all cases, the total number of endmembers to be extracted was set to $E = 16$ for all versions after estimating the virtual dimensionality (VD) of the data [6], although only seven endmembers were available for quantitative assessment in Table 2.3 due to the limited number of ground-truth signatures in our USGS library. Prior to a full examination and discussion of the results, it is also important to outline parameter values used for the PPI. It is worth noting that, in experiments with the AVIRIS scene, we observed that the PPI produced the same final set of experiments when the number of randomly generated skewers was set to $k = 10^4$ and above (values of $k = 10^3$, 10^5, and 10^6 were also tested). Based on the above simple experiments, we empirically set parameter t_v (threshold value) to the

TABLE 2.3 SAD-Based Spectral Similarity Scores Between Endmembers Extracted by Different Parallel Implementations of the PPI Algorithm and the USGS Reference Signatures Collected in the WTC Area

Dust/Debris Class	ENVI	Cluster-Based	Heterogeneous	FPGA
Gypsum Wall board – GDS 524	0.081	0.089	0.089	0.089
Cement – WTC01-37A(c)	0.094	0.094	0.099	0.099
Dust – WTC01-15	0.077	0.077	0.077	0.077
Dust – WTC01-36	0.086	0.086	0.086	0.086
Dust – WTC01-28	0.069	0.069	0.069	0.069
Concrete – WTC01-37Am	0.073	0.073	0.075	0.073
Concrete – WTC01-37B	0.090	0.090	0.090	0.090

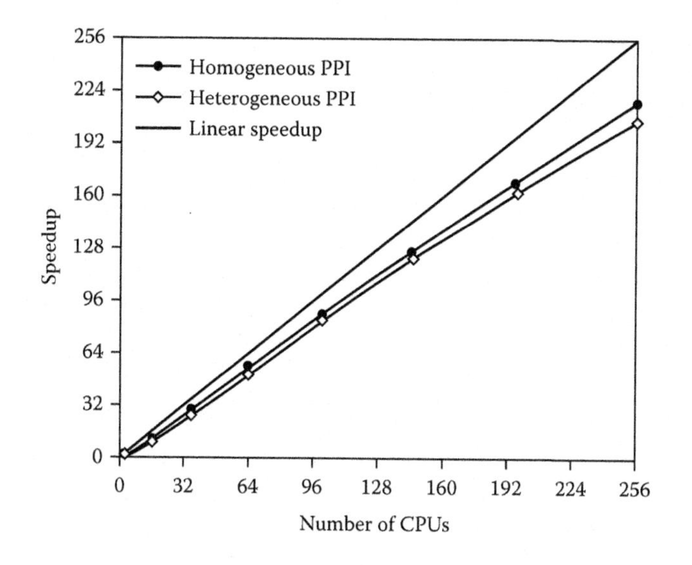

Figure 2.7 Scalability of the cluster-based and heterogeneous parallel implementations of PPI on Thunderhead.

mean of N_{PPI} scores obtained after $k = 1000$ iterations. In addition, we set the threshold angle value used to discard redundant endmembers during the process to $t_a = 0.01$. These parameter values are in agreement with those used before in the literature [32].

2.4.3 Performance Evaluation

To investigate the parallel properties of the parallel algorithms proposed in this chapter, we first tested the performance of the cluster-based implementation of PPI and its heterogeneous version on NASA's GSFC Thunderhead Beowulf cluster. For that purpose, Figure 2.7 plots the speedups achieved by multi-processor runs of the homogeneous and heterogeneous parallel versions of the PPI algorithm over the corresponding single-processor runs performed using only the Thunderhead processor. It should be noted that the speedup factors in Figure 2.7 were calculated as follows: the real time required to complete a task on p processors, $T(p)$, was approximated by $T(p) = A_p + \frac{B_p}{p}$, where A_p is the sequential (non-parallelizable) portion of the computation and B_p is the parallel portion. In our parallel codes, A_p corresponds to the *data partitioning* and *endmember selection* steps (performed by the master), while B_p corresponds to the *skewer generation*, *extreme projections*, and *candidate selection* steps, which are performed in "embarrasingly parallel" fashion at the different workers. With the above assumptions in mind, we can define the speedup for p processors, S_p, as follows:

$$S_p = \frac{T(1)}{T(p)} \approx \frac{A_p + B_p}{A_p + (B_p/p)}, \tag{2.5}$$

TABLE 2.4 Processing Times (Seconds) Achieved by the Cluster-Based and Heterogeneous Parallel Implementations of PPI on Thunderhead

Number of CPUs	1	4	16	36	64	100	144	196	256
Cluster-based PPI	2745	1012	228	94	49	30	21	16	12
Heterogeneous PPI	2745	1072	273	106	53	32	22	17	13

where $T(1)$ denotes single processor time. The relationship above is known as Amdahl's Law [46]. It is obvious from this expression that the speedup of a parallel algorithm does not continue to increase with increasing the number of processors. The reason is that the sequential portion A_p is proportionally more important as the number of processors increases, and, thus, the performance of the parallelization is generally degraded for a large number of processors. In fact, since only the parallel portion B_p scales with the time required to complete the calculation and the serial component remains constant, there is a theoretical limit for the maximum parallel speedup achievable for p processors, which is given by the following expression:

$$S_{\infty}^{p} = \lim_{p \to \infty} S_p = \frac{A_p + B_p}{A_p} = 1 + \frac{B_p}{A_p} \tag{2.6}$$

In our experiments, we have observed that although the speedup plots in Figure 2.7 flatten out a little for a large number of processors, they are very close to linear speedup, which is the optimal case in spite of equation 2.6. The plots also reveal that the scalability of the heterogeneous algorithm was esentially the same as that evidenced by its homogeneous version. For the sake of quantitative comparison, Table 2.4 reports the measured execution times by the tested algorithms on Thunderhead, using different numbers of processors. The results in Table 2.4 reveal that the heterogeneous implementation of PPI can effectively adapt to a massively parallel homogeneous environment, thus being able to produce a response in only a few seconds (12–13) using a relatively moderate number of processors.

After evaluating the performance of the proposed cluster-based implementation on a fully homogeneous cluster, a further objective was to evaluate how the proposed heterogeneous implementation performed on heterogeneous NOWs. For that purpose, we evaluated its performance by timing the parallel heterogeneous code using four (equivalent) networks of distributed workstations. Table 2.5 shows the measured execution times for the proposed heterogeneous algorithm and a homogeneous version

TABLE 2.5 Execution Times (Measured In Seconds) of the Heterogeneous PPI and its Homogeneous Version on the Four Considered Nows (16 Processors)

PPI Implementation	Fully Hetero	Fully Homo	Partially Hetero	Partially Homo
Heterogeneous	84	89	87	88
Homogeneous	667	81	638	374

that was directly obtained from the heterogeneous one by by simply replacing step 3 of the WEA algorithm with $\alpha_i = P/W$ for all $i \in \{1, 2, \cdots, P\}$.

As expected, the execution times reported in Table 2.5 show that the heterogeneous algorithm was able to adapt much better to fully (or partially) heterogeneous environments than the homogeneous version, which only performed satisfactorily on the fully homogeneous network. One can see that the heterogeneous algorithm was always several times faster than its homogeneous counterpart in the fully heterogeneous NOW, and also in both the partially homogeneous and the partially heterogeneous networks. On the other hand, the homogeneous algorithm only slightly outperformed its heterogeneous counterpart in the fully homogeneous NOW. Table 2.5 also indicates that the performance of the heterogeneous algorithm on the fully heterogeneous platform was almost the same as that evidenced by the equivalent homogeneous algorithm on the fully homogeneous NOW. This indicated that the proposed heterogeneous algorithm was always close to the optimal heterogeneous modification of the basic homogeneous one. On the other hand, the homogeneous algorithm performed much better on the partially homogeneous network (made up of processors with the same cycle-times) than on the partially heterogeneous network. This fact reveals that processor heterogeneity has a more significant impact on algorithm performance than network heterogeneity, a fact that is not surprising given our adopted strategy for data partitioning in the design of the parallel heterogeneous algorithm. Finally, Table 2.5 shows that the homogeneous version only slightly outperformed the heterogeneous algorithm in the fully homogeneous NOW. This clearly demonstrates the flexibility of the proposed heterogeneous algorithm, which was able to adapt efficiently to the four considered network environments.

To further explore the parallel properties of the considered algorithms in more detail, an in-depth analysis of computation and communication times achieved by the different methods is also highly desirable. For that purpose, Table 2.6 shows the total time spent by the tested algorithms in communications and computations in the four considered networks, where two types of computation times were analyzed, namely, sequential (those performed by the root node with no other parallel tasks active in the system, labeled as A_p in the table) and parallel (the rest of the computations, i.e., those performed by the root node and/or the workers in parallel, labeled as B_p in the table). The latter includes the times in which the workers remain idle. It can be seen from Table 2.6 that the A_p scores were relevant for both the heterogeneous and homogeneous implementations of PPI, mainly due to the final *endmember selection* step at is performed at the master node once the workers have finalized their parallel

TABLE 2.6 Communication (*com*), Sequential Computation (A_p), and Parallel Computation (B_p) Times Obtained on the Four Considered NOWs

	Fully Hetero			**Fully Homo**			**Partially Hetero**			**Partially Homo**		
	com	A_p	B_p	***com***	A_p	B_p	***com***	A_p	B_p	***com***	A_p	B_p
Heterogeneous	7	19	58	11	16	62	8	18	61	8	20	60
Homogeneous	14	19	634	6	16	59	9	18	611	12	20	342

TABLE 2.7 Load Balancing Rates for the Heterogeneous PPI and its Homogeneous Version on the Four Considered NOWs

	Fully Hetero		Fully Homo		Partially Hetero		Partially Homo	
	D_{all}	D_{minus}	D_{all}	D_{minus}	D_{all}	D_{minus}	D_{all}	D_{minus}
Heterogeneous	1.19	1.05	1.16	1.03	1.24	1.06	1.22	1.03
Homogeneous	1.62	1.23	1.20	1.06	1.67	1.26	1.41	1.05

computations. However, it can be seen from Table 2.6 that the A_p scores were not relevant when compared to the B_p scores, in particular, for the heterogeneous algorithm. This results in high parallel efficiency of the heterogeneous version. On the other hand, it can also be seen from Table 2.6 that the cost of parallel computations (B_p scores) dominated that of communications (labeled as *com* in the table) in the two considered parallel algorithms. In particular, the ratio of B_p to *com* scores achieved by the homogeneous version executed on the (fully or partially) heterogeneous network was very high, which is probably due to a less efficient workload distribution among the heterogeneous workers. Therefore, a study of load balance is highly required to fully substantiate the parallel properties of the considered algorithms.

To analyze the important issue of load balance in more detail, Table 2.7 shows the *imbalance* scores achieved by the parallel algorithms on the four considered NOWs. The imbalance is defined as $D = R_{max}/R_{min}$, where R_{max} and R_{min} are the maxima and minima processor run times, respectively. Therefore, perfect balance is achieved when $D = 1$. In the table, we display the imbalance considering all processors, D_{all}, and also considering all processors but the root, D_{minus}. As we can see from Table 2.7, the heterogeneous PPI was able to provide values of D_{all} close to 1 in all considered networks. Further, this algorithm provided almost the same results for both D_{all} and D_{minus} while, for the homogeneous PPI, load balance was much better when the root processor was not included. In addition, it can be seen from Table 2.7 that the homogeneous algorithm executed on the (fully or partially) heterogeneous networks provided the highest values of D_{all} and D_{minus} (and hence the highest imbalance), while the heterogeneous algorithm executed on the homogeneous network resulted in values of D_{minus} that were close to 1. It is our belief that the (relatively high) unbalance scores measured for the homogeneous PPI executed on the fully heterogeneous network are not only due to memory considerations or to an inefficient allocation of data chunks to heterogeneous resources, but to the impact of communications. As future research, we are planning to include considerations about the heterogeneous communication network in the design of the data partitioning algorithm.

Although the results presented above demonstrate that the proposed parallel implementations of the PPI algorithm are satisfactory from the viewpoint of algorithm scalability, code reusability, and load balance, there are many hyperspectral imaging applications that demand a response in real time. Although the idea of mounting clusters and networks of processing elements onboard airborne and satellite hyperspectral imaging facilities has been explored in the past, the number of processing elements in such experiments has been very limited thus far, due to payload requirements in

TABLE 2.8 Summary of Resource Utilization for the FPGA-based Implementation of the PPI Algorithm

Number of gates	Number of slices	Percentage of total	Operation frequency (MHz)
526,944	12,418	36%	18,032

most remote sensing missions. For instance, a low-cost, portable Myrinet cluster of 16 processors (with very similar specifications as those of the homogeneous network of workstations used in the experiments) was recently developed at NASA's GSFC for onboard analysis. The cost of the portable cluster was only $3,000. Unfortunately, it could still not facilitate real-time performance as indicated by Table 2.5, and the incorporation of additional processing elements to the low-scale cluster was reportedly difficult due to overheating and weight considerations. As an alternative to cluster computing, FPGA-based computing provides several advantages, such as increased computational power, adaptability to different applications via reconfigurability, and compact size. Also, the cost of the Xilinx Virtex-II XC2V6000-6 FPGA used for the experiments in this work is currently only slightly higher than that of the portable Myrinet cluster mentioned above.

In order to fully substantiate the performance of our FPGA-based implementation, Table 2.8 shows a summary of resource utilization by the proposed systolic array-based implementation of the PPI algorithm on the considered XC2V6000-6 FPGA, which was able to provide a response in only a few seconds for the considered AVIRIS scene. This result is even better than that reported for the cluster-based implementation of PPI executed on Thunderhead using 256 processors. Since the FPGA used in the experiments has a total of 33,792 slices available, the results addressed in Table 2.8 indicate that there is still room in the FPGA for implementation of additional algorithms. It should be noted, however, that the considered 614×512-pixel hyperspectral scene is just a subset of the total volume of hyperspectral data that was collected by the AVIRIS sensor over the Cuprite Mining District in a single pass, which comprised up to 1228×512 pixels (with 224 spectral bands). As a result, further experiments would be required in order to optimize our FPGA-based design to be able to process the full AVIRIS flight line in real time.

2.4.4 Discussion

This section has described different HPC-based strategies for a standard data processing algorithm in remote sensing, with the purpose of evaluating the possibility of obtaining results in valid response times and with adequate reliability in several HPC platforms where these techniques are intended to be applied. Our experiments confirm that the utilization of parallel and distributed computing paradigms anticipates ground-breaking perspectives for the exploitation of these kinds of high-dimensional data sets in many different applications.

Through the detailed analysis of the PPI algorithm, a well-known hyperspectral analysis method available in commercial software, we have explored different strategies to increase the computational performance of the algorithm (which can take up to several hours of computation to complete its calculations in latest-generation desktop computers). Two of the considered strategies, i.e., commodity cluster-based computing and distributed computing in heterogeneous NOWs, seem particularly appropriate for information extraction from very large hyperspectral data archives. Parallel computing architectures made up of homogeneous and heterogeneous commodity computing resources have gained popularity in the last few years due to the chance of building a high-performance system at a reasonable cost. The scalability, code reusability, and load balance achieved by the proposed implementations in such low-cost systems offer an unprecedented opportunity to explore methodologies in other fields (e.g. data mining) that previously looked to be too computationally intensive for practical applications due to the immense files common to remote sensing problems.

To address the near-real-time computational needs introduced by many remote sensing applications, we have also developed a systolic array-based FPGA implementation of the PPI. Experimental results demonstrate that our hardware version of the PPI makes appropriate use of computing resources in the FPGA and further provides a response in near-real-time that is believed to be acceptable in most remote sensing applications. It should be noted that onboard data processing of hyperspectral imagery has been a long-awaited goal by the remote sensing community, mainly because the number of applications requiring a response in realtime has been growing exponentially in recent years. Further, the reconfigurability of FPGA systems opens many innovative perspectives from an application point of view, ranging from the appealing possibility of being able to adaptively select one out of a pool of available data processing algorithms (which could be applied on the fly aboard the airborne/satellite platform, or even from a control station on Earth), to the possibility of providing a response in realtime in applications that certainly demand so, such as military target detection, wildland fire monitoring and tracking, oil spill quantification, etc. Although the experimental results presented in this section are very encouraging, further work is still needed to arrive at optimal parallel design and implementations for the PPI and other hyperspectral imaging algorithms.

2.5 Conclusions and Future Research

Remote sensing data processing exemplifies a subject area that has drawn together an eclectic collection of participants. Increasingly, this is the nature of many endeavors at the cutting edge of science and technology. However, a common requirement in most available techniques is given by the extremely high dimensionality of remote sensing data sets, which pose new processing problems. In particular, there is a clear need to develop cost-effective algorithm implementations for dealing with remote sensing problems, and the goal to speed up algorithm performance has already been identified in many on-going and planned remote sensing missions in order to satisfy the extremely high computational requirements of time-critical applications.

In this chapter, we have taken a necessary first step towards the understanding and assimilation of the above aspects in the design of innovative high-performance data processing algorithms and architectures. The chapter has also discussed some of the problems that need to be addressed in order to translate the tremendous advances in our ability to gather and store high-dimensional remotely sensed data into fundamental, application-oriented scientific advances through the design of efficient data processing algorithms. Specifically, three innovative HPC-based techniques, based on the well-known PPI algorithm, have been introduced and evaluated from the viewpoint of both algorithm accuracy and parallel performance, including a commodity cluster-based implementation, a heterogeneity-aware parallel implementation developed for distributed networks of workstations, and an FPGA-based hardware implementation. The array of analytical techniques presented in this work offers an excellent snapshot of the state-of-the-art in the field of HPC in remote sensing.

Performance data for the proposed implementations have been provided in the context of a real application. These results reflect the versatility that currently exists in the design of HPC-based approaches, a fact that currently allows users to select a specific high-performance architecture that best fits the requirements of their application domains. In this regard, the collection of HPC-based techniques presented in this chapter also reflects the increasing sophistication of a field that is rapidly maturing at the intersection of disciplines that still can substantially improve their degree of integration, such as sensor design including optics and electronics, aerospace engineering, remote sensing, geosciences, computer sciences, signal processing, and Earth observation related products. The main purpose of this book is to present current efforts towards the integration of remote sensing science with parallel and distributed computing techniques, which may introduce substantial changes in the systems currently used by NASA and other agencies for exploiting the sheer volume of Earth and planetary remotely sensed data collected on a daily basis.

As future work, we plan to implement the proposed parallel techniques on other massively parallel computing architectures, such as NASA's Project Columbia, the MareNostrum supercomputer at Barcelona Supercomputing Center, and several grid computing environments operated by the European Space Agency. We are also developing GPU-based implementations (described in detail in the last chapter of this book), which may allow us to fully accomplish the goal of real-time, onboard information extraction from hyperspectral data sets. We also envision closer multidisciplinary collaborations with environmental scientists to address global monitoring land services and security issues through carefully application-tuned HPC algorithms.

2.6 Acknowledgments

This research was supported by the European Commission through the Marie Curie Reseach Training Network project "Hyperspectral Imaging Network" (MRTN-CT-2006-035927). The authors gratefully thank John E. Dorband, James C. Tilton, and

J. Anthony Gualtieri for many helpful discussions, and also for their collaboration on experimental results using the Thunderhead Beowulf cluster at NASA's Goddard Space Flight Center.

References

[1] R. A. Schowengerdt, *Remote sensing, 3rd edition.* Academic Press: NY, 2007.

[2] C.-I Chang, *Hyperspectral data exploitation: theory and applications.* Wiley: NY, 2007.

[3] L. Chen, I. Fujishiro and K. Nakajima, Optimizing parallel performance of unstructured volume rendering for the Earth Simulator, *Parallel Computing*, vol. 29, pp. 355–371, 2003.

[4] G. Aloisio and M. Cafaro, A dynamic earth observation system, *Parallel Computing*, vol. 29, pp. 1357–1362, 2003.

[5] K. A. Hawick, P. D. Coddington and H. A. James, Distributed frameworks and parallel algorithms for processing large-scale geographic data, *Parallel Computing*, vol. 29, pp. 1297–1333, 2003.

[6] C.-I Chang, *Hyperspectral imaging: Techniques for spectral detection and classification.* Kluwer Academic Publishers: NY, 2003.

[7] R.O. Green et al., Imaging spectroscopy and the airborne visible/infrared imaging spectrometer (AVIRIS), *Remote Sensing of Environment*, vol. 65, pp. 227–248, 1998.

[8] A. Plaza, D. Valencia, J. Plaza and P. Martinez, Commodity cluster-based parallel processing of hyperspectral imagery, *Journal of Parallel and Distributed Computing*, vol. 66, no. 3, pp. 345–358, 2006.

[9] R. Brightwell, L. A. Fisk, D. S. Greenberg, T. Hudson, M. Levenhagen, A. B. Maccabe and R. Riesen, Massively parallel computing using commodity components, *Parallel Computing*, vol. 26, pp. 243–266. 2000.

[10] P. Wang, K. Y. Liu, T. Cwik, and R.O. Green, MODTRAN on supercomputers and parallel computers, *Parallel Computing*, vol. 28, pp. 53–64, 2002.

[11] S. Kalluri, Z. Zhang, J. JaJa, S. Liang, and J. Townshend, Characterizing land surface anisotropy from AVHRR data at a global scale using high performance computing, *International Journal of Remote Sensing*, vol. 22, pp. 2171–2191, 2001.

[12] J. C. Tilton, Method for implementation of recursive hierarchical segmentation on parallel computers, *U.S. Patent Office, Washington, DC, U.S. Pending Published Application 09/839147*, 2005. Available online: http://www.fuentek.com/technologies/rhseg.htm.

[13] J. Le Moigne, W. J. Campbell and R. F. Cromp, An automated parallel image registration technique based on the correlation of wavelet features, *IEEE Transactions on Geoscience and Remote Sensing*, vol. 40, pp. 1849–1864, 2002.

[14] M. K. Dhodhi, J. A. Saghri, I. Ahmad and R. Ul-Mustafa, D-ISODATA: A distributed algorithm for unsupervised classification of remotely sensed data on network of workstations, *Journal of Parallel and Distributed Computing*, vol. 59, pp. 280–301, 1999.

[15] T. Achalakul and S. Taylor, A distributed spectral-screening PCT algorithm, *Journal of Parallel and Distributed Computing*, vol. 63, pp. 373–384, 2003.

[16] A. Lastovetsky, *Parallel computing on heterogeneous networks*, Wiley-Interscience: Hoboken, NJ, 2003.

[17] K. Hawick, H. James, A. Silis, D. Grove, C. Pattern, J. Mathew, P. Coddington, K. Kerry, J. Hercus, and F. Vaughan, DISCWorld: an environment for service-based meta-computing, *Future Generation Computer Systems*, vol. 15, pp. 623–635, 1999.

[18] I. Foster and C. Kesselman, *The Grid: Blueprint for a New Computing Infrastructure*, Morgan Kaufman: San Francisco, CA, 1999.

[19] T. Vladimirova and X. Wu, On-board partial run-time reconfiguration for pico-satellite constellations, *First NASA/ESA Conference on Adaptive Hardware and Systems*, AHS, 2006.

[20] E. El-Araby, T. El-Ghazawi and J. Le Moigne, Wavelet spectral dimension reduction of hyperspectral imagery on a reconfigurable computer, *Proceedings of the 4th IEEE International Conference on Field-Programmable Technology*, 2004.

[21] D. Valencia and A. Plaza, FPGA-Based Compression of Hyperspectral Imagery Using Spectral Unmixing and the Pixel Purity Index Algorithm, *Lecture Notes in Computer Science*, vol. 3993, pp. 24–31, 2006.

[22] J. Setoain, C. Tenllado, M. Prieto, D. Valencia, A. Plaza and J. Plaza, Parallel hyperspectral image processing on commodity grahics hardware, *International Conference on Parallel Processing (ICPP)*, Columbus, OH, 2006.

[23] J. Boardman, F. A. Kruse and R.O. Green, Mapping target signatures via partial unmixing of AVIRIS data, *Summaries of the NASA/JPL Airborne Earth Science Workshop*, Pasadena, CA, 1995.

[24] D. A. Landgrebe, *Signal theory methods in multispectral remote sensing*, Wiley: Hoboken, NJ, 2003.

[25] J. Dorband, J. Palencia and U. Ranawake, Commodity computing clusters at Goddard Space Flight Center, *Journal of Space Communication*, vol. 1, no. 3, 2003. Available online: http://satjournal.tcom.ohiou.edu/pdf/Dorband.pdf.

[26] S. Tehranian, Y. Zhao, T. Harvey, A. Swaroop and K. McKenzie, A robust framework for real-time distributed processing of satellite data, *Journal of Parallel and Distributed Computing*, vol. 66, pp. 403–418, 2006.

[27] A. Plaza, J. Plaza and D. Valencia, AMEEPAR: Parallel Morphological Algorithm for Hyperspectral Image Classification in Heterogeneous Networks of Workstations, *Lecture Notes in Computer Science*, vol. 3391, pp. 888–891, Chapman Hall CRC Press: Boca Raton, FL, 2006.

[28] J. W. Larson et al., Components, the common component architecture, and the climate/weather/ocean community, *Proceeding of the 20th International Conference on Interactive Information and Processing Systems for Meteorology, Oceanography, and Hydrology*, Seattle, WA, 2004.

[29] P. Votava, R. Nemani, K. Golden, D. Cooke and H. Hernandez, Parallel distributed application framework for Earth science data processing, *IEEE International Geoscience and Remote Sensing Symposium*, Toronto, CA, 2002.

[30] T. W. Fry and S. Hauck, Hyperspectral image compression on reconfigurable platforms, *10th IEEE Symposium on Field-Programmable Custom Computing Machines*, Napa, CA, 2002.

[31] N. Keshava and J.F. Mustard, Spectral unmixing, *IEEE Signal Processing Magazine*, Vol. 19, pp. 44–57, 2002.

[32] A. Plaza, P. Martinez, R. Perez and J. Plaza, A quantitative and comparative analysis of endmember extraction algorithms from hyperspectral data, *IEEE Transactions on Geoscience and Remote Sensing*, vol. 42, pp. 650–663, 2004.

[33] C.-I Chang and A. Plaza, A Fast Iterative Implementation of the Pixel Purity Index Algorithm, *IEEE Geoscience and Remote Sensing Letters*, vol. 3 pp. 63–67, 2006.

[34] F. J. Seinstra, D. Koelma and J. M. Geusebroek, A software architecture for user transparent parallel image processing, *Parallel Computing*, vol. 28, pp. 967–993, 2002.

[35] B. Veeravalli and S. Ranganath, Theoretical and experimental study on large size image processing applications using divisible load paradigm on distributed bus networks, *Image and Vision Computing*, vol. 20, pp. 917–935, 2003.

[36] W. Gropp, S. Huss-Lederman, A. Lumsdaine and E. Lusk, *MPI: The complete reference, vol. 2, The MPI Extensions*, MIT Press: Cambridge, MA, 1999.

[37] A. Lastovetsky and R. Reddy, HeteroMPI: towards a message-passing library for heterogeneous networks of computers, *Journal of Parallel and Distributed Computing*, vol. 66, pp. 197–220, 2006.

[38] M. Valero-Garcia, J. Navarro, J. Llaberia, M. Valero and T. Lang, A method for implementation of one-dimensional systolic algorithms with data contraflow using pipelined functional units, *Journal of VLSI Signal Processing*, vol. 4, pp. 7–25, 1992.

[39] D. Lavernier, E. Fabiani, S. Derrien and C. Wagner, Systolic array for computing the pixel purity index (PPI) algorithm on hyperspectral images, *Proc. SPIE*, vol. 4321, 2000.

[40] Y. Dou, S. Vassiliadis, G. K. Kuzmanov and G. N. Gaydadjiev, 64-bit floating-point FPGA matrix multiplication, *ACM/SIGDA 13th Intl. Symposium on FPGAs*, 2005.

[41] Celoxica Ltd., *Handel-C language reference manual*, 2003. Available online: http://www.celoxica.com.

[42] Xilinx Inc. Available online: http://www.xilinx.com.

[43] Celoxica Ltd., *DK design suite user manual*, 2003. Available online: http://www.celoxica.com.

[44] *MPICH: a portable implementation of MPI*. Available online: http://www-unix.mcs.anl.gov/mpi/mpich.

[45] A. Lastovetsky and R. Reddy, On performance analysis of heterogeneous parallel algorithms, *Parallel Computing*, vol. 30, pp. 1195–1216, 2004.

[46] J. L. Hennessy and D. A. Patterson, *Computer architecture: a quantitative approach, 3rd ed.*, Morgan Kaufmann: San Mateo, CA, 2002.

Chapter 3

Computer Architectures for Multimedia and Video Analysis

Edmundo Sáez,
University of Córdoba, Spain

José González-Mora,
University of Málaga, Spain

Nicolás Guil,
University of Málaga, Spain

José I. Benavides,
University of Córdoba, Spain

Emilio L. Zapata,
University of Málaga, Spain

Contents

Multimedia processing involves applications with a high demanding computational power. New capabilities have been included in modern processors to cope with these new requirements, and specific architectures have been designed to increase the performance of different multimedia applications. Thus, multimedia extensions were included in general purpose processors to exploit the single-instruction multiple-data (SIMD) parallelism that appears in signal processing applications. Moreover, new architectures, such as Graphics Processing Units (GPU), are being successfully applied to general purpose computation (GPGPU) and, specifically, image and video processing. In this chapter we discuss the impact of the usage of multimedia extensions and GPUs in multimedia applications and illustrate this discussion with the study of two concrete applications involving video analysis.

3.1 Introduction

The growing importance of multimedia applications is greatly influencing current computer architectures. Thus, new extensions have been included in general purpose processors in order to exploit the subword parallelism (SIMD paradigm) that appears in some of these applications [1],[2]. Additionally, the capability of new processors designs, such as VLIW and SMT, to exploit the coarse and fine grain parallelism present at multimedia processing is being studied [3].

Also, specific architectures have been developed to speed up the performance of programs that process multimedia contents. Thus, specific implementations based on associative processors [4] and FPGAs [5] have been performed.

Graphics Processing Units (GPUs) were initially devoted to manage 3-D objects by including a powerful design with plenty of functional units able to apply geometric transformation and projections to a high number of spatial points and pixels describing the 3-D scenario. Nowadays, the two processors included in GPUs, named vertex and pixel processors, can be programmed using a high level language [6]. This allows other applications different from graphics to use the capabilities of those processing units, as, for example, video decoding [37].

Several specific benchmarks have been developed to measure the impact of architectural improvements on multimedia processing. This way, UCLA Media Bench is a suite of multimedia applications and data sets designed to represent the workload of emerging multimedia and communications systems [7]. Berkley multimedia workload [8] updates previous the benchmark by adding new applications (most notably MP3 audio) and modifying several data sets.

Attending to existing benchmarks, multimedia applications can be classified into the following three groups:

- *Coding/decoding algorithms.* Audio and/or video signals are digitized in order to obtain better quality and low bit-rates. Different efforts in the field of standardization have driven the development of specific standards for video transmission, such as MPEG 1, MPEG 2, and MPEG 4, and also for image compression, as JPEG. Signal coding and decoding are basic tools to generate multimedia contents. This fact, in addition to their high demanding computational requirements, have made these applications the most common used for testing the multimedia capabilities of a computer architecture.
- *Media content management and analysis.* The huge amount of digital content that is daily produced and stored increases the need for powerful tools for its manipulation. Thus, editing tools are used to cut and paste different pieces of sequences according to the audiovisual language. Simple editing effects can be applied in the compress domain. However, more complex effects must be generated in the uncompress domain, requiring decoding and encoding stages. As a result, an important challenge in this field is the topic of automatic indexing of the multimedia contents. It is based on the application of computational techniques to the digital audiovisual information in order to analyze those contents and extract some descriptors. These descriptors will be annotated in a database for indexing purposes, allowing the implementation of content-based retrieval techniques. Description of the multimedia content is a wide area. Currently, MPEG 7 is a standard that proposes both a set of multimedia descriptors and the relationship between them. It should be noted that the complexity of the analysis to be applied to the video content is related to the semantic level of the descriptors to be extracted. Thus, temporal video segmentation involves the comparison of low-level frame information to locate edition effects and camera movements. Object identification and tracking need more complex algorithms to improve the robustness.
- *3-D graphics.* Real-time 3D applications have become usual in multimedia applications. The use of more realistic scenarios implies the increase in the number of polygons needed to describe the objects. Thus, more computational requirements are needed for the rendering process.

Current multimedia benchmarks mainly focus on the computation of specific kernels or functions. However, we consider that the mapping of a complete application to a concrete computing architecture can help to extract interesting conclusions about the limits of the architecture and evaluate the impact of possible improvements.

Thus, in this chapter, we present two different applications involving video analysis: temporal video segmentation and object tracking. In the first application, multimedia extensions implemented in general purpose processors are used to increase the performance of a temporal video segmentation algorithm. Our study extracts the computational kernels that appear in the algorithm and analyzes how to map these kernels to the available SIMD instructions. Final speedup values consider the whole application, that is, optimized and non-optimized kernels. In the second application, we show the capabilities of the GPUs for general purpose processing and illustrate the performance that this architecture can achieve by implementing an application for object tracking in video. Reported performance is compared with that from general purpose processors showing that it depends on the problem size.

The rest of the chapter is organized as follows. In the next section, basic concepts about the two computing architectures are presented. Section 3.3 introduces the temporal video segmentation algorithm, including the analysis of its main computational kernels. Section 3.4 shows the optimization of the computational kernels in the proposed segmentation algorithm, and Section 3.5 analyses the achieved performance by using multimedia extensions. The following three sections illustrate the use of the GPU in object tracking in video. Hence, Section 3.6 introduces the techniques for object tracking using a second order approach while Section 3.7 studies the mapping of this algorithm to the GPU. Section 3.8 shows the performance of the GPU results compared with general purpose processors. Finally, in Section 3.9, main conclusions are summarized.

3.2 The Computing Architectures

Two different computing architectures have been proposed to implement multimedia applications. Following are the basic concepts of these architectures.

3.2.1 Multimedia Extensions

The multimedia extensions were proposed to deal with the demanding requirements of modern multimedia applications. They are based on the SIMD computation model and exploit the subword parallelism present in signal processing applications. Thus, arithmetic functions that operate with long words (e.g. 128 bits) can be subdivided to perform parallel computation with shorter words, as shown in Figure 3.1.

The major manufacturers have proposed their own set of multimedia extensions. Thus, AMD uses in their processors the 3DNow! technology. The processors PowerPC from Motorola include the AltiVec extensions. Finally, Sun Microsystems presented the VIS extensions in their UltraSparc processors. Nevertheless, the most famous extensions are those from Intel. In 1997, Intel introduced the first version of its multimedia extensions, the MMX technology. It consisted of 57 new instructions used to perform SIMD calculations on up to 8 integer operands. The FPU registers were used as MMX registers. Hence, combining floating-point and multimedia code was not advisable. The Pentium 3 processor introduced the SSE extensions, with 70 new instructions, some of which were used to perform SIMD calculations on four simple precision floating-point operands. Here, the set of registers, known as XMM registers, were independent of the FPU registers. Later, the Pentium 4 introduced the SSE2 extensions,

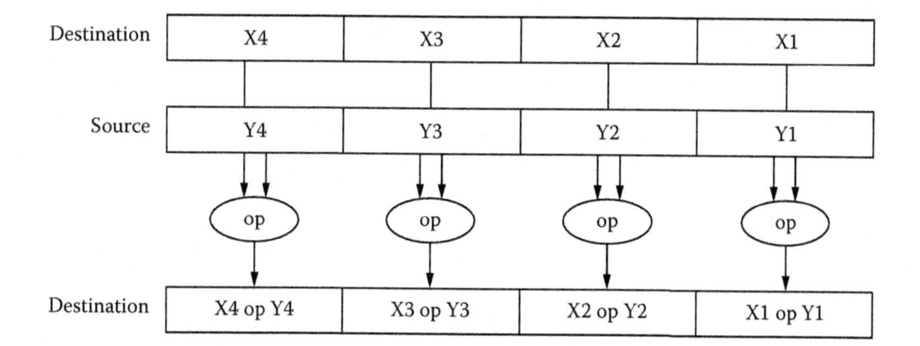

Figure 3.1 Typical SIMD operation using multimedia extensions.

containing 144 new instructions. The main innovation was the ability to work with up to two double precision floating-point operands. The last proposal from Intel is the SSE3 extensions, which includes instructions to perform horizontal and asymmetric computations, instead of vertical computations, as shown in Figure 3.1. A comparison of the performance of different sets of multimedia extensions is given in [23].

3.2.2 Graphics Processing Units

General purpose programming of Graphic Processing Units has become a topic of considerable interest in the past years due to the spectacular evolution of this kind of hardware in terms of capabilities to satisfy leisure and multimedia applications requirements. Nowadays it is common to find this kind of processor in almost every personal computer, and, in many occasions, their computing capabilities (in number of operations per second) are higher than those offered by the CPU, at a reasonable cost.

Initial GPU designs consisted of a specific fixed pipeline for graphic rendering. However, the increasing demand of customized graphic effects to fit different applications requirements motivated a fast evolution of its programming possibilities in subsequent generations. GPUs' desirable characteristics include a high memory bandwidth and a higher number of floating-point units than the CPUs. These facts make these platforms interesting candidates to be taken into account as efficient coprocessors of intensive computing applications in domestic platforms and workstations.

The computing capabilities of GPUs as intrinsic parallel systems have been shown in many recent works, including numerical computing applications, such as [35],[36].

Nowadays, the GPU is a deeply segmented architecture. The number of processing stages and its sequence greatly differ from one manufacturer to another, and even between the different models. However, a common high level structure can be distinguished, such as the one illustrated in Figure 3.2. Initial stages are related to the management of input vertex attributes defining the graphic objects feeding the graphic card. Ending stages focus on color processing of the pixels to be shown on the screen. In this design, *vertex processors* and *fragment processors* (darker in the figure) are the programmable stages of the pipeline.

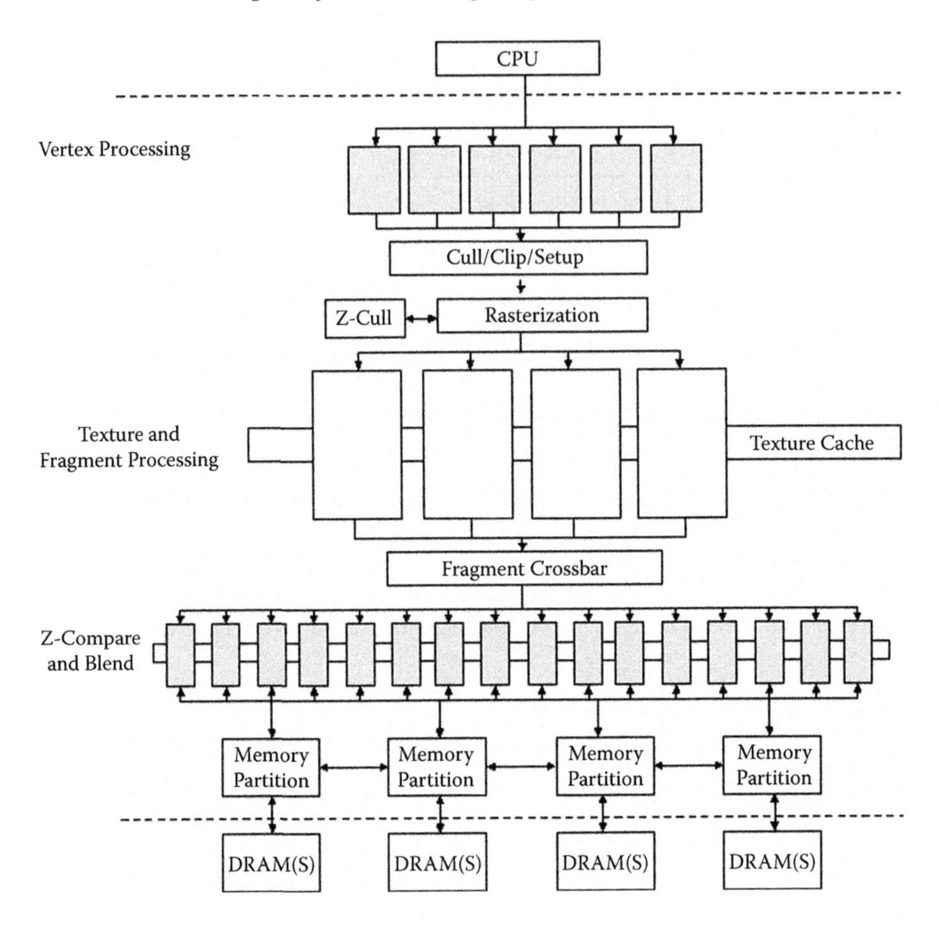

Figure 3.2 GPU pipeline.

The most distinctive characteristic is the main role of the data pipeline, in comparison to the instruction pipeline based CPU computing model (von-Neumann architecture). GPU instructions are employed to set up the hardware at a low level and to feed it with input data, instead of being stored in video memory. Data arrive at the pipeline as vertices and attributes and, after being transformed by the different execution units, they are represented as pixels in the framebuffer.

The graphic pipeline works on all the data simultaneously and the predominance of unary instructions allows it to simplify the hardware involved in risk management. No data forwarding or predictions units are used. That implies an important saving in the number of required transistors, which can be used to increment the computing resources and minimize the structural dependencies by means of a higher degree of parallelism.

Processing is organized using multiple uncoupled functional units working on data groups. Computations on the functional units generate resulting pixels in the framebuffer. Spatial data locality is exploited by assigning a rectangular area of adjacent pixels to a specific fragment processor in a so-called 'tiled' processing. In addition, the last GPU generations are able to exploit SIMD parallelism by computing simultaneously the four data associated to a pixel (three color components and one transparency value) as floating-point numbers.

In a computational model where data handling is so important, memory bandwidth has a fundamental role in the throughput. The GPU model looks for a fast memory access as uniform as possible in all its address spaces. To get this bandwidth, low latency memories with high bus dimensions are used; i.e. in 2002 they were equipped with 256-bit memories in addition to the 64 bits of many CPU architectures.

As a result of these characteristics, GPUs' theoretical capabilities are higher than those obtained from commodity CPUs. Furthermore, its design as a stream processing architectures scales well with respect to both the number of fragment processors and the core clock rate. However, real values in practical applications can be far away from the theoretical performance. In many cases, memory access is a major bottleneck that prevents the functional units from being active all the time. Thus, it is important to choose those designs that better map on these architectures, in order to take advantage of the GPU parallelism and make use of the spatial coherence of the geometric primitives.

3.3 Temporal Video Segmentation

Temporal video segmentation has become an active research field in the past years. As video archives continue growing, fast and reliable methods for indexing purposes are needed. One of the most useful techniques in video indexing is shot detection, where a shot is defined as a sequence of frames captured from a single camera operation. Shot detection is performed by means of shot transition detection algorithms.

Two different types of transitions are used to split a video into shots: abrupt and gradual transitions. Abrupt transitions, also referred to as cuts or straight cuts, occur when a sudden change from one shot to the next one is performed in just one frame. On the other hand, gradual transitions use several frames to link two shots together. Other authors refer to gradual transitions as optical cuts [9]. Depending on how the shots are mixed up in the transition, there are many different types of gradual transitions. Dissolves are the most common. Fades and wipes are also frequently used. What is most important about gradual transitions is that they are often used to establish some kind of semantic information in the video. For example, dissolves have been widely used to perform scene changes in video editing, where a scene is a set of shots closely related in terms of place and time.

While abrupt transitions detection is a relatively easy task [10], gradual transitions detection is still an open issue, as the amount of false positives reported by the algorithms is very high for certain sequences. The main problem in gradual transitions detection is that camera operation (pan, tilt, swing, zoom, etc.) originates similar patterns to those generated by gradual transitions [11]. Thus, a method to estimate global motion in video is needed in order to discard false positives induced by camera operations [12].

The first stage of shot transition detection algorithms is the extraction of characteristics from the video streams. One or more metrics are then used to compute several parameters from the characteristics. These metrics can be based on pixel luminance, contour information, block tracking, etc. Although most of the proposed methods make use of only one metric, using several of them is recommended as drawbacks from one metric could be compensated by the others [13], as long as the used metrics rely on different video characteristics.

The computed parameters are then used to determine the occurrence of a transition. Here, data driven methods address the problem from the data analysis point of view. On the other hand, model driven methods, based on mathematical models of video data, allow a systematic analysis of the problem and the use of domain-specific constraints, which helps to improve the efficiency [14],[15].

Other authors use non-deterministic classifiers to study the computed parameters in order to perform pattern recognition, as transitions generally resulting in a characteristic pattern in the parameters. Using a non-deterministic classifier makes unnecessary a specifically designed pattern recognition method, which usually needs several parameters to be tuned. Also, by using a supervised classification scheme, the system is able to learn the patterns generated by different types of gradual transitions. Some shot transition detection algorithms using neural networks [16],[17], hidden Markov models [18],[19], or support vector machines [20],[21] have been proposed.

3.3.1 Temporal Video Segmentation Algorithm

Many efforts have been devoted to developing reliable temporal video segmentation algorithms [22] running at real-time processing speed. However, this is not easy to achieve when different video features are computed; several video streams are processed in parallel; or even other tasks, such as a video decoder, are running at the same time. Thus, the optimization of temporal video segmentation applications is encouraged.

Certain characteristics of these algorithms, such as the calculation of numerical operations on large amounts of data stored in matrices, make them good candidates to be optimized by means of the multimedia extensions included in modern processors [23]. Thus, by optimizing these kinds of tasks, which are usually very time consuming, the algorithms are able to run faster.

The temporal video segmentation algorithm to be optimized was introduced in [24]. It is an algorithm designed to perform cut detection on MPEG compressed video using DC images. Two similarity values are computed for each pair of frames using a luminance based metric (LBM) and a contour based metric (CBM). Both sequences of similarity values are median filtered and then analyzed using a classifier to perform cut detection.

The LBM calculates the similarity using the expression

$$f_{LBM}(H_1, H_2) = \frac{\sum_b H_1[b] \cdot W_{1D}(H_2[b])}{\sqrt{H_1[b] \cdot W_{1D}(H_1[b])} \cdot \sqrt{H_2[b] \cdot W_{1D}(H_2[b])}} \tag{3.1}$$

where H_1 and H_2 are luminance histograms of DC images, and

$$W_{1D}(H[b]) = \sum_{i=-1}^{1} H[b+i] \tag{3.2}$$

On the other hand, the CBM calculates the similarity using the expression

$$f_{CBM}(OT_1, OT_2) = \frac{\sum_\alpha \sum_\theta OT_1[\alpha][\theta] \cdot W_{2D}(OT_2[\alpha][\theta])}{\sqrt{OT_1[\alpha][\theta] \cdot W_{2D}(OT_1[\alpha][\theta])} \cdot \sqrt{OT_2[\alpha][\theta] \cdot W_{2D}(OT_2[\alpha][\theta])}} \tag{3.3}$$

where OT_1 and OT_2 are the orientation tables calculated by the generalized Hough transform (GHT) [25] from the DC images, and

$$W_{2D}(OT[\alpha][\theta]) = \sum_{w_i=-2}^{2} \sum_{w_j=-2}^{2} OT[\alpha + w_i][\theta + w_j] \tag{3.4}$$

A block diagram of the algorithm is shown in Figure 3.3. The tasks used in this algorithm that are suitable to be optimized using the multimedia extensions are indicated by a dark circle.

Both LBM and CBM perform some of these tasks, such as sum of products and the functions W_{1D} and W_{2D}. The CBM also computes several arctangent values. Thus, a procedure to perform an arctangent calculation using the multimedia extensions should also be developed.

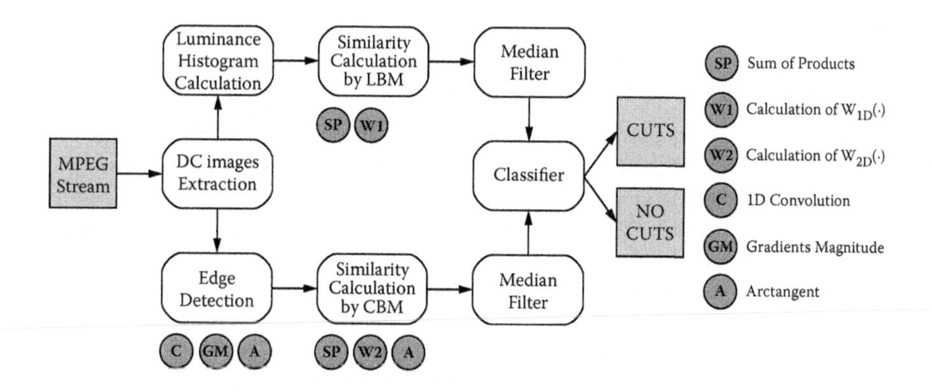

Figure 3.3 Temporal video segmentation algorithm to optimize.

The edge detection stage, implemented using the Canny Algorithm [26], includes some more operations suitable for the multimedia extensions, such as the 1D convolution, the gradients magnitude calculation, and the gradients direction calculation. This last operation is also based on an arctangent calculation.

Some other operations, such as the luminance histogram calculation or the creation of the orientation tables of the GHT, could not be optimized as the multimedia extensions are not applicable due to their memory access patterns.

3.4 Applying Multimedia Extensions to Temporal Video Segmentation

In this section, the most time-consuming tasks within temporal video segmentation are identified and their computation kernels optimized using the multimedia extensions provided by the Intel Pentium 4 processor.

3.4.1 Sum of Products

The sum of products is used by both LBM and CBM as a key operations to compute similarity values between frames. It is given by the expression

$$R = \sum_{i=0}^{n-1} U[i] \cdot V[i] \tag{3.5}$$

where U and V are the data vectors and R is the desired sum of products.

Vectors U and V are accessed in groups of 8 elements, 16 bits each, and stored in XMM registers. Then, the PMADDWD instruction is used to compute 4 sums of products (SP) in an XMM register. Each one is 32 bits wide. The PADDD instruction is used to accumulate sums of products during the iterations of the algorithm. Finally, 4 sums of sums of products (SSP) are computed in register XMM0. Then, shift instructions (PSRLDQ) as well as add instructions (PADDD) are used to calculate the value of R.

3.4.2 Calculation of $W_{ID}(\cdot)$

The function $W_{1D}(\cdot)$ is used by LBM to compute the similarity between two frames (see Section 3.4.1). In each iteration of the algorithm, a new group of 8 elements is accessed in memory and stored in register XMM2. Two more groups of 8 elements are already stored in registers XMM0 and XMM1 from previous iterations. The elements in these registers are arranged using shift instructions (PSLLDQ and PSRLDQ) and then added using the PADDW instruction. Thus, 8 elements are computed in parallel.

3.4.3 Calculation of $W_{2D}(\cdot)$

Similarly, the function $W_{2D}(\cdot)$ is used by CBM to compute the similarity between two frames. Its expression is also given in Section 3.4.1. In a first stage, horizontal sums are computed for each element in the orientation table using the expression

$$HS[\alpha][\theta] = \sum_{w_j=-2}^{2} OT[\alpha][\theta + w_j] \tag{3.6}$$

obtaining this way the horizontal sums table (HS). Then, the second stage performs vertical sums of elements in HS using the expression

$$W_{2D}(OT[\alpha][\theta]) = VS[\alpha][\theta] = \sum_{w_j=-2}^{2} HS[\alpha + w_i][\theta] \tag{3.7}$$

The process to calculate the horizontal sums is similar to that depicted in Section 3.3, which allows us to compute 8 elements in parallel. However, as 5 terms instead of 3 have to be added to obtain a horizontal sum, more instructions are needed.

To compute the vertical sums, the last expression is turned into

$$VS[\alpha][\theta] = VS[\alpha - 1][\theta] + HS[\alpha + 2][\theta] - HS[\alpha - 3][\theta] \tag{3.8}$$

In other words, each vertical sum is computed from the previous one, just using one PADDW and one PSUBW instruction. As in the previous stage, 8 elements are computed in parallel.

3.4.4 1-D Convolution

The algorithm to optimize implements horizontal and vertical 1-D convolution operations, using a five-element mask, in the edge detection stage. The source image elements I are multiplied by the mask elements M and added to obtain the elements of the convolved image C. Each image element is accessed only once and multiplied by every mask element. The computed products are accumulated using five registers, disposed following a top to bottom, right to left line. In each iteration, the leftmost accumulator produces one element of the convolved image.

This strategy has been implemented using the MULPS and ADDPS instructions, to obtain 4 single precision floating-point convolved terms in parallel, as shown in Figure 3.4 for the horizontal 1-D convolution. The implementation of the vertical 1-D convolution is highly similar.

3.4.5 Gradients Magnitude

The process to compute the magnitude of a gradient vector in the Canny Algorithm is also used in the edge detection stage. It is given by the expression

$$\|\nabla I\| = \sqrt{I_x^2 + I_y^2} \tag{3.9}$$

where I_x and I_y are the horizontal and vertical derivatives of the image. This expression is simple to optimize using the instructions ADDPS, MULPS, and SQRTPS, obtaining this way 4 terms per iteration.

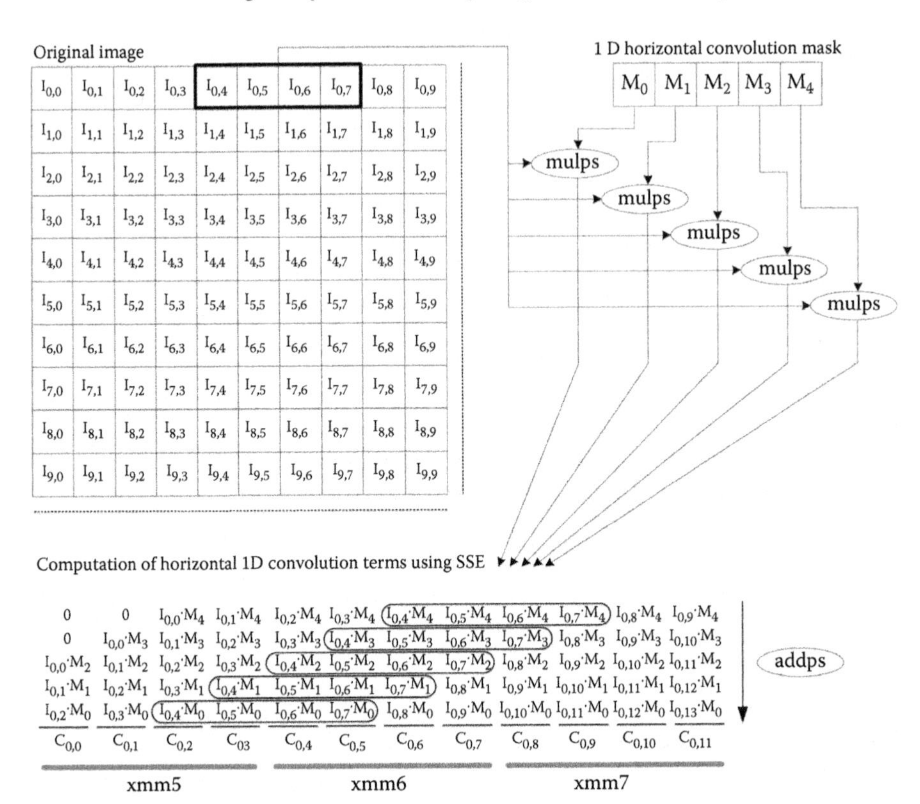

Figure 3.4 Implementation of the horizontal 1-D convolution.

3.4.6 Arctangent

The calculation of arctangent values is used in two stages of the algorithm to be optimized. It is used by the Canny Algorithm in the edge detection stage to compute the direction of a gradient vector. It is also used in the procedure to compute the similarity between two frames performed by the CBM.

The implementation of trigonometrical functions using multimedia extensions has been addressed in [27]. However, no method is proposed to perform arctangent calculations. A procedure to compute four arctangent in parallel has been developped. It is based on the five first terms of the Taylor's series, using the instructions MULPS, ADDPS, and SUBPS. This allows one to compute arctangent values with an error less than 3°.

As the Taylor's series to obtain an arctangent value is defined only within the interval [−45°, 45°], an extension has been implemented. This allows the method to work in parallel within the interval [0°, 360°]. In order to increase performance, no branch instructions have been used. Instead, a scheme using the instructions ANDPS and ORPS is proposed. This extension is shown in Figure 3.5.

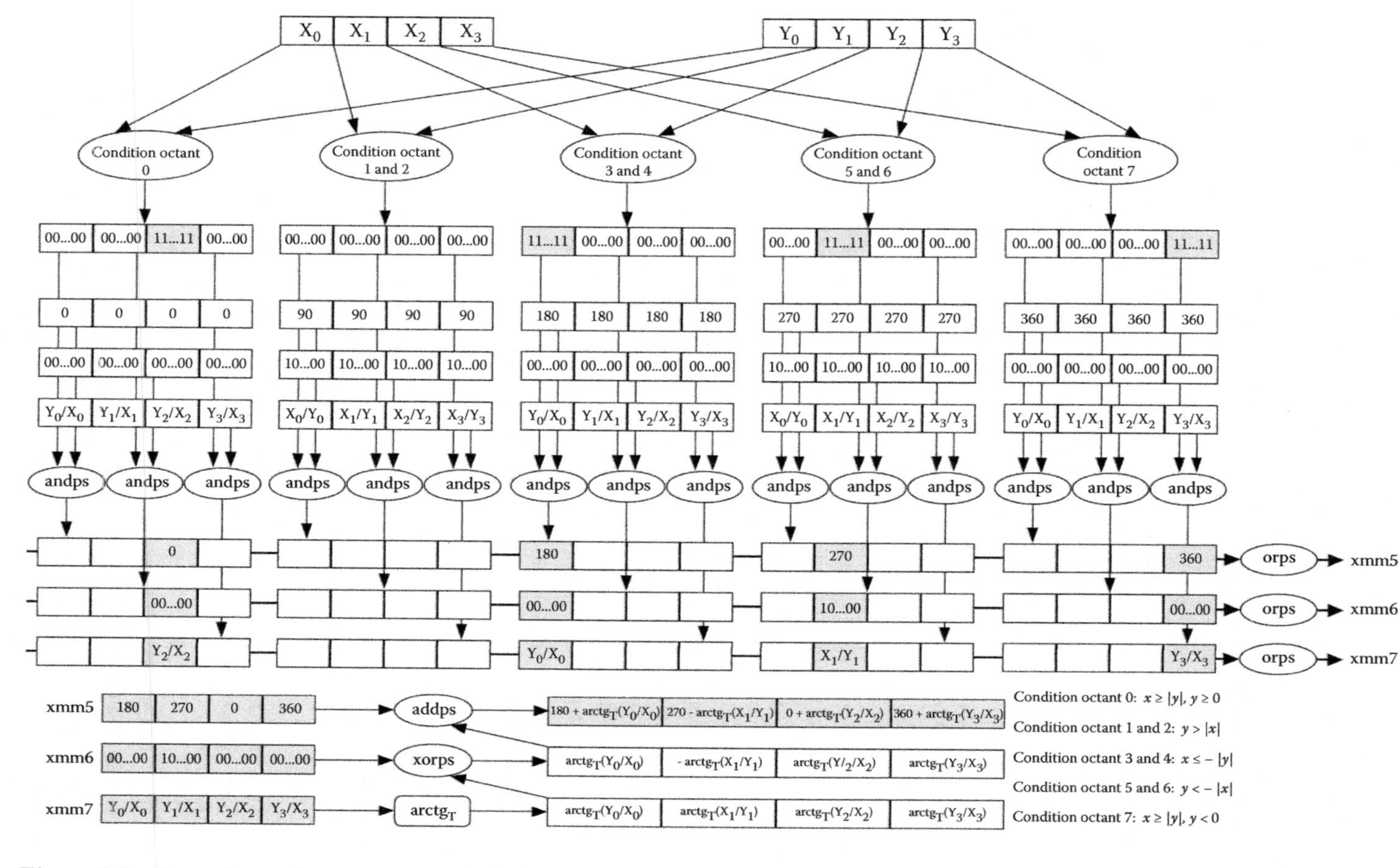

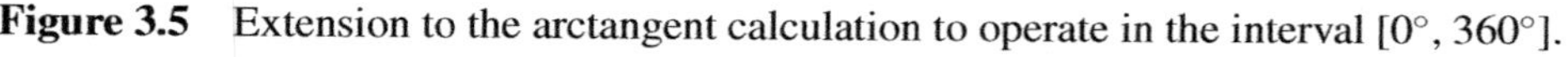

Figure 3.5 Extension to the arctangent calculation to operate in the interval [0°, 360°].

3.5 Performance of Multimedia Extensions

In order to implement the aforementioned optimized kernels, two key aspects have been studied. On one hand, it is important to choose the right compiler, as different compilers may produce different object codes, and thus different performance. On the other hand, it is even more important to make the correct decision about the implementation method (inlined assembler code, autovectorization or intrinsics).

The kernels have been implemented using the GNU C Complier (gcc) version 4.0.0 as well as the Intel C++ Compiler version 9.0. The obtained results show that the optimized versions produced by the latter are slightly faster. Regarding the implementation method, the results pointed out that the autovectorization is highly effective in simple algorithms as the sum of products, but inefficient in more complicated kernels, as the compiler is not able to design a suitable strategy. Good performance can be obtained using inlined assembler code. However, the compiler is unable to perform code reorganization in order to avoid structural risks. Thus, certain kernels exhibit low performance. Finally, the most efficient versions of the optimized kernels were those implemented using intrinsics. The use of intrinsics allows the programmer to design a parallelization strategy suitable for the kernel, as long as it lets the compiler reorganize the object code to obtain better performance.

Table 3.1 shows the clock cycles spent by the kernels in their sequential and optimized (parallel) implementations. The computed speedup for each kernel is also provided. To obtain these values the Intel C++ Compiler has been used. The algorithms have been implemented using intrinsics.

The speedup obtained by the optimized kernels is variable. Highest speedups are reported by those tasks that perform a vertical processing of a table, as the vertical sums in the calculation of $W_{2D}(\cdot)$ or the vertical 1-D convolution. This is due to the reduction of cache misses when reading data elements. When a table is accessed by columns, a cache miss will occur in each memory access if the table is large enough.

TABLE 3.1 Clock Cycles and Speedups for the Sequential/Optimized Kernel Implementations

Kernel Used	Sequential Clock Cycles	Optimized Clock Cycles	Speedup Measured
Sum of products (64 elements)	396	108	3.67
Sum of products (2025 elements)	9940	1084	9.17
$W_{1D}(\cdot)$	296	144	2.06
$W_{2D}(\cdot)$ (horizontal)	13100	7964	1.64
$W_{2D}(\cdot)$ (vertical)	53804	4064	13.24
1-D Convolution (horizontal)	30220	10168	2.97
1-D Convolution (vertical)	134436	6304	21.33
Gradients magnitude	49116	13824	3.55
Arctangent	144424	32684	4.42

TABLE 3.2 Percentage of Computation Time Spent by the Temporal Video Segmentation Algorithm in Different Tasks, Before and After the Optimization

Tasks Performed	**% Computation (Sequential)**	**% Computation (Optimized)**
Edge detection:		
1-D convolution (horizontal)	1.96	2.12
1-D convolution (vertical)	10.61	2.54
Gradients magnitude	3.35	2.12
Gradients direction	6.98	2.12
Similarity calculation by CBM:		
Contour points histogram calculation	2.79	3.81
Orientation tables calculation	33.24	48.31
Pairings insertion into orientation tables	1.12	5.93
Calculation of α angle	27.94	12.71
Calculation of $W_{2D}(\cdot)$ (horizontal)	1.12	0.85
Calculation of $W_{2D}(\cdot)$ (vertical)	1.68	0
Sum of Products (2025 elements)	2.51	0.42
Similarity calculation by LBM:		
Luminance histogram calculation	1.68	3.81
Others:	5.02	15.26

However, as data elements are read in parallel in the optimized versions, the memory accesses are reduced, and therefore the cache misses.

The proposed optimizations permit a speedup of 2.28 to be obtained. This speedup is noticeably lower than those from most of the optimized kernels. The fact that certain parts of the algorithm could not be optimized seriously limits the performance of the algorithm. Table 3.2 shows the percentage of computation time spent by the algorithm in different tasks before and after the optimization. Important tasks, such as the orientation tables calculation, contour points histogram calculation, and pairings insertion, as well as the luminance histogram calculation, could not be optimized due to their memory access patterns. In the sequential version of the algorithm, all of them represent 38.83% of time. Thus, as an important part of the algorithm could not be optimized, the speedup for the complete algorithm is not as high as those for the individual tasks. Nevertheless, thanks to the optimizations, the algorithm is able to run more than twice as fast before.

3.6 Object Tracking in Video

Tracking algorithms are a key element in many computer vision related applications, such as face recognition, augmented reality, and advanced computer interfaces. In many instances they are part of more complex systems requiring visual information processing, as, for example, video decoding. The use of GPUs in these systems as efficient processors for the tracking tasks can help to fulfill the temporal requisites. In these cases, the GPU can take the role of an efficient coprocessor for specific multimedia applications.

Some of the most studied algorithms for object tracking have been those based on template matching schemes using the sum of squared differences (SSD), such as the Lucas-Kanade algorithm [32]. Low-cost schemes have been implemented, reformulating the image registration process in a way that most of the required computations can be done offline, while still maintaining the same performance. There are also extensions that allow one to model appearance variations of the tracked objects [30],[31]. In addition, advanced optimization schemes based on second order approximations help to improve the convergence of the algorithm [29].

In this work, we use an adaptation of the second order approach to the inverse compositional image alignment algorithm [31]. Following is a brief overview of the technique.

Let us suppose that we have the template image $T(x)$ of an object whose evolution must be followed across an image sequence $I(x, t)$. The goal of the tracking algorithm is to determine the parameter value, p, of a warping function, $W(x, p)$, used to match the object pattern with its occurrence in the captured image. The easiest case, as illustrated in Figure 3.6, is where the warping function is a simple displacement of the image. In the implementation described here more complex warping functions are used, modeling projective transformations and image appearance changes.

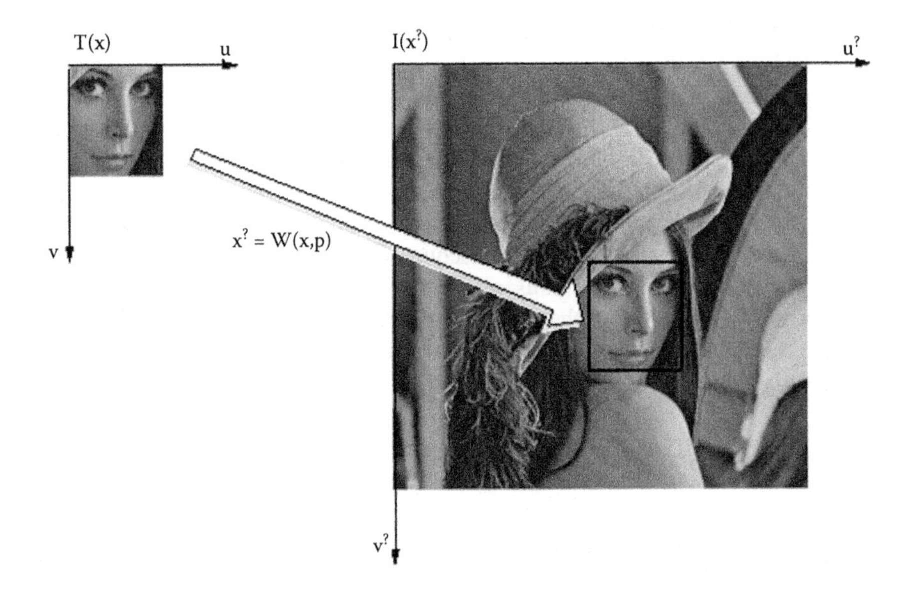

Figure 3.6 Tracking process: A warping function is applied to the template, $T(x)$, to match its occurrence in an image.

As all Lucas-Kanade [32] derived algorithms, the proposed implementation is an iterative technique to find previously mentioned parameters using a linear approximation of the problem that performs the following sum of squares differences minimization:

$$\sum_x \|T(W(x, \Delta p)) - I(W(x, p))\|^2 \tag{3.10}$$

Instead of the linear approximation proposed in previous approaches such as ICA [31], a second order approximation of the deformed template is used in the algorithm resulting in a higher convergence rate:

$$\begin{aligned} T(W(x, \Delta p)) \approx T(W(x, 0)) \quad + & \left.\frac{\partial T(W(x,p))}{\partial p}\right|_{p=0} + \cdots \\ & \cdots + \Delta p + \frac{1}{2}\Delta p^T \left.\frac{\partial^2 T(W(x,p))}{\partial p^2}\right|_{p=0} \Delta p \end{aligned} \tag{3.11}$$

To avoid the need to compute the second order partial derivative, the following approximation is employed (see [29] for a justification):

$$T(W(x, \Delta p)) \approx T(W(x, 0)) + \frac{1}{2}(\nabla_x T + \nabla_x I(W(x, p))\frac{\partial W(x, p)}{\partial}\Delta p \tag{3.12}$$

Using this approximation in 3.1, it results in the following expression to be minimized:

$$E(\Delta p) \approx \sum \left\| T(x) - I(W(x, p_c)) + \frac{1}{2}(\nabla_x T + \nabla_x I(W(x, p)))\frac{\partial W(x, p)}{\partial p}\Delta p \right\|^2 \tag{3.13}$$

Beginning with an initial pose defined by p_0 we find a new pose p by an incremental composition in successive iterations: $p = p_0 \cdot \Delta p$. This incremental value is a linear function of the image error e (difference between template and observed images in the initial pose) solved from 3.13:

$$\Delta p = -H_{esm}^{-1} \sum_x SD_{esm}^T e_{esm}(x) \tag{3.14}$$

where:

- $SD_{esm}^T = \frac{1}{2}(\nabla_x T(x) + \nabla_x I(W(x, p_c)))\left.\frac{\partial W(x,p)}{\partial p}\right|_{p=0}$ is the Jacobian of the warping function. This matrix can be obtained from the image intensity gradients and the warping function derivatives, $W'(x, p)$, with respect to the parameters p, evaluated at each point x.
- H, the Hessian approximation, is a matrix given by $H_{esm} = SD_{esm}^T SD_{esm}$.
- $e_{esm}(x) = T(x) - I(W(x, p_c))$.

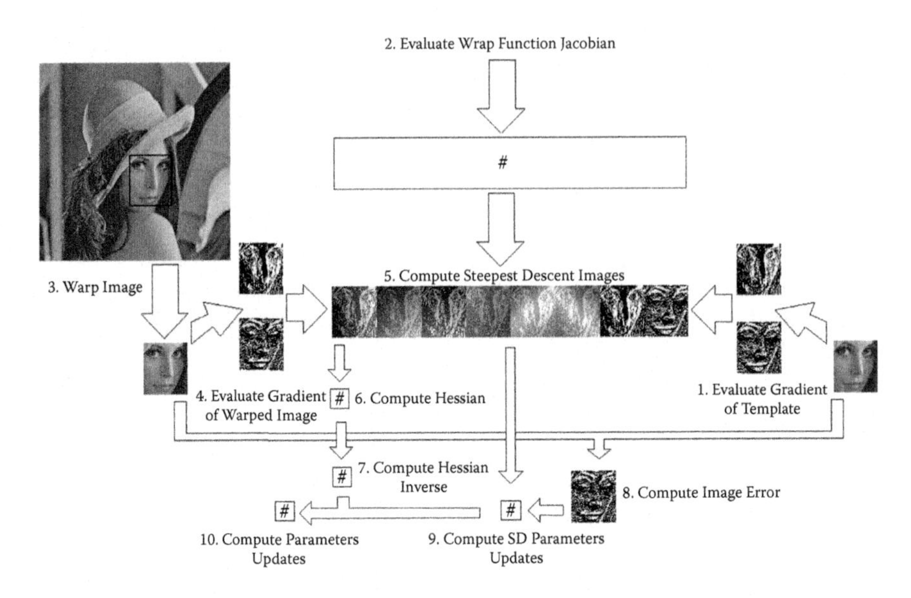

Figure 3.7 Steps of the tracking algorithm.

Figure 3.7 is a visual representation of the required steps of this algorithm in order to compute the pose parameter. Parameter increment requires the computation of the three terms defined in the right member of expression 3.14. The first, the Jacobian matrix SD_{esm}, also called, the *steepest descent matrix*, has a principal role in the algorithm (like other linear optimization techniques, such as the gradient descent method). It is formed by several components, each of them contributing to the increment of one pose parameter. Intuitively, the components can be interpreted as a set of variation modes of the template that occur by modifying its corresponding parameter (see step 5 of Figure 3.7). Its computation makes use of the template and warped image gradients (obtained at steps 1 and 4) and the warp function derivatives with respect to the pose parameters (step 2). Another required term, the *inverse Hessian matrix* H_{esm}^{-1}, is computed in step 6 and inverted in 7. Similarly, the *error matrix* e_{esm} is computed at step 8 by subtracting the template and the warped input image. Final computation of the parameter increment Δp is carried out at steps 9 and 10, by combining the three terms previously mentioned. More details about the tracking algorithm can be obtained in [28].

3.7 Mapping the Tracking Algorithm in GPUs

As opposed to other schemes based on meaningful characteristics matching (such as edges or corners), SSD algorithms make use of all the texture information inside the considered region of interest, by means of dense linear algebra operations over the contained pixels at this area. Furthermore, these tracking algorithms are organized in several well-defined stages, producing regular and predictable pattern accesses to the processed data. All of these make GPU an appropriate architecture to implement these kinds of algorithms in an efficient way. Following, the mapping in the GPU of each step of the algorithm is described.

Steps 1 and 2 in Figure 3.7 do not depend on the parameter values at the current iteration and can be precomputed in an initial step. *Gradient computation* can be approximated in the discrete domain by only using a finite convolution, involving the use of four adjacent pixels around the one considered. This scheme maps very well in GPU architectures because there is little reuse of input values and the same operation is applied to all of them.

At each iteration, we begin with the *capture of the interest region* (stage 3) using the initial parameters obtained at the previous iteration. This image rectification can be implemented in an efficient way in the GPU by using its specific texture interpolation hardware.

For the computation of step 5, we propose to organize the different steepest descent components in a packed form. This way, the resulting steepest descent matrix is composed of a set of submatrices, one per deformation mode parameter. This approach has two advantages:

1. Considering how task assignation is performed in fragment processors, the proposed data arrangement using the same dimensions than the template image allows a coherent access to input matrices (gradient and jacobian matrices). Thus the spatial locality is exploited in the cache associated to each fragment processor.
2. This 2-D distribution also allows one to efficiently assign workloads to different functional units following the 'tiled' task organization available in the GPU.

To *compute the parameters increment* (step 9) we have to multiply the steepest descent matrix by the image error. It is a costly computation because of the high input matrix dimensions. However, the resulting vector is quite small, as its dimensions coincide with the number of parameters of the warping function. Again, keeping in mind that GPU parallelism is achieved by dividing the computation of target elements between different functional units following their spatial distribution, a standard matrix multiplication algorithm, as in [35], will not provide good results in this case. In other words, the reduced dimensions of the rendering target do not allow one to efficiently assign the workload to all the processors. Thus, to take advantage of the GPU parallelism, we follow a different approach organized in two steps using the mentioned steepest descent matrix (see Figure 3.8):

- First, an element-wise multiplication for each Jacobian submatrix is carried out, producing a temporal matrix of high dimensions.

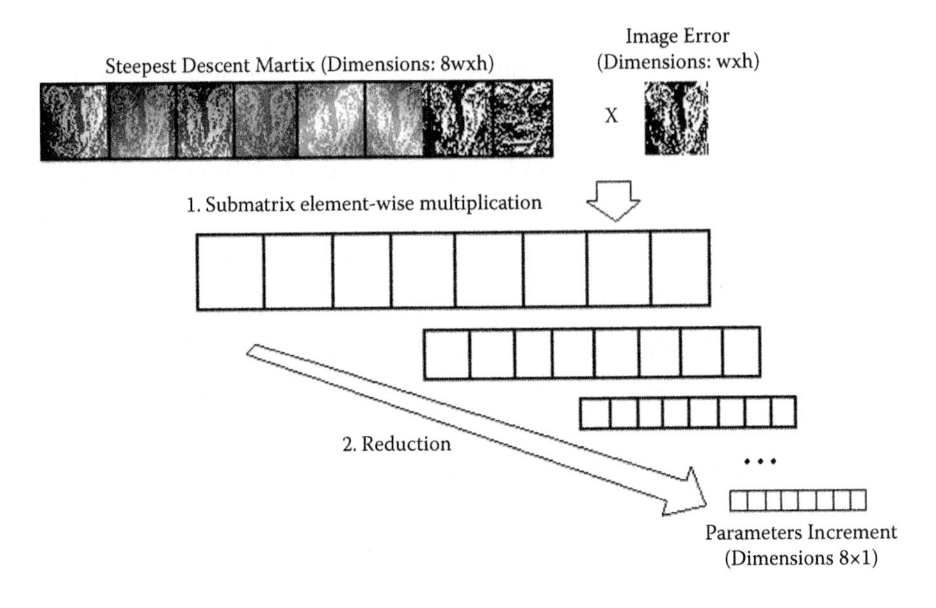

Figure 3.8 Efficient computation of a hessian matrix.

- Second, a reduction stage is done accumulating the temporal results previously obtained. This stage requires several iterations, each of them reducing the matrix dimensions by a 1/2 factor. To make this process as efficient as possible we restrict the algorithm input to multiples of 2 templates.

A similar technique has to be employed to compute the hessian matrix (step 6), because it requires a matrix multiplication between the high dimensions steepest descent matrix and its transpose, yielding a matrix of reduced dimensions.

Finally, the matrix-vector multiplication to obtain the parameter increment (step 10) is done using a multipass algorithm similar to those described in [35].

3.8 GPU Performance with the Tracking Algorithm

Multiple experiments were performed using different GPUs and commodity CPUs. Although there are useful libraries that abstract the GPU as a stream computing system (see Brook as an example), our algorithm was implemented upon OpenGL directly using C++ in order to have more control over the different code sections and to do the benchmarks; it was also necessary to introducte some specific OpenGL extensions to make the code more efficient. We employed Cg as a shader language.

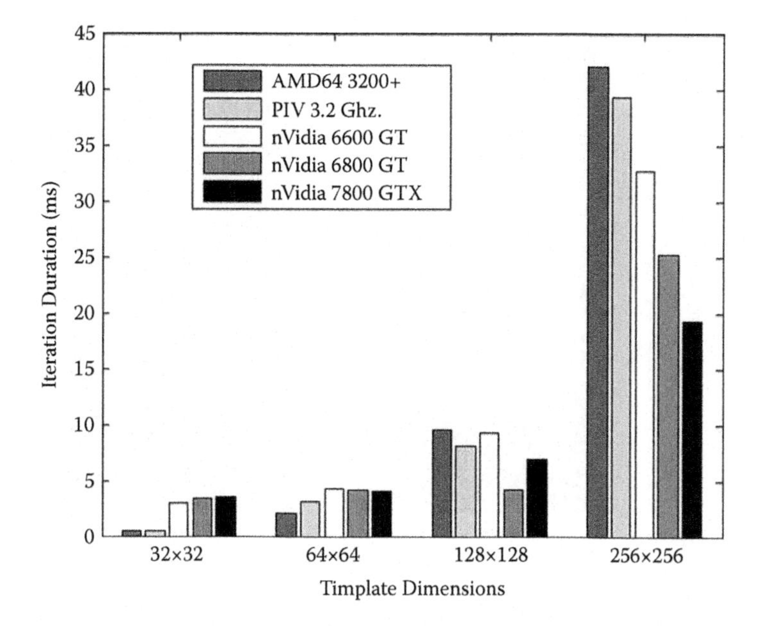

Figure 3.9 Time employed by a tracking iteration in several platforms.

The system was running a Linux operating system using nVidia proprietary drivers version 81.74; we made use of the recently introduced Framebuffer Objects extensions to avoid read/writing operations between target buffers and textures. We compared the performance of our algorithm against a CPU implementation based on the LTI-Lib [36] vision library.

In Figure 3.9, iteration times are represented for several GPUs and commodity CPUs. As mentioned in other applications [35], when the input data size is big enough, GPU functional units increment their workload, resulting in a better task assignment between different functional units. As a result, nVidia 7800 GTX tracking implementation results outperform those achieved by CPU implementations for template dimensions higher than 128×128. In any case, GPU versions appear as a competitive solution showing how this hardware platform can be used as an efficient coprocessor in those applications in which it is important to save CPU time for other higher level tasks in multimedia systems.

Figure 3.10 shows the relative computing time of different steps involved in the tracking process for the 128×128 template case. As an example, it can be seen that the required time for steepest descent computation is reduced in the GPU implementation. The reason is that it mostly requires element-wise operations that map well on these platforms. CPU efficiency in many intensive computing applications comes from exploiting data spatially located by using cache hierarchies; but in this case, with so little element reuse, it does not suppose a great advantage.

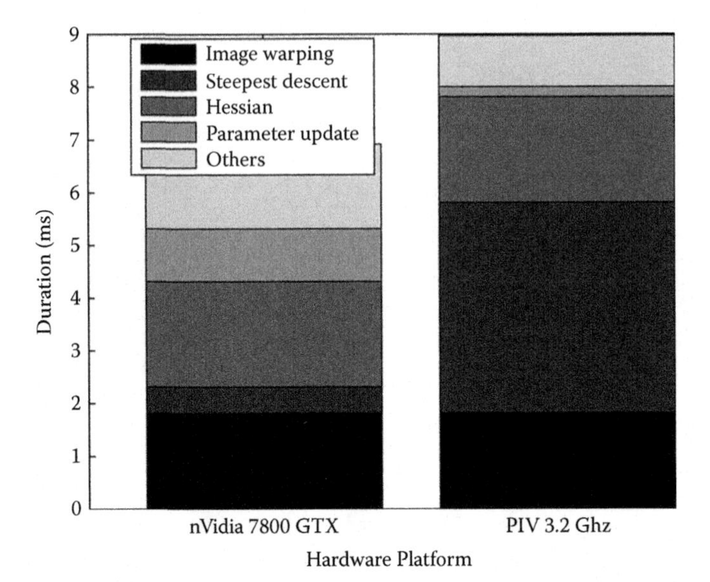

Figure 3.10 Time comparison for several stages.

3.9 Conclusions

In this chapter we have presented two approaches that implement multimedia applications using different computing architectures. On one hand, multimedia extensions have been used to increase the performance of a particular temporal video segmentation algorithm. It is based on luminance and contour information from MPEG video streams, being able to perform at real time. Several computation kernels used in this algorithm have been optimized using Intel's multimedia extensions, such as sum of products, 1-D convolution, and arctangent calculation, among others. The obtained results show that the reported speedup is variable from one kernel to another. Notwithstanding, the complete temporal video segmentation algorithm yielded a final speedup of 2.28. This means that the optimized algorithm is able to run more than twice as fast as the sequential version.

GPUs were used to perform a tracking algorithm based on sum-of-squared differences techniques. The results achieved confirm the great potential of these architectures in a multimedia application. We also have faced the limitations of current GPU designs, such as the inability to keep the computation units busy because of the limited bandwidth to the closest cache or the need of appropriate data organization to match efficient rasterization access patterns.

3.10 Acknowledgments

This work has been supported by the Spanish Ministry of Education, under contract TIC2003-06623, and the company Tecnologías Digitales AudioVisuales (TEDIAL), under contract 8.06/29.1821.

References

[1] N. Slingerland and A.J. Smith. Measuring the performance of multimedia instruction sets. *IEEE Transactions on Computers*, vol. 51, pp. 1317–1332, 2002.

[2] V. Lappalainen, T.D. Hamalainen and P. Liuha. Overview of research efforts on media ISA extensions and their usage in video coding. *IEEE Transactions on Circuits and Systems for Video Technology*, vol. 12, pp. 660–670, 2002.

[3] C. Limousin, J. Sebot, A. Vartanian and N. Drach, Architecture optimization for multimedia application exploiting data and thread-level parallelism, *Journal of Systems Architecture*, vol. 51, pp. 15–27, 2005.

[4] S. Balam and D. Schonfeld. Associative processors for video coding applications. *IEEE Transactions on Circuits and Systems for Video Technology*, vol. 16, pp. 241–250, 2006.

[5] B. Schoner, J. Villasenor, S. Molloy, and R. Jain, Techniques for FPGA implementation of video compression systems, *Third International ACM Symposium on Field-Programmable Gate Arrays* (FPGA'95), pp. 154–159, 1995.

[6] M. Pharr and R. Fernando, *GPU Gems 2: Programming Techniques for High-Performance Graphics and General-Purpose Computation*, Addison-Wesley: Reading, MA, 2005.

[7] C. L. Potkonjak, M. Mangione-Smith and W. H. MediaBench, A tool for evaluating and synthesizing multimedia and communications systems, *Proceedings Thirtieth Annual IEEE/ACM International Symposium on Microarchitecture*, pp. 330–335, 1997.

[8] N. T. Slingerland and A. J. Smith, Design and characterization of the Berkeley multimedia workload, *Multimedia Systems*, vol. 8, pp. 315–327, 2002.

[9] N. Patel and I. Sethi, Video shot detection and characterization for video databases, *Pattern Recognition*, vol. 30, pp. 583–592, 1997.

[10] G. Lupatini, C. Saraceno and R. Leonardi, Scene break detection: A comparison, *Proceedings International Workshop on Research Issues in Data Engineering*, Orlando, FL, USA, 1998, pp. 34–41.

[11] H. Lu, Y. Zhang and Y. Yao, Robust gradual scene change detection, *Proceedings IEEE International Conference on Image Processing*, vol. 3, Kobe, Japan, 1999, pp. 304–308.

[12] H. Zhang, A. Kankanhalli and S. Smoliar, Automatic partitioning of full-motion video, *ACM Journal on Multimedia Systems*, vol. 1, pp. 10–28, 1993.

[13] Y. Yuso, J. Kittler and W. Christmas, Combining multiple experts for classifying shot changes in video sequences, *Proceedings IEEE International Conference on Multimedia Computing and Systems*, vol. 2, Florence, Italy, 1999, pp. 700–704.

[14] P. Aigrain and P. Joly, The automatic real-time analysis of film editing and transition effects and its applications, *Computer and Graphics*, vol. 18, pp. 93–103, 1994.

[15] A. Hampapur, R. Jain and T. Weymouth, Production model based digital video segmentation, *International Journal of Multimedia Tools Applications*, vol. 1, pp. 9–46, 1995.

[16] R. Mallikarjuna, K. Ramakrishnan, N. Balakrishnan and S. Srinivasan, Neural net based scene change detection for video classification, *Proceedings IEEE Workshop on Multimedia Signal Processing*, Copenhagen, Denmark, 1999, pp. 247–252.

[17] S. Jun and S. Park, An automatic cut detection algorithm using median filter and neural network, *Proceedings International Technical Conference on Circuits/Systems, Computers and Communications*, Pusan, Korea, 2000, pp. 1049–1052.

[18] J. Boreczky and L. Wilcox, A hidden markov model framework for video segmentation using audio and image features, *Proceedings IEEE International Conference on Acoustics, Speech and Signal Processing*, vol. 6, Seattle, WA, USA, 1998, pp. 3741–3744.

[19] S. Eickeler and S. Muller, Content-based video indexing of tv broadcast news using Hidden Markov Models, *Proceedings IEEE International Conference on Acoustics, Speech and Signal Processing*, vol. 6, Phoenix, AR, USA, 1999, pp. 2997–3000.

[20] Y. Qi, A. Hauptmann and T. Liu, Supervised classification for video shot segmentation, *Proceedings IEEE International Conference on Multimedia and Expo*, vol. 2, Baltimore, MD, USA, 2003, pp. 689–692.

[21] R. Ewerth and B. Freisleben, Video cut detection without thresholds, *Proceedings Workshop on Systems, Signals and Image Processing (IWSSIP 2004)*, Poznan, Poland, pp. 227–230, 2004.

[22] I. Koprinska and S. Carrato, Temporal video segmentation: A survey. *Signal Processing: Image Communication*, vol. 16, pp. 477–500, 2001.

[23] N. Slingerland and A.J. Smith, Measuring the performance of multimedia instruction sets. *IEEE Transactions on Computers*, vol. 51, pp. 1317–1332, 2002.

[24] E. Saez, J. I. Benavides and N. Guil, Reliable real time scene change detection in MPEG compressed video. *Proceedings IEEE International Conference on Multimedia and Expo*, vol. 1, pp. 567–570, 2004.

[25] N. Guil, J. M. Gonzalez and E. L. Zapata, Bidimensional shape detection using an invariant approach. *Pattern Recognition*, vol. 32, pp. 1025–1038, 1999.

[26] J. Canny, A computational approach to edge detection. *IEEE Transactions on Pattern Analysis and Machine Intelligence*, vol. 8, pp. 679–698, 1986.

[27] L. Nyland and M. Snyder, Fast trigonometric functions using Intel's SSE2 instructions. *Technical Report TR03-041*, Dept. Computer Science, College of Arts and Sciences, University of North Carolina at Chapel Hill, NC, USA, 2003.

[28] S. Baker, R. Gross, Matthews and I. Lucas-Kanade, 20 Years on: A unifying framework: Part 3, *Technical Report CMU-RI-TR-03–35*, 2004.

[29] S. Benhimane and E. Malis, Real-time image-based tracking of planes using efficient second-order minimization, *Proc. IROS04*, 2004.

[30] S. Baker, I. Matthews, L. Lucas-Kanade, 20 years on: A unifying framework: Part 1. *International Journal of Computer Vision*, vol. 56, pp. 221–255, 2004.

[31] S. Baker and I. Matthews, Equivalence and efficiency of image alignment algorithms. *Proceedings International Conference on Computer Vision and Pattern Recognition*, pp. 1090–1097, 2001

[32] B. D. Lucas and T. Kanade, An iterative image registration technique with application to stereo vision, *IJCAI81*, pp. 674–679, 1981.

[33] G. Hager and P. Belhumeur, Efficient region tracking with parametric models of geometry and illumination. *IEEE Transactions on Pattern Analysis and Machine Intelligence*, pp. 1025–1039, 1998.

[34] M. Black and A. Jepson, Eigen-tracking: Robust matching and tracking of articulated objects using a view-based representation. *International Journal of Computer Vision*, vol. 26, pp. 63–84, 1998.

[35] K. Fatahalian et al., Understanding the efficiency of GPU algorithms for matrix-matrix multiplication, *Proceedings of the ACM SIGGRAPH/EUROGRAPHICS conference on Graphics hardware*, 2004.

[36] *LTI Computer Vision Library*, available online: http://ltilib.sourceforge.net.

[37] G. Shen, G. Gao, S. Li, H. Shum and Y. Zhang, Accelerate video decoding with generic GPU. *IEEE Transactions on Circuits and Systems for Video Technology*, vol. 15, pp. 685–693, 2005.

Chapter 4

Parallel Implementation of the ORASIS Algorithm for Remote Sensing Data Analysis

David Gillis,
Naval Research Laboratory

Jeffrey H. Bowles,
Naval Research Laboratory

Contents

ORASIS (the Optical Real-Time Adaptive Spectral Identification System) is a series of algorithms developed at the Naval Research Lab for the analysis of HyperSpectral Image (HSI) data. ORASIS is based on the Linear Mixing Model (LMM), which assumes that the individual spectra in a given HSI scene may be decomposed into a set of in-scene constituents known as endmembers. The algorithms in ORASIS are designed to identify the endmembers for a given scene, and to decompose (or demix) the scene spectra into their individual components. Additional algorithms may be used for compression and various post-processing tasks, such as terrain classification and anomaly detection. In this chapter, we present a parallel version of the ORASIS algorithm that was recently developed as part of a Department of Defense program on hyperspectral data exploitation.

4.1 Introduction

A casual viewing of the recent literature reveals that hyperspectral imagery is becoming an important tool in many disciplines. From medical and military uses to environmental monitoring and geological prospecting the power of hyperspectral imagery is being shown. From a military point of view, the primary use of hyperspectral data is for target detection and identification. Secondary uses include determination of environmental products, such as terrain classification or coastal bathymetry, for the intelligence preparation of the battlespace environment. The reconnaissance and surveillance requirements of the U.S. armed forces are enormous. Remarks at an international conference by General Israel put the requirements at a minimum of one million square kilometers per day that need to be analyzed. Usually, this work includes the use of high resolution panchromatic imagery, with analysts making determinations based on the shapes of objects in the image. Hyperspectral imagery and algorithms hold the promise of assisting the analyst by making determinations of areas of interest or even identification of militarily relevant objects using spectral information with spatial information being of secondary importance.

Both the power and the pitfalls of hyperspectral imaging originate with the vast amount of data that is collected. This data amount is a consequence of the detailed measurements being made. For example, given a sensor with a 2 meter ground sample distance (GSD) and a spectral range of 400 to 1000 nanometers (with a 5 nanometer spectral sampling), a coverage area of 1 square kilometer produces approximately 57 MB of hyperspectral data. In order to meet the million square kilometer requirement, a hyperspectral sensor would have to produce up to 57 *terabytes* per day. This is truly a staggering number. Only by automating the data processing, and by using state-of-the-art processing capability, will there be any chance of hyperspectral imagery making a significant contribution to military needs in reconnaissance and surveillance.

In order to deal with the large amounts of data in HSI, a variety of new algorithms have appeared in recent years. Additionally, advanced computing systems continue to improve processing speed, storage, and display capabilities. This is particularly true of the high-performance computing (HPC) systems.

One common technique used in hyperspectral data analysis is the Linear Mixing Model (LMM). In general terms (details are given in the next section), the LMM assumes that a given spectrum in a hyperspectral image is simply the weighted sum of the individual spectra of the components present in the corresponding image pixel. If we assume that the total number of major constituents in the scene (generally known as the scene endmembers) is smaller than the number of bands, then it follows that the original high-dimensional data can be projected into a lower-dimensional subspace (one that is spanned by the endmembers) with little to no loss of information. The projected data may then be used either directly by an analyst and/or fed to various other post-processing routines, such as classification or targeting.

In order to apply the LMM, the endmembers must be known. There have been a number of different methods for determining endmembers presented in the literature [1], including Pixel Purity [2], N-FINDR [3], and multidimensional morphological techniques [4]. The Optical Real-Time Adaptive Spectral Identification System (ORASIS) [5] is a series of algorithms that have been developed to find endmembers, using no *a priori* knowledge of the scene, capable of operating in (near) real-time. In addition to the main endmember selection algorithms, additional algorithms allow for compression, constrained or unconstrained demixing, and anomaly detection.

The original ORASIS algorithm was designed to run in scalar (single-processor) mode. Recently, we were asked to develop a parallel, scalable version of the ORASIS, as part of a Department of Defense Common High-Performance Computing Software Support Initiative (CHSSI) program [6]. In addition to ORASIS, this project included the development of parallel versions of N-FINDR and two LMM-based anomaly detection routines. In this chapter, we review the details of the algorithms involved in this project, and discuss the modifications that were made to allow them to be run in parallel. We also include the results of running our modified algorithms on a variety of HPC systems.

The remainder of this chapter is divided into six sections. In Section 4.2 we present the mathematical formalities of the linear mixing model. In Sections 4.3 and 4.4 we give a general overview of the (scalar) ORASIS and the anomaly detection and N-FINDR algorithms, respectively, used in this project. In Section 4.5 we discuss the modifications that were made to the scalar algorithms in order to be run in parallel mode, and present the computational results of our modifications in 4.6. We then present our conclusions in 4.7.

4.2 Linear Mixing Model

The linear mixing model assumes that each spectrum in a given hyperspectral image may be decomposed into a linear combination of the scene's constituent spectra, generally referred to as endmembers. Symbolically, let l be the number of spectral bands, and consider each spectrum as a vector in l-dimensional space. Let E_j be the

l-dimensional endmember vectors, k be the number of constituents in the scene, and $j = 1 \cdots k$. Then the model states that each scene spectrum s may be written as the sum

$$s = \sum_{j=1}^{k} \alpha_j E_j + N \tag{4.1}$$

where α_j is the abundance of the jth component spectrum E_j, and N is an l-dimensional noise vector. Intuitively, the α_j's represent the amount of each constituent that is in a given pixel, and are often referred to as the abundance (or mixing) coefficients. For physical reasons, one or both of the following constraints (respectively, sum-to-one and nonnegativity) are sometimes placed on the α_j's:

$$\sum_{j=1}^{k} \alpha_j = 1 \tag{4.2}$$

$$\alpha_j \geq 0 \tag{4.3}$$

Once the endmembers for a given scene are known (either by ORASIS or some other method), the abundance coefficients may be estimated using a least squares technique, a process generally known as demixing. If no constraints are placed on the coefficients, then this calculation reduces to a simple (and fast) matrix-vector product, as does the case involving the sum-to-one constraint (4.2). In the case of the nonnegativity constraint (4.3), the coefficients can only be found by using numerical optimization techniques. In this chapter, we consider only the unconstrained and nonnegative constrained problems.

After demixing, each of the l-dimensional spectra from the original scene may be replaced by the k-dimensional demixed spectra. In this way, a set of grayscale images (generally known as either *fraction planes* or *abundance planes*) is constructed, where each pixel in the image is given by the abundance coefficient of the corresponding spectra for the given endmember. As a result, the fraction planes serve to highlight groups of similar image spectra in the original scene. An example of this is given in Figure 4.1, which shows a single band of a hyperspectral image taken at Fort AP Hill with the NVIS sensor, along with two of the fraction planes created by ORASIS. Also, since the number of endmembers is generally much smaller than the original number of bands, the fraction planes retain the significant information in the scene but with a large reduction in the amount of data.

4.3 Overview of the ORASIS Algorithms

In its most general form, ORASIS is a collection of algorithms that work together to produce a set of endmembers. The first of these algorithms, the prescreener, is used to 'thin' the data; in particular, the prescreener chooses a subset of the scene

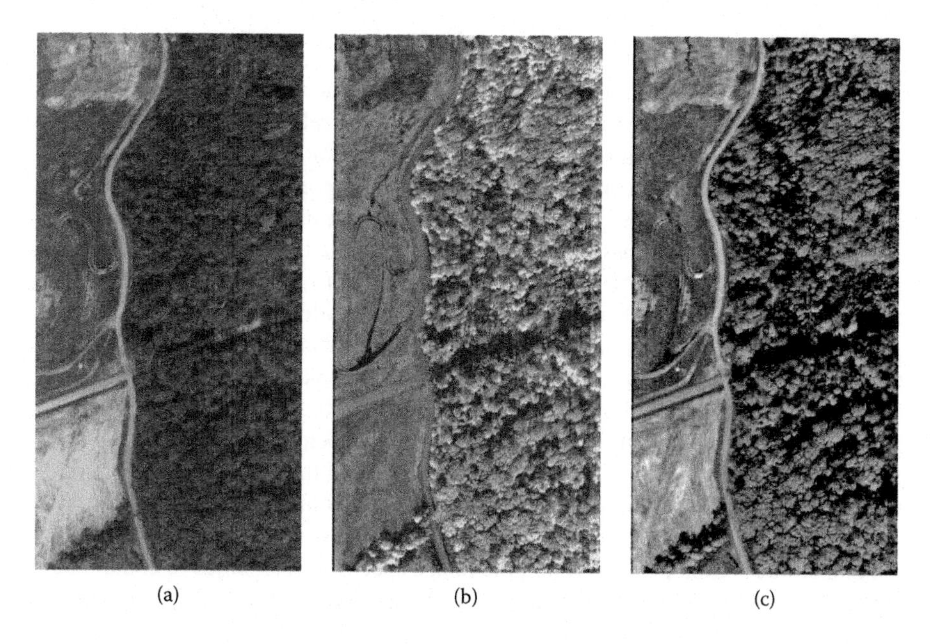

Figure 4.1 Data from AP Hill. (a) Single band of the original data. (b) (c) Fraction planes from ORASIS processing.

spectra (known as the exemplars) that is used to model the data. In our experience, up to 95% of the data in a typical scene may be considered redundant (adding no additional information) and simply ignored. The prescreener is used to reduce the complexity and computational requirements of the subsequent ORASIS processing, as well as acting as a compression algorithm. The second step is the basis selection module, which determines an optimal subspace that contains the exemplars. The existence of such a subspace is a consequence of the linear mixing model. Once the exemplars have been projected into the basis subspace, the endmember selection algorithm is used to actually calculate the endmembers for the scene. This algorithm, which we call the shrinkwrap, intelligently extrapolates outside the data set to find endmembers that may be closer to pure substances than any of the spectra that exist in the data. Large hyperspectral data sets provide the algorithm with many examples of the different mixtures of the materials present, and each mixture helps determine the makeup of the endmembers. The last step in ORASIS is the demixing algorithm, which decomposes each spectrum in the original scene into a weighted sum of the endmembers.

In this section we discuss the family of algorithms that make up ORASIS. This section is focused primarily on the original (scalar) versions of ORASIS; a discussion of the modifications made to allow the algorithms to run in parallel mode is given in Section 4.4.

4.3.1 Prescreener

The prescreener module in ORASIS has two separate but related functions. The first, which we denote "exemplar selection," is to replace the relatively large set of spectra in the original scene with a smaller representative set, known as the exemplars. The reason for doing this is that, by choosing a small set of exemplars that faithfully represents the image data, subsequent processing can be greatly sped up, often by orders of magnitude, with little loss in precision of the output. The second function of the prescreener, which we denote codebook replacement, is to associate each image spectrum with exactly one member of the exemplar set. This is done for compression. By replacing the original high-dimensional image spectra with an index to an exemplar, the total amount of data that must be stored to represent the image can be greatly reduced.

The basic concepts used in the prescreener are easy to understand. The exemplar set is initialized by adding the first spectrum in a given scene to the exemplar set. Each subsequent spectrum in the image is then compared to the current exemplar set. If the image spectrum is 'sufficiently similar' (meaning within a certain spectral 'error' angle), then the spectrum is considered redundant and is replaced, by reference, by a member of the exemplar set. If not, the image spectrum is assumed to contain new information and is added to the exemplar set. This process continues until every image spectrum has been processed.

The prescreener module can thus be thought of as a two-step problem; first, the exemplar selection process, is to decide whether or not a given image spectrum is 'unique' (i.e., an exemplar). If not, the second step (codebook replacement) is to find the best 'exemplar' to represent the spectrum. The trick, of course, is to perform each step as quickly as possible. Given the sheer size of most hyperspectral images, it is clear that a simple brute-force search and replace method would be quickly overwhelmed. The remainder of this subsection discusses the various methods that have been developed to allow the prescreener to run as quickly as possible (usually in near-real-time). In ORASIS, the two steps of the prescreener are intimately related; however, for ease of exposition, we begin by examining the exemplar selection step separately, followed by a discussion of the replacement process.

It is worth noting that the number of exemplars produced by the prescreener is a complicated function of instrument SNR, scene complexity (which might be viewed as a measure of how much hyperspectral 'space' the data fill), and processing error level desired (controlled by the error angle mentioned above). Figure 4.2 provides an example of how the number of exemplars scales with the error angle. This scaling is an important aspect of the porting of the ORASIS to the HPC systems. As discussed in later sections, the exponential increase in the number of exemplars as the error angle decreases creates problems with our ability to parallelize the prescreener.

4.3.1.1 Exemplar Selection

The exemplar selection algorithm selects a subset of spectra (known as the exemplars) from the image that is used to represent the image. Let $\{X_1, X_2, \cdots, X_N\}$ represent the image spectra, where N is the total number of pixels in the image. The exemplar set is initialized by setting the first exemplar E_1 equal to the first image spectrum X_1.

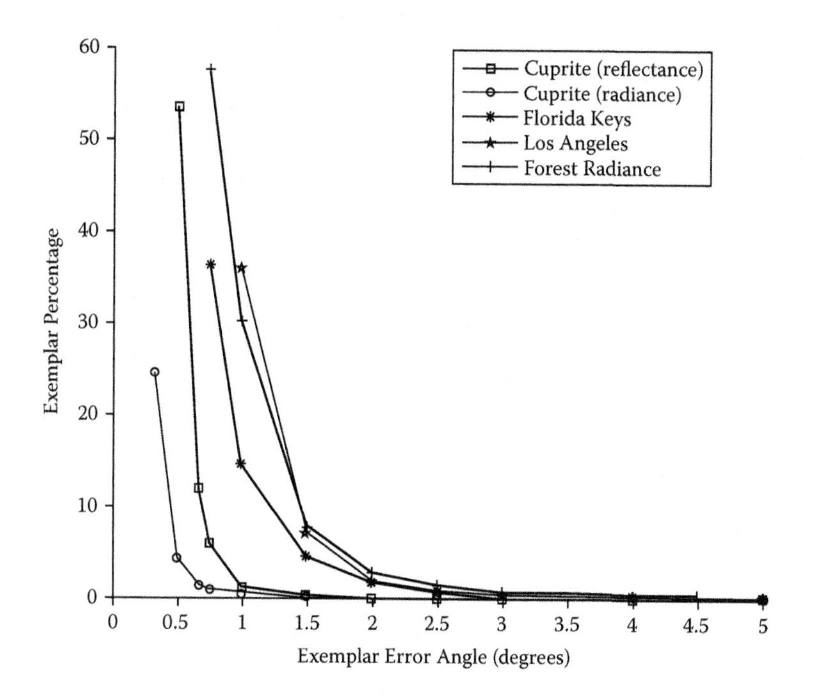

Figure 4.2 The number of exemplars as a function of the error angle for various hyperspectral images.

For each of the remaining image spectrum, the spectrum X_i is compared to the current set of exemplars $E_1, \cdots, E_m$ to see if it is 'sufficiently similar' (as defined below) to any member of the set. If not, the image spectrum is added to the exemplar set: $E_{m+1} = X_i$. Otherwise, the spectrum is considered to be spectrally redundant and is replaced by a reference to the matching exemplar. This process continues until every spectrum in the image has either been assigned to the exemplar set or given an index into this set.

By 'sufficiently similar' we simply mean that the angle $\theta(X_i, E_j)$ between the image spectrum X_i and the exemplar E_j must be smaller than some predetermined error angle θ_T. Recall that the angle between any two vectors is defined as $\theta(X_i, E_j) = \cos^{-1} \frac{|\langle X_i, E_j \rangle|}{\|X_i\| \cdot \|E_j\|}$, where $\langle X_i, E_j \rangle$ is the standard (Euclidean) vector inner (or dot) product, and $\|X_i\|$ is the standard (Euclidean) vector norm. It follows that an image spectrum is rejected (not added to the exemplar set) only if $\theta(X_i, E_j) \leq \theta_T$ for some exemplar E_j. If we assume that the vectors have been normalized to unit norm, then the rejection condition for an incoming spectrum becomes simply $|\langle X_i, E_j \rangle| \geq \cos^{-1} \theta_T$. Note that the inequality sign is reversed, since the cosine function is decreasing on the interval $(0, \pi)$.

The easiest approach to calculating the exemplar set would be a simple brute-force method where the entire set of angles between the candidate image spectrum and each member of the exemplar set is calculated and the minimum found. Given that the typical hyperspectral image contains on the order of 100,000 pixels (and growing), this approach would simply take far too long; thus, faster methods needed to be developed. The basic approach ORASIS uses to speed up the processing is to try to reduce the actual number of exemplars that must be checked in order to decide if a match is possible. To put this another way, instead of having to calculate the angle for each and every exemplar in the current set, we would like to be able to exclude as many exemplars as possible beforehand, and calculate angles only for those (hopefully few) exemplars that remain. In order to do this, we use a set of 'reference vectors' to define a test that quickly (i.e., in fewer processing steps) allows us to decide whether a given exemplar can possibly match a given image spectrum. All of the exemplars that fail this test can then be excluded from the search, without having to actually calculate the angle. Any exemplar that passes the test is still only a 'possible' match; the angle must still be calculated to decide whether the exemplar does actually match the candidate spectrum.

To define the reference vector test, suppose that we wish to check if the angle $\theta(X, E)$ between two unit normalized vectors, X and E, is below some threshold θ_T. Using the Cauchy-Schwarz inequality, it can be shown [5] that

$$\theta(X, E) \leq \theta_T \Leftrightarrow \sigma_{min} \leq \langle E, R \rangle \leq \sigma_{max} \tag{4.4}$$

where

$$\sigma_{min} = \langle X, R \rangle - \sqrt{2(1 - \cos(\theta_T))}$$
$$\sigma_{max} = \langle X, R \rangle + \sqrt{2(1 - \cos(\theta_T))}$$

and R is an arbitrary unit normalized vector. To put this another way, to test whether the angle between two given vectors is sufficiently small, we can choose some reference vector R, calculate σ_{min}, σ_{max} and $\langle E, R \rangle$, and check whether or not the rejection condition (Eq. 4.4) holds. If not, then we know that the vectors X and E cannot be within the threshold angle θ_T. We note that the converse does not hold.

Obviously, the above discussion is not of much use if only a single angle needs to be checked. However, suppose we are given two sets of vectors $X_1, \cdots, X_n$ (the candidates) and $E_1, \cdots, E_m$ (the exemplars), and assume that for each X_i we would like to see if there exists some E_j such that the angle between them is smaller than some threshold θ_T. Using the above ideas, we choose a reference vector R with $\|R\| = 1$ and define $\sigma_i = \langle \frac{E_j}{\|E_j\|}, R \rangle$, for each exemplar E_i. By renumbering the exemplars, if necessary, we may assume that $\sigma_1 \leq \sigma_2 \leq \cdots \leq \sigma_m$.

To test the candidate vector X_i we calculate

$$\sigma_{min}^i = \langle X_i, R \rangle - \sqrt{2 - (1 - \cos(\theta_T))}$$
$$\sigma_{max}^i = \langle X_i, R \rangle + \sqrt{2 - (1 - \cos(\theta_T))}$$

By the rejection condition (Eq. 4.4), it follows that the only exemplars that can be within the threshold angle are those whose sigma value σ_j lies in the interval

$[\sigma^i_{min}, \sigma^i_{max}]$; we call this interval the possibility zone for the vector X_i. All other exemplars can be immediately excluded. Assuming that the reference vector is chosen so that the sigma values are sufficiently spread out, and that the possibility zone for a given candidate is relatively small, then it is often possible using this method to significantly reduce the number of exemplars that need to be checked.

The preceding idea can be extended to multiple reference vectors as follows. Suppose that $R_1, \cdots, R_k$ is an orthonormal set of vectors, and let $\|X\| = \|E\| = 1$. Then X and E can be written as

$$X = \sum_{i=1}^{k} \alpha_i R_i + \alpha^{\perp} R^{\perp}$$

$$E = \sum_{i=1}^{k} \sigma_i R_i + \sigma^{\perp} S^{\perp}$$

where $\alpha_i = \langle X, R \rangle$, $\sigma_i = \langle E, R \rangle$, and $R^{\perp}$, $S^{\perp}$ are the residual vectors of X and E, respectively. In particular, $R^{\perp}$, $S^{\perp}$ have unit norm and are orthogonal to the subspace defined by the R_i vectors. It follows that the dot product of X and E is given by $\langle X, E \rangle = \sum \alpha_i \sigma_i + \alpha^{\perp} \sigma^{\perp} \langle R^{\perp}, S^{\perp} \rangle$.

By the Cauchy-Schwartz inequality, $\langle R^{\perp}, S^{\perp} \rangle \leq \|R^{\perp}\| \cdot \|S^{\perp}\| = 1$, and by the assumption that X and E have unit norm

$$\begin{aligned} \alpha^{\perp} &= \sqrt{1 - \sum \alpha_i^2} \\ \sigma^{\perp} &= \sqrt{1 - \sum \sigma_i^2} \end{aligned} .$$

If we define the projected vectors $\alpha_p = (\alpha_1, \cdots, \alpha_k, \alpha^{\perp})$ and $\sigma_p = (\sigma_1, \cdots, \sigma_k, \sigma^{\perp})$, then the full dot product satisfies $\langle X, E \rangle \leq \sum \alpha_i \sigma_i + \alpha^{\perp} \sigma^{\perp} \equiv \langle \alpha_p, \sigma_p \rangle$.

This allows us to define a multizone rejection condition that, as in the single reference vector case, allows us to exclude a number of exemplars without having to do a full dot product comparison. The exemplar search process becomes one of first checking that the projected dot product $\langle \alpha_p, \sigma_p \rangle$ is below the rejection threshold. If not, there is no need to calculate the full dot product, and we move on to the next exemplar. The trade-off is that each of the reference vector dot products must be taken before using the multizone rejection test. In our experience, the number of reference zone dot products (we generally use three or four reference vectors) is generally much smaller than the number of exemplars that are excluded, saving us from having to calculate the full band exemplar/image spectra dot products, and thus justifying the use of the multizone rejection criterion. However, the overhead does limit the number of reference vectors that should be used.

We note that the choice of reference vectors is important in determining the size of the possibility zone, and therefore in the overall speed of the prescreener. The principal components of the exemplars tend to give the best results, which is not surprising since the PCA eigenvectors provide by construction the directions that

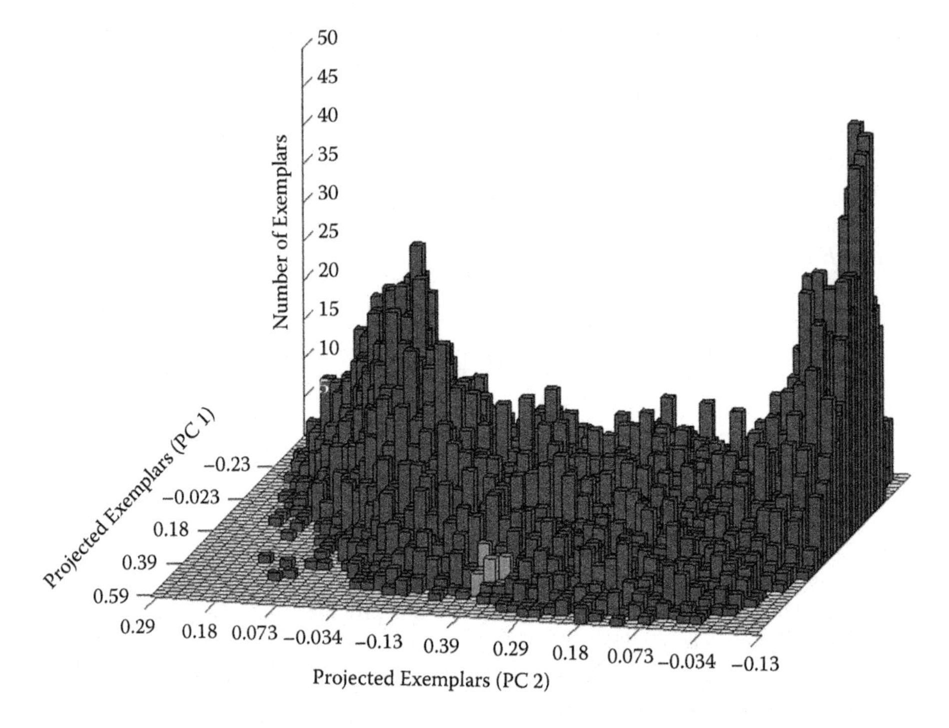

Figure 4.3 Three-dimensional histogram of the exemplars projected onto the first two reference vectors.

maximize the variance of the projected exemplars. In the prescreener, the PCA vectors are calculated on the fly using a weighted exemplar substitution method to calculate the (noncentered) covariance matrix and then the eigenvectors. Experience has shown that sufficiently accurate directions can be determined using only the first hundred or so exemplars. Conceptually, the use of PCA eigenvectors for the reference vectors assures that a grass spectrum is compared only to exemplars that look like grass and not to exemplars that are mostly water, for example.

An example of the power of the possibility zone is given in Figure 4.3, which shows a histogram of a set of exemplars projected onto two reference vectors (in this example the reference vectors are the first two principal components of the exemplars). Using the multizone rejection condition, only the highlighted (lighter colored) exemplars need to be fully tested for the given candidate image spectrum. All other exemplars can be immediately excluded, without having to actually calculate the angle between them and the candidate.

The single and multizone rejection conditions allow us to quickly reduce the number of exemplars that must be compared to an incoming image spectrum to find a match. We note that each test uses only the spectral information of the exemplars and image spectra; however, hyperspectral images typically exhibit a large amount of spatial homogeneity. As a result, neighboring pixels tend to be spectrally similar. In terms of exemplar selection, this implies that if two consecutive pixels are rejected, then there is a reasonable chance that they both matched the same exemplar. For this reason, we keep a dynamic list (known as the popup stack) of the exemplars that were most recently matched to an image spectrum. Before applying the rejection conditions, a candidate image spectrum is compared to the stack to see if it matches any of the recent exemplars. This list is continuously updated, and should be small enough to be quickly searched but large enough to capture the natural scene variation. In our experience, a size of four to six works well; the current version of ORASIS uses a five-element stack.

4.3.1.2 Codebook Replacement

In addition to exemplar selection, the second major function of the prescreener is the codebook replacement process, which substitutes each redundant (i.e., non-exemplar) spectrum in a given scene with an index to one of the exemplar spectra. By doing so, the high-dimensional image spectra may be replaced by a simple scalar (the index), thus greatly reducing the amount of data that must be stored. In the compression community, this is known as a vector quantization compression scheme. We note that this process only affects how the image spectra pair up with the exemplars, and does not change the spectral content of the exemplar set. Thus, it does not affect any subsequent processing, such as the endmember selection stage.

In exemplar selection, each new candidate image spectrum is compared to the list of 'possible' matching exemplars. A few of these candidate spectra will not 'match' any of the exemplars and will become new exemplars. However, the majority of the candidates will match at least one of the exemplars and be rejected as redundant. In these cases, we would like to replace the candidate with a reference to the 'best' matching exemplar, for some definition of best.

In ORASIS, there are a number of different ways of doing this replacement. For this project, we implemented two replacement strategies, which we denote 'first match' and 'best fit.' We note for completeness that other replacement strategies are available; however, they were not implemented in this version of the code.

The 'first match' strategy simply replaces the candidate spectrum with the first exemplar within the possibility zone that it matches. This is by far the easiest and fastest method, and is used by default.

The trade-off for the speed of the first match method is that the first matching exemplar may not be the best, in the sense that there may be another exemplar that is closer (in terms of difference angles) to the candidate spectrum. Since the search is stopped at the first matching exemplar, the 'better' matching exemplar will never be found. In a compression scenario, this implies that the final amount of distortion from using the first match is higher than it could be if the better matching exemplar was used.

To overcome the shortcomings of the first match method, the user has the option of the 'best fit' strategy, which simply checks every single exemplar in the possibility zone and chooses the exemplar that is closest to the candidate. This method guarantees that the distortion between the original and compressed images will be minimized.

The obvious drawback is that this approach can take much longer than the simple first match method. Since, as we noted earlier, the codebook replacement does not affect any steps later in the program, we use the best fit strategy only when compression is a major concern in the processing.

4.3.2 Basis Selection

Once the prescreener has been run and the exemplars calculated, the next step in the ORASIS algorithm is to define an appropriate, low-dimensional subspace that contains the exemplars. One way to interpret the linear mixing model (Eq. 4.1) is that, if we ignore noise, then every image spectrum may be written as a linear combination of the endmember vectors. It follows that the endmembers define some subspace within band space that contains the data. Moreover, the endmembers are, in mathematical terms, a basis for that subspace. Reasoning backwards, it follows that if we can find some low-dimensional subspace that contains the data, then we simply need to find the 'right' basis for that subspace to find the endmembers. Also, by projecting the data into this subspace, we can reduce both the computational complexity (by working in lower dimensions) as well as the noise.

The ORASIS basis selection algorithm constructs the desired subspace by building up a set of orthonormal basis vectors from the exemplars. At each step, a new dimension is added until the subspace contains the exemplar set, up to a user-defined error criterion. The basis vectors are originally chosen from the exemplar set, and then they orthonormalized using a Gramm-Schmidt-like procedure (we note for completeness that earlier ORASIS publications have referred to the basis selection algorithm as a 'modified Gramm-Schmidt procedure.' We have since learned that this term has a standard meaning in mathematics that is unrelated to our procedure, and we have stopped using this phrase to describe the algorithm.)

The algorithm begins by finding the two exemplars $E_{i(1)}$, $E_{i(2)}$ that have the largest angle between them. These exemplars become known as 'salients,' and the indices $i(1)$ and $i(2)$ are stored for use later in the endmember selection stage. The first two salients are then orthonormalized (via Gramm-Schmidt) to form the first two basis vectors B_1 and B_2. Next, the set of exemplars is projected down into the two-dimensional subspace (plane) spanned by B_1 and B_2, and the residual (distance from the original to the projected spectrum) is calculated for each exemplar. If the value of the largest residual is smaller than some predefined error threshold, then the process terminates. Otherwise, the exemplar $E_{i(3)}$ with the largest residual is added to the salient set, and the index is saved. This exemplar is orthonormalized to the current basis set to form the third basis vector B_3. The exemplars are then projected into the three-dimensional subspace spanned by $\{B_1, B_2, \cdots, B_k\}$ and the process repeated. Additional basis vectors are added until either a user-defined error threshold is reached or a predetermined maximum number of basis vectors has been chosen.

At the end of the basis selection process, there exists a k-dimensional subspace that is spanned by the basis vectors $\{B_1, B_2, \cdots, B_k\}$, and all of the exemplars have been projected down into this subspace. As we have noted, under the assumptions of the linear mixing model, the endmembers must also span this same space. It follows that we are free to use the low-dimensional projected exemplars in order to find the endmembers. The salients $\{E_{i(1)}, E_{i(2)}, \cdots, E_{i(k)}\}$ are also saved for use in the next step, where they are used to initialize the endmember selection algorithm.

It is worth noting that the basis algorithm described above guarantees that the largest residual (or error) is smaller than some predefined threshold. In particular, ORASIS will generally include all outliers, by increasing the dimensionality of the subspace until it is large enough to contain them. This is by design, since in many situations (e.g., target and/or anomaly detection) outliers are the objects of most interest. By comparison, most statistically based methods (such as Principal Component Analysis) are designed to exclude outliers (which by definition lie in the tails of the distribution). One problem with our inclusive approach is that it can be sensitive to noise effects and sensor artifacts; however, this is usually avoided by having the prescreener remove any obviously 'noisy' spectra from the scene.

We note for completeness that newer versions of ORASIS include options for using principal components as a basis selection scheme, as well as an N-FINDR-like algorithm for improving the original salients. Neither of these modifications were used in this version of the code.

4.3.3 Endmember Selection

The next stage in the ORASIS processing is the endmember selection algorithm, or the 'shrinkwrap.' As we have discussed in previous sections, one way to interpret the linear mixing model (Eq. 4.1) is that the endmember vectors define some k-dimensional subspace (where k is equal to the number of endmembers) that contains the data. If we apply the sum-to-one (Eq. 4.2) and nonnegativity (Eq. 4.3) constraints, then a slightly stronger statement may be made; the endmembers are in fact the vertices of a $(k-1)$ simplex that contains the data. Note that this simplex must lie within the original k-dimensional subspace containing the data.

ORASIS uses this idea by defining the endmembers to be the vertices of some 'optimal' simplex that encapsulates the data. This is similar to a number of other 'geometric' endmember algorithms, such as Pixel Purity Index (PP) and N-FINDR, and is a direct consequence of the linear mixing model. We note that, unlike PP and N-FINDR, ORASIS does not assume that the endmembers are necessarily in the data set. We believe this is an important point. By assuming that each endmember must be one of the spectra in the given scene, there is an implicit assumption that there exists at least one pixel that contains only the material corresponding to the endmember. If this condition fails, then the endmember will only appear as a mixture (mixed pixel), and will not be present (by itself) in the data. This can occur, for example, in scenes with a large GSD (where the individual objects may be too small to fill an entire pixel). One of the goals of ORASIS is to be able to detect these 'virtual'-type endmembers (i.e. those not in the data), and to estimate their signature by extrapolating from the mixtures those that are present in the data.

From the previous subsection, the inputs to the endmember module are the exemplars from the prescreener, projected down into some k-dimensional subspace, as well as an initial set of k vectors known as the salients. By construction, the salients form an initial $(k-1)$ simplex within the subspace. The basic idea behind the shrinkwrap is to systematically 'push' the vertices of this simplex outwards. At each step, the vertices of the simplex are adjusted and a new simplex is formed. This process continues until every exemplar lies within the new simplex.

To begin the shrinkwrap, we check to see if all the exemplars are already inside the simplex defined by the salients. If so, then we assume that the salients are in fact the endmembers, and we are done. In almost every case, however, there will be at least one point outside of the initial simplex, and it must be expanded in order to encapsulate the exemplars. To do so, we find the exemplar E_{max} that lies the furthest distance outside of the current simplex. This is easily done by using the current endmembers (the vertices of the current simplex) to demix the data and search for the most negative abundance coefficient. The vertex V_{max} that is the furthest from the most outlaying exemplar E_{max} is held stationary, and the remaining vertices are moved outward (using steps of convex combinations) until the E_{max} exemplar lies inside the new simplex. The process is then simply repeated until all exemplars are within the simplex. The final endmembers are then defined to be the vertices of this final encompassing simplex.

4.3.4 Demixing

The final step in ORASIS is to decompose each of the scene spectra into a weighted sum of the endmembers. In the HSI literature this process is commonly referred to as demixing the data. Note that, in almost all cases, the measured image spectra will not lie exactly in the subspace defined by the endmembers; this is due to both modeling error and various types of sensor noise. It follows that the demixing process will not be exactly solvable, and the abundance coefficients must be estimated. The process of estimating the coefficients will differ depending on whether or not either (or both) of the constraints given in Eqs. 4.2 and 4.3 are applied. In this subsection, we discuss the two demixing algorithms (constrained and unconstrained) that are available in ORASIS.

The demixed data (with or without constraints) produced by the linear mixing model have a number of useful properties. For example, demixing allows the original high-dimensional image spectra to be replaced with the lower-dimensional demixed data, with little loss of information. This reduction, typically on the order of 10 to 1, can greatly simplify and speed up further processing. Also, demixing the data produces 'maps' of the abundance coefficients $\alpha_{i,j}$. By replacing each image spectrum with its demixed version, a series of k (= number of endmembers) grayscale images can be created. Each image will highlight only those pixels that contain the given endmember. For example, in a scene containing water, grass, and dirt elements, the pixels that contain water will be bright (have high abundance coefficients) only in the water endmember image, and will be dark in the remaining grass and dirt endmember images. Remembering that physically the $\alpha_{i,j}$'s represent the abundance of material j in image spectrum i, the images produced in this way are often referred to as abundance planes (or maps). Exactly what the abundance planes measure physically depends on what calibrations/normalizations have been performed during the processing. If the data have been calibrated and the endmembers are normalized, then the abundance maps represent the radiance associated with each endmember. Other interpretations are possible, such as relating the abundance maps to the fraction of total radiance from each endmember. In this case, the abundance maps are sometimes called the fraction planes.

4.3.4.1 Unconstrained Demix

The easiest method for demixing the data occurs when no constraints are placed on the abundance coefficients. If we let P be the $k \times n$ matrix (where k is the number of endmembers and n is the number of spectral bands) defined by $P = (X^t \cdot X)^{-1} X^t$, where $|X_1 X_2 \cdots X_k|$ is the $n \times k$ matrix whose columns are the endmembers, then it is straightforward to show that the least squares estimate $\hat{\alpha}$ to the true unknown mixing coefficients α for a given image spectrum Y is given by $\hat{\alpha} = PY$.

Note that the matrix P depends only on the endmembers. It follows that once P has been calculated, the unconstrained demixing process reduces to a simple matrix-vector product, which can be done very quickly.

4.3.4.2 Constrained Demix

The constrained demixing algorithm is used when the nonnegativity constraints (Eq. 4.3) are applied to the abundance coefficients. In this case, there is no known analytical solution, and numerical methods must be used. Our approach is based on the well-known Non-Negative Least Squares (NNLS) method of Lawson and Hanson [7]. The NNLS algorithm is guaranteed to converge to the unique solution that is closest (in the least squares sense) to the original spectrum. The FORTRAN code for the NNLS algorithm is freely available from Netlib [8]. We note that, compared to the unconstrained demixing algorithm, the NNLS can be significantly (orders of magnitude) slower. At the current time, ORASIS does not implement the sum-to-one constraint, either with or without the nonnegativity constraint.

4.4 Additional Algorithms

While the main focus in this chapter is the ORASIS algorithm, we include for completeness a brief description of the other major algorithms that were implemented in this project. This section discusses the algorithms in their original scalar form; we discuss the modifications made to run them in parallel in the next section.

4.4.1 ORASIS Anomaly Detection

The ORASIS Anomaly Detection (OAD) algorithm [9], originally developed as part of the Adaptive Spectral Reconnaissance Program (ASRP), is a method for using the ORASIS outputs (exemplars and endmembers) to identify potential objects of interest within hyperspectral imagery. The term 'anomaly' is generally used in the HSI literature to refer to objects that are significantly different (generally in a spectral sense, though spatial context is also used) from the background clutter of the scene. Generally speaking, anomaly detection algorithms do not attempt to identify (in a material sense) the detected anomalies; in contrast, target detection algorithms attempt to find those spectra in the image containing specific materials (targets).

The first step of OAD is to simply run ORASIS to create a set of exemplars and to identify endmembers. Next, each exemplar is assigned an 'anomaly measure' as defined below. An initial target map is then created by assigning to each image spectrum a score equal to that of its corresponding exemplar. A threshold is applied to the target map and the surviving spectra are segmented to create a list of distinct objects. Finally, the various spatial properties (e.g., width, height, aspect ratio) of the objects are calculated and stored. Spatial filters may then be applied to reduce false alarms by removing those objects that are not relevant.

The OAD anomaly measure attempts to define how spectrally different a given exemplar is from the general background of the scene. To do so, OAD first separates the set of endmembers into 'target' and 'background' classes. Intuitively, background endmembers are those endmembers that appear as a mixture element in a large number of the exemplars; conversely, target endmembers are those that appear in only a small number of exemplars. To put it another way, the abundance coefficient corresponding to a background endmember will be relatively large for a majority of the exemplars in a given scene, while the abundance coefficient of a target endmember should be relatively small for almost all exemplars. In statistical terms, the histogram of abundance coefficients for a background endmember will be relatively wide (high standard deviation) with a relatively large mean value (see Figure 4.4(a)), while target endmembers will have relatively thin (low standard deviation) histograms, with small means and a few pixels with more extreme abundance values (Figure 4.4(b)).

After the endmembers have been classified, the OAD algorithm discards the background endmembers and uses only the target dimensions. A number of partial measures are calculated, including measures of how 'target-like' (i.e., how much target abundance is present) a given exemplar is, and how 'isolated' or unique (i.e., how many other exemplars are nearby, in target space) that exemplar is. The partial measures are then combined into a single scalar anomaly measure.

As an example, Figure 4.5 shows the results of applying the OAD algorithm (with spatial filters) to the HYDICE Forest Radiance I data set.

4.4.2 N-FINDR

The N-FINDR algorithm is an alternative endmember selection algorithm developed by Technical Research Associates, Inc. As with ORASIS, N-FINDR uses the structure imposed on the data by the linear mixture model to define endmembers. In geometrical terms, the LMM (Eq. 4.1) states that the endmembers form a k-dimensional subspace that contains the image spectra (ignoring noise). If the sum-to-one and non-negativity constraints (Eqs. 4.2 and 4.3, respectively) are enforced, then the linear

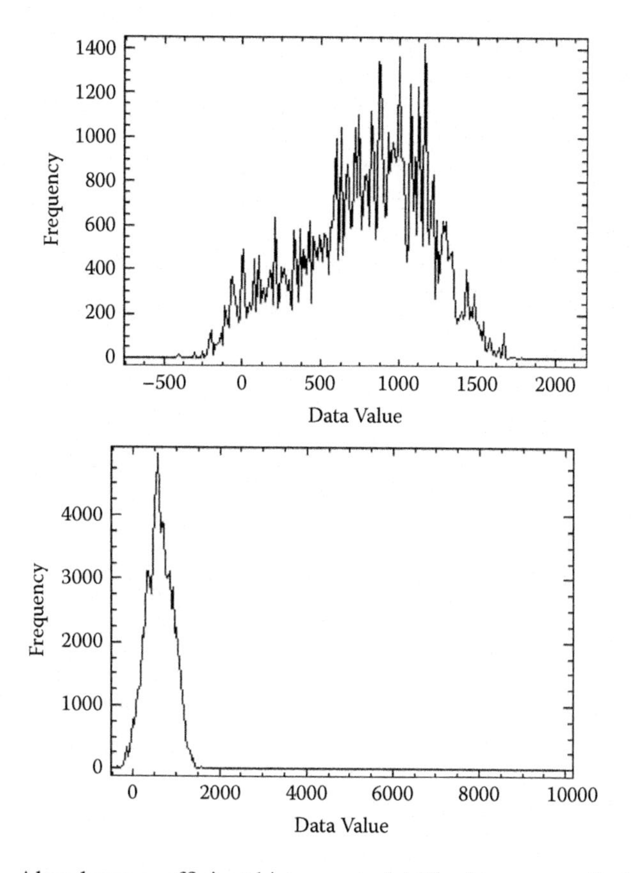

Figure 4.4 Abundance coefficient histograms. (a) The histogram of a background endmember. (b) The histogram of a target endmember.

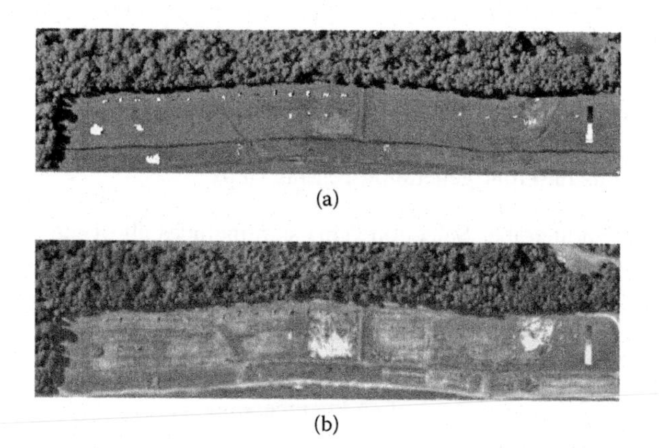

Figure 4.5 HYDICE data from Forest Radiance. (a) A single band of the raw data. (b) Overlay with the results of the OAD.

mixing model implies that the endmembers are in fact the vertices of a $(k-1)$ simplex that encapsulates the data. Roughly speaking, N-FINDR uses this approach to determine endmembers by choosing the k image spectra that define a solid whose volume is maximized. The number of endmembers k that are chosen may be defined by the user, or determined autonomously using a principal components-like analysis. We note that, in contrast to ORASIS, N-FINDR will only choose spectra that are in the image to be endmembers, and thus implicitly assumes that full-pixel examples of each endmember exist in the data. On the other hand, unlike ORASIS, this approach guarantees that each endmember is physically meaningful.

4.4.3 The Stochastic Target Detector

The Stochastic Target Detector (STD) [10] is an anomaly detection algorithm that was originally developed by Technical Research Associates, Inc. STD is similar to OAD in that endmembers are divided into target and background groups, although STD also includes a 'neutral' endmember type, which includes endmembers that are neither target or background endmembers (e.g. noise artifacts, shading effects, etc.). The algorithm then uses various statistical measures on the target and background endmember planes to determine how 'anomalous' a given input spectra is. The final output is a target map, assigning a single detection measure to each image spectra. As with OAD, the target map can then be thresholded, segmented, spatially filtered, etc.

4.5 Parallel Implementation

The algorithms presented in the two previous sections were all originally designed to run in scalar (single processor) environments. In this section, we discuss the changes that were made to run the various algorithms in parallel.

One of the goals of our project was to present the algorithms in such a way as to allow the user to be able to 'mix-n-match' various pieces in order to obtain a specific result as quickly as possible. For example, a user interested only in anomaly detection has no need for compression. Similarly, a user may want to be able to compare results using different endmember selection schemes, etc. With this in mind, we divided the various algorithms into four general processing steps:

1. ORASIS Endmember Selection. This step includes all of the major ORASIS algorithms (prescreener, basis selection, shrinkwrap) and outputs a set of endmembers. In addition, the user may optionally select the 'compression' option, which runs the best-fit module of the prescreener.
2. N-FINDR Endmember Selection. This step simply runs the parallel version of N-FINDR and outputs a set of endmembers.
3. Demixing. This step uses the endmembers from either of the first two steps to demix the image spectra. The user has the option of using either unconstrained or nonnegatively constrained demixing.

4. Anomaly Detection. This step uses the endmembers from either of the first two steps to run the OAD or STD algorithms.

Each of the preceding four steps was modified to be able to run in parallel mode. Note that the steps themselves must be run serially, since each step depends on the outcome of the earlier steps in order to be run. Also, in general only one of the endmember selection schemes will be used.

The general strategy we used for parallelization was the well-known master-slave paradigm, in which one process (the master) acts as a scheduler, distributing data and tasks to the remaining processes (slaves), which in turn send results back to the master for consolidation and output. Inter-process communication was done using the Message Passing Interface (MPI) [11], a communications library that has been implemented by all major HPC hardware manufacturers. We note that, at the time of this project, the full MPI-2 standard had not yet been implemented by all vendors. Since one of the requirements for this project was to be as portable as possible, we decided to use the older MPI-1.1 standard. In particular, we did not have access to the parallel Input/Output (I/O) routines of the MPI-2 standard. As a result, the master was tasked with reading the input image files and then distributing the data to each of the slaves. Given the size of typical hyperspectral images, this presented a rather large bottleneck, and forced us to structure the code in ways that were perhaps not ideal. As the MPI-2 standard and Parallel I/O become better implemented, it is hoped that these bottlenecks can be removed.

4.5.1 ORASIS Endmember Selection

The first step in the ORASIS program, the exemplar selection part of the prescreener, turned out to be quite difficult to parallelize. For a number of reasons, which we discuss in more detail later in this subsection, we decided to use only a limited number (typically three or four) of slave processes in this step. The master began by sending to each slave a fixed number of lines from the beginning of the cube; for example, the first slave received the first 100 lines of the cube, the second slave received the next 100 lines, etc. The individual slaves would then run their own (scalar) version of the prescreener, with each slave keeping a 'personal' list of exemplars. Once a slave had finished the initial block of data, it would ask the master for a new block to process, using its own internal list of exemplars, until the entire cube had been processed. In this way, each slave process builds up a set of exemplars, which is then sent to the master. The master then consolidates the individual lists into a 'full' set of exemplars, which it then broadcasts to all of the slave processes (including those processes not involved with the exemplar selection).

If the compression option is selected by the user, then the next step is the codebook replacement module, which replaces each image spectrum with the closest member of the exemplar list. We again use a master-slave formulation; the master begins by sending each slave an image frame to process. The slave then runs the (scalar) 'best-fit' codebook replacement algorithm on the frame. As the slave finishes, it sends back to the master the codebook for that frame, and receives a new frame. This process simply continues until the entire image has been processed. The frame-by-frame approach is needed since the amount of time needed to find the best matching exemplar can vary widely among the incoming image spectra, which tended to lead to serious load balancing issues.

The two remaining modules, basis determination and endmember selection, operate only on the exemplars. At this point, each process has a copy of the exemplar list, and the master distributes the computing load by assigning each process a subgroup of exemplars to work on. The slaves send the results of their computations back to the master, who first selects the basis vectors and then (after further processing by the slaves) determines the endmembers.

We conclude this subsection with a discussion of why we decided to cap the number of slave processes used during the exemplar selection process. The reason for doing so was due to the fact that the prescreener is based on accumulated knowledge. In particular, each incoming spectrum must be compared to the entire list of already known exemplars. Since each process contains its own list of exemplars, it is possible that a process could encounter an incoming spectrum that appears to be a new exemplar (since it does not match any exemplar on its list), while in reality it should be discarded (since it matches an exemplar found on another process's list). As a result, each process contains a number of 'redundant' exemplars, and the total number of exemplars will increase with the number of processes. As an extreme example, if the number of slave processes was equal to the number of image spectra, then no spectrum would be considered redundant, and every image spectrum would become an exemplar. Since the computation time of the remaining modules scales approximately as the square of the number of exemplars, the speedup resulting in parallelizing the prescreener (using all available slaves) was quickly nullified by the increase in computing time needed for the rest of the algorithm. For this reason, it was decided to cap the number of processes used in the prescreener.

We also note that the reason for sending 'blocks' of data, instead of simply partitioning the scene into equal areas and then assigning each area to a slave, was a load balancing problem. The actual amount of processing time needed to run the prescreener varies directly with the number of exemplars, which is itself closely tied to the scene. For example, a given image may contain both large, homogenous areas (grass fields, say) as well as areas with very diverse spectra (e.g. urban areas). In this type of image, the homogeneous areas will have only a relatively few spectra, and thus run very quickly, since very few comparisons are needed to see if a match occurs. Conversely, the urban areas will contain relatively many exemplars, and consequently take much longer to run than the grass fields.

4.5.2 N-FINDR Endmember Selection

To parallelize N-FINDR, a master-slave formulation was again used, and the scene partitioned spatially. To run the algorithm, the master process reads in the data cube and sends a given number of spatially contiguous frames to each process. The individual processes then run a (slightly modified) version of the scalar N-FINDR algorithm to determine a set of 'possible' endmembers for that section of the data. Each set of possible endmembers is returned to the master process, which then consolidates the combined list of possible endmembers to create a final set of endmembers. We note that the design and implementation of the parallel N-FINDR algorithm was done by Michael Winter of the University of Hawaii and TRA.

4.5.3 Spectral Demixing

Once the endmembers have been calculated, either by ORASIS or by N-FINDR, the next step is to estimate the abundance coefficients, or demix, the individual spectra. Two demixing routines are available: an unconstrained demix, which places no restrictions on the abundance coefficients, and a constrained demix, which requires that the abundance coefficients be strictly nonnegative. We note that either demixing routine operates strictly on a spectrum-by-spectrum case and is therefore embarrassingly parallel.

In the parallel version of the demixing algorithm, the master begins by sending the endmembers to each of the slave processes, which then calculates the correct (constrained or unconstrained) demixing matrix. The master then sends out a single image frames to each of the slaves. Each slave demixes the individual pixels in the frame and returns the demixed frame to the master, who then sends out a new frame to the slave. Once each frame has been demixed, the master then writes out the demixed cube to disk.

4.5.4 Anomaly Detection

After the demixing process, the next (optional) step in the program is anomaly detection. The user may choose one of two anomaly detection routines: the ORASIS Anomaly Detection (OAD) algorithm or the Stochastic Target Detector (STD) algorithm.

In the parallel version of OAD, the master separates the endmembers into target/background classes and broadcasts that information to the slaves. At this point in the processing, each slave already contains a copy of the exemplars and the endmembers. The exemplars are then partitioned by the master into subgroups, and each slave calculates the anomaly measures for each of the exemplars in its group. The master then receives back each of the anomaly measures, creates a target image, and writes out the target image to disk.

In the STD algorithm, the actual calculation times are dominated by matrix-matrix multiplications. Most of these multiplications involve small matrices (on the order of the number of endmembers) with a few relatively large ones (on the order of the number of pixels in the scene). Using platform-specific optimized linear algebra libraries, the total running time of the (scalar) STD algorithm for the images we used was on the order of a few seconds; therefore, we did not attempt to develop a parallel version of this algorithm.

4.6 Results

In this section, we discuss the timing and validation results of the parallel algorithms presented in the last section. We begin with a discussion of the hardware used in the test procedure. As we have noted earlier, this work was done as part of a CHSSI project on hyperspectral data validation. One of the requirements of this project was to design the code to be completely portable, while also allowing for the use of optimized, hardware-specific libraries. To meet this requirement, we ran our algorithms on three different HPC systems representing a variety of architectures and operating systems. The details of the hardware are summarized in Table 4.1. The base code was written in ISO compatible C++ (except the public domain WNNLS routine used in the constrained demixing algorithm, which is written in FORTRAN). All message passing and numerical linear algebra calculations were done using vendor-specific implementations of the MPI and BLAS (including LAPACK) libraries, respectively. The code compilation and linking of the various libraries was handled through the use of machine-specific make files.

To test our algorithms, we ran a series of experiments on each of the three test machines. The major goals of the test procedure were to show that the algorithms performed well on a variety of data cubes and under various user configurations. With that in mind, we constructed a series of five test runs for each machine, meant to model typical hyperspectral linear modeling tasks. In particular, we developed tests for compression, terrain categorization, and anomaly detection [12].

The first test, compression, used only the endmember selection and unconstrained demixing modules. Since N-FINDR is not designed for compression, we used only the ORASIS endmember selection module (including the 'best-fit' codebook replacement algorithm) for this test.

The second test, terrain categorization (TerrCat), used only the endmember selection and constrained demixing algorithms. This test was subdivided into two parts, one for each of the two endmember selection algorithms.

The third test, anomaly detection (ATR), used the endmember selection and unconstrained demixing algorithms, as well as one of the anomaly detection algorithms. As in TerrCat, this test was subdivided into two parts, one for each of the endmember selection algorithms. For the ORASIS endmember test, the OAD anomaly detection algorithm was used; the STD algorithm was used for the N-FINDR endmember tests.

Each of the three test scenarios was applied to different image cubes, to verify that the code was able to handle data from various sensors and in various formats. The specifications for each of the image cubes used are summarized in Table 4.2.

TABLE 4.1 Summary of HPC Platforms

Machine	Location	Machine Type	Operating System	Processors
Longview	SPAWAR	HP Superdome	HP-UX	48
Huinalu	MHPCC	IBM Netfinity Supercluster	Linux	512
Shelton	ARL	IBM P5	AIX	512

TABLE 4.2 Summary of Data Cubes

Test	Sensor	Cube Name	Samples	Lines	Bands	Data Type
Compression	AVIRIS	Cuprite VNIR	610	1024	64	16 bit integer
TerrCat	AVIRIS	Cuprite	610	1024	224	16 bit integer
ATR	NVIS	AP Hill	244	512	293	32 bit float

In Tables 4.3–4.5, we summarize the timing results for each of the three HPC platforms. Each table contains the results of running each of the five test runs, with the times given in seconds. Each test was run on a number of different processors (4, 8, 16, and 64) to test scalability. A Not Applicable (NA) score in the tables indicates that the given test was not run on the specified number of processors.

A second series of tests was developed to test the validity of the results. This test was somewhat of a challenge, since there is no 'correct' answer to the endmember selection process. However, we can use the compression test from above to verify that the endmember selection and demixing algorithms are working correctly, by first compressing the image cube and then comparing the decompressed cube with the original input cube. Since we can derive theoretical limits on how much distortion can occur, it is reasonable to assume that if the actual measured distortion is within the theoretical bounds, then the algorithms are performing correctly. Or, to put in another way, when the algorithms are not performing correctly, the distortion is much higher than it should be.

TABLE 4.3 Timing Results for the Longview Machine (in seconds)

		Number of Processes			
Test	**Algorithm**	**4**	**8**	**32**	**64**
Compression	ORASIS	83	NA	32	NA
TerrCat	ORASIS	152	NA	35	NA
TerrCat	N-FINDR	168	NA	41	NA
ATR	ORASIS	48	NA	15	NA
ATR	N-FINDR	25	NA	11	NA

TABLE 4.4 Timing Results for the Huinalu Machine (in seconds)

		Number of Processes			
Test	**Algorithm**	**4**	**8**	**32**	**64**
Compression	ORASIS	111	80	32	53
TerrCat	ORASIS	128	78	57	51
TerrCat	N-FINDR	140	82	60	48
ATR	ORASIS	77	47	32	31
ATR	N-FINDR	24	21	9	19

TABLE 4.5 Timing Results for the Shelton Machine (in seconds)

		Number of Processes			
Test	**Algorithm**	**4**	**8**	**32**	**64**
Compression	ORASIS	97	70	46	NA
TerrCat	ORASIS	197	111	59	6
TerrCat	N-FINDR	11	10	11	NA
ATR	ORASIS	158	94	35	25
ATR	N-FINDR	13	9	7	NA

TABLE 4.6 Statistical Tests used for Compression. X = original Spectrum, Y = Reconstructed Spectrum, n =Number of Bands

Measure	Formula
Absolute error	$\frac{1}{n}\sum_{i=1}^{n} \lvert X_i - Y_i \rvert$
Relative error	$\frac{1}{n}\sum_{i=1}^{n} \frac{\lvert X_i - Y_i \rvert}{X_i}$
RMS	$\sqrt{\sum_{i=1}^{n} \frac{(X_i - Y_i)^2}{n}}$
SNR	$10 \cdot \log_{10}\left(\frac{(max Y_i)^2}{RMS}\right)$
Error angle	$\cos^{-1}(\frac{X \cdot Y}{\lVert X_i \rVert \cdot \lVert Y_i \rVert})$

To verify the compression results, we calculated a variety of statistical measures to compare the original input cube and the decompressed image cube. For our test cube, we used the Cuprite image from Table 4.2, and the error angle in the prescreener was set to 0.5 degrees. Each measure was calculated on a pixel-by-pixel case, by comparing the original and reconstructed spectra. We then calculated the minimum, maximum, and mean results (among all pixels) for each measure. The details of the individual measures are given in Table 4.6, and the results for each of the three platforms are summarized in Tables 4.7–4.9. We note that the results were consistent among each of the three platforms, implying that the code was running correctly on each machine.

TABLE 4.7 Compression Results for the Longview Machine

Measure	Mean	Min	Max
Absolute error	541.8	374.6	1140.3
Relative error	0.75	0.73	0.75
RMS	592.9	390.6	1341.5
SNR	23.2	19.7	27.9
Error angle	0.38	0.05	0.73

TABLE 4.8 Compression Results for the Huinalu Machine

Measure	Mean	Min	Max
Absolute error	541.8	374.6	1140.2
Relative error	3.0	2.9	3.01
RMS	592.9	390.6	1341.4
SNR	35.2	31.8	39.9
Error angle	0.38	0.05	1.4

4.7 Conclusions

Given the size of most hyperspectral images, it is clear that automated, and efficient, processing algorithms are needed in order to keep up with the flow of data. Modern high-performance systems appear to offer the best hope of doing so, but a number of issues remain.

In the particular case of ORASIS, these issues include the data passing overhead, as the master process needs to send large chunks of data to each of the slaves. Better implementation of the MPI-2 standard, including the use of Parallel I/O, should remove most of that overhead. A bigger issue in our case was our lack of success in completely parallelizing the prescreener. This was offset by the near perfect speedup of the demixing routines, which, especially in the case of the constrained demix, tends to dominate the total processing time. In compression tasks, the best-fit algorithm performed well up to about 32 processes; for reasons we do not yet fully understand, increasing the number above that led to a decrease in performance. The remaining algorithms (basis determination, endmember selection, and OAD) also performed reasonably well, but, given the performance of the scalar versions, the speedup that results from parallelization is fairly slight. We note that N-FINDR, which was better able to take advantage of parallel processing by partitioning the scene spatially, performed very well.

One last issue, which we did not discuss in the text, is the question of how best to process the data spatially. Modern pushbroom sensors, which take data on a line-by-line basis, are capable of producing images that are many thousands of lines long (and

TABLE 4.9 Compression Results for the Shelton Machine

Measure	Mean	Min	Max
Absolute error	541.8	374.6	1140.3
Relative error	3.0	2.9	3.0
RMS	592.9	390.6	1341.4
SNR	35.2	31.8	39.9
Error angle	0.39	0.05	0.74

many gigabytes in size). ORASIS, like most linear mixing model-based algorithms, tends to do better on relatively small (1000 lines or so) images. The reason for this is simply that larger scenes will contain a larger number of endmembers. As the number of endmembers starts to reach the number of bands, the advantages of using linear mixing quickly diminishes. The question of how to best partition these very long data sets is a question we hope to pursue in the future.

4.8 Acknowledgments

This research was supported by the Office of Naval Research and the Common High Performance Computing Software Support Initiative. The parallel N-FINDR algorithm was designed and implemented by Michael Winter of the University of Hawaii. Michael Bettenhausen of the Naval Research Laboratory also contributed to the parallel versions of ORASIS described in this chapter.

References

[1] A. Plaza, P. Martinez, R. Perez and J. Plaza. A quantitative and comparative analysis of endmember extraction algorithms from hyperspectral data. *IEEE Transactions on Geoscience and Remote Sensing*, vol. 43, pp. 650-663, 2004.

[2] J. Boardman, F. Kruse and R. Green. Mapping target signatures via partial unmixing of AVIRIS data. *Summaries of Fifth Annual JPL Airborne Earth Science Workshop*, pp. 23-26, 1995.

[3] M. Winter. N-FINDR: an algorithm for fast autonomous spectral end-member determination in hyperspectral data. *Proceedings of SPIE*, vol. 3753, pp. 266-275, 1999.

[4] A. Plaza, P. Martinez, R. Perez and J. Plaza. Spatial/spectral endmember extraction by multidimensional morphological operations. *IEEE Transactions on Geoscience and Remote Sensing*, vol. 40, pp. 2025-2041, 2002.

[5] J. Bowles and D. Gillis. An optical real-time adaptive spectral identification system (ORASIS). *Hyperspectral Data Exploitation: Theory and Applications*. C.-I. Chang, Ed. John Wiley and Sons: Hoboken, NJ., 2007.

[6] *Department of Defense High Performance Computing Modernization Program (HPCMP) Main Page*. http://www.hpcmo.hpc.mil/index.html (accessed May 31, 2006).

[7] C. Lawson and R. Hanson. *Solving Least Squares Problems. Classics in Applied Mathematics 15*. SIAM: Philadelphia, PA. 1995.

[8] *The Netlib repository at UTK and ORNL.* http://www.netlib.org (accessed May 31, 2006). The FORTRAN code for NNLS can be found at http://www.netlib.org/lawson-hanson/all (accessed May 31, 2006).

[9] J. M. Grossmann, J. H. Bowles, D. Haas, J. A. Antoniades, M. R. Grunes, P. J. Palmadesso, D. Gillis, K. Y. Tsang, M. M. Baumback, M. Daniel, J. Fisher and I. A. Triandaf. Hyperspectral analysis and target detection system for the Adaptive Spectral Reconnaissance Program (ASRP). *Proceedings of SPIE*, vol. 3372, pp. 2-13, 1998.

[10] Hoff, L. E. and E. M. Winter. Stochastic target detection, *proceedings of the MSS on CC&D*, 2001.

[11] W. Gropp, E. Lusk and A. Skjellum. *Using MPI, Second Edition.* MIT Press: Cambridge, MA. 1999.

[12] D. Gillis, J. Bowles, M. Bettenhausen and M. Winter. Endmember selection and demixing in hyperspectral Imagery. *MHPCC Application Briefs*, 2003.

Chapter 5

Parallel Implementation of the Recursive Approximation of an Unsupervised Hierarchical Segmentation Algorithm

James C. Tilton,
NASA Goddard Space Flight Center

Contents

The hierarchical image segmentation algorithm (referred to as HSEG) is a hybrid of hierarchical step-wise optimization (HSWO) and constrained spectral clustering that produces a hierarchical set of image segmentations. HSWO is an iterative approach to region growing segmentation in which the optimal image segmentation is found at N_R regions, given a segmentation at $N_R + 1$ regions. HSEG's addition of constrained spectral clustering makes it a computationally intensive algorithm, for all but the smallest of images. To counteract this, a computationally efficient recursive approximation of HSEG (called RHSEG) has been devised. Further improvements in processing speed are obtained through a parallel implementation of RHSEG. This chapter describes this parallel implementation and demonstrates its computational efficiency on a Landsat Thematic Mapper test scene.

5.1 Introduction

Image segmentation is the partitioning of an image into related sections or regions. For remotely sensed images of the earth, an example of an image segmentation would be a labeled map that divides the image into areas covered by distinct earth surface covers such as water, snow, types of natural vegetation, types of rock formations, types of agricultural crops, and types of other man created development. In unsupervised image segmentation, the labeled map may consist of generic labels such as region 1, region 2, etc., which may be converted to meaningful labels by a post-segmentation analysis.

Segmentation is a key first step for a number of approaches to image analysis and compression. In image analysis, the group of image points contained in each region provides a good statistical sampling of image values for more reliable labeling based on region mean feature values. In addition, the region shape can be analyzed for additional clues to the appropriate labeling of the region. In image compression, the regions form a basis for compact representation of the image. The quality of the prerequisite image segmentation is a key factor in determining the level of performance for these image analysis and compression approaches.

A segmentation hierarchy is a set of several image segmentations of the same image at different levels of detail in which the segmentations at coarser levels of detail can be produced from simple merges of regions at finer levels of detail. This is useful for applications that require different levels of image segmentation detail depending on the particular image objects segmented. A unique feature of a segmentation hierarchy that distinguishes it from most other multilevel representations is that the segment or region boundaries are maintained at the full image spatial resolution for all levels of the segmentation hierarchy.

In a segmentation hierarchy, an object of interest may be represented by multiple image segments in finer levels of detail in the segmentation hierarchy, and may be merged into a surrounding region at coarser levels of detail in the segmentation hierarchy. If the segmentation hierarchy has sufficient resolution, the object of interest will be represented as a single region segment at some intermediate level of segmentation detail. The segmentation hierarchy may be analyzed to identify the hierarchical level at which the object of interest is represented by a single region segment. The object may then be identified through its spectral and spatial characteristics. Additional clues for object identification may be obtained from the behavior of the image segmentations at the hierarchical segmentation levels above and below the level at which the object of interest is represented by a single region.

Segmentation hierarchies may be formed through a region growing approach to image segmentation. In region growing, spatially adjacent regions iteratively merge through a specified merge selection process. Hierarchical step-wise optimization (HSWO) is a form of region growing segmentation in which the iterations consist of finding the best segmentation with one region less than the current segmentation [1, 2, 3]. The best segmentation is defined through a mathematical criterion such as a minimum vector norm or minimum mean squared error. An augmentation of HSWO, called HSEG (for hierarchical segmentation), was introduced by Tilton [4] in which an option is provided for merging spatially non-adjacent regions as controlled by a threshold based on previous merges of spatially adjacent regions. This can be thought of as a form of constrained spectral clustering.

The introduction of constrained spectral clustering in HSEG makes it a computationally intensive algorithm, for all but the smallest of images. This is because of a requirement to evaluate the dissimilarity between all pairs of regions, rather than just spatially adjacent regions. For a 1024×1024 pixel image, this leads to the order of 10^6 dissimilarity evaluations per iteration in the initial processing stages.

This computational difficulty is overcome by a recursive approximation of HSEG, called RHSEG. An early version of RHSEG was discussed in Tilton [5]. This recursive formulation not only limits the number of comparisons between spatially non-adjacent regions to a more reasonable number, but also lends itself to a straightforward and efficient implementation on parallel computing platforms. The current parallel implementation is similar to the implementation first disclosed in a NASA internal document [6] and U.S. Patent No. 6, 895, 115 B2. This implementation for two-dimensional data has been recently extended to accommodate one- and three-dimensional data [7].

This chapter is organized as follows. A high-level description of HSEG is followed by a more detailed description of RHSEG. Then a description of the parallel implementation of RHSEG is provided. Finally, timing comparisons are provided for several degrees of parallelism, from 256 CPUs down to 1 CPU.

5.2 Description of the Hierarchical Segmentation (HSEG) Algorithm

The hierarchical image segmentation algorithm, HSEG, is based upon the relatively widely utilized hierarchical step-wise optimization (HSWO) region growing approach of Beaulieu and Goldberg [3], which can be summarized as follows:

1. Initialize the segmentation by assigning each image pixel a region label. If a pre-segmentation is provided, label each image pixel according to the pre-segmentation. Otherwise, label each image pixel as a separate region.
2. Calculate the dissimilarity criterion value between all pairs of spatially adjacent regions, find the pair of spatially adjacent regions with the smallest dissimilarity criterion value, and merge that pair of regions.
3. Stop if no more merges are required. Otherwise, return to step 2.

HSEG differs from HSWO in one major aspect. The HSEG algorithm allows for the merging of spatially non-adjacent regions, as controlled by the S_{wght} parameter. For $S_{wght} = 0.0$, only spatially adjacent regions are allowed to merge, as in HSWO. However, for $S_{wght} > 0.0$, HSEG allows merges between spatially non-adjacent regions. For $S_{wght} = 1.0$, merges between spatially adjacent and non-adjacent regions are given equal weight. For values of S_{wght} between 0.0 and 1.0, merges between spatially adjacent regions are favored by a factor of $1.0/S_{wght}$. Allowing for a range of merge priorities for spatially non-adjacent regions provides HSEG with a great deal of flexibility in tailoring the segmentation results to a particular need.

HSEG also provides a selection of dissimilarity functions for determining most similar pairs of regions for merging. The available selection of dissimilarity functions is based on vector norms, mean-squared error, entropy, spectral information divergence (SID), spectral angle mapper (SAM), and normalized vector distance (NVD). See the RHSEG and HSEGViewer User's Manual [8] for the mathematical definitions of these dissimilarity functions. Options for other dissimilarity functions can be easily added.

5.3 The Recursive Formulation of HSEG

The merging of spatially non-adjacent regions in HSEG leads to heavy computational demands. These demands are significantly reduced through a recursive approximation of HSEG, called RHSEG, which recursively subdivides the imagery data into smaller sections to limit to a manageable number the number of regions considered at any point in the algorithm (usually in the range of 1000 to 4000 regions). RHSEG includes a provision to blend the results from the subsections to avoid processing window artifacts. This recursive approximation also leads to a very efficient parallel implementation. This parallel implementation of RHSEG is so efficient that full Landsat Thematic Mapper (TM) scenes (approximately 7000 by 6500 pixels) can be processed in 2 – 8 minutes on a Beowulf cluster consisting of 256 2.4GHz CPUs (http://thunderhead.gsfc.nasa.gov). This is only 10 to 20 times the amount of time that the Landsat TM sensor takes to collect this amount of data.

The two spatial dimensional version of RHSEG was described in [5] and [9]. A description of RHSEG, generalized to N_D spatial dimensions, follows (see also [7]):

1. Given an input image X, specify the number levels of recursion (L_r) required and pad the input image, if necessary, so that each spatial dimension of the data set can be evenly divided by $2^{(L_r-1)}$. (A good value for L_r results in an image section at recursive level L_r consisting of roughly 1000 to 4000 pixels.) Set $L = 1$.
2. Call $rhseg(L, X)$, where $rhseg(L, X)$ is as follows:
   - 2.1 If $L = L_r$, go to step 2.3. Otherwise, divide the image data into 2^{N_D} equal subsections and call $rhseg(L + 1, X/2^{N_D})$ for each image section (represented as $X/2^{N_D}$).
   - 2.2 After all 2^{N_D} calls to $rhseg()$ from step 2.1 complete processing, reassemble the image segmentation results.
   - 2.3 If $L < L_r$, initialize the segmentation with the reassembled segmentation results from step 2.2. Otherwise, initialize the segmentation with one pixel per region. Execute the HSEG algorithm on the image X with the following modification: Terminate the algorithm when the number of regions reaches the preset value N_{min}.

3. Execute the HSEG algorithm (per Section 5.2) on the image X using as a pre-segmentation the segmentation output by the call to $rhseg()$ in step 2.

The defaults for the user specifiable parameters L_r and N_{min} depend on the size of the image data and are calculated internally by RHSEG.

Under a number of circumstances, the segmentations produced by the RHSEG algorithm exhibit processing window artifacts. These artifacts are region boundaries that are along the processing window seams, even though the image pixels across the seams are very similar. Processing window artifacts are usually minor but can be more noticeable, depending on the image. They tend to be more noticeable and prevalent in larger images. However, all processing window artifacts can be completely eliminated by adding a fourth step to the definition of $rhseg(L, X)$ given above (following [10] and [11]):

2.4. If $L = L_r$, exit. Otherwise do the following (and then exit):

(a) For each region, identify other regions that may contain pixels that are more similar to it than the region that they are currently in. These regions are placed in a *candidate_region_label* set for each region. This is done by:

i. scanning the processing window seam between sections processed at the next deeper level of recursion for pixels that are more similar (by a factor of F_{seam}) to the region existing across the processing window seam.

ii. for $S_{wght} > 0.0$ identifying regions that have a dissimilarity between each other less than $F_{region} \times S_{wght} \times T_{max}$ (T_{max} is the maximum of the merge threshold encountered so far in HSEG).

(b) For each region with a $candidate\_region\_label$ set of size greater than zero, identify pixels in the region that are more similar by a factor of F_{split} to regions in the $candidate\_region\_label$ set than to the region they are currently in. If $S_{wght} = 1.0$, simply switch the region assignment of these pixels to the more similar region. Otherwise, split these pixels out of their current regions and remerge them through a restricted version of HSEG in which region growing is performed with these split-out pixels and merging is restricted to neighboring regions, the region from which the pixel was split out from, and regions in the $candidate\_region\_label$ set of the region from which the pixel was split out from.

Processing window artifact elimination as introduced here not only eliminates the processing window artifacts, but does so with minimal computational overhead. The computation time no more than doubles for a wide range of image sizes [11]. The default value of 1.5 for the parameters values F_{seam}, F_{region}, and F_{split} works well for a wide range of images.

A demonstration version of the sequential implementation of RHSEG is available from http://ipp.gsfc.nasa.gov/RHSEG/. This demonstration version is full featured, limited only by a three month time limit.

5.4 The Parallel Implementation of RHSEG

An earlier parallel implementation of RHSEG using PVM is described in Tilton [5]. The current parallel implementation of RHSEG using MPI is similar to the PVM implementation disclosed earlier in Tilton [6] and U.S. Patent No. 6, 895, 115 B2. A key difference between the current MPI implementation and the earlier implementations is the manner in which processing tasks are allocated amongst the available CPUs (see below).

The recursive division of the data, which is at the core of RHSEG, divides the data into $2^{N_D(L_r-1)}$ sections, where N_D is the spatial dimensionality of the data. A naive parallelization approach would be to process each these sections on a separate CPU. However, for large images, the number of CPUs required for this approach becomes unrealistic, as shown in Table 5.1.

The practical solution to this problem is to determine the number of recursive levels, designated as $L_i (\leq L_r)$, that divides the data into a number of sections less than or equal to the available number of CPUs (P), by solving $P \geq 2^{N_D(L_i-1)}$. Then RHSEG is run sequentially for the recursive levels $> L_i$.

After the sequential recursive levels complete processing, and parallel processing is completed at recursive level L_i, one must choose whether or not the input data and pixel-based results data, such as the current region label map, are passed back to recursive levels $< L_i$ or kept at recursive level L_i. Such pixel-based data can be kept at recursive level L_i and retrieved as necessary by recursive levels $< L_i$. Since the region merging decisions involve only region-based information, which is always passed back to recursive levels $< L_i$, communications are, for the most part, only required back to recursive level L_i to update the region label map. The only other communications required back to recursive level L_i are at recursive level 1 for the computation of the value of the global dissimilarity function value, if requested, for the connected component labeling step in processing window artifact elimination, which is required at all recursive levels $< L_r$ when $S_{wght} = 0.0$, and when outputting the segmentation hierarchy results.

For parallel computing systems with slower interprocessor communications, keeping the pixel-based data at processing level L_i may lead to an increase in processing time. However, for large images, bringing the pixel-based data all the way back to recursive level 1 can easily lead to a large of amount of disk swapping as the program memory required exceeds the available RAM memory. A compromise is to bring the pixel-based data back up to a recursive level L_o, which is less than or equal to L_i but greater than or equal to 1. This is illustrated in Figure 5.1. The case illustrated is for a 512×512 pixel image with $L_r = 5$, $L_i = 3$, and $L_o = 2$. A total of 16 tasks (on 16

TABLE 5.1 The Number of CPUs Required for a Naive Parallelization of RHSEG with One CPU per 4096 Pixel Data Section for Various Dimensionalities

L_r	Image size ($N_D = 1$)	# CPUs	L_r	Image size ($N_D = 2$)	# CPUs
1	4096	1	1	64^2	1
2	8192	2	2	128^2	4
3	16,384	4	3	256^2	16
4	32,768	8	4	512^2	64
5	65,536	16	5	$1,024^2$	256
6	131,072	32	6	$2,048^2$	1,024
7	262,144	64	7	$4,096^2$	4,096
8	524,288	128	8	$8,192^2$	16,384
9	1,048,576	256	9	$16,384^2$	65,536
10	2,097,152	512	—	—	—
L_r	**Image size ($N_D = 3$)**	**# CPUs**	L_r	**Image size ($N_D = 4$)**	**# CPUs**
1	16^3	1	1	8^4	1
2	32^3	8	2	16^4	16
3	64^3	64	3	32^4	256
4	128^3	512	4	64^4	4096
5	256^3	4,096	5	128^4	65,536
6	512^3	32,768	—	—	—
L_r	**Image size ($N_D = 6$)**	**# CPUs**	L_r	**Image size ($N_D = 12$)**	**# CPUs**
1	4^6	1	1	2^{12}	1
2	8^6	64	2	4^{12}	4,096
3	16^6	4,096	—	—	—

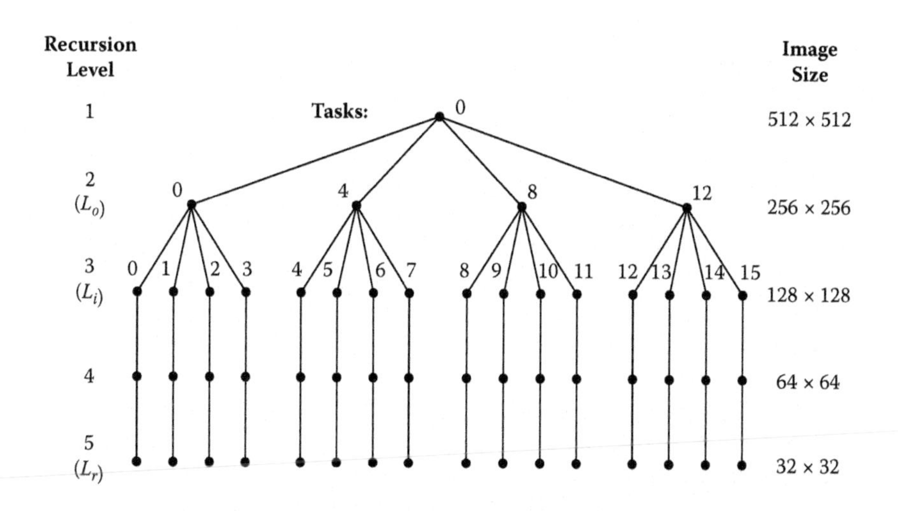

Figure 5.1 Graphical representation of the recursive task distribution for RHSEG on a parallel computer.

CPUs) are utilized. The data are input at recursive level 3 (L_i), and then subdivided and fed down to recursive level 5 (L_r), where HSEG processing is initiated. All results are fed back up through recursive level 2 (L_o). Only non-pixel-based results are fed back up to recursive level 1, and the pixel-based results are maintained and output from recursive level 2 (L_o). Task 0 is active at all recursive levels. At recursive level 1, it covers the entire 512×512 image; at recursive level 2, it covers a 256×256 portion of the image; and at recursive levels 3 through 5 it covers a 128×128 portion of the image. Tasks 4, 8, and 12 are active for recursive levels 2 through 5. At recursive level 2, they each cover a 256×256 portion of the image; and at recursive levels 3 through 5 they cover a 128×128 portion of the image. Tasks 1, 2, 3, 5, 6, 7, 9, 10, 11, 13, 14, and 15 are active at recursive levels 3 through 5, and each cover a 128×128 portion of the image.

The scheme of utilizing these three recursive processing levels in this way is the subject of U.S. Patent No. 6, 895, 115 B2 (see also [6]). However, the current implementation features a more efficient allocation of processing tasks amongst the available CPUs. The previous implementation restricted each CPU to processing only at a particular recursive level. However, the current implementation reuses certain CPUs at multiple recursive levels. In Figure 5.1, the numbered tasks correspond to individual CPUs. Thus task 0 is performed on CPU No. 0 and is active at all recursive levels. Similarly, tasks 4, 8, and 12 are performed on CPU numbers 4, 8, and 12, respectively, and are active at recursive levels 2 and above. Finally, tasks 1, 2, 3, 5, 6, 7, 9, 10, 11, 13, 14, and 15 are performed on CPU numbers 2, 3, 5, 6, 7, 9, 10, 11, 13, 14, and 15, respectively, and are active at recursive levels 3 and above.

5.5 Processing Time Performance

Processing time performance is reported here for a subset of a Landsat Thematic Mapper (TM) image that was collected May 28, 1999 from over Maryland and Virginia. The test scene contains 2048 columns and 2048 rows with six spectral bands. The computing platform utilized is a 256 node Beowulf cluster equipped with dual processor 2.4 GHz Pentium 4 Xeons, 256 Gbyte DDR memory (1.0 GBytes RAM available per CPU), 20 Tbyte disk space, and a 2.2 Gbyte/sec Myrinet fiber interconnection system. The interconnection system is sufficiently fast so that no processing time improvement is obtained by setting L_o less than L_i. When run with only one CPU (i.e., in serial), L_i also always equals L_o (and is called L_{io}) and specifies the recursive level at which image, region label map, and other pixel-oriented intermediate results are swapped in and out of the program in a scheme to reduce program RAM requirements (see [12]).

Table 5.2 compares the processing times for RHSEG for three settings of the S_{wght} parameter for 1, 4, 16, 64, and 256 CPUs. As can be seen from the speedup factors, the parallel implementation is very efficient. For $S_{wght} = 0.0$, the speedup factor even exceeds the number of CPUs for 16 and 64 CPUs. This extra speedup in the parallel implementation versus the serial can be explained by inefficiencies in the serial

TABLE 5.2 RHSEG Processing Time Results for a Six-Band Landsat Thematic Mapper Image with 2048 Columns and 2048 Rows. (For the 1 CPU case, the processing time shown is for the values of L_i and L_o that produce the smallest processing time.) Processing Time Shown as Hours:Minutes:Seconds

		$S_{wght} = 0.0$		$S_{wght} = 0.1$		$S_{wght} = 1.0$	
# of CPUs	**L_r:L_i: L_o**	**Processing Time**	**Measured Speedup**	**Processing Time**	**Measured Speedup**	**Processing Time**	**Measured Speedup**
1	7:1:1	—	—	1:52:53	—	2:45:50	—
1	7:4:4	0:25:39	—	—	—	—	—
4	7:2:2	0:06:58	3.7	0:31:44	3.6	0:45:19	3.7
16	7:3:3	0:01:29	17.3	0:09:13	12.2	0:13:41	12.1
64	7:4:4	0:00:24	64.1	0:02:34	44.0	0:04:11	39.6
256	7:5:5	0:00:11	139.9	0:00:50	135.4	0:01:31	109.3

implementation. It is clearly more efficient to swap the pixel-oriented intermediate results back and forth between parallel tasks than it is to swap this information in and out of disk files, as required in the serial version.

It is interesting to investigate the amount of time spent in the parallel implementation in actual computation versus transferring data between parallel tasks and waiting for other tasks. Table 5.3 shows the percentage of time task 0 spent in these activities for 4, 16, 64, and 256 CPUs, plus the percentage of time spent in setup (processing input parameters, calculating other program parameters and inputting distributing the input image data, etc.) and in other activities for the 2048 × 2048 Landsat TM test scene for $S_{wght} = 0.1$.

For the test scene, the percentage of time task 0 spent in actual computation ranges from about 84% for 256 CPUs up to over 98% for 16 CPUs. The wait time ranges from about 10.5% for 256 CPUs and 4 CPUs down to just over 1% for 16 CPUs. The percentage of time spent in setup, data transfer, and other activities is very small compared to the compute and wait times.

It should be noted that the relative percentages task 0 spends in computation versus waiting is very data dependent. In the case of 256 CPUs, task 0 processes the upper

TABLE 5.3 The Percentage of Time Task 0 of the Parallel Implementation of RHSEG Spent in the Activities of Set-up, Computation, Data Transfer, Waiting for Other Tasks, and Other Activities for the 2048 × 2048 Landsat TM Test Scene

# CPUs	Set-up	Computation	Data transfer	Waiting	Other
256	0.82%	84.42%	1.38%	10.44%	2.94%
64	0.33%	95.88%	0.32%	2.57%	0.90%
16	0.12%	98.46%	0.07%	1.26%	0.09%
4	0.06%	89.37%	0.01%	10.48%	0.08%

left 128 × 128 pixel portion of the image. For 64 CPUs, task 0 processes the upper left 256 × 256 pixel portion, for 16 CPUs task 0 processes the upper left 512 × 512 portion, and for 4 CPUs task 0 processes the upper left 1024 × 1024 portion. The wait time would be 0% if the image data are very homogeneous outside of these upper left portions of the image and relatively heterogeneous inside of these upper left portions.

5.6 Concluding Remarks

The hierarchical image segmentation algorithm (referred to as HSEG) was described as a hybrid of hierarchical step-wise optimization (HSWO) and constrained spectral clustering that produces a hierarchical set of image segmentations. HSEG's addition of constrained spectral clustering makes it a computationally intensive algorithm, for all but the smallest of images. This chapter described a computationally efficient recursive approximate implementation of HSEG (called RHSEG) designed to reduce the computational requirements of HSEG and provided a description of the parallel implementation of RHSEG. The speedup provided by the parallel implementation was shown for a Landsat TM test scene to range from 42% to 108% of the number of CPUs, depending on the number of CPUs utilized and the setting of a program parameter. For the same Landsat TM test scene, the percentage of time spent in actual computation was shown to range from about 84% to over 98%, depending on the number of CPUs utilized.

References

[1] J.-M. Beaulieu, M. Goldberg, Hierarchical picture segmentation by approximation, *Proceedings Canadian Communications and Energy Conference*, Montreal, P.Q., Canada, pp. 393–396, 1982.

[2] J. C. Tilton and S. C. Cox, Segmentation of remotely sensed data using parallel region growing, *1983 International Geoscience and Remote Sensing Symposium (IGARSS'83) Digest*, San Francisco, CA, vol. 1, Section WP-4, paper 9, Aug. 31–Sep. 2, 1983.

[3] J.-M. Beaulieu and M. Goldberg, Hierarchy in picture segmentation: a step-wise optimal approach, *IEEE Transactions on Pattern Analysis and Machine Intelligence*, Vol. 11, No. 2, pp. 150–163, Feb. 1989.

[4] J. C. Tilton, image segmentation by region growing and spectral clustering with a natural convergence criterion, *Proceedings of the 1998 International Geoscience and Remote Sensing Symposium*, Seattle, WA, pp. 1766–1768, July 6–10, 1998.

[5] J. C. Tilton, A recursive PVM implementation of an image segmentation algorithm with performance comparisons in-between the HIVE and Cray T3E, *Proceedings of the Seventh Symposium on the Frontiers of Massively Parallel Computation*, Annapolis, MD, pp. 146–153, Feb. 23–25, 1999.

[6] J. C. Tilton, Method for implementation of recursive hierarchical segmentation on parallel computers, *Disclosure of Invention and New Technology (Including Software): NASA Case No. GSC 14,305-1*, NASA Goddard Space Flight Center, Feb. 2, 2000. NOTE: Patent No. 6,895,115 B2 was issued for this technology on May 17, 2005 by the United States Patent and Trademark Office.

[7] J. C. Tilton, D-dimensional formulation and implementation of recursive hierarchical segmentation, *Disclosure of Invention and New Technology: NASA Case No. GSC 15199-1*, May 26, 2006.

[8] J. C. Tilton, RHSEG and HSEGViewer User's Manual, provided with the demonstration version of RHSEG and HSEGViewer available from http://ipp.gsfc.nasa.gov/RHSEG (version 1.25 released Dec. 14, 2006).

[9] J. C. Tilton, Method for recursive hierarchical segmentation by region growing and spectral clustering with a natural convergence criterion, *Disclosure of Invention and New Technology: NASA Case No. GSC 14,328-1*, Feb. 28, 2000. See also http://cisto.gsfc.nasa.gov/TILTON/publications/RHSEG_disclosure/NF1679.html.

[10] J. C. Tilton, Method for recursive hierarchical segmentation which eliminates processing window artifacts, *Disclosure of Invention and New Technology: NASA Case No. GSC 14,681-1*, Oct. 11, 2002 (revised Jan. 24, 2003). NOTE: U.S. Patent Application Serial No. 10/845,419 was filed on this technology on May 11, 2004.

[11] J. C. Tilton, A split-remerge method for eliminating processing window artifacts in recursive hierarchical segmentation, *Disclosure of Invention and New Technology: NASA Case No. GSC 14,994-1*, Mar. 9, 2005. NOTE: U.S. Patent Application Serial No. 11/251,530 was filed on this technology on September 29, 2005 (this is a continuation in part of U.S. Patent Application Serial No. 10/845,419).

[12] J. C. Tilton, Memory efficient serial implementation of recursive hierarchical segmentation, *Disclosure of Invention and New Technology: NASA Case No. GSC 15198-1*, May 23, 2006.

Chapter 6

Computing for Analysis and Modeling of Hyperspectral Imagery

Gregory P. Asner,
Carnegie Institution of Washington

Robert S. Haxo,
Carnegie Institution of Washington

David E. Knapp,
Carnegie Institution of Washington

Contents

Hyperspectral remote sensing is increasingly used for Earth observation and analysis, but the large data volumes and complex analytical techniques associated with imaging spectroscopy require high-performance computing approaches. In this chapter, we highlight several analytical methods employed in vegetation and ecosystem studies using airborne and space-based imaging spectroscopy. We then summarize the most common high-performance computing approaches used to meet these analytical demands, and provide examples from our own work with computing clusters. Finally, we discuss several emerging areas of high-performance computing, including data processing onboard aircraft and spacecraft and distributed Internet computing, that will change the way we carry out computations with high spatial and spectral resolution observations of ecosystems.

6.1 Introduction

There is an increasing demand for high spatial and spectral resolution remote sensing data for environmental studies ranging from ecological dynamics of terrestrial and aquatic systems to urban development. A good example is hyperspectral remote sensing, also called imaging spectroscopy, which is rapidly advancing from the remote sensing research arena to a mapping and analysis science in support of conservation, management, and policy development. Hyperspectral remote sensing is the measurement, in narrow contiguous wavelength bands, of solar radiation reflected by materials in the environment (Figure 6.1). These measurements express the chemical composition and structural properties of the materials of interest. In industrial operations, spectroscopy is used for material identification, manufacturing, and quality assurance. In Earth observation, imaging spectroscopy is used to estimate chemical concentrations and structures in vegetation, phytoplankton, soils, rocks, and a wide range of synthetic materials [1, 2].

The advent of hyperspectral remote sensing technology represents a progression from basic panchromatic and multispectral camera-like imaging of the past to a more data-rich and physically-based imaging and analysis arena for 21st century science. Field, airborne, and even space-based hyperspectral sensors are available today to government, commercial, and private organizations, yet the collection and analysis of imaging spectrometer data continue to be a challenge. Both the data volume and the processing techniques currently require a level of technological and scientific investment that is beyond the reach of many agencies and organizations. Continued effort is thus needed to advance the science of imaging spectroscopy from an esoteric specialty area to a mainstream set of applied methods for earth science. This is particularly true today as the earth science community is challenged to demonstrate the societal benefit of its observations and studies, especially from expensive investments such as remote sensing.

In this chapter, we summarize the major processing challenges and steps involved in hyperspectral image data collection and analysis, and we provide examples of

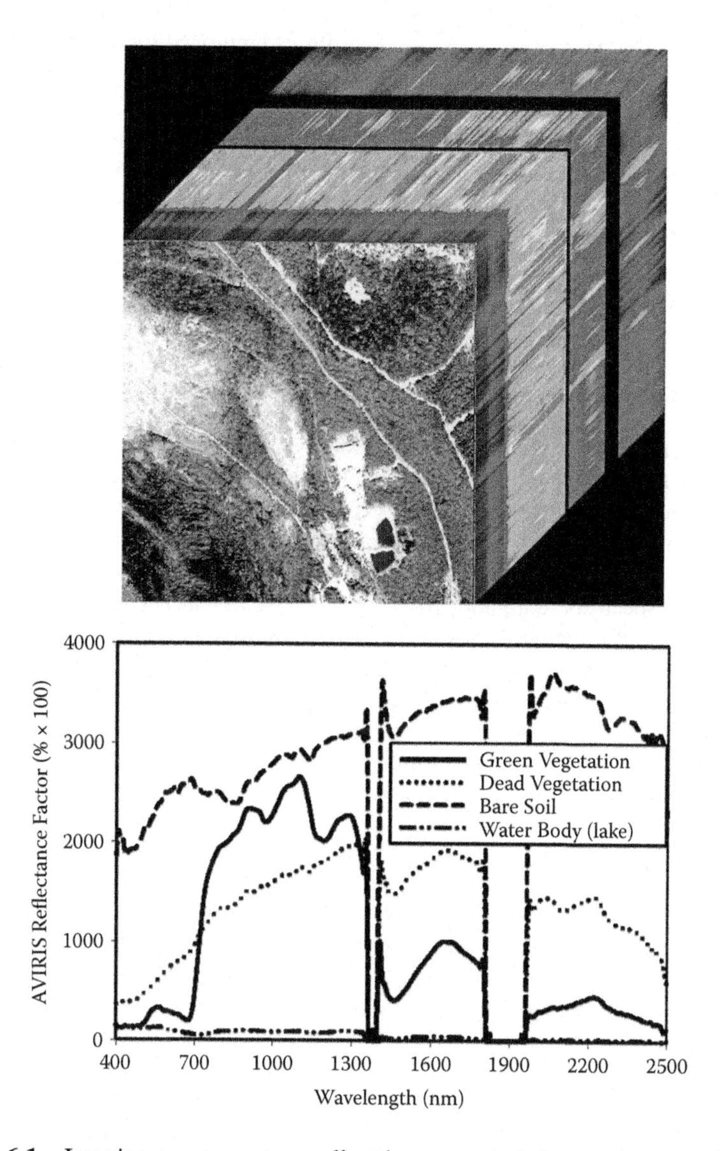

Figure 6.1 Imaging spectrometers collect hyperspectral data such that each pixel contains a spectral radiance signature comprised of contiguous, narrow wavelength bands spanning a broad wavelength range (e.g., 400–2500 nm). Top shows a typical hyperspectral image cube; each pixel contains a detailed hyperspectral signature such as those shown at the bottom.

how high-performance computing is used to meet these challenges. We highlight why specific types of high-performance computing approaches are matched to the demands of different types of scientific algorithms employed for hyperspectral data analysis. Whereas some analytical methods require true parallel processing, others benefit from the strategic use of distributed computing techniques. We also look into the future by outlining a framework for processing hyperspectral data onboard aircraft and spacecraft. Near-real-time processing of data is the next frontier in bringing hyperspectral imaging from a specialty to a mainstream science for environmental research and monitoring.

6.2 Hyperspectral Imagery and Analysis

The term hyperspectral is used in a wide variety of ways by remote sensing practitioners, and there is some confusion in the literature as to what hyperspectral really means. Some call any imaging system with more than about 5–10 channels 'hyperspectral.' However, a hyperspectral sensor is a system that collects images with pixels containing a series of contiguous, narrowband spectral channels covering a particular region of the spectrum (e.g., 400–1050 nm or 400–2500 nm). The spectrum in each pixel provides information on absorption and scattering features of materials in that pixel. Because the data are collected in image format with each spectrum spatially located, the measurements are organized as 3-D 'cubes' that allow analysis of remotely sensed materials in a geographic context (Figure 6.1). In this section, we discuss some of the most common hyperspectral imaging systems and analytical techniques that subsequently demand high-performance computing techniques.

6.2.1 Typical Imaging Systems

Imaging spectrometers are used to collect hyperspectral data from field, airborne, or space-based vantage points. Imaging spectrometers vary in design, but the two most common systems employ either scanning 'whiskbroom' sensors or pushbroom arrays (Table 6.1). The data vary in spatial resolution (ground instantaneous field-of-view;

TABLE 6.1 The Basic Characteristics of Several Well-Known Imaging Spectrometers

Sensor	Bands	Range (nm)
Airborne Visible/Infrared Imaging Spectrometer	220	360–2510
Compact Airborne Spectrographic Imager-1500	288	400–1050
Digital Airborne Imaging Spectrometer	79	400–12,000
Earth Observing-1 Hyperion	220	400–2500
HyMap Imaging Spectrometer	128	400–2500
PROBE-1	100–200	400–2400

GIFOV) based on flying altitude, aircraft speed, scan and data rate, and desired signal-to-noise (SNR) properties of the imagery. Smaller GIFOV measurements from aircraft (e.g. < 5 m) require slower speed over ground, high sensor SNR, and fast data rates. Many applications require these high spatial resolution measurements, but few airborne and no space-based spectrometers can deliver the information. Probably the most well-known airborne sensor to do so is the NASA Airborne Visible and Infrared Imaging Spectrometer (AVIRIS), which can fly sufficiently low and slow to acquire data at about 1.8 m resolution; otherwise, this sensor is most often flown at higher altitudes to obtain data in the 4–20 m resolution range. AVIRIS is a very high fidelity spectrometer that collects spectra in 9.6 nm wide bands (full width at half maximum; FWHM) spanning the 360–2510 nm wavelength range and with a cross-track swath of 614 pixels [3, 4]. AVIRIS has been continuously improved over the past 15 or more years and is now in its fifth major version for use in Earth observation and analysis [4].

Another imaging spectrometer that has undergone continual improvements over the years is the Compact Airborne Spectral Imager (CASI) [5]. CASI collects spectral data at 2.4 nm FWHM sampling across a wavelength range of about 400–1050 nm. The SNR and overall fidelity of CASI are now roughly similar to those of AVIRIS, but CASI stands as a unique instrument because of its programmability and very high spectral resolution. Its pushbroom array allows the CASI to be operated at low altitude to achieve spatial resolutions of less than 1 m. The most recent version of CASI has 1500 cross-track pixels. Other spectrometers such as HyMap and PROBE-1 provide a near-contiguous spectral sampling of the 400–2500 nm wavelength range, but at differing operational spatial resolutions, SNRs, and swath widths.

There are very few spaceborne imaging spectrometers available for scientific use; the space-based technology has not been made operational for the earth sciences. The earth Observing-1 (EO-1) satellite does carry the Hyperion imaging spectrometer (Table 6.1), which was placed in low Earth orbit in December 1999. EO-1 is a technology demonstration and thus does not provide large-scale coverage of the Earth's surface; however, Hyperion data can be requested from the U.S. Geological Survey. The Hyperion imagery has a spatial resolution (GIFOV) of 30 m, and the spectra cover the 400–2500 nm wavelength region in 220 channels [6]. Hyperion is a pushbroom imager with relatively low signal-to-noise and image uniformity as compared to systems such as AVIRIS and HyMap, but it does provide a chance to test imaging spectroscopy concepts and analysis methods just about anywhere in the world.

The data volumes associated with airborne hyperspectral data are significantly larger than typical multispectral data for a given spatial resolution and coverage. For a given length of distance flown, the width of the sensor scan can also affect the data volume. AVIRIS collects 640 pixels per scan, whereas the CASI-1500 collects 1500 pixels across its linear array. In the spectral domain, the AVIRIS and the CASI-1500 sensors contain 220 and 288 spectral bands per pixel, respectively, while the Landsat Thematic Mapper (TM) and Enhanced Thematic Mapper Plus (ETM+) sensors contain only seven multispectral bands. Furthermore, many multispectral sensors only have an 8-bit dynamic range of intensity per pixel per band, whereas AVIRIS and CASI data have a greater dynamic range and are stored as 14- and 16-bit values per

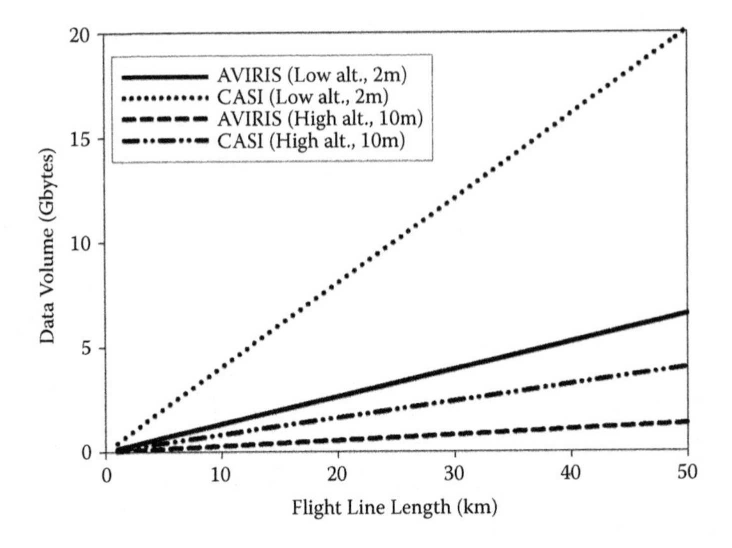

Figure 6.2 Change in data volume with flight distance for two common imaging spectrometers, AVIRIS and CASI-1500, flown at 2 m and 10 m GIFOV.

pixel per band, respectively. Figure 6.2 shows how the data volume increases dramatically as these spectral and spatial parameters increase. If a line were drawn on this graph to show the data volume for a seven-band sensor with similar spatial coverage characteristics of the AVIRIS or CASI sensor, the data volumes would be so low that the line would not be distinguishable from the x-axis.

6.2.2 Typical Analysis Steps

Calibrating imagery to radiometric units is a necessary step to compare the image data to field spectra and data collected from other sensors. With multispectral data, calibration is typically done relative to other images with known or well-modeled atmospheric conditions. For example, a time series of images has one image converted to surface reflectance, while the other images have their radiometry adjusted to the surface reflectance image by regressing the radiometric values based on relatively unchanging surfaces in the image [7]. Unfortunately, this usually involves careful analysis and selection of calibration targets by a human analyst. Other relative calibration methods, such as dark object subtraction (DOS), have been developed by Chavez [8, 9] to minimize the effects of atmospheric scattering on multispectral data.

To take full advantage of the precision of the radiance information inherent in the narrow spectral bands of hyperspectral sensors, it is critical that the data be calibrated to land surface reflectance (Figure 6.3). Hyperspectral data contain a wealth of information that makes it possible to remove the effects of the most variable constituents of the atmosphere (principally aerosols and water vapor) based on the relationship between the numerous spectral bands. These relationships have been used to develop

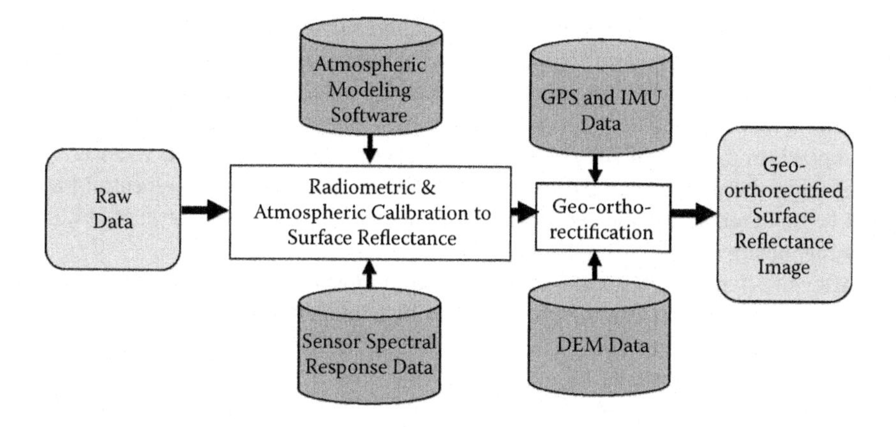

Figure 6.3 Major processing steps used to derive calibrated, geo-referenced surface reflectance spectra for subsequent analysis of hyperspectral images.

atmospheric correction algorithms for the MODIS program [10, 11, 12]. The higher spectral resolution of instruments such as AVIRIS and Hyperion are making atmospheric correction an automated process based on the spectral data content and not on human-selected calibration targets [3, 13]. This makes the calibration process easy to parallelize and distribute across many computer processors [14].

Many airborne hyperspectral sensors are used in conjunction with Global Positioning System and Inertial Motion Units to precisely geo-locate each pixel based on the location and attitude of the sensor. A separate digital elevation model makes it possible to remove any parallax in the image resulting from variations in scan angle and surface elevation. The combination of these data makes it possible to map the precise coordinate of each pixel and render a geo-orthorectified image [15].

Although we are used to seeing an organized grid of pixels when images are displayed, the precise location of each pixel does not fall on regularly spaced intervals. To create a visually pleasing, geo-orthorectified rendering of an image, the nearest neighbor (NN) resampling kernel is typically used. This resampling method has distinct advantages for hyperspectral data. First, most hyperspectral remote sensing is done to find subtle variations in the spectra of different land surfaces. Resampling methods such as bilinear interpolation and cubic convolution perform a weighted averaging of the closest 8 and 16 pixels, respectively, for each geo-orthorectified pixel. These two resampling methods have been used in multispectral imagery for image interpretation and to avoid the 'block' appearance that the NN kernel can produce. The second reason that the NN kernel is superior for hyperspectral remote sensing is that it requires much less computation per pixel.

Although the resampling method can impact computation time, something as basic as the organization of the imagery on a disk can also affect the computation time. Most multi-band imagery is organized in one of three different ways: band sequential (BSQ), band interleaved by line (BIL), and band interleaved by pixel (BIP). While the organization of the data does not significantly affect the processing speed for multispectral imagery, the organization of the large number of bands in hyperspectral imagery can have important implications for processing an image. Since most processing of hyperspectral data occurs on a pixel-by-pixel basis (discussed later), it is more efficient to organize the data in BIL or BIP format, so that the entire spectrum for a pixel (in the case of BIP) or a line of pixels (in the case of BIL) is located close together in the data stream. If BSQ is used, the data for the various spectral bands will be scattered across the entire length of the data file on the disk or in RAM, thus slowing the computations on the data.

After the hyperspectral data are calibrated, atmospherically corrected, and georeferenced, the spectra can then be analyzed to estimate the structural and/or chemical composition of materials in the image. For our purposes, we focus on terrestrial vegetation and ecosystems because it presents one of the most challenging areas for imaging spectroscopy. Whereas geological applications of imaging spectroscopy rely heavily on analysis of spectral absorption features (e.g., rock minerals) [16], vegetation spectroscopy places equal emphasis on light scattering and absorption because the three-dimensionality of vegetation canopies creates a solar-reflected radiation field that can be dominated by photon scattering rather than solely by absorption [17, 18].

A wide variety of techniques are available for the analysis of hyperspectral imagery. The simplest approaches employ 'vegetation indices' that take combinations of a few narrow wavelength bands to estimate canopy structural or biochemical properties [19]. These approaches do not necessarily require high-performance computing environments unless the data are collected over large geographic regions and/or at high spatial resolution. A more computer-intensive processing effort is required for methods that utilize the shape of the spectrum, which necessarily requires many wavebands and more advanced methods. For example, spectral mixture analysis is very commonly used to estimate the fractional contribution of materials within image pixels [20]. This type of sub-pixel analysis is made robust by hyperspectral imaging that resolves the most unique reflectance features of different materials; it is a method often used to quantify the fractional cover of live and dead vegetation and bare soils in hyperspectral data [21]. Finally, the most computer-intensive methods for hyperspectral data analysis of vegetation involve the use of mechanistic models that are numerically inverted to derive model parameters leading to the best estimate of the observed pixel spectrum [22]. Inverse modeling can employ simple or very complex simulations, depending upon the application and goal of the study [23]. At times, these approaches may use information from neighboring pixels, which leads to even higher computational demand.

The products derived from hyperspectral remote sensing can be used as observational constraints in Earth system models. For example, estimates of vegetation canopy nitrogen from hyperspectral observations can provide inputs to ecosystem models seeking to simulate plant growth [24]. Likewise, remotely sensed estimates of vegetation cover and density can play a key role in constraining biosphere-atmosphere model simulations of carbon storage and fluxes, and hyperspectral data are just now being considered for use in data assimilation approaches for ecological forecasting [25]. The computational demand increases with higher and higher levels of data analysis and use. From vegetation indices to carbon cycle models, the computing approaches must scale with the intensity of the effort. In the following sections, we describe how these different analytical methods set the computing requirements.

6.2.3 Computing Challenges in Hyperspectral Analysis

The large computational demand of hyperspectral image analysis stems from two particular issues. First, the images can be extremely data-rich because they have a large number of spectral bands. Second, and more importantly, the analysis techniques often utilize the shape of the spectra, not just the discrete bands, to remotely sense the chemical and structural attributes of materials. Typical methods for spectral analysis include the use of spectral derivatives, Monte Carlo simulations, and the inversion of physically-based reflectance models. Each of these approaches requires a different type and level of computational power, and can benefit from a well-matched computing architecture, to achieve the analytical speeds needed to make hyperspectral imaging truly operational or 'science ready.' Here, we use a couple of examples from our own work to highlight the different types of computational needs.

Spectral mixture analysis (SMA) is a common approach to decomposing image pixels into fractional cover estimates of various Earth surface materials. The diversity of materials to be estimated depends upon the spectral and spatial resolution of the imagery, the fidelity of the spectral measurements, and the uniqueness of the spectral properties of each material. In ecosystem research, the spectral 'endmembers' tend to be a combination of biotic and abiotic materials such as plant canopies (or species), dead vegetation, soils, and rocks. In our work in arid and semi-arid regions, as well as in humid tropical forests and savannas, we have found that no single spectral signature can represent a generic vegetation canopy; patch of dried, non-photosynthetic vegetation (NPV); or bare soil. Instead, a wide range of spectral signatures can depict the presence of these materials. A subsequent finding, however, was that the shapes (or derivatives) of high spectral resolution signatures in the combined 700–750 nm and 2000–2400 nm wavelength ranges can best quantify the fractional cover of vegetation, NPV, and soils [26, 27]. Although this wavelength range best serves to separate the materials within each image pixel, the variability within each endmember class nonetheless persists, and thus a Monte Carlo approach, AutoMCU©, was developed to unmix each image pixel iteratively until a stable solution could be found (Figure 6.4). The computational demand for this type of per-pixel, Monte Carlo mixture analysis is high and thus requires distributed computing methods described in Section 6.3.

Imaging spectroscopy can also be used to estimate the chemical composition of vegetation canopies. This type of analysis employs canopy radiative transfer (CRT) models that simulate top-of-canopy reflectance signatures based on a set of scale-dependent soil, vegetation, solar, and sensor input parameters. The CRT models range in complexity from simple 2-stream, turbid medium simulations to complex 3-D models that combine both radiative transfer and geometric-optical approaches [17, 28]. These models are often run in a forward mode to study the effects of differing inputs (often measured in the field) on predicted top-of-canopy reflectance signatures (Figure 6.5). Run iteratively in this mode, the input parameters and output spectral

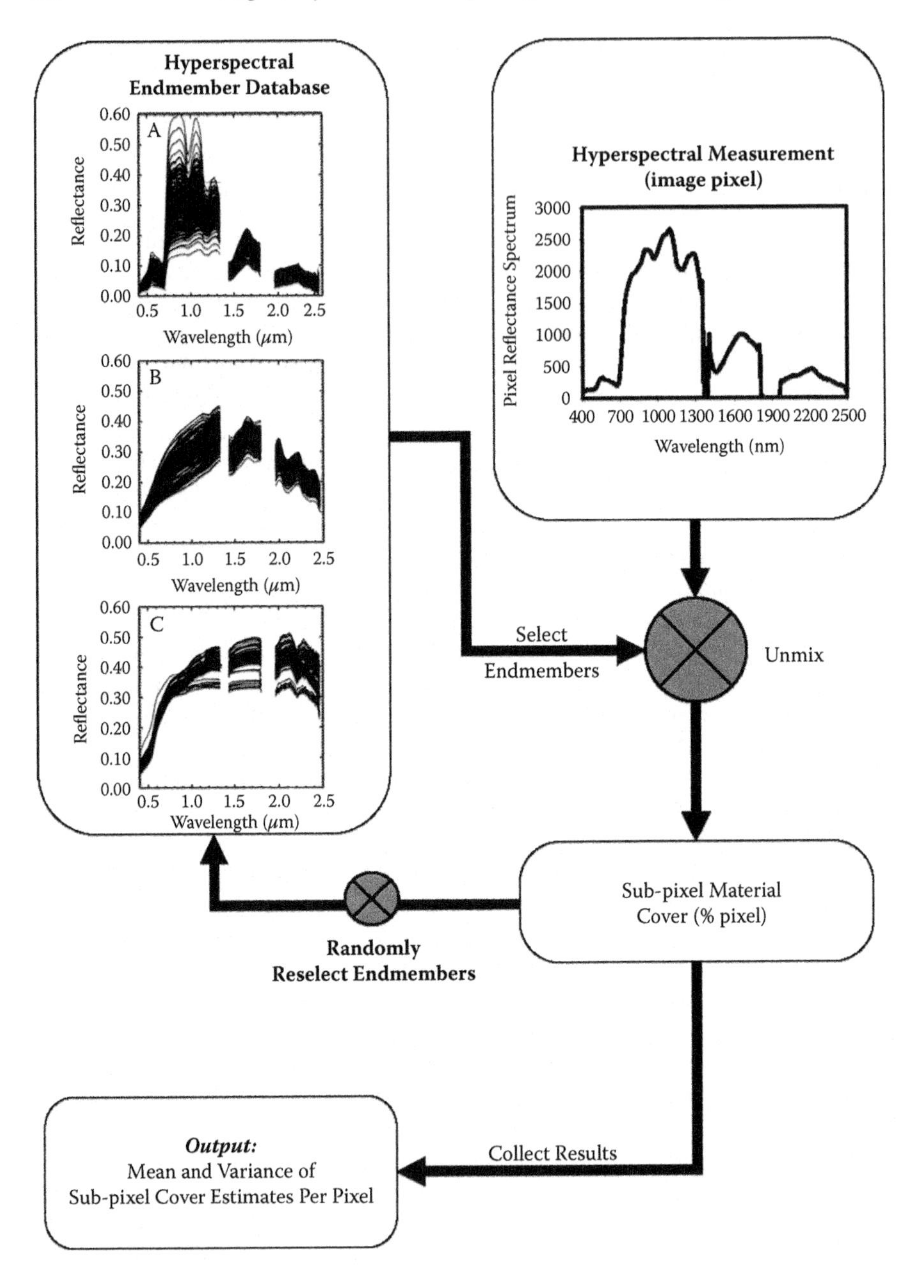

Figure 6.4 A per-pixel, Monte Carlo mixture analysis model used for automated, large-scale quantification of fractional material cover in terrestrial ecosystems [18, 21]. A spectral endmember database of (A) live, green vegetation; (B) non-photosynthetic vegetation; and (C) bare soil is used to iteratively decompose each pixel spectrum in an image into constituent surface cover fractions.

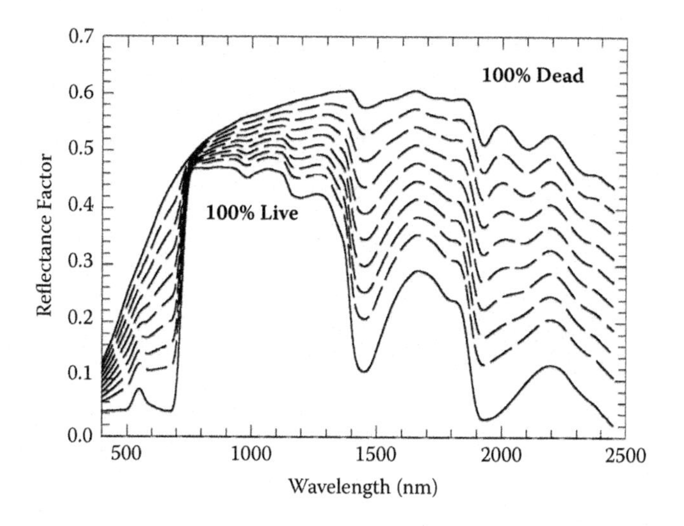

Figure 6.5 Example forward canopy radiative transfer model simulations of how a plant canopy hyperspectral reflectance signature changes with increasing quantities of dead leaf material.

signatures can be compiled and used to understand the relative importance of parameters in the models, which sets the physical basis for parameter estimation via model inversion [18].

Inverse CRT modeling usually requires the use of numerical inversion techniques to match a measured hyperspectral signature acquired from an airborne or space-based imaging spectrometer to a set of model parameters. Numerical inversion techniques can range from Newton minimization functions to simulated annealing, neural networks, and genetic algorithms [29]. All of these methods are computationally demanding; the inversion of a single pixel can take minutes of CPU time depending upon the spectral resolution of the data, the complexity of the CRT model, and the sophistication of the inversion technique. There is also an ecological component to the inverse modeling process; some canopy or ecosystem parameters represented in the CRT models co-vary in biochemically or biophysically predictable ways. This is critically important because a model inversion can be made more efficient if the solution domain is narrowed via knowledge of these ecological covariations among parameters [30]. However, these covariates are often fuzzy and thus a Monte Carlo technique can be employed to accommodate the uncertainty in how model parameters interrelate. In this case, a Monte Carlo routine is used to iteratively run a numerical model inversion on a pixel-by-pixel basis (Figure 6.6). This creates an extremely high computational demand. In our work using 3-D CRT models with Newton minimization codes and Monte Carlo techniques, we find that a combination of distributed and true parallel computing techniques provide the best performance in processing hyperspectral data. The distributed computing component breaks images into subsets for subsequent processing on different computer nodes, whereas the per-pixel inversion

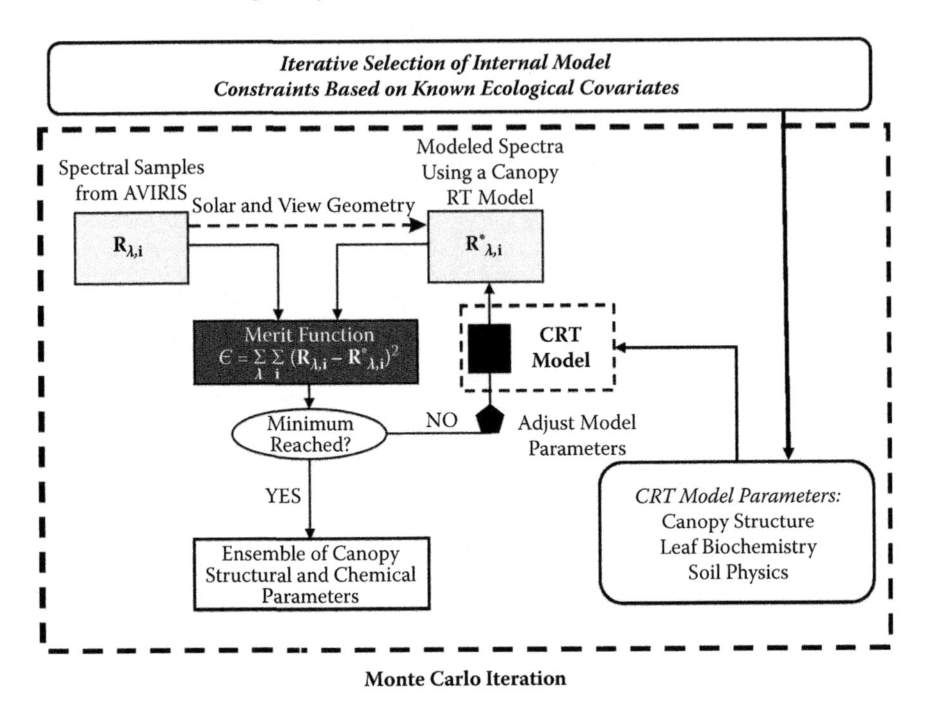

Figure 6.6 Schematic of a typical canopy radiative transfer inverse modeling environment, with Monte Carlo simulation over a set of ecologically-constrained variables. This example mentions AVIRIS as the hyperspectral image data source.

utilizes 2–4 processors per node with message passing between these processors (discussed in Section 6.3).

We are developing new inverse modeling algorithms that require information from neighboring pixels. Very high spatial resolution spectrometers can provide multiple hyperspectral signatures per vegetation canopy (e.g., a set of 0.5 m resolution signatures spanning the top of a single tree crown). In this case, the model parameter estimates of any given pixel can benefit from information obtained through the parameter estimation of neighboring pixels. To maximize the use of information among pixels, a fully parallel computing approach is required to simultaneously solve for all model parameters among all image pixels or groups of pixels.

6.3 Meeting the Computing Demands

The data volumes, processing, steps, and analytical methods described above require high-performance computing (HPC) techniques. For our purposes, we define an HPC cluster as a set of networked computers that are dedicated to a given cluster. These computers typically are identical in configuration and are in close physical proximity.

The cluster network or networks are private; that is, there is very limited direct access from systems not associated with the cluster.

Software resides on the HPC cluster computers that enables them to work in a unified fashion. Clustering software includes software for monitoring and controlling the computers in the cluster; software for scheduling and controlling jobs running on the cluster; and software needed by the jobs running on the cluster, in particular, libraries that enable the sub-job running on each computer to communicate with the related sub-jobs running on other computers in the cluster.

6.3.1 High-Performance Computing Jobs

The jobs appropriate for an HPC cluster can be grouped in three classes:

- Independent, simultaneous calculations involving different data sets.
- Simultaneous calculations involving the same data set in which the calculations running on the compute nodes are independent.
- Simultaneous calculations involving the same data set in which the results of the calculations running on a compute node are dependent upon the calculations run on another compute node

6.3.1.1 Job Class 1 – Independent Calculations on Different Data Sets

When performing independent calculations on different data sets, the HPC cluster can simply be a group of networked workstations on which individual jobs are run. As the jobs are independent, there is no network needed for communication between the workstations. An example of this job class is simultaneously running an atmospheric correction code (Figure 6.3) on a number of remote sensing data sets. This case is simply one or more workstations providing the independent CPU cycles needed to process more imagery.

In this computing class, the independent jobs are initiated with a startup script running on a control computer having network connectivity to all workstations. The startup script assigns individual calculations to the workstations, with the number of jobs assigned to each workstation being the same as the number of CPUs on it. If clustering control software is added to a network of workstations, the software assigns tasks to available workstations, which can vary dynamically, and enables the user to queue jobs to be executed as other jobs finish.

6.3.1.2 Job Class 2 – Independent Calculations on a Single Data Set

Because the calculations are independent, this class is often termed trivial parallel processing. The compute nodes do not communicate between themselves so, as with job class 1, there are no networking needs for communication between the nodes. An example of this job class is a calculation performed independently on each pixel of a remote sensing data set (e.g., the AutoMCU spectral mixture model; Figure 6.4). Like most spectral mixture models, the AutoMCU code operates on a per-pixel basis without a need to gather information from neighboring pixels. In this particular example, the code is computationally expensive because of the Monte Carlo routine, matrix inversions, and overall image data volume involved. The calculation begins by sub-dividing the data set, then each compute node loops through the pixels of an assigned data subset, and finally the subset results are recombined once all are received from the nodes.

6.3.1.3 Job Class 3 – Dependent Calculations on a Single Data Set

This class includes distributed parallel computing that requires significant networking resources for communication between compute nodes. The addition of a high-speed, low-latency network for communication between compute nodes may significantly improve performance. At the application level, communication is handled by the Messages Passing Interface (MPI). Examples of this job class include general circulation models (GCM) and canopy radiative transfer inverse modeling methods (Figure 6.5). In this case, the program assigns the calculations for a number of pixels (or cells) to each processor in the cluster. Since MPI uses shared memory for communication between processes running on the same compute node, performance enhancements can be realized when cells that directly communicate with each other are run on the same compute node.

6.3.2 Computing Architectures

HPC clusters can be divided into two subcategories: (a) clusters that use commodity hardware, for example, Ethernet for the cluster network; and (b) clusters that use specialized hardware to achieve higher performance, for example, low-latency, high-speed networks (Figure 6.7).

These architectures have some common components and considerations:

6.3.2.1 Cluster Computer Nodes

Typical computer nodes for clusters have more than one processor. Processor technology for HPC clusters has recently transitioned from 32-bit to 64-bit, with a 32-bit compatibility mode (e.g., $\times 86 - 64$) and pure 64-bit processors. Multi-core CPUs are being delivered, but currently there is a lack of understanding as to whether CPU bus bandwidths associated with multiple cores will support full use of multiple cores for HPC clusters. Finally, the memory in an HPC cluster is distributed on the compute nodes. With dual or quad-processor nodes (and soon with dual and multi-core CPUs), the memory on each of the compute nodes can be shared among the processors/cores on the node.

6.3.2.2 Cluster Front-End Computer

The cluster front-end computer is used as the portal for users to submit jobs to the cluster. This particular computer thus requires both private and outside-of-network

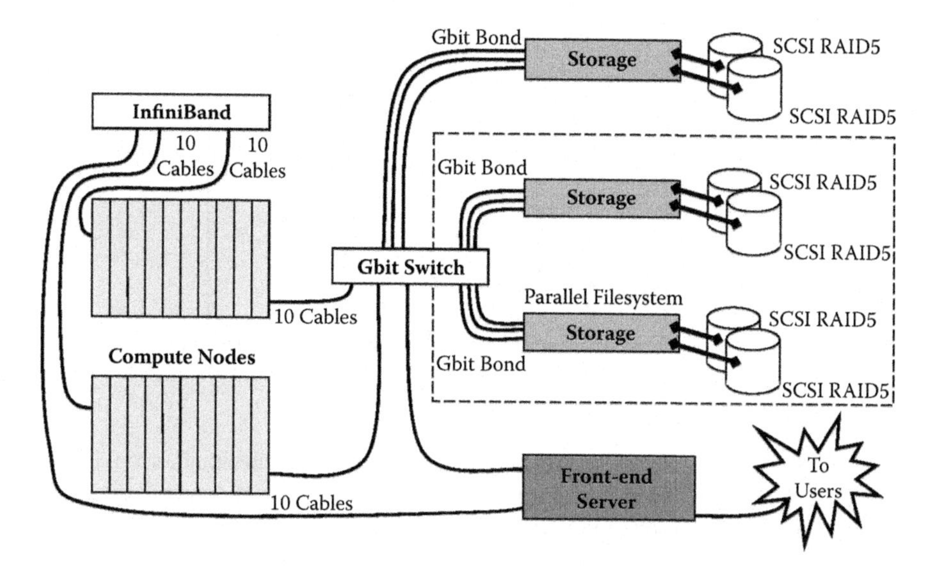

Figure 6.7 Schematic of a small HPC cluster showing 20 compute nodes, front-end server, InfiniBand high-speed/low-latency network, Gigabit Ethernet management network, and storage in parallel and conventional file systems on SCSI RAID-5 drive arrays.

access. The front-end computer can be a different configuration from the compute nodes and is generally not used for computations.

6.3.2.3 Networking

Gigabit Ethernet is often employed for user communication with the front-end computer and for management of the nodes in the cluster. In addition, we use Gigabit Ethernet for accessing an IBRIX Fusion (www.ibrix.com) parallel file system and the conventional network file system (NFS) data storage. Since several nodes accessing a server can saturate a single Gigabit network interface card, bandwidth is increased by binding three Gigabit Ethernet interfaces on each storage server into a single-channel bonding interface. The resulting channel bonding interface has a bandwidth of 3 × 100 MB/sec, or about 300MB/sec. A Gigabit network is sufficient for a cluster in which node-to-node communication is not needed for calculations. However, with a latency of 150 msec per message, Ethernet is a poor choice for calculations in which significant communication is required, i.e., for distributed parallel MPI computations. Newer networks such as InfiniBand (www.InfiniBandTA.org) and Myrinet (www.myricom.com) provide not only increased bandwidth over Gigabit Ethernet (900 MB/sec and 450 MB/sec, respectively), but also have a significantly lower latency (10 sec and 7 sec, respectively).

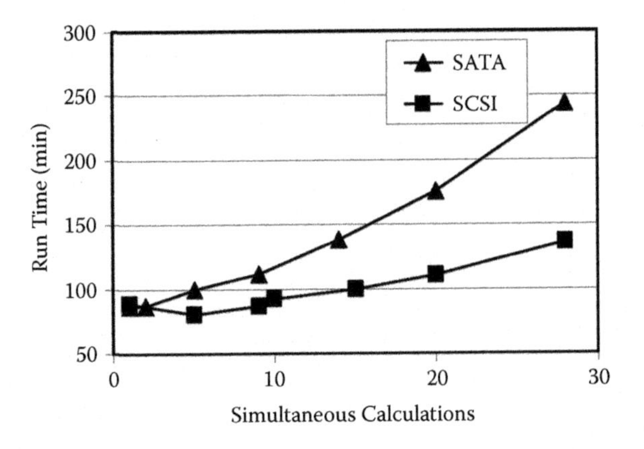

Figure 6.8 Effect of storage RAID-5 subsystem on independent simultaneous calculations, with storage systems accessed via NFS. With multiple simultaneous accesses, the SCSI array outperforms the SATA array.

6.3.2.4 Data Storage

Simultaneous access to a data set or multiple data sets on a single RAID array by many compute nodes can seriously limit the performance of an HPC cluster. Figure 6.8 shows times for independent calculations on our HPC cluster with data accessed via standard NFS from SATA and SCSI disk arrays. Notice that the run-time for an individual computation increases with an increasing number of simultaneous

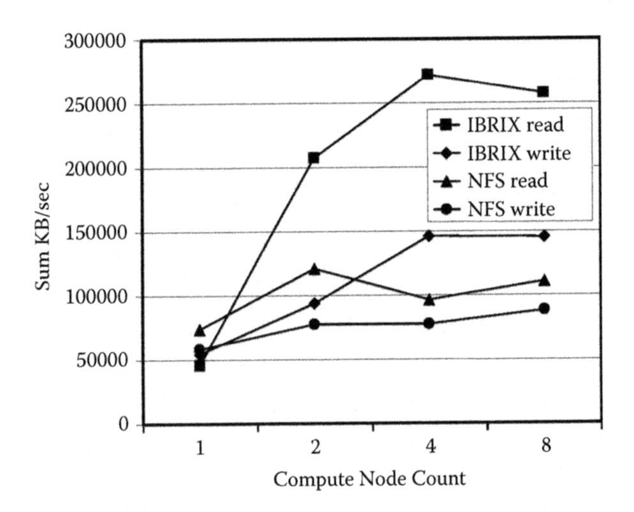

Figure 6.9 Performance comparison of multiple computer node access to a data storage system using the traditional NFS or newer IBRIX parallel file system.

calculations. This problem is especially noticeable when the data are accessed from a SATA RAID-5 array. Ideally, the curves would be flat, and the time required for an individual calculation would not increase as more calculations were performed simultaneously.

These types of storage hardware limitations require the use of SCSI arrays for hyperspectral data analyses. They also require a parallel file system to handle the multiple accesses that simultaneously occur on a disk array (e.g., www.ibrix.com). We compared the performance of multiple computer nodes accessing a Linux ext3 file system via NFS versus accessing an IBRIX Fusion parallel file system (Figure 6.9). The parallel file system provides significantly better throughput when accessed by multiple nodes.

6.4 Future Possibilities of Hyperspectral Computing

Hyperspectral signatures contain detailed information on the chemical and structural composition of materials. In this chapter, we have discussed some of the algorithms, computing methods, and hardware used for analysis of imaging spectrometer data. Our examples were specific to vegetation and terrestrial ecosystems, but many of the same principles apply to analyses of aquatic ecosystems and the built environment. Future remote sensing developments will probably bring the type of shortwave reflectance spectroscopy presented here together with hyperspectral-thermal and/or active light detection and ranging (LiDAR) data, which will greatly increase the computer processing requirements for Earth systems analysis. Meanwhile, the spatial and spectral resolutions of the sensors will continue to increase, allowing access to smaller surface features and more chemical determinations. How will we cope with such enormous data sets? The capabilities of even the largest computers are not likely to keep pace with the expected data stream from hyperspectral imaging. The solutions may lie in acquiring only the data needed for subsequent analysis and Internet-based distributed computing techniques.

6.4.1 Onboard Processing and Data Compression

To date, the Earth observing community has mostly operated within the framework that the data are collected, delivered to the analyst, and then processed to derive products. This approach may no longer work as the data volume increases with future sensors. An inherently important quality of high spectral resolution data is its ability to quantify land, aquatic, and atmospheric constituents 'on the fly.' For example, clouds, aerosol, and cirrus are readily detected using the detailed spectroscopic signatures provided by full-range (400–2500 nm) hyperspectral sensors [31]. Likewise, water bodies, snow, rocks, soils, and vegetation are very easily identified in the data. It is thus possible to program a given flight or satellite overpass to collect only the spectra that pass a set of tests to remove features of low interest during the image acquisition.

For instance, in coastal zone mapping missions, land and clouds can easily be detected in-flight, and those pixels can be set to null. The data can then be compressed using a wide range of high-performance compression algorithms now available or soon to be available.

Data compression has traditionally been a subject to avoid in imaging spectroscopy because of the unrecoverable loss of critical information during the compress-uncompress process. However, innovations in lossless or low-loss compression techniques are paving the way for reduced data volumes onboard both aircraft and spacecraft. An example algorithm developed at the NASA Goddard Space Flight Center was used to compress AVIRIS hyperspectral imagery at different compression ratios. The data were then uncompressed and analyzed using the AutoMCU algorithm described in Section 6.2. The output from different compression tests were then compared to those derived from data that had never been compressed. The results show that compression ratios of up to 3:1 lead to no degradation of a very sensitive mixture modeling algorithm; compression as high as 8:1 results in only a 10–15% decrease in algorithm accuracy. These findings are sensor and algorithm specific, but they nonetheless demonstrate the value of data compression to reduce volume, telemetry time, and potentially even processing time.

As remote sensing systems continue to provide ever increasing amounts of spectral data, this information could facilitate pre-processing steps beyond that of data removal and compression. What is the value of onboard processing of the data to science-ready format? In many practical applications of remote sensing, the image products are quickly needed for conservation and management purposes. For ecosystem analysis, processing onboard an aircraft could allow for rapid determination of fire fuel loads, vegetation stress (e.g., precise agriculture), and biodiversity. To do so, the sensors must provide data-rich, high-fidelity spectral information; combinations of hyperspectral and LiDAR technology may be best suited for such an effort. It is unlikely that spacecraft would carry sufficient payload for onboard science-product generation, but it is possible that the most important processing step of atmospheric correction could be carried out in orbit.

6.4.2 Distributed Internet Computing

Computing has undergone a revolution from the days of large, mainframe facilities to relatively low-cost distributed 'Beowulf' computing clusters that can be built or purchased by even a small agency or organization. An even more revolutionary step was taken in 1999 when a group from the Search for Extraterrestrial Intelligence (SETI) organization and the University of California, Berkeley, developed software that uses the vast amount of 'idle time' on computers connected to the Internet. When their project, SETI@home (setiathome. sslberkeley.edu), was first established, about 200,000 PCs were linked, providing analysis of radio signals collected by telescopes like the Arecibo Observatory in Puerto Rico [32]. By 2001, more than 1 million computers were analyzing 100-sec segments of the radio signals, all while the PCs would have been otherwise inactive. In the years since that historic beginning in massive distributed Internet computing, additional projects have been initiated, such as on protein folding (folding.stanford.edu), climate prediction (climateprediction.net), and drug analysis for AIDS (fightaidsathome.scripps.edu).

Surprisingly, we could find no examples of distributed Internet computing for Earth remote sensing image analysis. Given the per-pixel or per-image nature of most processing algorithms, it is likely that Internet computing could easily advance the computational power needed for the operational analysis of hyperspectral and similar data. Given the great public interest in accessing images of Earth, such as through Google Earth (earth.google.com), people would likely be interested in processing Earth observing data for their particular geographic area or for another area of interest to them. Participants could then receive direct access to processed image products, providing a natural conduit for environmental education. The possibilities are real, and the software structures are available (e.g., www.mithral.com). The future of high-performance remote sensing may rest in the willingness of a community to provide large-scale, distributed computing power.

6.5 Acknowledgments

This work was supported by the NASA Terrestrial Ecology Program and The Carnegie Institution. Discussion of instrumentation, computing hardware, software, and Internet websites by trade name does not constitute any endorsement by the authors or The Carnegie Institution.

References

[1] G. Vane and A. F. H. Goetz. *Terrestrial imaging spectroscopy.* Remote Sensing of Environment, 24:1–29, 1988.

[2] S. L. Ustin, D. A. Roberts, J. A. Gamon, G. P. Asner, and R. O. Green. *Using imaging spectroscopy to study ecosystem processes and properties.* BioScience, 54:523–534, 2005.

[3] R. O. Green, M. L. Eastwood, C. M. Sarture, T. G. Chrien, M. Aronsson, B. J. Chippendale, J. A. Faust, B. E. Pavri, C. J. Chovit, M. S. Solis, M. R. Olah, and O. Williams. *Imaging spectroscopy and the Airborne Visible Infrared Imaging Spectrometer (AVIRIS).* Remote Sensing of Environment, 65:227–248, 1998.

[4] R. O. Green, M. L. Eastwood, and C. M. Sarture. *Advances in the airborne visible and infrared imaging spectrometer (AVIRIS), 1987–2005.* Proceedings of the NASA Airborne Earth Science Workshop. Jet Propulsion Laboratory, Pasadena, CA, 2005.

[5] S. K. Babey and C. D. Anger. *Compact airborne spectrographic imager (CASI): a progress review*. SPIE, 1937:152–163, 1993.

[6] S. G. Ungar, J. S. Pearlman, J. A. Mendenhall, and D. Reuter. *Overview of the Earth Observing-1 (EO-1) mission*. IEEE Transactions on Geoscience and Remote Sensing, 41:1149–1160, 2003.

[7] F. G. Hall, D. E. Strebel, J. E. Nickeson, and S. J. Goetz. *Radiometric rectification: toward a common radiometric response among multidate, multisensor images*. Remote Sensing of Environment, 35:11–27, 1991.

[8] P. S. Chavez 1988. *Improved dark-object subtraction technique for atmospheric scattering correction of multispectral data*. Remote Sensing of Environment, 24:459–479, 1988.

[9] P. S. Chavez. *Image-based atmospheric corrections revisited and improved*. Photogrammetric Engineering and Remote Sensing, 62:1025–1036, 1996.

[10] E. F. Vermote, N. El Saleous, C. O. Justice, Y. J. Kauffman, J. L. Privette, L. Remer, J. C. Roger, and D. Tanre. *Atmospheric correction of visible to middle-infrared EOS-MODIS data over land surfaces: background, operational algorithm and validation*. Journal of Geophysical Research, 102:17,131–17,141, 1997.

[11] E. F. Vermote, N. Z. El Saleous, and C. O. Justice. *Atmospheric correction of MODIS data in the visible to middle infrared: First results*. Remote Sensing of Environment 83:97–111, 2002.

[12] L. A. Remer, Y. J. Kauffman, D. Tanre, S. Mattoo, D. A. Chu, J. V. Martins, R. R. Li, C. Ichoku, R. C. Levy, and R. G. Kleidman. *The MODIS aerosol algorithm, products, and validation*. Journal of Atmospheric Sciences, 62:947–973, 2005.

[13] R. O. Green, J. E. Conel, J. S. Margolis, C. J. Bruegge, and G. L. Hoover. *An inversion algorithm for retrieval of atmospheric and leaf water absorption from AVIRIS radiance compensation for atmospheric scattering*. Proceedings of the NASA Airborne Earth Science Workshop. Jet Propulsion Laboratory, Pasadena, CA, 1991.

[14] P. Wang, K. Y. Liu, T. Cwik, and R. O. Green. *MODTRAN on supercomputers and parallel computers*. Parallel Computing, 28:53–64, 2002.

[15] J. W. Boardman. *Precision geocoding of low-altitude AVIRIS data: lessons learned in 1998*. Proceedings of the NASA Airborne Earth Science Workshop. Jet Propulsion Laboratory, Pasadena, CA, 1999.

[16] R. N. Clark. *Spectroscopy of rocks and minerals, and principles of spectroscopy*. Remote Sensing for the Earth Sciences, pages 3-58, Wiley & Sons Inc., New York, 1999.

[17] N. S. Goel. *Models of vegetation canopy reflectance and their use in estimation of biophysical parameters from reflectance data*. Remote Sensing Reviews, 4:1–212, 1988.

[18] G. P. Asner. *Biophysical and biochemical sources of variability in canopy reflectance*. Remote Sensing of Environment, 64:234–253, 1998.

[19] B. Zhang, X. Zhang, T. Lie, G. Xu, L. Zheng, and Q. Tong. *Dynamic analysis of hyperspectral vegetation indices*. SPIE, 4548:32–38, 2001.

[20] M.O. Smith, J. B. Adams, and D. E. Sabol. *Spectral mixture analysis—new strategies for the analysis of multispectral data*. Imaging spectrometry—a tool for environmental observations. Kluwer Academic Publishers, Dordrecht, Netherlands, 1994.

[21] D. A. Roberts, M. O. Smith, and J. B. Adams. *Green vegetation, non-photosynthetic vegetation, and soils in AVIRIS data*. Remote Sensing of Environment, 44:255–269, 1993.

[22] S. Jacquemoud, C. Bacour, H. Poilve, and J. P. Frangi. *Comparison of four radiative transfer models to simulate plant canopies reflectance: direct and inverse mode*. Remote Sensing of Environment, 74:471–481, 2000.

[23] G. P. Asner, J. A. Hicke, and D. B. Lobell. *Per-pixel analysis of forest structure: Vegetation indices, spectral mixture analysis and canopy reflectance modeling*. Methods and Applications for Remote Sensing of Forests: Concepts and Case Studies. Kluwer Academic Publishers, New York.

[24] M. L. Smith, S. V. Ollinger, M. E. Martin, J. D. Aber, R. A. Hallett, and C. L. Goodale. *Direct estimation of aboveground forest productivity through hyperspectral remote sensing of canopy nitrogen*. Ecological Applications, 12:1286–1302, 2002.

[25] M. Hill, G. P. Asner, and A. Held. *A biogeophysical approach to remote sensing of vegetation in coupled human-environment systems: societal benefits and global context*. Submitted to the Journal of Spatial Science.

[26] G. P. Asner and D. B. Lobell. *A biogeophysical approach to automated SWIR unmixing of soils and vegetation*. Remote Sensing of Environment, 74:99–112, 2000.

[27] G. P. Asner and K. B. Heidebrecht. *Spectral unmixing of vegetation, soil and dry carbon in arid regions: comparing multi-spectral and hyperspectral observations*. International Journal of Remote Sensing, 23:3,939–3,958, 2002.

[28] R. B. Myneni, R. N. Ramakrishna, and S. W. Running. *Estimation of global leaf area index and absorbed PAR using radiative transfer models*. IEEE Transactions on Geoscience and Remote Sensing, 35:1380–1393, 1997.

[29] M. Cheney. *Inverse boundary-value problems*. American Scientist, 85:448–455, 1997.

[30] G. P. Asner, C. A. Wessman, D. S. Schimel, and S. Archer. *Variability in leaf and litter optical properties: Implications for BRDF model inversions using AVHRR, MODIS, and MISR*. Remote Sensing of Environment, 63:243–257, 1998.

[31] B. C. Gao, A. F. H. Goetz, and J. A. Zamudio. *Removing atmospheric effects from AVIRIS data for surface reflectance retrievals*. Proceedings of the NASA Airborne Earth Science Workshop. Jet Propulsion Laboratory, Pasadena, CA, 1991.

[32] J. Bohannon. *Distributed computing: grassroots computing*. Science, 308: 810–813, 2005.

Chapter 7

Parallel Implementation of Morphological Neural Networks for Hyperspectral Image Analysis

Javier Plaza,
University of Extremadura, Spain

Rosa Pérez,
University of Extremadura, Spain

Antonio Plaza,
University of Extremadura, Spain

Pablo Martínez,
University of Extremadura, Spain

David Valencia,
University of Extremadura, Spain

Contents

Improvement of spatial and spectral resolution in latest-generation Earth observation instruments is introducing extremely high computational requirements in many remote sensing applications. While thematic classification applications have greatly benefited from this increasing amount of information, new computational requirements have been introduced, in particular, for hyperspectral image data sets with hundreds of spectral channels and very fine spatial resolution. Low-cost parallel computing architectures such as heterogeneous networks of computers have quickly become a standard tool of choice for dealing with the massive amount of image data sets. In this chapter, a new parallel classification algorithm for hyperspectral imagery based on morphological neural networks is presented and discussed. The parallel algorithm is mapped onto heterogeneous and homogeneous parallel platforms using a hybrid partitioning scheme. In order to test the accuracy and parallel performance of the proposed approach, we have used two networks of workstations distributed among different locations, and also a massively parallel Beowulf cluster at NASA's Goddard Space Flight Center in Maryland. Experimental results are provided in the context of a real agriculture and farming application, using hyperspectral data acquired by the Airborne Visible Infra-Red Imaging Spectrometer (AVIRS), operated by the NASA Jet Propulstion Laboratory, over the valley of Salinas in California.

7.1 Introduction

Many international agencies and research organizations are currently devoted to the analysis and interpretation of high-dimensional image data collected over the surface of the Earth [1]. For instance, NASA is continuously gathering hyperspectral images using the Jet Propulsion Laboratory's Airborne Visible-Infrared Imaging Spectrometer (AVIRIS) [2], which measures reflected radiation in the wavelength range from 0.4 to 2.5 μm using 224 spectral channels at a spectral resolution of 10 nm. The incorporation of hyperspectral instruments aboard satellite platforms is now producing a near-continual stream of high-dimensional remotely sensed data, and cost-effective techniques for information extraction and mining from massively large hyperspectral data repositories are highly required [3]. In particular, although it is estimated that several Terabytes of hyperspectral data are collected every day, about 70% of the collected data is never processed, mainly due to the extremely high computational requirements.

Several challenges still remain open in the development of efficient data processing techniques for hyperspectral image analysis [1]. For instance, previous research has demonstrated that the high-dimensional data space spanned by hyperspectral data sets is usually empty [4], indicating that the data structure involved exists primarily in a subspace. A commonly used approach to reduce the dimensionality of the data is the principal component transform (PCT) [5]. However, this approach is characterized by its global nature and cannot preserve subtle spectral differences required to obtain a good discrimination of classes [6]. Further, this approach relies on spectral properties of the data alone, thus neglecting the information related to the spatial arrangement of the pixels in the scene. As a result, there is a need for feature extraction techniques able to integrate the spatial and spectral information available from the data simultaneously [5].

While such integrated spatial/spectral developments hold great promise in the field of remote sensing data analysis, they introduce new processing challenges [7, 8]. The concept of Beowulf cluster was developed, in part, to address such challenges [9, 10]. The goal was to create parallel computing systems from commodity components to satisfy specific requirements for the earth and space sciences community. Although most dedicated parallel machines employed by NASA and other institutions during the last decade have been chiefly homogeneous in nature, a current trend is to utilize heterogeneous and distributed parallel computing platforms [11]. In particular, computing on heterogeneous networks of computers (HNOCs) is an economical alternative that can benefit from local (user) computing resources while, at the same time, achieving high communication speed at lower prices. The properties above have led HNOCs to become a standard tool for high-performance computing in many ongoing and planned remote sensing missions [3, 11].

To address the need for cost-effective and innovative algorithms in this emerging new area, this chapter develops a new parallel algorithm for the classification of hyperspectral imagery. The algorithm is inspired by previous work on morphological neural networks, such as autoassociative morphological memories and morphological perceptrons [12], although it is based on different concepts. Most importantly, it can be tuned for very efficient execution on both HNOCs and massively parallel, Beowulf-type commodity clusters. The remainder of the chapter is structured as follows.

- Section 7.2 describes the proposed heterogeneous parallel algorithm, which consists of two main processing steps: 1) a parallel morphological feature extraction taking into account the spatial and spectral information, and 2) robust classification using a parallel multi-layer neural network with back-propagation learning.
- Section 7.3 describes the algorithm's accuracy and parallel performance. Classification accuracy is discussed in the context of a real application that makes use of hyperspectral data collected by the AVIRIS sensor, operated by NASA's Jet Propulsion Laboratory, to assess agricultural fields in the valley of Salinas, California. Parallel performance in the context of the above-mentioned application is then assessed by comparing the efficiency achieved by an heterogeneous parallel version of the proposed algorithm, executed on a fully heterogeneous network, with the efficiency achieved by its equivalent homogeneous version, executed on a fully homogeneous network with the same aggregate performance as the heterogeneous one. For comparative purposes, performance data on Thunderhead, a massively parallel Beowulf cluster at NASA's Goddard Space Flight Center, are also given.
- Finally, Section 7.4 concludes with some remarks and hints at plausible future research, including implementations of the proposed parallel algorithm on specialized hardware architectures.

7.2 Parallel Morphological Neural Network Algorithm

This section describes a new parallel algorithm for the analysis of remotely sensed hyperspectral images. Before describing the two main steps of the algorithm, we first formulate a general optimization problem in the context of HNOCs, composed of different-speed processors that communicate through links at different capacities [11]. This type of platform can be modeled as a complete graph, $G = (P, E)$, where each node models a computing resource p_i weighted by its relative cycle-time w_i. Each edge in the graph models a communication link weighted by its relative capacity, where c_{ij} denotes the maximum capacity of the slowest link in the path of physical communication links from p_i to p_j. We also assume that the system has symmetric costs, i.e., $c_{ij} = c_{ji}$. Under the above assumptions, processor p_i will accomplish a share of $\alpha_i \times W$ of the total workload W, with $\alpha_i \geq 0$ for $1 \leq i \leq P$ and $\sum_{i=1}^{P} \alpha_i = 1$. With the above assumptions in mind, an abstract view of our problem can be simply stated in the form of a client-server architecture, in which the server is responsible for the efficient distribution of work among the P nodes, and the clients operate with the spatial and spectral information contained in a local partition. The partitions are then updated locally and the resulting calculations may also be exchanged between the clients, or between the server and the clients. Below, we describe the two steps of our parallel algorithm.

7.2.1 Parallel Morphological Algorithm

The proposed feature extraction method is based on mathematical morphology [13] concepts. The goal is to impose an ordering relation (in terms of spectral purity) in the set of pixel vectors lying within a spatial search window (called a structuring element) designed by B [5]. This is done by defining a cumulative distance between a pixel vector $f(x, y)$ and all the pixel vectors in the spatial neighborhood given by B (B-neighborhood) as follows: $D_B[f(x, y)] = \sum_i \sum_j \text{SAD}[f(x, y), f(i, j)]$, where (x, y) refers to the spatial coordinates in the B-neighborhood and SAD is the spectral angle distance [1]. From the above definitions, two standard morphological operations called erosion and dilation can be respectively defined as follows:

$$(f \otimes B)(x, y) = argmin_{(s,t)\in Z^2(B)} \sum_s \sum_t \text{SAD}(f(x, y), \ f(x + s, y + t)) \tag{7.1}$$

$$(f \oplus B)(x, y) = argmax_{(s,t)\in Z^2(B)} \sum_s \sum_t \text{SAD}(f(x, y), \ f(x - s, y - t)) \tag{7.2}$$

Using the above operations, the opening filter is defined as $(f \circ B)(x, y) = [(f \otimes C) \oplus B](x, y)$ (erosion followed by dilation), while the closing filter is defined as $(f \bullet B)(x, y) = [(f \oplus C) \otimes B](x, y)$ (dilation followed by erosion). The composition of the opening and closing operations is called a spatial/spectral profile, which is defined as a vector that stores the relative spectral variation for every step of an increasing series. Let us denote by $\{(f \circ B)^{\lambda}(x, y)\}$, $\lambda = \{0, 1, ..., k\}$, the *opening series* at $f(x, y)$, meaning that several consecutive opening filters are applied using the same window B. Similarly, let us denote by $\{(f \bullet B)^{\lambda}(x, y)\}$, $\lambda = \{0, 1, ..., k\}$, the *closing series* at $f(x, y)$. Then, the spatial/spectral profile at $f(x, y)$ is given by the following vector:

$$p(x, y) = \{\mathrm{SAD}((f \circ B)^{\lambda}(x, y), (f \circ B)^{\lambda-1}(x, y))\} \cup \{\mathrm{SAD}((f \bullet B)^{\lambda}(x, y), (f \bullet B)^{\lambda-1}(x, y))\} \quad (7.3)$$

Here, the step of the opening/closing series iteration at which the spatial/spectral profile provides a maximum value gives an intuitive idea of both the spectral and spatial distributions in the B-neighborhood [5]. As a result, the profile can be used as a feature vector on which the classification is performed using a spatial/spectral criterion.

In order to implement the algorithm above in parallel, two types of partitioning can be exploited:

- Spectral-domain partitioning subdivides the volume into small cells or sub-volumes made up of contiguous spectral bands, and assigns one or more sub-volumes to each processor. With this model, each pixel vector is split amongst several processors, which breaks the spectral identity of the data because the calculations for each pixel vector (e.g., for the SAD calculation) need to originate from several different processing units.
- Spatial-domain partitioning provides data chunks in which the same pixel vector is never partitioned among several processors. With this model, each pixel vector is always retained in the same processor and is never split.

In this work, we adopt a spatial-domain partitioning approach for several reasons:

- A first major reason is that the application of spatial-domain partitioning is a natural approach for morphological image processing, as many operations require the same function to be applied to a small set of elements around each data element present in the image data structure, as indicated in the previous subsection.
- A second reason has to do with the cost of inter-processor communication. In spectral-domain partitioning, the window-based calculations made for each hyperspectral pixel need to originate from several processing elements, in particular, when such elements are located at the border of the local data partitions (see Figure 7.1), thus requiring intensive inter-processor communication.

However, if redundant information such as an overlap border is added to each of the adjacent partitions to avoid access from outside the image domain, then boundary data to be communicated between neighboring processors can be greatly minimized. Such an overlapping scatter would obviously introduce redundant computations, since the intersection between partitions would be non-empty. Our implementation makes

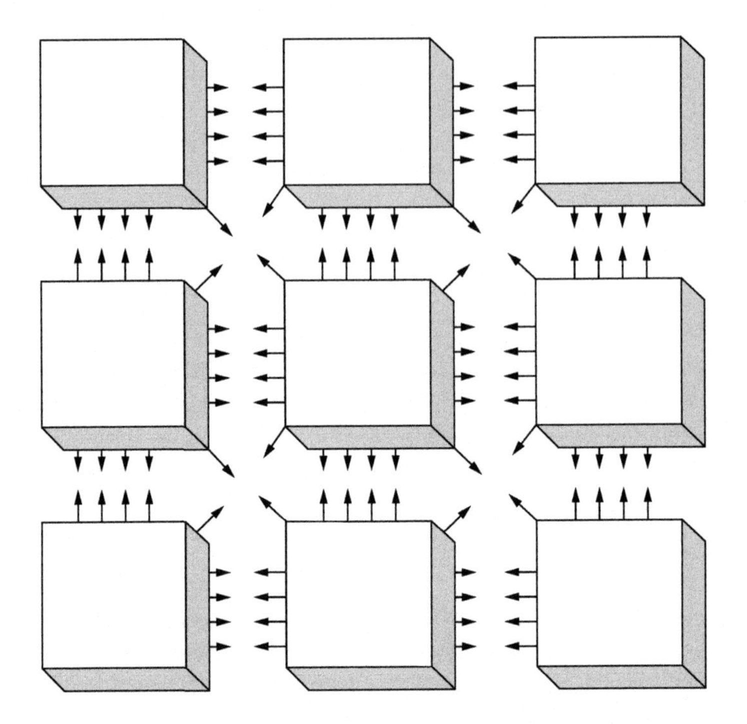

Figure 7.1 Communication framework for the morphological feature extraction algorithm.

use of a constant structuring element B (with size of 3×3 pixels) that is repeatedly iterated to increase the spatial context, and the total amount of redundant information is minimized. To do so, we have implemented a special 'overlapping scatter' operation that also sends out the overlap border data as part of the scatter operation itself (i.e., redundant computations replace communications).

To implement the algorithm, we made use of MPI *derived datatypes* to directly scatter hyperspectral data structures, which may be stored non-contiguously in memory, in a single communication step. A comparison between the associative costs of redundant computations in overlap with the overlapping scatter approach, versus the communications costs of accessing neighboring cell elements outside of the image domain, has been presented and discussed in previous work [7].

A pseudo-code of the proposed HeteroMORPH parallel algorithm, specifically tuned for HNOCs, is given below:

Inputs: N-dimensional cube f, structuring element B.

Output: Set of morphological profiles for each pixel.

1. Obtain information about the heterogeneous system, including the number of processors, P; each processor's identification number, $\{p_i\}_{i=1}^{P}$; and processor cycle-times, $\{w_i\}_{i=1}^{P}$.

2. Using B and the information obtained in step 1, determine the total volume of information, R, that needs to be replicated from the original data volume, V, according to the data communication strategies outlined above, and let the total workload W to be handled by the algorithm be given by $W = V + R$.
3. Set $\alpha_i = \lfloor \frac{(P/w_i)}{\sum_{i=1}^{P}(1/w_i)} \rfloor$ for all $i \in \{1, ..., P\}$.
4. For $m = \sum_{i=1}^{P} \alpha_i$ to $(V + R)$, find $k \in \{1, .., P\}$ so that $w_k \cdot (\alpha_k + 1) = \min\{w_i \cdot (\alpha_i + 1)\}_{i=1}^{P}$ and set $\alpha_k = \alpha_k + 1$.
5. Use the resulting $\{\alpha_i\}_{i=1}^{P}$ to obtain a set of P spatial-domain heterogeneous partitions (with overlap borders) of W, and send each partition to processor p_i, along with B.
6. Calculate the morphological profiles $p(x, y)$ for the pixels in the local data partitions (in parallel) at each heterogeneous processor.
7. Collect all the individual results and merge them together to produce the final output.

A homogeneous version of the HeteroMORPH algorithm above can be simply obtained by replacing step 4 with $\alpha_i = P/w_i$ for all $i \in \{1, ..., P\}$, where w_i is the communication speed between processor pairs in the network, which is assumed to be homogeneous.

7.2.2 Parallel Neural Algorithm

In this section, we describe a supervised parallel classifier based on a multi-layer perceptron (MLP) neural network with back-propagation learning. This approach has been shown in previous work to be very robust for the classification of hyperspectral imagery [14]. However, the considered neural architecture and back-propagation-type learning algorithm introduce additional considerations for parallel implementations on HNOCs.

The architecture adopted for the proposed MLP-based neural network classifier is shown in Figure 7.2. As shown in the figure, the number of input neurons equals the number of spectral bands acquired by the sensor. In the case of PCT-based pre-processing or morphological feature extraction commonly adopted in hyperspectral analysis, the number of neurons at the input layer equals the dimensionality of feature vectors used for classification. The second layer is the hidden layer, where the number of nodes, M, is usually estimated empirically. Finally, the number of neurons at the output layer, C, equals the number of distinct classes to be identified in the input data. With the above architecture in mind, the standard back-propagation learning algorithm can be outlined by the following steps:

1. *Forward phase*. Let the individual components of an input pattern be denoted by $f_j(x, y)$, with $j = 1, 2, ..., N$. The output of the neurons at the hidden layer is obtained as: $H_i = \varphi(\sum_{j=1}^{N} \omega_{ij} \cdot f_j(x, y))$ with $i = 1, 2, ..., M$, where $\varphi(\cdot)$ is the activation function and ω_{ij} is the weight associated to the connection between the i-th input node and the j-th hidden node. The outputs of the MLP

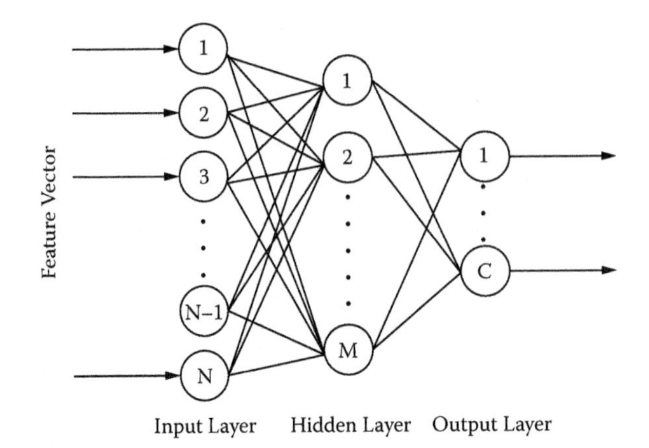

Figure 7.2 MLP neural network topology.

are obtained using $O_k = \varphi(\sum_{i=1}^{M} \omega_{ki} \cdot H_i)$, with $k = 1, 2, ..., C$. Here, ω_{ki} is the weight associated to the connection between the i-th hidden node and the k-th output node.

2. *Error back-propagation.* In this stage, the differences between the desired and obtained network outputs are calculated and back-propagated. The *delta* terms for every node in the output layer are calculated using $\delta_k^o = (O_k - d_k) \cdot \varphi'(\cdot)$, with $i = 1, 2, ..., C$. Here, $\varphi'(\cdot)$ is the first derivative of the activation function. Similarly, *delta* terms for the hidden nodes are obtained using $\delta_i^h = \sum_{k=1}^{C}(\omega_{ki} \cdot \delta_i^o) \cdot \varphi(\cdot))$, with $i = 1, 2, ..., M$.
3. *Weight update.* After the back-propagation step, all the weights of the network need to be updated according to the *delta* terms and to η, a learning rate parameter. This is done using $\omega_{ij} = \omega_{ij} + \eta \cdot \delta_i^h \cdot f_j(x, y)$ and $\omega_{ki} = \omega_{ki} + \eta \cdot \delta_k^o \cdot H_i$. Once this stage is accomplished, another training pattern is presented to the network and the procedure is repeated for all incoming training patterns.

Once the back-propagation learning algorithm is finalized, a classification stage follows, in which each input pixel vector is classified using the weights obtained by the network during the training stage [14].

Two different schemes can be adopted for the partitioning of the multi-layer perceptron classifier:

- The exemplar partitioning scheme, also called training example parallelism, explores data level parallelism and can be easily obtained by simply partitioning the training pattern data set. Each process determines the weight changes for a disjoint subset of the training population, and then changes are combined and applied to the neural network at the end of each epoch. This scheme requires a suitable large number of training patterns to take advantage of it, which is not a very common situation in most remote sensing applications, as long as it is a very hard task to get ground-truth information for regions of interest in a hyperspectral scene.

- The hybrid partition scheme, on the other hand, relies on a combination of neuronal level as well as synaptic level parallelism [15], which allows one to reduce the processors' intercommunications at each iteration. In the case of neuronal parallelism (also called vertical partitioning), all the incoming weights to the neurons local to the processor are computed by a single processor. In synaptic level parallelism, each workstation will compute only the outgoing weight connections of the nodes (neurons) local to the processor. In the hybrid scheme, the hidden layer is partitioned using neuronal parallelism while weight connections adopt the synaptic scheme.

The parallel classifier presented in this section is based on a hybrid partitioning scheme, where the hidden layer is partitioned using neuronal level parallelism and weight connections are partitioned on the basis of synaptic level parallelism [16]. As a result, the input and output neurons are common to all processors, while the hidden layer is partitioned so that each heterogeneous processor receives a number of hidden neurons, which depends on its relative speed. Each processor stores the weight connections between the neurons local to the processor. Since the fully connected MLP network is partitioned into P partitions and then mapped onto P heterogeneous processors using the above framework, each processor is required to communicate with every other processor to simulate the complete network. For this purpose, each of the processors in the network executes the three phases of the back-propagation learning algorithm described above. The HeteroNEURAL algorithm can be summarized as follows:

Inputs: N-dimensional cube f, training patterns $f_j(x, y)$.

Output: Set of classification labels for each image pixel.

1. Use steps 1–4 of the HeteroMORPH algorithm to obtain a set of values $(\alpha_i)_{i=1}^{P}$, which will determine the share of the workload to be accomplished by each heterogeneous processor.
2. Use the resulting $(\alpha_i)_{i=1}^{P}$ to obtain a set of P heterogeneous partitions of the hidden layer and map the resulting partitions among the P heterogeneous processors (which also store the full input and output layers along with all connections involving local neurons).
3. *Parallel training.* For each considered training pattern, the following three parallel steps are executed:
   (a) *Parallel forward phase.* In this phase, the activation value of the hidden neurons local to the processors are calculated. For each input pattern, the activation value for the hidden neurons is calculated using $H_i^P = \varphi(\sum_{j=1}^{N} \omega_{ij} \cdot f_j(x, y))$. Here, the activation values and weight connections of neurons present in other processors are required to calculate the activation values of output neurons according to $O_k^P = \varphi(\sum_{i=1}^{M/P} \omega_{ki}^P \cdot H_i^P)$,

with $k = 1, 2, ..., C$. In our implementation, broadcasting the weights and activation values is circumvented by calculating the partial sum of the activation values of the output neurons.

(b) *Parallel error back-propagation.* In this phase, each processor calculates the error terms for the local hidden neurons. To do so, *delta* terms for the output neurons are first calculated using $(\delta_k^o)^P = (O_k - d_k)^P \cdot \varphi'(\cdot)$, with $i = 1, 2, ..., C$. Then, error terms for the hidden layer are computed using $(\delta_i^h)^P = \sum_{k=1}^{P}(\omega_{ki}^P \cdot (\delta_k^o)^P) \cdot \varphi'(\cdot)$, with $i = 1, 2, ..., N$.

(c) *Parallel weight update.* In this phase, the weight connections between the input and hidden layers are updated by $\omega_{ij} = \omega_{ij} + \eta^P \cdot (\delta_i^h)^P \cdot f_j(x, y)$. Similarly, the weight connections between the hidden and output layers are updated using the expression $\omega_{ki}^P = \omega_{ki}^P + \eta^P \cdot (\delta_k^o)^P \cdot H_i^P$.

4. *Classification.* For each pixel vector in the input data cube f, calculate (in parallel) $\sum_{j=1}^{P} O_k^j$, with $k = 1, 2, ..., C$. A classification label for each pixel can be obtained using the winner-take-all criterion commonly used in neural networks by finding the cumulative sum with maximum value, say $\sum_{j=1}^{P} O_{k^*}^j$, with $k^* = \arg\{\max_{1 \leq k \leq C} \sum_{j=1}^{P} O_k^j\}$.

7.3 Experimental Results

This section provides an assessment of the effectiveness of the parallel algorithms described in the previous section. The section is organized as follows. First, we describe a framework for the assessment of heterogeneous algorithms and provide an overview of the heterogeneous and homogeneous networks used in this work for evaluation purposes. Second, we briefly describe the hyperspectral data set used in the experiments. Performance data are given in the last subsection.

7.3.1 Performance Evaluation Framework

Following a recent study [17], we assess the proposed heterogeneous algorithms using the basic postulate that they cannot be executed on a heterogeneous network faster than its homogeneous prototype on an equivalent homogeneous cluster network. Let us assume that a heterogeneous network consists of $\{p_i\}_i^P$ heterogeneous workstations with different cycle-times w_i, which span m communication segments $\{s_j\}_{j=1}^m$, where $c^{(j)}$ denotes the communication speed of segment s_j. Similarly, let $p^{(j)}$ be the number of processors that belong to s_j, and let $w_t^{(j)}$ be the speed of the t-th processor connected to s_j, where $t = 1, ..., p^{(j)}$. Finally, let $c^{(j,k)}$ be the speed of the communication link between segments s_j and s_k, with $j, k = 1, ..., m$. According to [17], the above network can be considered equivalent to a homogeneous one made up of $\{q_i\}_{i=1}^P$ processors with a constant cycle-time and interconnected through a homogeneous communication network with speed c if, and only if, the following expressions are satisfied:

$$c = \frac{\sum_{j=1}^{m} c^{(j)} \cdot [\frac{p^{(j)}(p^{(j)}-1)}{2}] + \sum_{j=1}^{m} \sum_{k=j+1}^{m} p^{(j)} \cdot p^{(k)} \cdot c^{(j,k)}}{\frac{P(P-1)}{2}} \tag{7.4}$$

and

$$w = \frac{\sum_{j=1}^{m} \sum_{t=1}^{p^{(j)}} w_t^{(j)}}{P} \tag{7.5}$$

where the first expression states that the average speed of point-to-point communications between processors $\{p_i\}_{i=1}^{P}$ in the heterogeneous network should be equal to the speed of point-to-point communications between processors $\{q_i\}_{i=1}^{P}$ in the homogeneous network, with both networks having the same number of processors. On the other hand, the second expression simply states that the aggregate performance of processors $\{p_i\}_{i=1}^{P}$ should be equal to the aggregate performance of processors $\{q_i\}_{i=1}^{P}$.

We have configured two networks of workstations to serve as sample networks for testing the performance of the proposed heterogeneous hyperspectral imaging algorithm. The networks are considered approximately equivalent under the above framework. Their description follows:

- *Fully heterogeneous network.* This network, already described and used in Chapter 2 of the present volume, consists of 16 different workstations and 4 communication segments, where processors $\{p_i\}_{i=1}^{4}$ are attached to communication segment s_1, processors $\{p_i\}_{i=5}^{8}$ communicate through s_2, processors $\{p_i\}_{i=9}^{10}$ are interconnected via s_3, and processors $\{p_i\}_{i=11}^{16}$ share the communication segment s_4. The communication links between the different segments $\{s_j\}_{j=1}^{4}$ only support serial communication. The communication network of the fully heterogeneous network consists of four relatively fast homogeneous communication segments, interconnected by three slower communication links with capacities $c^{(1,2)} = 29.05$, $c^{(2,3)} = 48.31$, $c^{(3,4)} = 58.14$ in milliseconds, respectively. Although this is a simple architecture, it is also a quite typical and realistic one as well.
- *Fully homogeneous network.* Consists of 16 identical Linux workstations $\{q_i\}_{i=1}^{16}$ with a processor cycle-time of $w = 0.0131$ seconds per megaflop, interconnected via a homogeneous communication network where the capacity of links is $c = 26.64$ milliseconds.

Finally, in order to test the proposed algorithm on a large-scale parallel platform, we have also experimented with Thunderhead, a massively parallel Beowulf cluster at NASA's Goddard Space Flight Center. The system is composed of 256 dual 2.4 GHz Intel Xeon nodes, each with 1 GB of memory and 80 GB of main memory. The total peak performance of the system is 2457.6 GFlops. Along with the 512-processor computer core, Thunderhead has several nodes attached to the core with 2 Ghz optical fibre Myrinet. In all considered platforms, the operating system used at the time of the experiments was Linux Fedora Core, and MPICH was the message-passing library used (see http://www-unix.mcs.anl.gov/mpi/mpich).

7.3.2 Hyperspectral Data Sets

Before empirically investigating the performance of the proposed parallel hyperspectral imaging algorithms in the five considered platforms, we first describe the hyperspectral image scene that will be used in the experiments. The scene was collected by the 224-band AVIRIS sensor over Salinas Valley, California, and is characterized by high spatial resolution (3.7-meter pixels). The relatively large area covered (512 lines by 217 samples) results in a total image size of more than 1 GB. Figure 7.3(a) shows the spectral band at 587 nm wavelength and a sub-scene (called hereinafter Salinas A), which comprises 83 × 86 pixels and is dominated by directional features. Figure 7.3(b) shows the ground-truth map, in the form of a class assignment for each labeled pixel with 15 mutually exclusive ground-truth classes. As shown by Figure 7.3(b), ground truth is available for nearly half of the Salinas scene. The data set above represents a very challenging classification problem (due to the spectral similarity of most classes, discriminating among them is very difficult). This fact has made the scene a universal and widely used benchmark to validate the classification accuracy of hyperspectral algorithms [5].

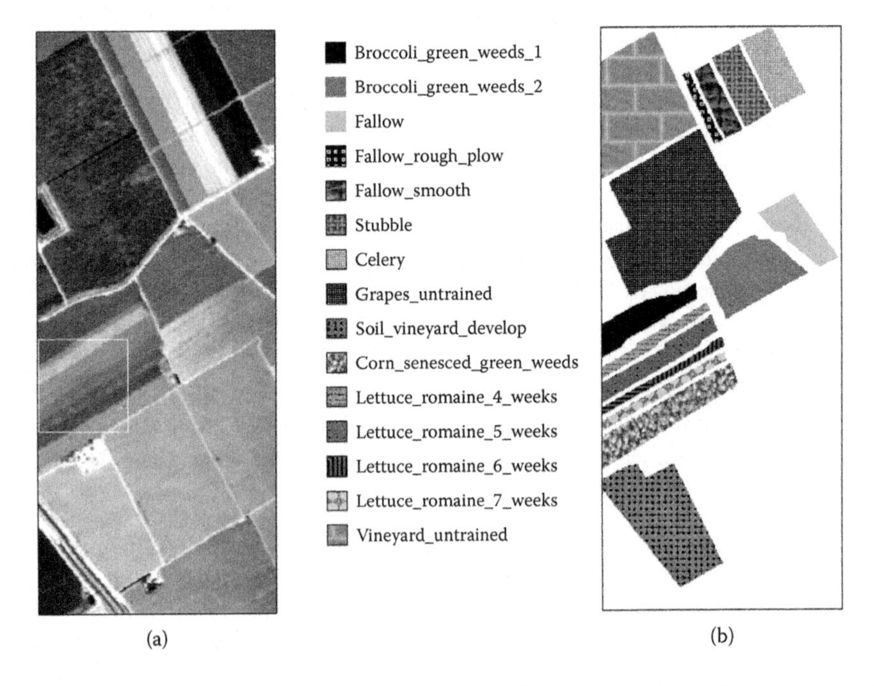

Figure 7.3 AVIRIS scene of Salinas Valley, California (a), and land-cover ground classes (b).

TABLE 7.1 Classification Accuracies (in Percentage) Achieved by The Parallel Neural Classifier for the AVIRIS Salinas Scene Using Morphological Features, PCT-Based Features, and the Original Spectral Information (Processing Times in a Single Thunderhead Node are Given in the Parentheses)

AVIRIS Salinas Class Label	Spectral Information (2981)	PCT-Based Features (3256)	Morphological Features (3679)
Fallow rough plow	96.51	91.90	96.78
Fallow smooth	93.72	93.21	97.63
Stubble	94.71	95.43	98.96
Celery	89.34	94.28	98.03
Grapes untrained	88.02	86.38	95.34
Soil vineyard develop	88.55	84.21	90.45
Corn senesced green weeds	82.46	75.33	87.54
Lettuce romaine 4 weeks	78.86	76.34	83.21
Lettuce romaine 5 weeks	82.14	77.80	91.35
Lettuce romaine 6 weeks	84.53	78.03	88.56
Lettuce romaine 7 weeks	84.85	81.54	86.57
Vineyard untrained	87.14	84.63	92.93
Overall accuracy	87.25	86.21	95.08

In order to test the accuracy of the proposed parallel morphological/neural classifier, a random sample of less than 2% of the pixels was chosen from the known ground-truth of the Salinas scene described above. Morphological profiles were then constructed in parallel for the selected training samples using 10 iterations, which resulted in feature vectors with dimensionality of 20 (i.e., 10 structuring element iterations for the *opening series* and 10 iterations for the *closing series*). The resulting features were then used to train the parallel back-propagation neural network classifier with one hidden layer, where the number of hidden neurons was selected empirically as the square root of the product of the number of input features and information classes (several configurations of the hidden layer were tested and the one that gave the highest overall accuracies was reported). The trained classifier was then applied to the remaining 98% of the labeled pixels in the scene, yielding the classification accuracies shown in Table 7.1.

For comparative purposes, the accuracies obtained using the full spectral information and PCT-reduced features as input to the neural classifier are also reported in Table 7.1. As shown in the table, morphological input features substantially improve individual and overall classification accuracies with regard to PCT-based features and the full spectral information (e.g., for the directional 'lettuce' classes contained in the Salinas A subscene). This is not surprising since morphological operations use both spatial and spectral information as opposed to the other methods, which rely on spectral information alone. For illustrative purposes, Table 7.1 also includes (in the parentheses) the algorithm processing times in seconds for the different approaches tested, measured on a single processor in the Thunderhead system. Experiments were performed using the GNU-C/C++ compiler in its 4.0 version. As shown in table,

TABLE 7.2 Execution Times (in Seconds) and Performance Ratios Reported for the Homogeneous Algorithms Versus The Heterogeneous Ones on the Two Considered Networks

	Homogeneus Network		Heterogeneus Network	
Algorithm	**Time**	**Homo/Hetero**	**Time**	**Homo/Hetero**
HeteroMORPH	221	1.11	206	10.98
HomoMORPH	198		2261	
HeteroCOM	289	1.12	242	11.86
HomoCOM	258		2871	
HeteroNEURAL	141	1.12	130	9.70
HomoNEURAL	125		1261	

the computational cost was slightly higher when morphological feature extraction was used.

7.3.3 Assessment of the Parallel Algorithm

To investigate the properties of the parallel morphological/neural classification algorithm developed in this work, the performance of its two main modules (HeteroMORPH and HeteroNEURAL) was first tested by timing the program using the heterogeneous network and its equivalent homogeneous one. For illustrative purposes, an alternative implementation of HeteroMORPH without 'overlapping scatter' was also tested; i.e., in this implementation the overlap border data are not replicated between adjacent processors but communicated instead. This approach is denoted as HeteroCOM, with its correspondent homogeneous version designated by HomoCOM.

As expected, the execution times reported in Table 7.2 for the three considered heterogeneous algorithms and their respective homogeneous versions indicate that the heterogeneous implementations were able to adapt much better to the heterogeneous computing environment than the homogeneous ones, which were only able to perform satisfactorily on the homogeneous network. For the sake of comparison, Table 7.2 also shows the performance ratios between the heterogeneous algorithms and their respective homogeneous versions (referred to as Homo/Hetero ratio in the table and simply calculated as the execution time of the homogeneous algorithm divided by the execution time of the heterogeneous algorithm).

From Table 7.2, one can also see that the heterogeneous algorithms were always several times faster than their homogeneous counterparts in the heterogeneous network, while the homogeneous algorithms only slightly outperformed their heterogeneous counterparts in the homogeneous network. The Homo/Hetero ratios reported in the table for the homogeneous algorithms executed on the homogeneous network were indeed very close to 1, a fact that reveals that the performance of heterogeneous algorithms was almost the same as that evidenced by homogeneous algorithms when they were run in the same homogeneous environment. The above results demonstrate

TABLE 7.3 Communication (COM), Sequential Computation (SEQ), and Parallel Computation (PAR) Times for the Homogeneous Algorithms Versus the Heterogeneous Ones on the Two Considered Networks After Processing the AVIRIS Salinas Hyperspectral Image

	Homogeneous Network			**Heterogeneous Network**		
	COM	SEQ	PAR	COM	SEQ	PAR
HeteroMORPH	7	19	202	11	16	190
HomoMORPH	14	18	180	6	16	2245
HeteroCOM	57	16	193	52	15	182
HomoCOM	64	15	171	69	13	2194
HeteroNEURAL	4	27	114	7	24	106
HomoNEURAL	9	27	98	3	24	1237

the flexibility of the proposed heterogeneous algorithms, which were able to adapt efficiently to the two considered networks.

Interestingly, Table 7.2 also reveals that the performance of the heterogeneous algorithms on the heterogeneous network was almost the same as that evidenced by the equivalent homogeneous algorithms on the homogeneous network (i.e., the algorithms achieved essentially the same speed, but each on its network). This seems to indicate that the heterogeneous algorithms are very close to the optimal heterogeneous modification of the basic homogeneous ones. Finally, although the Homo/Hetero ratios achieved by HeteroMORPH and HeteroCOM are similar, the processing times in Table 7.2 seem to indicate that the data replication strategy adopted by HeteroMORPH is more efficient than the data communication strategy adopted by HeteroCOM in our considered application.

To further explore the above observations in more detail, an in-depth analysis of computation and communication times achieved by the different methods is also highly desirable. For that purpose, Table 7.3 shows the total time spent by the tested algorithms in communications (labeled as COM in the table) and computations in the two considered networks, where two types of computation times were analyzed, namely, sequential (those performed by the root node with no other parallel tasks active in the system, labeled as SEQ in the table) and parallel (the rest of the computations, i.e., those performed by the root node and/or the workers in parallel, labeled as PAR in the table). The latter includes the times in which the workers remain idle. It is important to note that our parallel implementations have been carefully designed to allow overlapping of communications and computations when no data dependencies are involved.

It can be seen from Table 7.3 that the COM scores were very low when compared to the PAR scores in both HeteroMORPH and HeteroNEURAL. This is mainly due to the fact that these algorithms involve only a few inter-processor communications, which leads to almost complete overlapping between computations and communications in most cases. In the case of HeteroMORPH, it can be observed that the SEQ and PAR scores are slightly increased with regard to those obtained for HeteroCOM

TABLE 7.4 Load-Balancing Rates for the Parallel Algorithms on the Homogeneous and Heterogeneous Network

	Homogeneus Network		Heterogeneus Network	
Algorithm	D_{All}	D_{Minus}	D_{All}	D_{Minus}
HeteroMORPH	1.03	1.02	1.05	1.01
HomoMORPH	1.05	1.01	1.59	1.21
HeteroCOM	1.06	1.04	1.09	1.03
HomoCOM	1.07	1.03	1.94	1.52
HeteroNEURAL	1.02	1.01	1.03	1.01
HomoNEURAL	1.03	1.01	1.39	1.19

as a result of the the data replication strategy introduced by the former algorithm. However, Table 7.3 also reveals that the COM scores measured for HeteroCOM were much higher than those reported for HeteroMORPH, and could not be completely overlapped with computations due to the high message traffic resulting from communication of full hyperspectral pixel vectors across the heterogeneous network. This is the main reason why the execution times measured for HeteroCOM were the highest in both networks, as already reported by Table 7.2. Finally, the fact that the PAR scores produced by the homogeneous algorithms executed on the heterogeneous network are so high is likely due to a less efficient workload distribution among the heterogeneous workers. Therefore, a study of load balance is highly required to fully substantiate the parallel properties of the considered algorithms.

In order to measure load balance, Table 7.4 shows the imbalance scores achieved by the parallel algorithms on the two considered networks. The imbalance is defined as $D = R_{max}/R_{min}$, where R_{max} and R_{min} are the maxima and minima processor run-times, respectively. Therefore, perfect balance is achieved when $D = 1$. In the table, we display the imbalance considering all processors, D_{All}, and also considering all processors but the root, D_{Minus}. As we can see from Table 7.4, both the HeteroMORPH and HeteroNEURAL algorithms were able to provide values of D_{All} close to 1 in the two considered networks, which indicates that the proposed heterogeneous data partitioning algorithm is effective. Further, the above algorithms provided almost the same results for both D_{All} and D_{Minus} while, for the homogeneous versions, load balance was much better when the root processor was not included. While the homogeneous algorithms executed on the heterogeneous network provided the highest values of D_{All} and D_{Minus} (and hence the highest imbalance), the heterogeneous algorithms executed on the homogeneous network resulted in values of D_{Minus} that were close to optimal.

Despite the fact that conventional feature extraction algorithms (such as those based on PCT) do not take into account the spatial information explicitly into the computations—a fact that has traditionally been perceived as an advantage for the development of parallel implementations—and taking into account that both HeteroMORPH and HeteroNEURAL introduce redundant information expected to slow

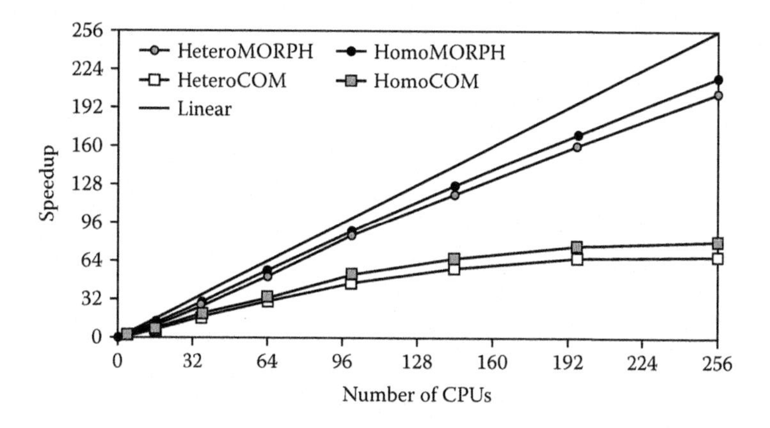

Figure 7.4 Scalability of parallel morphological feature extraction algorithms on Thunderhead.

down the computation a priori, the results in Table 7.4 indicate that the two heterogeneous algorithms are effective in finding an appropriate workload distribution among the heterogeneous processors. On the other hand, the higher imbalance scores measured for HeteroCOM (and its homogeneous version) are likely due to the impact of inter-processor communications. In this case, further research is required to adequately incorporate the properties of the heterogeneous communication network into the design of the heterogeneous algorithm.

Taking into account the results presented above, and with the ultimate goal of exploring issues of scalability (considered to be a highly desirable property in the design of heterogeneous parallel algorithms), we have also compared the performance of the heterogeneous algorithms and their homogeneous versions on the Thunderhead Beowulf cluster. Figure 7.4 plots the speedups achieved by multi-processor runs of the heterogeneous parallel implementations of the morphological feature extraction algorithm over the corresponding single-processor runs of each considered algorithm on Thunderhead. For the sake of comparison, Figure 7.4 also plots the speedups achieved by multi-processor runs of the homogeneous versions on Thunderhead. On the other hand, Figure 7.5 shows similar results for the parallel neural network classifier. As Figure 7.4 and 7.5 show, the scalability of heterogeneous algorithms was essentially the same as that evidenced by their homogeneous versions, with both HeteroNEURAL and HeteroMORPH showing scalability results close to linear in spite of the fact that the two algorithms introduce redundant computations expected to slow down the computation a priori. Quite opposite, Figure 7.4 shows that the speedup plot achieved by HeteroCOM flattens out significantly for a high number of processors, indicating that the ratio of communications to computations is progressively more significant as the number of processors is increased, and parallel performance is significantly degraded. The above results clearly indicate that the proposed data replication strategy is more appropriate than the tested data communication strategy in the design of a

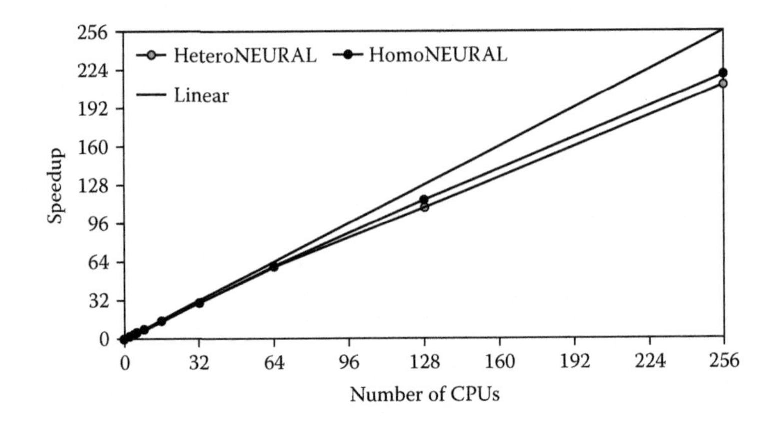

Figure 7.5 Scalability of parallel neural classifier on Thunderhead.

parallel version of morphological feature extraction in the context of remote sensing applications.

Overall, experimental results in our study reveal that the proposed heterogeneous parallel algorithms offer a relatively platform-independent and highly scalable solution in the context of realistic hyperspectral image analysis applications. Contrary to common perception that spatial/spectral feature extraction and back-propagation learning algorithms are too computationally demanding for practical use and/or (near) real-time exploitation in hyperspectral imaging, the results in this chapter demonstrate that such approaches are indeed appealing for parallel implementation, not only because of the regularity of the computations involved in such algorithms, but also because they can greatly benefit from the incorporation of redundant information to reduce sequential computations at the master node and involve minimal communication between the parallel tasks, namely, at the beginning and ending of such tasks.

7.4 Conclusions and Future Research

In this chapter, we have presented an innovative parallel algorithm for hyperspectral image analysis based on morphological neural networks, and implemented several variations of the algorithm on both heterogeneous and homogeneous networks and clusters. The parallel performance evaluation strategy conducted in this work was based on experimentally assessing the heterogeneous algorithm by comparing its efficiency on a fully heterogeneous network (made up of processing units with different speeds and highly heterogeneous communication links) with the efficiency achieved by its equivalent homogeneous version on an equally powerful homogeneous network. Scalability results on a massively parallel commodity cluster are also provided.

Experimental results in this work anticipate that the (readily available) computational power offered by heterogeneous architectures offers an excellent alternative for the efficient implementation of hyperspectral image classification algorithms based on morphological neural networks, which can successfully integrate the spatial and spectral information in the data in simultaneous fashion. In future research, we are planning on implementing the proposed parallel neural algorithm using hardware architectures taking advantage of the efficient systolic array design already conducted by the morphological and neural stages of the algorithm [18].

7.5 Acknowledgment

The authors, thank J. Dorband, J. C. Tilton, and J. A. Gualtieri for their support with experiments on NASA's Thunderhead system. They also acknowledge their appreciation for Profs. M. Valero and F. Tirado.

References

[1] C.-I. Chang. *Hyperspectral imaging: Techniques for spectral detection and classification*. Kluwer: New York, 2003.

[2] R. O. Green. Imaging spectroscopy and the airborne visible/infrared imaging spectrometer (AVIRIS). *Remote Sensing of Environment,* vol. 65, pp. 227–248, 1998.

[3] G. Aloisio and M. Cafaro. A dynamic earth observation system. *Parallel Computing*, vol. 29, pp. 1357–1362, 2003.

[4] D. A. Landgrebe. *Signal theory methods in multispectral remote sensing*. Wiley: Hoboken, 2003.

[5] A. Plaza, P. Martinez, J. Plaza, and R. M. Perez. Dimensionality reduction and classification of hyperspectral image data using sequences of extended morphological transformations. *IEEE Transactions on Geoscience and Remote Sensing*, vol. 43, pp. 466–479, 2005.

[6] T. El-Ghazawi, S. Kaewpijit, and J. L. Moigne. Parallel and adaptive reduction of hyperspectral data to intrinsic dimensionality. *Proceedings of the IEEE International Conference on Cluster Computing,* pp. 102–110, 2001.

[7] A. Plaza, D. Valencia, J. Plaza, and P. Martinez. Commodity cluster-based parallel processing of hyperspectral imagery. *Journal of Parallel and Distributed Computing*, vol. 66, pp. 345–358, 2006.

[8] P. Wang, K. Y. Liu, T. Cwik, and R. O. Green. MODTRAN on supercomputers and parallel computers. *Parallel Computing*, vol. 28, pp. 53–64, 2002.

[9] T. Sterling. Cluster computing. *Encyclopedia of Physical Science and Technology,* vol. 3, 2002.

[10] J. Dorband, J. Palencia, and U. Ranawake. Commodity clusters at Goddard Space Flight Center. *Journal of Space Communication,* vol. 3, pp. 227–248, 2003.

[11] A. Lastovetsky. *Parallel computing on heterogeneous networks.* Wiley-Interscience: Hoboken, NJ, 2003.

[12] G. X. Ritter, P. Sussner, and J. L. Diaz. Morphological associative memories. *IEEE Transactions on Neural Networks*, vol. 9, pp. 281–293, 2004.

[13] P. Soille. *Morphological image analysis: Principles and applications.* Springer: Berlin, 2003.

[14] J. Plaza, A. Plaza, R. M. Perez, and P. Martinez. Automated generation of semi-labeled training samples for nonlinear neural network-based abundance estimation in hyperspectral data. *Proceedings of the IEEE International Geoscience and Remote Sensing Symposium*, pp. 345–350, 2005.

[15] S. Suresh, S. N. Omkar, and V. Mani. Parallel implementation of back-propagation algorithm in networks of workstations. *IEEE Transactions on Parallel and Distributed Systems*, vol. 16, pp. 24–34, 2005.

[16] J. Plaza, R. M. Perez, A. Plaza, P. Martinez and D. Valencia. Parallel morphological/neural classification of remote sensing images using fully heterogeneous and homogeneous commodity clusters. *Proceedings of the IEEE International Conference on Cluster Computing*, pp. 328–337, 2006.

[17] A. Lastovetsky and R. Reddy. On performance analysis of heterogeneous parallel algorithms. *Parallel Computing*, vol. 30, pp. 1195–1216, 2004.

[18] D. Zhang and S. K. Pal. *Neural Networks and Systolic Array Design.* World Scientific: Singapore, 2002.

Chapter 8

Parallel Wildland Fire Monitoring and Tracking Using Remotely Sensed Data

David Valencia,
University of Extremadura, Spain

Pablo Martínez,
University of Extremadura, Spain

Antonio Plaza,
University of Extremadura, Spain

Javier Plaza,
University of Extremadura, Spain

Contents

Predicting the potential behavior and effects of wildland fires using remote sensing technology is a long-awaited goal. The role of high-performance computing in this task is essential since fire phenomena often require a response in (near) real-time. Several studies have focused on the potential of hyperspectral imaging as a baseline technology to detect and monitor wildland fires by taking advantage of the rich spectral information provided by imaging spectrometers. The propagation of fires is a very complex process that calls for the integrated use of advanced processing algorithms and mathematical models in order to explain and further characterize the process. In this chapter, we describe several advanced hyperspectral data processing algorithms that are shown to be useful in the task of detecting/tracking wildland fires and further study how such algorithms can be integrated with mathematical models, with the ultimate goal of designing an integrated system for surveillance and monitoring of fires.

8.1 Introduction

Many efforts have been conducted by international organizations to deal with natural and human-induced disasters through the use of remote sensing technology. Many of them are focused on post-evaluation and management of the disaster as a way to improve future evaluation and prediction. Several missions operated by international agencies are designed to produce a great amount of image data, which can be processed to evaluate and track these disasters, but this approach introduces strong computational requirements that are always challenging in terms of budget and, in some cases, lack of knowledge of the remote sensing community on high-performance computing solutions.

For instance, wildland fires represent one of the most important sources of biodiversity loss on our planet and introduce important requirements from the viewpoint of algorithm design and high-performance implementations. This is a general problem that causes important environmental risks. In particular, the importance of preserving forests is enormous since they are multifunctional in their outcome, from economical (resources), social (recreational), and environmental perspectives (protection against atmospheric contamination and wildfires, climate control, mitigation of climate change, and water/soil preservation).

In this chapter, we evaluate different possibilities to approach the problem of monitoring and tracking wildland fires using remotely sensed hyperspectral imagery. We also outline the design of a system to monitor and track wildland fires using image data sets produced by both airborne and satellite hyperspectral sensors. The chapter is organized as follows:

- Section 8.2 introduces the requirements for real-time response on hazards and disasters, using wildland fires as a potential case study.
- Section 8.3 describes the model adopted in this chapter for characterization of wildland fires. Ideally, this model should be integrated with advanced data processing algorithms to produce advanced fire characterization products.
- Section 8.4 describes a collection of hyperspectral data processing algorithms that may be used for detecting and monitoring wildland fires. Different approaches are evaluated in this section, including detection, classification, segmentation, and spectral unmixing.
- Section 8.5 details parallel implementations of some of the proposed algorithms, including morphological techniques and neural networks for classification and spectral mixture analysis of hyperspectral data sets.
- Section 8.6 outlines a high-performance system that integrates the above-mentioned parallel algorithms with mathematical models for potential prevention and response to wildland fires. The proposed system integrates several computer architectures, such as homogeneous and heterogeneous networks of computers including grid environments, and specialized hardware platforms such as those based on programmable hardware.
- Section 8.7 provides experimental results to evaluate the accuracy and parallel efficiency of the proposed parallel algorithms. Due to the lack of hyperspectral images of fires with reliable ground-truth, we use standard hyperspectral data sets collected in the framework of other applications to provide our experimental assessment.

8.2 Real-Time Response to Hazards and Disasters: Wildland Fires as a Case Study

Many remote sensing missions have been defined with the ultimate goal of monitoring natural disasters, e.g., detection of red tides using Envisat's Medium Resolution Imaging Spectrometer Instrument (MERIS) [16]. In the next decade, and thanks to the

Figure 8.1 MERIS hyperspectral image of the fires that took place in the summer of 2005 in Spain and Portugal.

expected improvements in the spatial and spectral resolutions of hyperspectral imaging instruments, it is anticipated that several missions will be focused on the detection, monitoring, and tracking of hazards. For instance, Figure 8.1 shows a MERIS image obtained during the summer of 2005 over Spain and Portugal. The figure reveals the fires that occurred in the area, which caused the loss of many forest areas [17]. The main disadvantage of monitoring fires using the MERIS instrument is that the revisit time of MERIS is 3 days, thus limiting its exploitation for active fire tracking.

Fire is an important and recurrent phenomenon in all forested and non-forested regions of the Earth. In some ecosystems, fire plays an ecologically significant role in biochemical cycles and disturbance dynamics. In other ecosystems, fire may lead to the destruction of forests or to long-term site degradation. As a consequence of demographic and land use changes, and the cumulative effects of anthropogenic disturbances, many forest types adapted to fire are becoming more vulnerable to high-intensity wildfires. For several years the number of fires in European forests has been increasing [17]. Forest fires are usually a result of the simultaneous existence of several phenomena: droughts, the effect of air pollution (decline and decay of trees, the formation of loose canopy and lush growth of grasses, all resulting in large amounts of flammable material), pyromaniacs, *illegal* selective burning of areas for construction, and so on.

Decades of research to understand and quantify fire, together with revolutionary advances in computer technology and computational algorithms, have lead to the development of sophisticated mathematical models that can predict and visualize growth

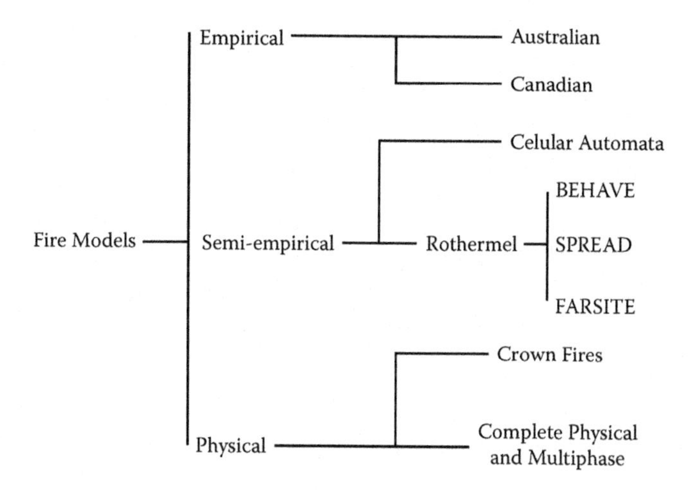

Figure 8.2 Classification of fire spread models.

and spread of fire under a variety of conditions. However, these models generally require substantial computational and data resources. Numerous fire spread models have been proposed following several methods (see Figure 8.2) that can be grouped into:

Empirical (or statistical). Statistical, stochastic, also called empirical models are predicting more probable fire behavior from average conditions and accumulated knowledge obtained from laboratory and outdoor experimental fires, or historical fires. There are two empirical models widely in use, Australian and Canadian [18, 19]. These models make no attempt to include any physical mechanisms for fire spread; they are purely statistical descriptions of test fires of such spreads. For example, the Canadian Forest Service has integrated 25 years of researching experimental and real scenario fires to develop the Canadian Forest Fire Behavior Prediction System, which is now available in book and electronic forms. It consists of 89 formulas developed empirically.

Semi-empirical (semi-physical or laboratory models). Semi-empirical models are based on a global energy balance and on the assumption that the energy transferred to the unburned fuel is proportional to the energy released by the combustion of the fuel. Several terms of the model must be fitted from laboratory fire experimental results and field campaign data [1]. The simplicity of this approach has allowed the development of operational tools.

Physical (theoretical or analytical). Models based on physical principles, have the potential to accurately predict the parameters of interest over a broader range of input variables than empirically based models. Physics based models can also provide the basic information needed for the proper description of physical processes (i.e., fluid flow, heat transfer, and chemical kinetics). But physics-based models [20] imply that the developer has an adequate understanding of the underlying physical relations sufficient to achieve the desired objectives, that the underlying physics can be represented mathematically in a manner that permits numerical solution. Improved models are needed for increased accuracy in fire behavior prediction, fire danger rating calculations, and fuel hazard assessment. Models with the goal to predict 3-dimensional fire shapes are often referred to as crown fire models.

The amount of living vegetation, and its moisture content, have a strong effect on the propagation and severity of wildland fires, and they are key components for all models types mentioned above. The direct observation of vegetation greenness is, therefore, essential for any fire-spread model (fuel bed component of the models), along with other information that can be retrieved using remote sensing. Current assessment of living vegetation moisture relies on various methods of manual sampling. While these measurements are quite accurate, they are difficult to obtain over broad areas, so they fail to portray changes in the pattern of vegetation greenness and moisture across the landscape.

The current polar orbiting satellites provide the potential for delivering information about the greenness and moisture of vegetation, along with other parameters for fire management. Some initiatives focused on the usage of constellations of satellites, as the FUEGO mission of the European Space Agency (http://http://virtual.vtt.fi/space/firealert/space_segment_detailed.html), in order to collect data about fires over a particular site in very small intervals of time. This will provide the scientists investigating forest fires with a great amount of data for their models, but again this wealth of information will be very difficult to manage and compute without the aid of high-performance computing.

In this chapter, we outline the requirements and preliminary design of a system for forest fire monitoring and tracking, combining advanced hyperspectral data processing techniques (implemented on high-performance computing platforms) with a well-known semi-empirical mathematical model for forest fire spread. Although the proposed system is aimed at wildland fire characterization, experimental results are given using data sets in other application domains for preliminary validation purposes.

8.3 Mathematical Model for Wildland Fire Characterization

In the past years, several models and systems to evaluate and predict fire risks have been made, as was pointed out before. The most interesting ones are the group of models that are based on the use of Rothermel's equations [1]. The main equations provided by the model allow the calculation of the rate of spread (ROS) and fire intensity I_b as follows:

$$\mathrm{ROS} = \frac{I_R \xi (1 + \phi_w + \phi_s)}{\rho_b \varepsilon Q_{ig}} \quad (8.1)$$

where ROS is the heading fire steady state spread rate ($\frac{m}{min}$), I_R is the reaction intensity ($\frac{kJ}{min \cdot m^2}$), ξ is the propagating flux ratio, ϕ_w is the wind factor (dimensionless), ϕ_s is the slope factor (dimensionless), ρ_b is the ovendry bulk density ($\frac{kg}{m^3}$), ε is the effective heating number (dimensionless), and Q_{ig} is the heat of pre-ignition ($\frac{kJ}{kg}$).

$$I_b = h \cdot w \cdot \frac{R}{60} \tag{8.2}$$

where I_b is the fire line intensity (kW/m) that describes the rate of energy release per unit length in the fire front, h is the heat yielded by the fuel (kJ/kg) or the total heat less the energy required for vaporizing moisture, w is the weight of fuel per area (kg/m^2) burned in the flaming front, and $R/60$ is the fire spread rate converted to units of m/sec.

Rothermel's model is the basis for many systems in the U.S., including the BEHAVE fire behavior prediction system, the FARSITE fire area simulator, the National Fire Danger Rating System (NFDRS), the National Fire Management Analysis System (NFMAS) for economic planning, the Rare Event Risk Assessment Process (RERAP) [22], and many more.

To facilitate the use in models and systems, fuelbed inputs have been formulated into fuel models. A fuel model is a set of fuelbed inputs needed by a particular fire behavior or fire effects model. Although a fuel model technically includes all fuel inputs to the model, several fuel inputs have never been subject to control by a user when creating a custom fuel model, leaving them static, and loosing, then, some accuracy in the predictions. The fuel models have worked well for predicting spread rate and intensity of active fires at the peak of the fire season, in part because the associated dry conditions lead to a more uniform fuel complex. However, they have deficiencies for other purposes, including prescribed fire, wildland fire use, simulating the effects of fuel treatments on potential fire behavior, and simulating the transition to crown fire using crown fire initiation models, along with comparisons of behaviors with similar species in different latitudes.

To solve the above-mentioned problems it is necessary to change the static nature of the models, allowing the change of some characteristics that can be known at the beginning of the fire season, increasing, thus, the accuracy of the predictions and simulations. To tackle this from the point of view of the field work is impossible, so it is necessary to work together with images provided by hyperspectral sensors and techniques able to obtain abundance maps, which label the vegetal species in the areas and extract further information on physical, biological, and characteristics of plants present in the image. This can be done by developing advanced hyperspectral data processing algorithms, as will be described in the following section.

8.4 Advanced Hyperspectral Data Processing Algorithms

In the present section, we introduce several advanced hyperspectral data processing algorithms intended to be used as potential building blocks for an integrated system to monitor wildland fires through the combination of field and ground-truth information, hyperspectral and additional remote sensing data (i.e. fuelbeds characteristics), and detailed mathematical models.

8.4.1 Morphological Algorithm for Endmember Extraction and Classification

The ultimate goal of endmember extraction algorithms is to select the spectrally purest constituent spectra in a scene. These endmembers are assumed to interact following a linear or non-linear spectral mixing model, and such models can be used to identify land cover within a fine spatial resolution scene. For this reason, it is very important to accurately obtain the signatures of the endmembers present in the scene. In the following, we describe a morphological algorithm for endmember extraction that integrates the spatial and spectral information in the analysis. Mathematical morphology [12] is a classic non-linear spatial processing technique that provides a remarkable framework to achieve the desired integration of spatial and spectral information. Parallelization of this technique for efficient execution in multiprocessors will be detailed in subsequent sections of this chapter.

Before describing our proposed approach, let us denote by $\mathbf{F}$ a hyperspectral data set defined on an N-dimensional (N-D) space, where N is the number of channels or spectral bands. The main idea of the algorithm is to impose an ordering relation, in terms of spectral purity, in the set of pixel vectors lying within a spatial search window or structuring element (SE) around each image pixel vector [12]. To do so, we first define a cumulative distance between one particular pixel $\mathbf{f}(x, y)$, where $\mathbf{f}(x, y)$ denotes an N-D vector at discrete spatial coordinates $(x, y) \epsilon Z^2$, and all the pixel vectors in the spatial neighborhood one given by B (B-neighborhood) as:

$$D_B[\mathbf{f}(x, y)] = \sum_i \sum_j \text{SID}[\mathbf{f}(x, y), \mathbf{f}(i, j)] \tag{8.3}$$

where (i, j) are the spatial coordinates in the B-neighborhood and SID is the spectral information divergence, a commonly used distance in remote sensing applications [15] defined as follows:

$$\text{SID}[\mathbf{f}(x, y), \mathbf{f}(i, j)] = \sum_{l=1}^{N} p_l \cdot \log\left(\frac{p_l}{q_l}\right) + \sum_{l=1}^{N} q_l \cdot \log\left(\frac{q_l}{p_l}\right) \tag{8.4}$$

where

$$p_l = \frac{f_l(x, y)}{\sum_{k=1}^{N} f_k(x, y)} \tag{8.5}$$

$$q_l = \frac{f_l(i, j)}{\sum_{k=1}^{N} f_k(i, j)} \tag{8.6}$$

Based on the distance above, we calculate the extended morphological erosion of $\mathbf{F}$ by B for each pixel in the input data scene as follows [4]:

$$(\mathbf{f} \Theta B)(x, y) = argmin_{(i,j)}\{D_B[\mathbf{f}(x + i, y + j)]\} \tag{8.7}$$

where the $argmin$ operator selects the pixel vector that is most highly similar, spectrally, to all the other pixels in the B-neighborhood. On the other hand, the extended morphological dilation of **f** by B is calculated as follows [4]:

$$(\mathbf{f} \oplus B)(x, y) = argmax_{(i,j)}\{D_B[\mathbf{f}(x + i, y + j)]\} \tag{8.8}$$

where the $argmax$ operator selects the pixel vector that is most spectrally distinct to all the other pixels in the B-neighborhood. With the above definitions in mind, we provide below an unsupervised classification algorithm for hyperspectral imagery called the Automatical Morphological Classification (AMC), which relies on extended morphological operations. The inputs to the algorithm are a hyperspectral data cube **f**, a structuring element B, and the number of classes to be detected c. The output of the algorithm is a 2-D matrix that contains a classification label for each pixel $f(x, y)$ in the input image. The algorithm can be summarized by the following steps:

1. Initialize a morphological eccentricity index score MEI$(x, y) = 0$ for each pixel.
2. Move B through all the pixels of **F**, defining a local spatial search area around each $\mathbf{f}(x, y)$, and calculate the maximum and minimum pixel, at each B-neighborhood using dilation and erosion, respectively. Update the MEI at each pixel using the SID between the maximum and the minimum.
3. Select the set of c pixel vectors in **F** with a higher associated score in the resulting MEI image.
4. Estimate the sub-pixel abundance $\alpha_i(x, y)$ of the pure pixels selected in the previous stage within $\mathbf{f}(x, y)$, using the standard linear mixture model described in [10].
5. Obtain a classification label for each pixel $\mathbf{f}(x, y)$ by assigning it to the class with the highest sub-pixel fractional abundance score in that pixel. This is done by comparing all estimated abundance fractions $\{\alpha_1(x, y), \alpha_2(x, y), \ldots, \alpha_c(x, y)\}$ and finding the one with the maximum value, say $\alpha_{i^*}(x, y)$, with $i^* = arg\{max_{1 \leq i \leq c}\{\alpha_i(x, y)\}\}$.

One of the main features of the algorithm is the regularity of its computations. As shown in previous work [4], its computational complexity is $O(p_f \times p_B \times N)$, where $p_{\mathbf{F}}$ is the number of pixels in the hyperspectral image **F** and p_B is the number of pixels in the structuring element B. This results in high computational cost in real applications.

8.4.2 Orthogonal Subspace Projection Algorithm for Target Detection

As opposed to classification algorithms, which aim at assigning appropriate class labels to each pixel, target detection algorithms aim at finding a collection of significant pixel vectors in accordance with different criteria.

One of the most effective target detection algorithms for hyperspectral image analysis is the automatic target generation process (ATGP), developed to find potential target pixels that can be used to generate a signature matrix using the concept of orthogonal subspace projection (OSP) [10]. The algorithm makes use of an OSP projector defined defined by

$$P_{\mathbf{U}}^{\perp} = \mathbf{I} - \mathbf{U}(\mathbf{U}^T\mathbf{U})^{-1}\mathbf{U}^{\mathrm{T}} \quad (8.9)$$

The ATGP repeatedly makes use of equation 8.9 to find target pixel vectors of interest from the data without prior knowledge, regardless of the types of pixels that the targets one. It can be briefly described as follows. Let's assume that $\mathbf{t}_0$ is an initial target pixel vector. The algorithm begins with the initial target pixel vector $\mathbf{t}_0$ by applying an orthogonal subspace projector $P_{\mathbf{t}_0}^{\perp}$ specified by equation 8.9 with $\mathbf{U} = \mathbf{t}_0$ to all image pixel vectors. It then finds a target pixel vector, denoted by $\mathbf{t}_1$, with the maximum orthogonal projection in the orthogonal complement space, denoted by $< \mathbf{t}_0 >^{\perp}$, that is orthogonal to the space $< \mathbf{t}_0 >$ linearly spanned by $\mathbf{t}_0$. The reason for this selection is that the selected t_1 generally has the most distinct features from $\mathbf{t}_0$ in the sense of orthogonal projection, because $\mathbf{t}_1$ has the largest magnitude of the projection in $< \mathbf{t}_0 >^{\perp}$ produced by $P_{\mathbf{t}_0}^{\perp}$. A second target pixel vector $\mathbf{t}_2$ can be found by applying an orthogonal subspace projector $P_{[\mathbf{t}_0\mathbf{t}_1]}^{\perp}$ with $\mathbf{U} = [\mathbf{t}_0\mathbf{t}_1]$ to the original image, and the target pixel vector that has the maximum orthogonal projection in $< \mathbf{t}_0, \mathbf{t}_1 >^{\perp}$ is selected as $\mathbf{t}_2$. The above procedure is repeated until a certain stopping rule is satisfied, usually determined by an estimated number of target pixel vectors required to generate, using several different procedures for the determination of the number of target pixel vectors. If we consider p as the number of target pixel vectors to generate the stopping criterion, the ATGP can be summarized by the following steps:

1. *Initial condition.* Select an initial target pixel vector of interest denoted by $\mathbf{t}_0$. In order to initialize the ATGP without knowing $\mathbf{t}_0$, we select a target pixel vector with the maximum length as the initial target $\mathbf{t}_0$, namely, $\mathbf{t}_0 = arg\{max\{\mathbf{f}(x, y) \cdot \mathbf{f}(x, y)^T\}\}$, which has the highest intensity, i.e., the brightest pixel vector in the image scene. Set $k = 1$ and $\mathbf{U} = [\mathbf{t}_0]$. It is worth noting that this selection may not necessarily be the best selection. However, according to the experiments, it was found that the brightest pixel vector was always extracted later on, if it was not used as an initial target pixel vector in the initialization.
2. *Iterative target detection.* At the k-th iteration, apply $P_{\mathbf{t}_0}^{\perp}$ via equation 8.9 to all image pixels $\mathbf{f}(x, y)$ in the image and find the k-th target, $\mathbf{t}_k$, generated at the k-th stage, which has the maximum orthogonal projection.
3. *Stopping rule.* If $k < p - 1$, let $\mathbf{U}_k = [\mathbf{U}_{k-1}t_k] = [t_1, t_2, \cdots, t_k]$ be the k-th target matrix and go to step 2. Otherwise, continue. At this point, the ATGP is terminated. The resulting target matrix is $\mathbf{U}_{p-1}$, which contains $p - 1$ target pixel vectors as its column vectors, which do not include the initial target pixel vector $\mathbf{t}_0$. The final set of target pixel vectors produced comprises p target pixel vectors, $\{\mathbf{t}_0, \mathbf{t}_1, \cdots, \mathbf{t}_{p-1}\} = \{\mathbf{t}_0\} \bigcup \{\mathbf{t}_1, \mathbf{t}_2, \cdots, \mathbf{t}_{p-1}\}$, that where found by repeatedly using equation 8.9. These target pixel vectors are the resulting targets for the hyperspectral image $\mathbf{F}$.

8.4.3 Self-Organizing Map for Neural Network Classification

A further approach for advanced classification of remotely sensed data consists of using a self-organizing map (SOM) [5]. This neural network architecture has been demonstrated in previous work to be very useful for analyzing hyperspectral images [8]. The main objective of the SOM model is to transform a given N-D signal or input pattern into a multidimensional map and the adaptive refining of such transformation using different topological criteria.

The SOM-based classification technique proposed in this chapter consists of N input neurons and M output neurons, where N is the dimensionality of input pixel vectors and M is the number of endmembers obtained using an endmember extraction algorithm (for instance, the morphological endmember extraction procedure described in this chapter). Therefore, our SOM neural network is composed of two layers, with forward connections from the input layer towards the output layer and the set of weights associated distributed in a matrix, named in the following $W_{M \times N}$. The analysis process performed by the network is divided in two phases: training and classification:

- In the training phase, the different patterns are presented to the neural network in a way that makes the forward connections change to adapt to the information contained in the training data.
- In the classification phase, the forward connections project the input patterns, i.e., the pixel vectors are classified in the feature space using the Euclidean distance to identify the winning neuron.

The whole procedure can be summarized in the following steps:

1. *Initialization of weights.* Random values are normalized to initialize the weight vectors: $w_i^{(0)}$, with $i = 1, 2, \cdots, M$.
2. *Training.* This step makes use of the endmembers obtained using some endmember extraction algorithm, e.g., the morphological endmember extraction procedure described in this chapter, which is used to provide input training patterns for the neural network.
3. *Clustering.* For each input pattern x, a winning neuron i^* is obtained in time t using a similarity criteria based on the Euclidean distance, i.e., $i^*[x] = \min_{1<=j<=M} ||x - w_j||^2$.
4. *Adjustment of weights.* The winning neuron and those in its neighborhood change their weights following the expression $w_i^{(t+1)} = w_i^t + \sum_{t'=t_0}^{t_{max}} \sum \alpha(t') \sum \sigma(t') \sum \left(x - w_i^{(t)}\right)$, where $\alpha(t)$ and $\sigma(t)$ are learning and neighborhood functions, respectively. One must take into account that the weights associated to i^* are modified proportionally to the learning frequency.
5. *Stopping condition.* The SOM algorithm ends as soon as a predetermined number of iterations, t_{max}, is completed.

From the steps indicated, it is clear that the SOM algorithm is naturally sequential. As a result, the parallelization strategies for this algorithm must particularly address the problem of data dependencies.

8.4.4 Spectral Mixture Analysis Algorithms

To conclude this section on advanced data processing techniques, we address the important issue of mixed pixels in hyperspectral image analysis. In several works, it has been demonstrated that the knowledge on the endmembers present in the image is not sufficient to characterize the different properties of plants, biomass, etc. In other words, once a set of endmembers has been extracted, it is important to be able to express the mixed pixels as a decomposition of a collection of 'pure' spectral signatures and a set of 'abundances' that indicate the individual proportion or contribution of each of the pure spectra to the mixed pixel [14].

The model used to properly describe the above situation is denominated as a 'mixture model,' which often assumes that the scene is composed of a limited set of endmembers with unique spectral features and a majority of mixed pixels with different endmembers participating in different proportions. In the models, two possibilities are observed:

- *Linear mixture model.* It considers that each incident beam of solar radiation only interacts with a single component or endmember, so that the total radiation reflected by a pixel can be decomposed proportionally to the abundance of each of the endmembers in the pixel.
- *Non-linear mixture model.* It supposes that the endmembers interact following a non-linear model. The non-linear effects appearing in this case are due, fundamentally, to multiple scattering effects in the light reflected by different materials.

In any case, one should note that the linear mixture model is not intended to fully describe the mixture problem since it is mainly a simplification. The real situation is dominated by secondary effects due to light multiply scattered by several covers, absorption and diffusion effects, shadows, etc. These effects pose an interesting question on which one of the mixture models is predominant in the mixture pixel spectra obtained by the sensors.

In the particular case of the wildland fires, the main goal is to estimate the temperature and quantity of fire appearing in each pixel of the image in order to produce maps of fires and hot spots that initiated the fires. For this purpose, several investigations have been carried out on the use of linear unmixing for the detection of the fraction of a fire endmember inside a pixel. The model proposed in [3] is the one used as a baseline in our investigations, and its basic idea is to use the linear spectral mixing model as an initial estimation, and then use the best fit linear spectral mixing model to identify fire temperature and land cover within a fine spatial resolution image scene. To accomplish this goal, we propose the use of artificial neural networks for linear unmixing of hyperspectral data.

8.5 Parallel Implementations

In the present section, we explore the application of parallel techniques in order to implement the different methods described in Section 8.4. First, we provide a parallel version of the AMC classification algorithm for both homogeneous and heterogeneous platforms, and then we describe a parallel version of the proposed SOM-based classification algorithm.

8.5.1 Parallelization of Automated Morphological Classification (AMC) Algorithm for Homogeneous Clusters

The parallel implementation of AMC is based on the definition of an efficient data partitioning scheme for hyperspectral imagery. We have considered two different approaches to the problem: partitioning in the spatial domain and partitioning in the spectral domain:

- The spatial-domain partitioning option divides the hyperspectral image in multiple blocks, in a way that the pixels for each block preserve its entire spectral identity.
- The spectral-domain partitioning option divides the original image in blocks constituted by several bands, in a way that we can preserve the spatial identity for each band but all the pixels in each block lose their spectral identity.

In other words, in spatial-domain partitioning the information of a single pixel in the image would be scattered across several different processing units. The selection of a partitioning scheme in the spectral domain is critical and could substantially increase the costs of communication and/or coordination between processors [7]. Besides, the overhead introduced by the communication increases with the number of processors, thus introducing problems in the load-balancing accomplished by the designed algorithms [25].

At this point, we introduce the concept of a parallelizable spatial/spectral pattern (PSSP), which is defined as the maximum amount of information that the parallel system can process without the need for additional communication and/or coordination between processors [24]. Such patterns are automatically generated by a spatial domain partitioning (SDP) module.

Figure 8.3 describes the partitioning framework using two computing units. In the example, the SDP divides the image into two PSSPs. The values of the MEI index for two pixels of the original hyperspectral image are calculated in parallel by each of the processors, using a square-shaped SE of 3 $\times$ 3 pixels. Such values are then updated in a local 2-D image. At the end of the process, the SDP integrates the various local images, obtaining a resulting 2-D image that is used as a baseline to extract a final set of endmembers, which lead to a final classification result. An issue of major importance in the design of SE-based parallel image processing applications is the possibility of

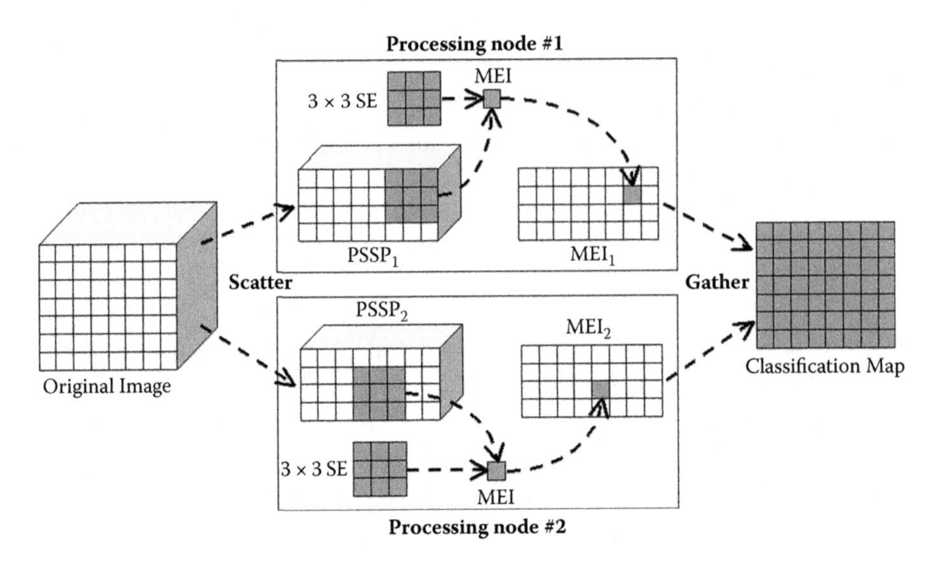

Figure 8.3 Concept of parallelizable spatial/spectral pattern (PSSP) and proposed partitioning scheme.

accessing pixels out of the spatial domain of the partition available in the processor. This is normally managed by a determined border-handling strategy. In our parallel implementation, two such strategies have been implemented, as described below:

- *Border-handling strategy relative to the pixels out of the domain of the original image.* This strategy is necessary in the situation illustrated in Figure 8.4. In this case, only the pixels of the SE that fall inside the image domain are used for the MEI computation. This strategy is similar to the mirroring technique commonly used in kernel-based image processing applications.

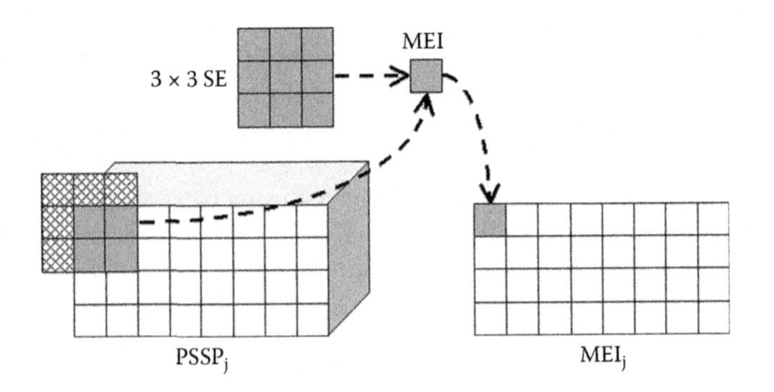

Figure 8.4 Problem of accessing pixels outside the image domain.

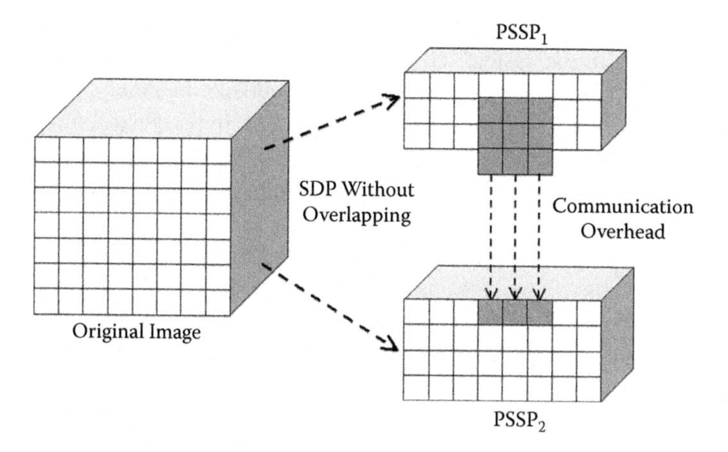

Figure 8.5 Additional communications required when the SE is located around a pixel in the border of a PSSP.

- *Border-handling strategy relative to the pixels out of the domain of the PSSP.* This strategy is applied when the pixel located in a remote processor is required in the calculation of the MEI index associated with another pixel in a given processor (see Figure 8.5). To resolve this issue, we introduce redundant computations to minimize the communication/coordination between processors. In other words, we replicate the information necessary to avoid border effects between different processors, as shown in Figure 8.6.

According to our experiments [11], the cost of processing the information resulting from the procedure above is sensibly inferior to dealing with the overhead

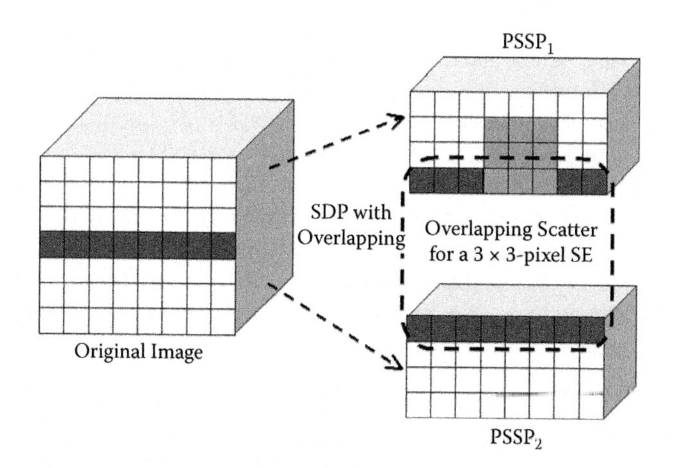

Figure 8.6 Border-handling strategy relative to pixels in the border of a PSSP.

introduced by communication among different processors if no redundant information is introduced in the system. Given the characteristics of the AMC algorithm, which relies on the utilization of an SE of 3×3 pixels iteratively, the number of redundant pixels p_R introduced in the processing of a hyperspectral image is given by $p_R = 2 \times [(2^{\lceil \frac{\log_2 N}{2} \rceil} t) - 1] \times I_R + 2 \times [(2^{\lfloor \frac{\log_2 N}{2} \rfloor}) - 1] \times I_C$, where N is the number of processors, I_R is the number of rows in the original image, and I_C is the number of columns in the original image.

All the algorithms in this section have been implemented in the C++ programming language, using calls to a message passing interface (MPI). Specifically, we used the MPICH-1.2.5 version due the demonstrated flexibility of this version in order to migrate the code to different parallel architectures, which increases code reusability and portability.

8.5.2 Parallelization of Automated Morphological Classification (AMC) Algorithm for Heterogeneous Clusters

As has been mentioned before, spatial/spectral morphological algorithms are particularly suitable for being implemented on heterogeneous architectures. Our efforts in this area have been directed toward the minimization of the execution time of the algorithms in order to provide (near) real-time responses. To describe and calculate the optimal data partitioning and the best data communication scheme for heterogeneous networks of computers, we resort to the HeteroMPI [2] library.

The first step to accomplish the HeteroMPI-based implementation is to define a performance model that is able to capture the data partitioning and communication framework for the heterogeneous platform [24]. Listing 1 shows the most important fragments of the code that describes the adopted performance model, which has six input parameters. Specifically, parameter m specifies the number samples of the data cube, while parameter n specifies the number of lines. Parameters *se_size* and *iter* respectively denote the size of the SE and the number of iterations executed by the algorithm. Finally, parameters p and q indicate the dimensions of the computational processor grid (in columns and rows, respectively) used to map the spatial coordinates of the individual processors within the processor grid layout. Finally, parameter *partition_size* is an array that indicates the size of the local PSSPs (calculated automatically using the computing power of the heterogeneous processors).

It should be noted that some of the definitions have been removed from Listing 1 for simplicity. However, some of the most representative sections are included. The *coord* section defines the mapping of individual abstract processors performing the algorithm onto the grid layout using variables I and J. The node primitive defines the amount of computations that will be made by every processor, which depends on its spatial coordinates in the grid as indicated by I and J and the computing power of the individual processors as indicated by *partition_size*. Finally, the parent directive indicates the spatial localization of the master processor in the grid. An additional link section is used to define the individual communications that every processor carries out based on its position in the grid. Further information on performance model definition is available in [2].

Once a performance model for the parallel algorithm has been defined, implementation using the standard HeteroMPI is quite straightforward [26]. Listing 2 shows the most interesting fragments of the HeteroMPI-based code of our parallel implementation. The HeteroMPI runtime system is initialized using operation HeteroMPI_Init. Then, operation HeteroMPI_Recon updates the estimation of performances of processors. This is followed by the creation of a group of processes using operation HeteroMPI_Group_create. The members of this group then perform the computations of the heterogeneous parallel algorithm using standard MPI mechanisms. This is followed by freeing the group using operation HeteroMPI_Group_free and the finalization of the HeteroMPI run-time system using operation HeteroMPI_Finalize. In the code described above, the benchmark function used to measure the processing power of the processors in HeteroMPI_Recon is essential, mainly because a poor estimation of the power and memory capacity of processors may result in load unbalancing problems. In our implementation, the benchmark function is based on a simple 3×3 morphological computation to calculate the MEI index, as described in Figure 8.3.

Listing 1 Typical performance model for hyperspectral analysis algorithms.

```
algorithm hhaa_ rend(int m, int n, int iter, int p, int q, int partition_size[p * q]) {
  coordI = p, J = q;
  node {I >= 0 && J >= 0: benchmark*(partition_size[I * q + J]*iter);};
  parent[0, 0];
}
```

Listing 2 Parallel implementation of AMC algorithm with HeteroMPI directives to calculate the load balance based on a benchmark function containing the core computations of the method.

```
main(int argc, char *argv[]){
  HeteroMPI_Init(&argc,&argv);
  If(HeteroMPI_Is_member(HMPI_COMM_WORLD_GROUP)){
    HeteroMPI_Recon(benchmark_function,dims,15,&output_p);
  }
  HeteroMPI_Group_create(&gid,&MPC_NetType_hhaa_rend,modelp,num_param);
  If(HeteroMPI_Is_free()){
    HeteroMPI_Group_create(&gid,&MPC_NetType_hhaa_rend,NULL,0);
  }
  if(HeteroMPI_Is_free()){
    HeteroMPI_Finalize(0);
  }
  If(HeteroMPI_Is_member(&gid)){
    HeteroMPI_Group_performances(&gid,speeds);
    //Calculations of the size and communications based on the obtained performance
    //Definition of the best size and distribution of communications
    Read_image(name,image,lin,col,bands,data_type,init);
    for( i = I_max ; i > 1 ; i - - ){
      AMC_algorithm(image,lin,col,bands,sizeofB,res);
    }
    if(HeteroMPI_Is_member(&gid)){
      free(image);
    }
    HeteroMPI_Group_free(&gid);
    HeteroMPI_Finalize(0);
  }
}
```

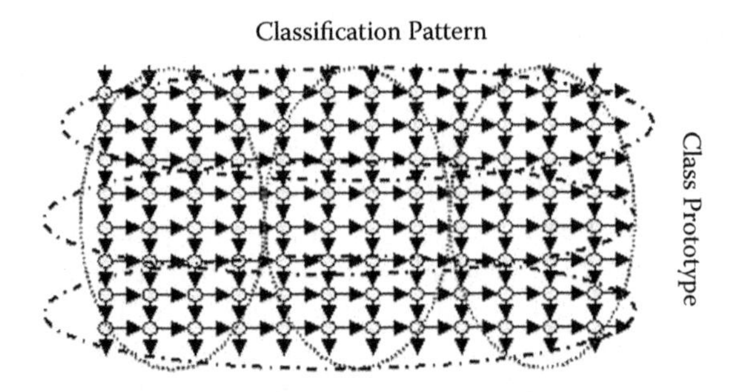

Figure 8.7 Partitioning options for the considered neural algorithm.

8.5.3 Parallelization of the SOM-based Classification Algorithm for Homogeneous Clusters

In order to parallelize the SOM algorithm, we face similar problems to those already addressed for the AMC algorithm in previous subsections. A straightforward approach to parallelization of the neural algorithm is to simply replicate the whole neural network architecture, which is a feasible approach due to the random nature of the initial weights of the network. However, this option results in the need for very complex rules of reduction, and integrity hazards.

Taking into account our previous studies [27] and considering the relatively small size of the training set [28], we can state that the overhead of the neural network is mainly located in the training process (in the form of Euclidean distance calculations and adjustment of weight factors). This fact makes partitioning of the neural network (weight factors matrix) an appealing solution in order to reduce the processing load and time. Again, two main alternatives can be adopted to carry out such partitioning: (1) division by input neurons (endmembers/training patterns); or (2) division by output neurons (class prototypes). The two options are simply illustrated in Figure 8.7.

It should be noted that, in the latter case, the parallelization strategy is very simple. Quite opposite, when the former approach is adopted, there is a need to communicate both calculations and intermediate results among different processors. This introduces an overhead in communications that may significantly slow down the algorithm: According to our preliminary experiments, this option could even result in worse results than those found by the sequential version of the algorithm. On the other hand, the partitioning scheme based on dividing by class prototype only introduces a minor communication overhead, i.e., that created by the need to obtain the winner class. To do so, a protocol similar to $\log_2 N$ synchronizing barriers is adopted. Also, there is a need to introduce a broadcast/all-reduce protocol to obtain the class prototype through local minimum calculations in a batch SOM processing way.

The winner neuron for each pattern needs to be tailored, and subsequent modifications for the weighting factor need to be stored for later addition/subtraction. This approach also allows us to directly obtain the winner neuron at each iteration without the need for any further calculations. It also facilitates a more pleasingly parallel solution, aimed at taking full advantage of the processing power available in the considered parallel architecture while minimizing the communication overhead.

At this point, we must emphasize that the proposed scheme still introduces the need to replicate calculations in order to reduce communications, as was the case with the AMC algorithm. However, the amount of replicated data is limited to the presence of the complete training pattern set at each processor, along with administrative information, i.e., which processor holds the winner neuron, which processor holds the neurons in the neighborhood of the winner neuron, etc. Such administrative information can be used to reduce the communication overhead even further. For instance, using the above information, we consider two implementations of the neighborhood modification function $\sigma(t')$, where the first one is applied when a node is in the neighborhood of the winner neuron and the second is considered when the node is outside the domain of that processor. To assess the integrity of the considered neighborhood function, a look-up table is locally created at each processor so that the value of $\sigma(t)$ is stored for every pair of neurons. While in the present work the function selected is gaussian, i.e., $\sigma(t) = e^{\frac{\|i^* - i\|}{t}}$, other neighborhood functions may also be considered [8].

In any regard, we emphasize that when the neighborhood function is applied to the processor that holds the winner neuron, it is used in a traditional way. On the contrary, when the function is applied to other processors, a modified version is implemented to average the distances with all possible winners. There are two main reasons for this decision: (1) First and foremost, this approach significantly reduces the amount of communications; and (2) it represents a more meaningful and robust neighborhood function [6]. As a final major remark, we must point out that our MPI-based implementation makes use of blocking primities, thus ensuring that all processors are synchronized and preventing integrity problems in the calculations with the matrices of weights $W_{M \times N}$.

8.6 Architectural Outline of an Advanced System for Management of Wildland Fires Using Hyperspectral Imagery

In the present section, we outline the system proposed for the surveillance and management of wildland fires. This system is built on the algorithms and models described in Sections 8.4 and 8.5, and also on the mathematical model for characterization of wildland fires described in Section 8.3. Figure 8.8 provides a flowchart of the proposed design, in which each processing component and the interconnections between them are displayed. It can be seen that the first stage of the flowchart is the extraction of the endmembers of the image that are to be used, first, as inputs to the linear unmixing process; second, as relevant spectra of the classes available on the scene; and, finally, as input of the classification stage that will provide the distribution of fuels, along with any other elements present in the image. The spectra obtained can be compared with information about laboratory and field measurements of specimens (bulk density, humidity, etc.) available in the database. The database gives relevant characteristics

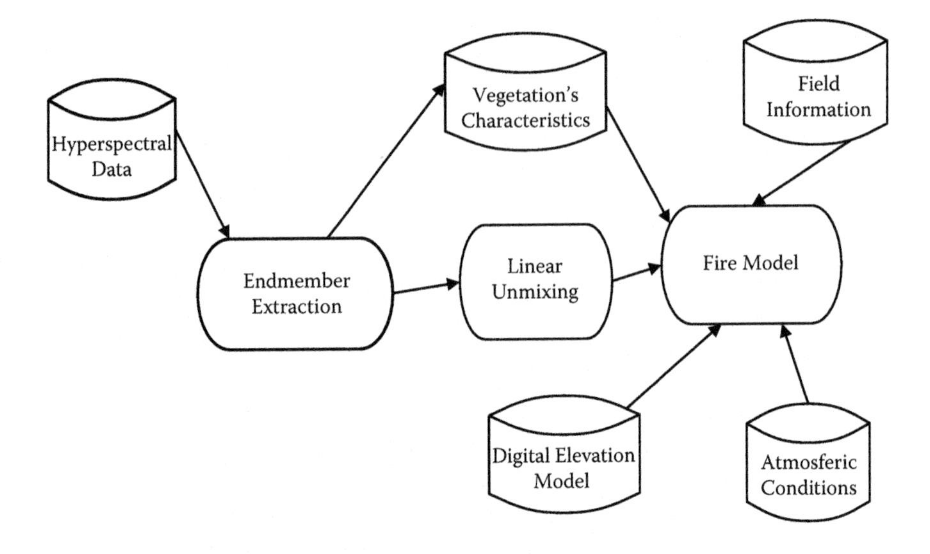

Figure 8.8 Functional diagram of the system design model.

of the vegetation that allows one to change the statical nature of the fuel models, thus providing the mathematical model with more precise data, and increasing its accuracy.

After this stage, the set of endmembers obtained used as a learning pattern for the neural network-based linear unmixing algorithm. In this particular case, we propose the use of a Hopfield neural network to perform the unmixing [6]. This stage will provide the abundance fractions of each endmember at each pixel in the image. In the system, these abundances indicate the precise distribution of fuels inside each pixel in the scene (i.e., mixture of fuelbeds, different stages of senescence for a plant, etc.). These abundances, combined with the experimental laboratory data adjusted to the actual conditions in the studied area, provide us with knowledge to define a particular fire risk index. These abundances may also be combined with distribution maps (such as those produced by the proposed SOM-based classifier), giving important information for the mathematical models to operate.

The model can be improved for realistic modeling of the behavior of the fire by additional information sources; i.e., the slope factor can be calculated using a digital elevation model (DEM) of the area, the wind factor can be obtained from meteorological measures, etc. Further, we emphasize that the proposed system adopts a component-based approach, in which each module can be replaced with different algorithms or enhanced versions, thus providing greater flexibility and accuracy on the particular results obtained.

In the following, we introduce the computational resources in which the proposed system can be implemented, including *homogeneous platforms, heterogeneous platforms*, and *specialized programmable hardware*, listing the advantages and disadvantages of each considered architecture from the viewpoint of their incorporation to the system.

8.6.1 Homogeneous Parallel Platforms

Homogeneous resources are the most widely used high-performance computer architectures. In this category, we can include architectures with shared memory multiprocessors (SMPs), vector processors, and homogeneous clusters.

- *Vector Processors*. The vector processor provides a single control flow with serially executed instructions operating on both vector and scalar operands. The parallelism of this architecture is at the instruction level [2].
- *Shared Memory Multiprocessors* (SMPs). The shared memory multiprocessor architecture consists of a number of identical processors sharing a global main memory. In general, the processors of an SMP computer framework are of the vector or superscalar architecture [2].
- *Homogeneous Clusters*. A computer system of the distributed memory multiprocessor architecture consists of a number of identical processors not sharing global main memory and interconnected via a communication network. The architecture is also called the massively parallel processors (MPP) architecture [2].

In our opinion, the main advantages of homogeneous resources from the viewpoint of their incorporation to a remote sensing data processing system are

- Regularity of computations in the processing and data access leads to a better understanding for new users.
- In the case of vector processors, if the size of the vector register is greater than or equal to the number of bands of the spectral signature, one can achive a very good speedup.
- Due to computer market sales and trends, it is common to buy a large number of computers, making homogeneous clusters a reasonable solution for high-performance computing on a large scale.
- Homogeneous platforms often allow abstractions from more complex questions (most of them at a hardware level), and allow the programmer to concentrate only on the development of applications.

On the other hand, we believe that the main drawbacks of homogeneous platforms for their incorporation to a remote sensing data processing development are

- The SMP and vector processors architectures tend to be very expensive and closed solutions for high-performance computing.
- Due to the price of supercomputers, they are usually only available in large computing facilities.
- The large size of hyperspectral data forces the allocation of great amounts of data, and a bad partitioning policy may result in too many cache misses and inefficient memory usage.

8.6.2 Heterogenous Parallel Platforms

New research trends in high-performance computing tend to the use of heterogeneous computing resources, such as heterogeneous networks and Grid architectures, covered extensively in subsequent chapters of this volume. In the following, we outline the platforms that could be integrated in our proposed system.

- *Heterogeneous networks of computers* (HNOCs). With the commercial availability of networking hardware, it soon became obvious that networked groups of machines distributed among different locations could be used together by one single parallel remote sensing code as a distributed-memory machine. As a result, HNOCs have now become a very popular tool for distributed computing with essentially unbounded sets of machines, in which the number and location of machines may not be explicitly known.
- *Grid architectures*. The grid is a new class of infrastructure. By providing scalable, secure, high-performance mechanisms for discovering and negotiating access to remote resources, the grid promises to make it possible for scientific collaborations to share resources on an unprecedented scale, and for geographically distributed groups to work together in ways that were previously impossible.

In our opinion, the main advantages of heterogeneous platforms for their incorporation to a remote sensing data processing system are

- Distribution of data based on availability processing power, which produce a better utilization of the resources in terms of usage and accessibility for different users.
- Greater supercomputing structures with different functionalities can be created with the use of grid infrastructure, allowing the interconnection of systems geographically distributed and the sharing of functionalities, databases, etc.
- Due to the inclusion of some homogeneous resources, even in the case of having them distributed in several centers or laboratories, one can still make use of the advantages indicated for homogeneous resources locally.

The main disadvantages of this platforms in the context of our considered application are, in our opinion

- The cost is still an issue, even though reusability of components allows integration of existing resources.
- There is a deep need for knowledge of compilers and libraries, although access to large computing infrastructures is often provided at a high level.
- The control of load balance and correct communication between several configurations is, in some situations, not fully defined and correctly exploited by general-purpose libraries, which forces the developers to explicitly expose the parallelism through compiler directives.
- The programming paradigms are generally more complex.

These architectures are especially indicated for embarrassingly parallel applications, which need great processing power and small communication, allowing data scattering in terms of availability of processing resources. This is indeed the case for the AMC, ATGP, and SOM algorithms described in Section 8.5 and also for the mathematical model in Section 8.6.

8.6.3 Programmable Hardware

To conclude this section, we emphasize the importance of specialized hardware for onboard processing as part of any integrated system for remote sensing data analysis. By means of onboard processing, one can drastically reduce the size of the data to be communicated by means of compression (e.g., for real-time surveillance).

In order to consider the design of a specialized electronic system for remote sensing data processing, there are several possibilities for implementation, each with their own advantages and disadvantages in terms of cost, flexibility, performance, and complexity. On the one hand, the best performance is usually given by full-custom designs. On the other hand, semi-custom designs based on programmable hardware provide a very good compromise in terms of cost-performance ratio. Examples include field-programmable gate arrays (FPGAs) and graphics processing units (GPUs). The application of these specialized hardware platforms for remote sensing data analysis are extensively covered in several chapters of the present volume and therefore are out of the scope of this chapter.

8.7 Experimental Results

In the present section, we illustrate the classification accuracy and parallel performance of the parallel algorithms described in Section 8.5. First, a brief overview of the parallel architectures used in this study is provided. Then, performance data for the parallel algorithm are given and discussed in light of realistic hyperspectral imaging applications.

8.7.1 Parallel Computer Architectures

Two parallel computers have been used to evaluate the computational performance of the morphological algorithm proposed. The first parallel computer used in the study is a Beowulf-type cluster called Thunderhead, located at NASA's Goddard Space Flight Center in Maryland (a detailed system description is available at http://thunderhead.gsfc.nasa.gov). The second architecture used for experiments is available at the Heterogeneous Computing Laboratory (HCL), University College Dublin (UCD). This heterogeneous cluster is composed of 16 heterogeneous nodes of two types (Xeon and CSUltra), running Fedora Core Linux and Sun Os operating systems. The MPI implementation is MPICH-1.2.5 with the HeteroMPI library installed. The nodes are interconnected via a 100 Mbit Ethernet communication network with a switch enabling parallel communication among the processors. Although this is a simple configuration, it is also a quite typical and realistic heterogeneous platform as well.

8.7.2 Hyperspectral Data Sets

The hyperspectral scene used for the experiments in this chapter was collected by the NASA Jet Propulsion Laboratory Airborne Visible Infra-Red Imaging Spectrometer (AVIRIS) sensor and is characterized by very high spectral resolution (1939×614 pixels with 224 narrow spectral bands in the range $0.4\mu m - 2.5\mu m$ and moderate spatial resolution 20-m pixels). It was gathered over the Indian Pines test site in Northwestern Indiana, a mixed agricultural/forested area, early in the growing season.

The data set represents a very challenging classification problem. The primary crops of the area, mainly corn and soybeans, were very early in their growth cycle with only about 5% canopy cover. This fact makes most of the scene pixels highly mixed in nature. Discriminating among the major crops under these circumstances can be very difficult, a fact that has made this scene a universal and extensively used benchmark to validate classification accuracy of hyperspectral imaging algorithms [29]. Figure 8.9 shows the spectral band at 587 nm of the original scene. Part of these data, including ground-truth, are available online (from http://dynamo.ecn.purdue.edu/̃biehl/MultiSpec). Apart from the availability of detailed ground-truth information, we have particularly selected this scene because it contains several agricultural classes at different stages of growth and senescence and

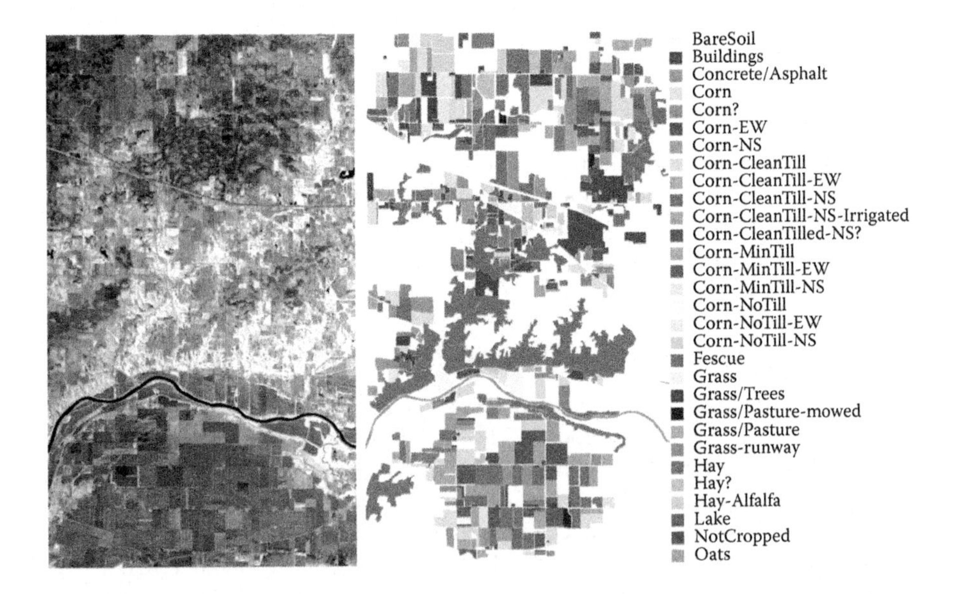

Figure 8.9 (Left) Spectral band at 587 nm wavelength of an AVIRIS scene comprising agricultural and forest features at Indian Pines, Indiana. (Right) Ground-truth map with 30 mutually exclusive land-cover classes.

with variable water content, which makes it a perfect scenario for testing the ability of the algorithms presented in this chapter in the task of generating accurate biomass fuel maps.

8.7.3 Performance Evaluation

Before analyzing its parallel performance, we first briefly discuss the classification accuracy obtained by the AMC algorithm with the different ground-truth classes available for the AVIRIS Indian Pines scene. For this purpose, Table 8.1 shows the

TABLE 8.1 Classification Accuracy Obtained by the Proposed Parallel AMC Algorithm for Each Ground-Truth Class in the AVIRIS Indian Pines Data

Class	Classification Accuracy (%)
BareSoil	98.05
Buildings	30.43
Concrete/Asphalt	96.24
Corn	99.37
Corn?	86.77
Corn-EW	37.01
Corn-NS	91.50
Corn-CleanTill	65.39
Corn-CleanTill-EW	69.88
Corn-CleanTill-NS	71.64
Corn-CleanTill-NS-Irrigated	60.91
Corn-CleanTilled-NS	70.27
Corn-MinTill	79.71
Corn-MinTill-EW	65.51
Corn-MinTill-NS	69.57
Corn-NoTill	87.20
Corn-NoTill-EW	91.25
Corn-NoTill-NS	44.64
Fescue	42.37
Grass	70.15
Grass/Trees	51.30
Grass/Pasture-mowed	79.87
Grass/Pasture	66.40
Grass-runway	60.53
Hay	62.13
Hay?	61.98
Hay-Alfalfa	83.35
Lake	83.41
NotCropped	99.20
Oats	78.04
Road	86.60
Woods	88.89
Overall:	72.35

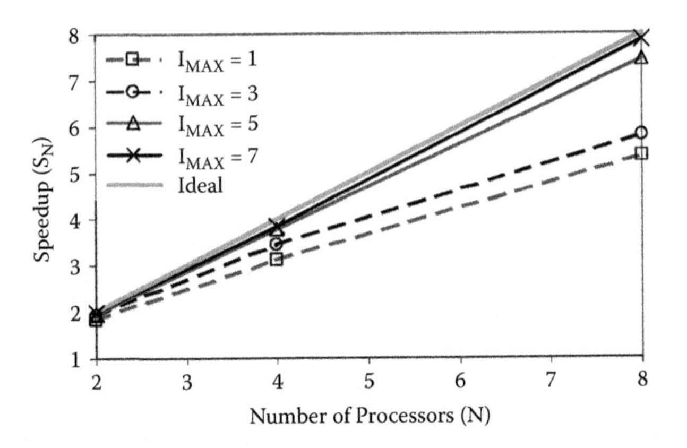

Figure 8.10 Speedups achieved by the parallel AMC algorithm using a limited number of processors on Thunderhead.

overall and individual classification accuracies (in percentages) achieved by the proposed parallel AMC algorithm, using a 3×3 structuring element for the construction of morphological operations and $I_{MAX} = 7$ algorithm iterations (the classification accuracies obtained for less iterations were lower).

As shown by Table 8.1, the lowest classification accuracies were reported for the buildings class, due to the presence of mixed pixels in this area as a result of the coarse spatial resolution of the scene, and for some of the corn classes due to the early growth stage of the crops in these areas, which also results in heavily mixed pixels. Quite opposite, very high classification accuracies were reported for macroscopically pure classes such as BareSoil, Concrete/Asphalt, and Woods. The measured overall accuracy of 72.35% is a very good classification result given the extremely high complexity of the data set, as reported in previous studies [15]. In addition, Table 8.1 also reveals that the results obtained are sufficient to meet the requirements of the system proposed in terms of accuracy, for the generation of biomass maps, and the capacity to discriminate between several vegetation canopies at different stages of growth and senescence.

Figure 8.10 shows the speedup achieved by the parallel AMC algorithm using a limited number of processors on Thunderhead. As we can see in the figure, the speedups achieved were better when the number of iterations (parameter I_{MAX}) were increased, thus revealing that the algorithm scales better when the amount of data to be processed is higher. On the other hand, Figure 8.11(a)–(b) reports the speedup and parallel efficiency achieved by the first three steps of the AMC algorithm (which extract the most relevant endmembers from the Indian Pines scene), using much larger numbers of processors on the Thunderhead system. As we can see, the speedup and parallel efficiency were better when the number of iterations (parameter I_{MAX}) was increased, thus revealing that the algorithm scales better when the amount of data to be processed is higher. On the other hand, Figure 8.11(c) and (d) illustrates the performance of the

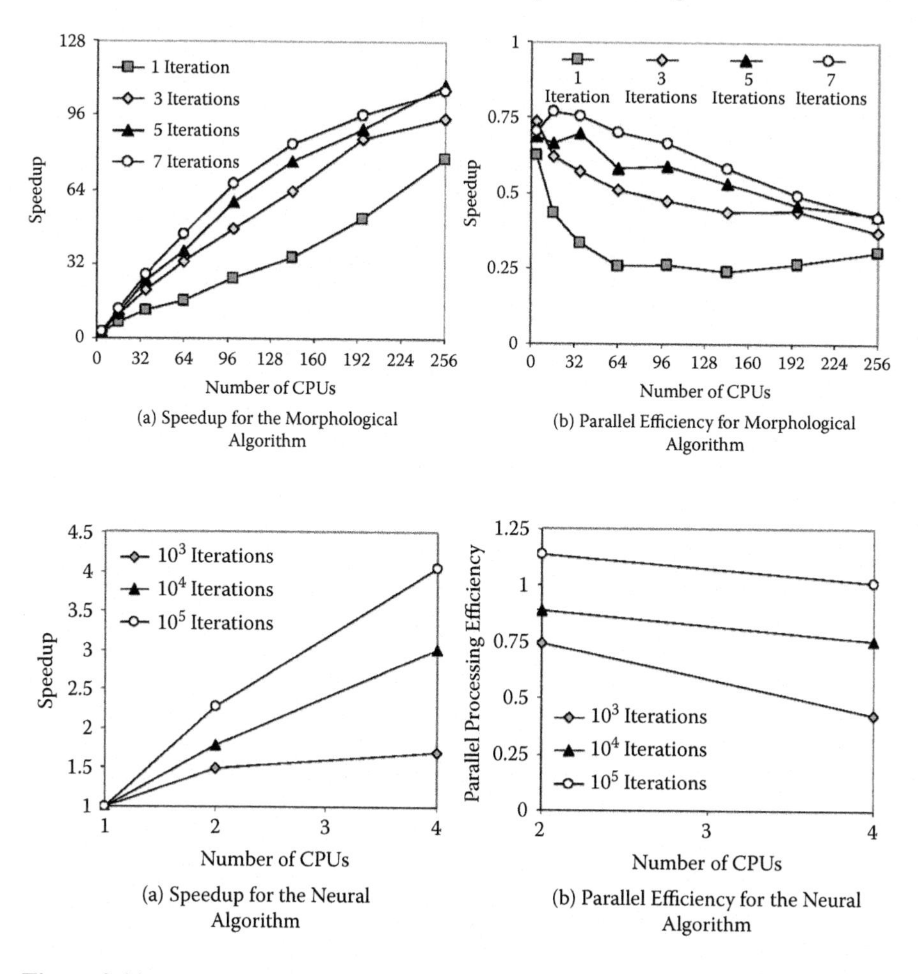

Figure 8.11 Speedups achieved by the parallel SOM-based classification algorithm (using endmembers produced by the first three steps of the AMC algorithm) using a large number of processors on Thunderhead.

parallel SOM-based classification stage (only a maximum of four Thunderhead processors were required to complete all calculations), which also scales well and even results in super-linear scalability. Finally, Figure 8.12 shows performance results for a straightforward master-slave parallelization of the ATGP algorithm, in which the data one partitioned and processed locally at each worker and the targets identified locally at each processor are gathered at the master processor. This implementation suffers from unbalanced communications, which lead to low speedups. Here, the lack of an appropriate benchmark function for dynamic distributions may also be an issue affecting parallel performance.

In order to illustrate the performance of our proposed HeteroMPI-based implementation of the AMC algorithm on the HCL heterogeneous cluster at UCD, Table 8.2

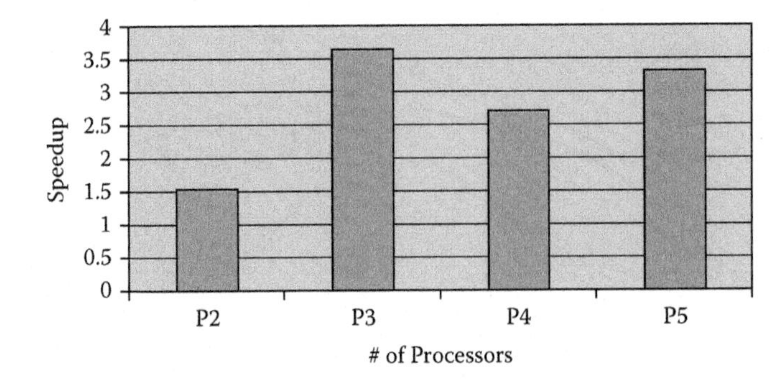

Figure 8.12 Speedups achieved by the parallel ATGP algorithm using a limited number of processors on Thunderhead.

shows the execution times obtained for each heterogeneous processor. The processing times reported in Table 8.2 are well balanced as indicated by Table 8.3, which shows the minima (t_{min}) and maxima (t_{max}) processing times (in seconds) and the load imbalance (defined as $R = \frac{t_{max}}{t_{min}}$).

To conclude this chapter, we provide a preliminary evaluation (in terms of elapsed time) of the different stages in order to evaluate the global response time that we expect for our proposed fire tracking/surveillance system. We emphasize that the fire model used in our experimentation is the one available in the Firelib software package [22],

TABLE 8.2 Execution Times (Seconds) of the HeteroMPI-Based Parallel Version of AMC Algorithm on the Different Heterogeneous Processors of the HCL Cluster

I_{MAX}	1	2	3	4	5	6	7
0	45.37	92.81	135.15	180.23	226.41	266.37	311.95
1	43.31	92.63	135.42	180.15	225.79	279.03	315.19
2	46.84	92.15	134.23	178.81	226.79	272.02	314.95
3	45.73	92.16	139.06	185.82	226.41	271.77	329.92
4	45.55	92.79	137.15	184.96	221.29	265.24	318.17
5	44.93	93.65	135.11	185.14	224.56	274.07	323.19
6	46.86	90.55	135.79	180.23	228.80	278.85	315.95
7	46.38	91.82	137.82	187.15	229.43	274.45	325.80
8	54.57	107.41	161.06	204.08	268.01	314.47	357.69
9	54.90	108.67	158.50	201.71	266.86	315.72	350.59
10	54.43	105.76	158.10	201.46	262.69	315.14	356.83
11	53.28	105.94	158.48	199.51	262.71	315.30	349.34
12	52.49	106.81	157.97	198.88	259.77	311.86	346.56
13	49.98	102.27	158.63	193.45	257.12	308.38	347.03
14	50.10	87.64	158.68	196.05	250.73	308.77	328.95

TABLE 8.3 Minima (t_{min}) and Maxima (t_{max}) Processor Run-Times (In Seconds) and Load Imbalance (R) of the Heterompi-Based Implementation of AMC Algorithm on the HCL Cluster

I_{MAX}	2	3	4	5	6	7
t_{max}	54.90	108.67	161.06	204.08	268.01	315.72	357.69
t_{min}	43.31	87.64	134.23	178.81	221.29	265.24	311.95
$R = \frac{t_{max}}{t_{min}}$	1.26	1.23	1.19	1.14	1.21	1.19	1.13

with the inclusion of the results from previous hyperspectral analysis stages in order to refine and complement the static nature of the method.

Specifically, the execution of the Firelib software [22] (software based on the Rothermel model described in Section 8.6) resulted in an average elapsed time of 60 seconds for an image grid of a size similar to the hyperspectral scene considered in Figure 8.9. The execution time for the parallel AMC method is 18 seconds when 256 processors are used on Thunderhead (it should be noted that the HeteroMPI-based version of AMC took 358 seconds to be executed on the HCL heterogeneous cluster with 16 processors). On the other hand, the proposed parallel SOM-based classification algorithm resulted in an execution time of 560 seconds on Thunderhead. This algorithm employs the endmembers provided by the parallel AMC algorithm as input. Therefore, the estimated execution time of the whole system (including the parallel version of Firelib and the combined AMC/SOM classification), which would be able to accurately characterize the fire risk, is around 660 seconds. This response time can be further reduced by incorporating specialized hardware implementations of AMC and SOM on GPUs and reconfigurable hardware architectures, as indicated by subsequent chapters.

8.8 Conclusions

In this chapter, we have outlined the preliminary design of a fire tracking/surveillance system based on the integration of remotely sensed hyperspectral imagery and advanced processing algorithms implemented on a parallel computing infrastructure. The system is inspired by semi-empirical fire models such as the one proposed by Rothermel. In particular, this chapter conducts a preliminary evaluation on how the proposed parallel processing algorithms can be used to transform the (generally) static nature of fire models into a more dynamic framework, able to incorporate features that are highly relevant to characterize the evolution of wildland fires, such as automatic characterization of water content and distribution of biomass using spectral mixture analysis techniques. Further, we evaluated the performance of the proposed parallel algorithms using high-performance architectures such as homogeneous and heterogeneous supercomputers, and also studied their scalability and parallel efficiency from a computational standpoint. From our experimental assessment, we conclude that the strong real-time restrictions required for a complex system of this kind are still subject to future developments and investigations. Our future research will be directed toward the optimization of the parallel algorithms proposed in this chapter for more efficient execution on distributed platforms and specialized hardware architectures.

8.9 Acknowledgments

The authors thank Drs. Robert Green, Anthony Gualtieri, James Tilton, John Dorband, and Alexey Lastovetsky for many helpful discussions, and also for their collaboration in the development of experiments on the Thunderhead Beowulf cluster and the HCL cluster at University College Dublin.

References

[1] R. C. Rothermel, A mathematical model for Predicting fire spread in wildland fuels, Research Paper INT-115. Ogden, UT: U.S. Department of Agriculture, Forest Service, Intermountain Forest and Range Experiment Station, 1972.

[2] A. Lastovetsky, *Parallel Computing on Heterogeneous Networks*, John Wiley & Sons, Inc., Hoboken, New Jersey, 2003.

[3] R. O. Green et al., Imaging spectroscopy and the airborne visible/infrared imaging spectrometer (AVIRIS), *Remote Sensing of Environment*, vol. 65, pp. 227–248, 1998.

[4] A. Plaza, P. Martinez, R. Perez and J. Plaza, Spatial/spectral endmember extraction by multidimensional morphological operations, *IEEE Transactions on Geoscience and Remote Sensing,* vol. 40, no. 9, pp. 2025–2041, Sept. 2002.

[5] T. Kohonen, *Self-Organizing Maps,* Springer, Berlin, Heidelberg, 1995. (Second Edition 1997).

[6] P. Martinez, P. L. Aguilar, R. Perez and A. Plaza, Systolic SOM neural network for hyperspectral image classification. *Neural Networks and Systolic Array Design,* edited by D. Zhang and S. K. Pal. World Scientific, pp. 26–43, 2002.

[7] A. Plaza, Extended morphological profiles for hyperspectral image analysis on parallel computers, *Neural Information Processing Systems Conference,* Vancouver, 2003.

[8] P. L. Aguilar, A. Plaza, R. M. Perez, P. Martinez, Morphological endmember identification and its systolic array design. In *Neural Networks and Systolic Array Design,* pp. 47–69, 2002.

[9] G. Aloisio and M. Cafaro, A dynamic earth observation system, *Parallel Computing,* vol. 29, pp. 1357–1362, 2003.

[10] C.-I Chang, *Hyperspectral Imaging: Spectral Detection and Classification,* Kluwer Academic/Plenum Publishers, New York, 2003.

[11] A. Plaza, D. Valencia, J. Plaza and C.-I Chang, Parallel implementation of endmember extraction algorithms from hyperspectral data. *IEEE Geoscience and Remote Sensing Letters,* vol. 3, no. 3, pp. 334–338, 2006.

[12] P. Soille, *Morphological Image Analysis, Principles and Applications (2nd ed.),* Springer, Berlin, 2003.

[13] R. Brightwell, L. A. Fisk, D. S. Greenberg, T. Hudson, M. Levenhagen, A. B. Maccabe and R. Riesen, Massively parallel computing using commodity components, *Parallel Computing,* vol. 26, pp. 243–266, 2000.

[14] N. Keshava and J. F. Mustard, Spectral unmixing, *IEEE Signal Processing Magazine,* vol. 19, pp. 44–57, 2002.

[15] A. Plaza, *Development, validation and testing of a new morphological method for hyperspectral image analysis that integrates spatial and spectral information,* Ph.D. Dissertation, Computer Science Department, University of Extremadura, Spain, 2002.

[16] ENVISAT's MERIS webpage: http://envisat.esa.int/instruments/meris/.

[17] Spanish National Association of Forest Entreprises webpage: http://www.asemfo.org.

[18] GISCA webpage: http://www.gisca.adelaide.edu.au/.

[19] Canadian Fire model webpage: http://www.firegrowthmodel.com/index.cfm.

[20] D. D. Evans, R. G. Rehm, E. S. Baker, E. G. McPherson and J. B. Wallace, Physics-based modeling of community fires. *International Interflam Conference, 10th Proceedings,* vol. 2, 2004.

[21] C.-I. Chang and A. Plaza, A fast iterative implementation of the pixel purity index algorithm. *IEEE Geoscience and Remote Sensing Letters,* vol. 3, pp. 63–67, 2006.

[22] Public Domain Sotfware for the Wildland Fire Community webpage: http://www.fire.org.

[23] J. Setoain, C. Tenllado, M. Prieto, D. Valencia, A. Plaza and J. Plaza, Parallel hyperspectral image processing on commodity grahics hardware, *Proceedings of International Conference on Parallel Processing (ICPP'2006),* Columbus, OH, 2006.

[24] D. Valencia and A. Plaza, *FPGA-Based Compression of Hyperspectral Imagery Using Spectral Unmixing and the Pixel Purity Index Algorithm,* vol. 3993, Lecture Notes in Computer Science, Springer, New York, pp. 24–31, 2006.

[25] D. Valencia, A. Plaza, P. Martinez and J. Plaza, On the use of cluster computing architectures for implementation of hyperspectral analysis algorithms, *Proceedings of the 10th IEEE Symposium on Computers and Commuications (ISCC),* Cartagena, Spain, pp. 995–1000, 2005.

[26] D. Valencia, A. Lastovetsky and A. Plaza, Design and implementation of a parallel heterogeneous algorithm for hyperspectral image analysis using HeteroMPI, *Proceedings of 5th International Symposium on Parallel and Distributed Computing (ISPDC),* Timisoara, Romania, 2006.

[27] J. Plaza, R. Perez, A. Plaza, P. Martinez and D. Valencia. Parallel morphological/neural classification of remote sensing images using fully heterogeneous and homogeneous commodity clusters, *Proceedings of IEEE International Conference on Cluster Computing (Cluster'2006),* Barcelona, Spain, 2006.

[28] T. D. Hamalainen, Parallel implementation of self-organizing maps. In *Self-Organizing Neural Networks: Recent Advances and Applications,* U. Seiffert and L. C. Jain, Eds. Springer, Berlin, pp. 245–278, 2002.

[29] J. A. Gualtieri, Hyperspectral analysis, the support vector machine, and land and benthic habitats, *IEEE Workshop on Advances in Techniques for Analysis of Remotely Sensed Data,* pp. 354–364, 2003.

Chapter 9

An Introduction to Grids for Remote Sensing Applications

Craig A. Lee,
The Aerospace Corporation

Contents

This chapter introduces fundamental concepts for grid computing, Web services, and service-oriented architectures. We briefly discuss the drawbacks of previous approaches to distributed computing and how they are addressed by service architectures. After presenting the basic service architecture components, we discuss current Web service implementations, and how grid services are built on top of them to enable the design, deployment,.and management of large, distributed systems. After discussing best practices and emerging standards for grid infrastructure, we also discuss some end-user tools. We conclude with a short review of some scientific grid projects whose science goals are directly relevant to remote sensing.

9.1 Introduction

The concept of distributed computing has been around since the development of networks and many computers could interact. The current notion of *grid computing*, however, as a field of distributed computing, has been enabled by the pervasive availability of these devices and the resources they represent. In much the same way that the World Wide Web has made it easy to distribute Web content, even to PDAs and cell phones, and engage in user-oriented interactions, grid computing endeavors to make distributed computing resources easy to utilize for the spectrum of application domains [26].

Managing distributed resources for any computational purpose, however, is much more difficult than simply serving Web content. In general, grids and grid users require information and monitoring services to know what machines are available, what current loads are, and where faults have occurred. Grids also require scheduling capabilities, job submission tools, support for data movement between sites, and notification for job status and results. When managing sets of resources, the user may need workflow management tools. When managing such tasks across administrative domains, a strong security model is critical to authenticate user identities and enforce authorization policies.

With such fundamental capabilities in place, it is possible to support many different styles of distributed computing. Data grids will be used to manage access to massive data stores. Compute grids will connect supercomputing installations to allow coupled scientific models to be run. Task farming systems, such as SETI@Home [29], and Entropia [9], will be able to transparently distribute independent tasks across thousands of hosts. Supercomputers and databases will be integrated with cell phones to allow seamless interaction.

This level of *managed resource sharing* will enable *resource virtualization*. That is to say, computing tasks and computing resources *will not have to be hard-wired in a fixed configuration to fixed machines to support a particular computing goal.* Such flexibility will also support the dynamic construction of groups of resources and institutions into *virtual organizations*.

This flexibility and wide applicability to the scientific computing domain means that grids have clear relevance to *remote sensing* applications. From a computational viewpoint, remote sensing could have a broad interpretation to mean both on-orbit sensors that are remote from the natural phenomena being measured, and in-situ sensors that could be in a sensor web. In both cases, the sensors are remote from the main computational infrastructure that is used to acquire, disseminate, process, and understand the data.

This chapter seeks to introduce grid computing technology in preparation for the chapters to follow. We will briefly review previous approaches to distributed computing before introducing the concept of *service architectures*. We then introduce current Web and grid service standards, along with some end-user tools for building grid applications. This is followed by a short survey of current grid infrastructure and science projects relevant to remote sensing. We conclude with a discussion of future directions.

9.2 Previous Approaches

The origins of the current grid computing approach can be traced to the late 1980's and early 1990's and the tremendous amounts of research being done on parallel programming and distributed systems. Parallel computers in a variety of architectures had become commercially available, and networking hardware and software were becoming more widely deployed. To effectively program these new parallel machines, a long list of parallel programming languages and tools were being developed and evaluated to support both *shared-memory* and *distributed-memory machines* [30]. With the commercial availability of networking hardware, it soon became obvious that networked groups of machines could be used together by one parallel code as a distributed-memory machine. NOWs (network of workstations) became widely used for parallel computation. Such efforts gave rise to the notion of *cluster computing*, where commodity processors are connected with commodity networks. Dedicated networks with private IP addresses are used to support parallel communication, access to files, and booting the operating system on all cluster nodes from a single OS "image" file. A special, front-end machine typically provides the public interface.

Of course, networks were originally designed and built to connect heterogeneous sets of machines. Indeed, the field of *distributed computing* deals with essentially unbounded sets of machines where the number and location of machines may not be explicitly known. This is, in fact, a fundamental difference between clusters and grids, as a distributed computing infrastructure. In a cluster, the number and location of nodes are known and relatively fixed, whereas in a grid, this information may be relatively dynamic and have to be discovered at run-time. Indeed, early distributed computing focused on basic capabilities such as algorithms for consensus, synchronization, and distributed termination detection, using whatever programming models were available.

At that time, systems such as the Distributed Computing Environment (DCE) [44] were built to facilitate the use of groups of machines, albeit in relatively static, well-defined, closed configurations. DCE used the notion of *cells* of machines in which users could run codes. Different mechanisms were used for inter-cell and intra-cell communication.

The Common Object Request Broker Architecture (CORBA) managed distributed systems by providing an object-oriented, client-side API that could access other objects through an Object Request Broker (ORB) [43]. CORBA is genuinely object-oriented and supports the key object-oriented properties such as encapsulation of state, inheritance, and methods that separate interfaces from implementations. To manage interfaces, CORBA used the notion of an Interface Definition Language (IDL), which could be used to produce *stubs* and *skeletons*. To use a remote object, a client would have to compile-in the required stub. If an object interface changed, the client would have to be recompiled. Interoperability was not initially considered by the CORBA standards and many vendor ORBs were not compatible. Hence, deploying a distributed system of any size required deploying the same ORB everywhere. Interoperability was eventually addressed by the Internet Inter-ORB Protocol (IIOP) [42].

While CORBA provided novel capabilities at the time, some people argued that the CORBA paradigm was not sufficiently loosely coupled to manage open-ended distributed systems. Interfaces were brittle, vendor ORBs were non-interoperable, and no real distinction was made between objects and object instances.

At roughly the same time, the term *metacomputing* was being used to describe the use of aggregate computing resources to address application requirements. Research projects such as Globus [38], Legion [41], Condor [33], and UNICORE [46] were underway and beginning to provide initial capabilities using 'home-grown' implementations. The Globus user, for instance, could use the *Globus Monitoring and Discovery Service (MDS)* to find appropriate hosts. The *Globus Resource Access Manager (GRAM)* client would then contact the *Gatekeeper* to do authentication, and request the local job manager to allocate and create the desired process. The *Globus Access to Secondary Storage (GASS)* (now deprecated) could be used to read remote files. Eventually the term *metacomputing* was replaced by *grid computing* by way of the analogy with the electrical power grid, where power is available everywhere on demand.

While these early grid systems also provided novel capabilities, they nonetheless had design issues, too. Running a grid task using Globus was still oriented toward running a pre-staged binary identified by a known path name. Legion imposed an object model on all aspects of the system and applications regardless of whether it was necessary or not. The experience gained with these systems, however, was useful since they generated widespread interest and motivated further development.

9.3 The Service-Oriented Architecture Concept

With the rapid growth of interest in grid computing, it quickly became clear that the most widely supported models and best practices needed to be standardized. The lessons learned from these earlier approaches to distributed computing motivated the adoption and development of even more loosely coupled models that required even less a priori information about the computing environment. This approach is generally called *service-oriented architectures*. The fundamental notion is that hosts interact via *services* that can be dynamically *discovered* at run-time.

Of course, this requires some conventions and at least one well-known service to enable the discovery process. This is illustrated in Figure 9.1. First, an available service must *register* itself with a well-known *registry service*. A client can then query the registry to find appropriate services that match the desired criteria. One or more *service handles* are returned to the client, which may select among them. The service is then invoked with any necessary input data and the results are eventually returned.

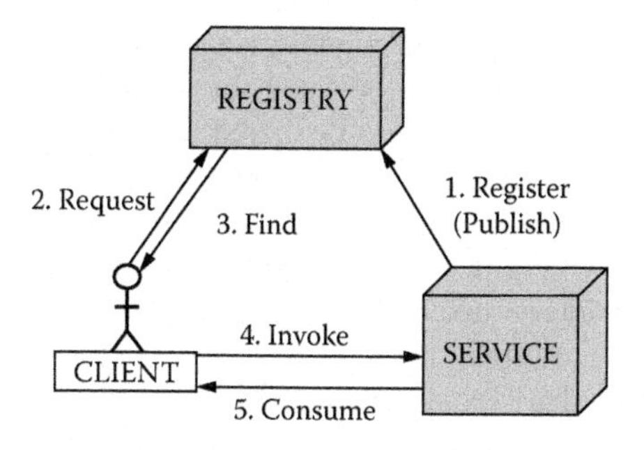

Figure 9.1 The service architecture concept.

This fundamental interaction identifies several key components that must all be clearly defined to make this all work:

- **Representation.** Service-related network messages and services must observe some common, base-level representation. In many cases, the eXtensible Markup Language (XML) is used.
- **Transport Protocol.** A network transport protocol is necessary for simply transferring service-related messages between the source and destination.
- **Interaction Protocol.** Each service invocation must observe a protocol whereby the requester makes the request, the service acknowledges the request, and eventually returns the results, assuming no failures occur.
- **Service Description.** For each service that is available, there needs to be an on-line service description that captures all relevant properties of the service, e.g., the service name, what input data are required, what type of output data is produced, the running time, the service cost, etc.
- **Service Discovery.** In a loosely coupled, open-ended environment, it is untenable that every client must know a priori of every service it needs to use. Both client and service hosts may change for a variety of reasons, and being able to automatically discover and reconfigure the interactions is a key capability.

The clear benefit of this architectural concept is that service selection, location, and execution *do not have to be hard-wired.* The proper representations, descriptions, and protocols enable services to be potentially hosted in multiple locations, discovered and utilized as necessary. Service discovery enables a system to improve flexibility and fault tolerance by dynamically finding service providers when necessary. Of course, a service does not have to be looked up every time it is used. Once discovered, a service handle could be cached and used many times. It is also possible to *compose* multiple services into a single, larger, composite service.

Another extremely important concept is that service architectures enable the *management of shared resources.* We use the term *resource*, in its most general sense, to mean all manners of machines, processes, networks, bandwidth, routing, files, data, databases, instruments, sensors, signals, events, subsystems comprised of sets of resources, etc. Such resources can be made accessible as services in a distributed environment, which also means they can be shared among multiple clients. With multiple clients potentially competing for a shared resource, there must be well-defined security models and policies in place to establish client identity and determine who, for example, gets to use certain machines, what services or processes they can run there, who gets to read/write databases containing sensitive information, who gets notified about available information, etc.

Being able to manage shared resources means that we can *virtualize* those resources. Systems that are 'stovepiped' essentially have system functions that are all hard-wired to specific machines, and the interactions among those machines are also hard-wired. Hence, it is very difficult to modify or extend those systems to interact, or interoperate, with systems that they were not designed for. If specific system functions are not hard-wired to certain machines, and can be scheduled on different hosts while still functioning within the larger system, we have virtualized that system function. Hence, if system functions are addressed by *logical name*, they can be used regardless of what physical machine they are running on.

Note that data can also be virtualized. Data storage systems can become 'global' if the file name space is mapped across multiple physical locations. Again, by using a logical name, a file can be read or written without the client having to know where the file is physically stored. In fact, files could be striped or replicated across multiple locations to enable improved performance and reliability.

9.4 Current Approaches

This tremendous flexibility of virtualization is enabled by the notion of a service architecture, which is clearly embodied by design in *Web services*. As we shall see, each of the key components are represented. While the basic Web services stack provides the fundamental mechanism for discovery and client-server interaction, there are many larger issues associated with managing sets of distributed resources that are not addressed. Such distributed resource management, however, was the original focus of grid computing to support large-scale scientific computations. After introducing the Web services stack, we shall look at how grid services build on top of them, and some of the current and emerging end-user grid tools.

9.4.1 Web Services

Web services are not to be confused with Web pages, Web forms, or using a Web browser. Web services essentially define how a client and service can interact with a minimum of a priori information. The following Web service standards provide the necessary key capabilities [47]:

- **XML (eXtensible Markup Language).** XML is a common representation using content-oriented markup symbols that has gained wide-spread use in many applications. Besides providing basic structuring for attribute values, XML namespaces and schemas can be defined for specific applications.

- **HTTP (Hyper Text Transport Protocol).** HTTP is the transport protocol developed for the World Wide Web to move structured data from point A to B. While its use for Web services is not mandatory, it is nonetheless widely used.
- **SOAP (Simple Object Access Protocol).** SOAP provides a request-reply interaction protocol with an XML-based message format. At the top level, SOAP messages consist of an envelope with delivery information, a header, and a message body with processing instructions. (We note that while SOAP was originally an acronym, it no longer is since technically it is not used for objects.)
- **WSDL (Web Services Description Language).** WSDL provides an XML-based service interface description format for interfaces, attributes, and other properties.
- **UDDI (Universal Description, Discovery and Integration).** UDDI provides a platform-independent framework for publishing and discovering services. A WSDL service description may be published as part of a service's registration that is provided to the client upon look-up.

9.4.2 Grid Architectures

While Web services were originally motivated to provide basic client-server discovery and interaction, grids were originally motivated by the need to manage groups of machines for scientific computation. Hence, from the beginning, the grid community was concerned about issues such as the scheduling of resources, advance reservations (scheduling resources ahead of time), co-scheduling (scheduling sets of resources), workflow management, virtual organizations, and a distributed security model to manage access across multiple administrative domains. Since scientific computing was the primary motivation, there was also an emphasis on *high performance* and managing *massive data.*

With the rapid emergence of Web services to address simple interactions in support of e-commerce, however, it quickly became evident that they would provide a widely accepted, standardized infrastructure on which to build. What Web services did not initially address adequately, however, was *state management, lifetime management,* and *notification.*

State management determines how data, i.e., state, are handled across successive service invocations. Clearly a client may need to have a sequence of related interactions with a remote service. The results of any one particular service invocation may depend on results from the previous invocations. This would suggest that, in general, services may need to be *stateful.* However, a service may be servicing multiple clients with separate invocation streams. Also, successive service invocations may, in fact, involve the interaction of two different services in two different locations. Hence, given these considerations, managing the *context* or *session state* separately from the service, such that the service itself is *stateless* rather than *stateful*, has distinct advantages such as supporting workflow management and fault tolerance.

There are several design choices for how to go about this. A simple avenue is to carry all session states on each service invocation message. This would allow services to be stateless, enabling better fault tolerance because multiple servers could be used since they don't encapsulate any state. However, this approach is only reasonable for applications with small data sets, such as simple e-commerce interactions. For scientific applications where data sets may involve megabytes, gigabytes, or more, this is simply not scalable.

Another approach is to use a *service factory* to create a *transient service instance* that manages all states relevant to a particular client and invocation context. After the initial call, a new service handle is returned to the client that is used for all subsequent interactions with the transient service. While this approach may be somewhat more scalable, it means that all data are hidden in the service instance.

WS-Context [48] is yet another approach where explicit *context structures* can be defined that capture all relevant context and session information for a set of related services and calls. While context structures can be passed by value (as part of a message), they can also be referenced by a URI (Uniform Resource Identifier) and accessed through a separate *Context Service* that manages a store of context structures. Contexts are created with a *begin* command and destroyed with a *complete* command. A *timeout* value can also be set for a context.

A fourth, similar, approach is the Web Services Resource Framework (WSRF) [12], where all relevant data, local or remote, can be managed as *resources*. Resources are accessed through a WS-Resource-qualified endpoint reference that is essentially a 'network-wide' pointer to a WS-Resource. Such endpoints may be returned as a reply to a service request, returned from a query to a service registry, or from a request to a resource factory to create a new WS-Resource. In fact, WSRF does not define specific mechanisms for creating resources. Nonetheless, the lifetime of a resource can be explicitly managed. A resource may be immediately destroyed with an explicit *destroy* request message, or through a *scheduled* destruction at a future time.

Equally important as the management of states is the management of services themselves. Services can be manually installed on particular hosts and registered with a registry, but this can become untenable for even moderately sized systems. If a process or host should crash, identifying the problem and manually rebooting the system can be tedious. Hence, there is clearly a need to be able to dynamically install, boot, and terminate services. For this reason, the concept of *service containers* was developed. Services are typically deployed within a *container* and have a specific *time-to-live*. If a service's time-to-live is not occasionally extended, it will eventually be terminated by the container, thereby reclaiming the resources (host memory and cycles) for other purposes without having to have a distributed garbage collection mechanism.

Event notification is an essential part of any distributed system [19]. Events can be considered to be simply small messages that have the semantics of being delivered and acted up on as quickly as possible. Hence, events are commonly used to asynchronously signal any system occurrences that have a time-sensitive nature. Simple, atomic events can be used to represent occurrences such as process completion, failure, or heartbeats during execution. Events could also be represented by attribute-value pairs with associated metadata, such as changes in a sensor value that exceeds some threshold. Events could also have highly structured, compound attributes, such as the 'interest regions' in a distributed simulation. (Interest regions can be used by simulated entities to 'advertise' what types of simulated events are of interest.)

Regardless of the specific representation, events have producers and consumers. In simple systems, event producers and consumers may be explicitly known to one another and be connected point-to-point. Many event systems offer *topic-oriented publish/subscribe* where producers (publishers) and consumers (subscribers) use a *named channel* to distribute events related to a well-known topic. In contrast, *content-oriented publish/subscribe* delivers events by matching an event's content (attributes and values) to content-based subscriptions posted by consumers.

In the context of grid service architectures, WSRF uses WS-Notification [28], which supports event publishers, consumers, topics, subscriptions, etc., for XML-based messages. Besides defining *NotificationProducers* and *NotificationConsumers* that can exchange events, WS-Notification also supports the notion of subscriptions to WSRF *Resource Properties*. That is to say, a client can subscribe to a remote (data) resource and be notified when the resource value changes. In conjunction with WS-Notification, WS-Topics presents XML descriptions of topics and associated meta-data, while WS-BrokeredNotification defines an intermediary service to manage subscriptions.

Along side all of these capabilities is an integral security model. This is critical in a distributed computing infrastructure that may span multiple administrative domains. Security requires that *authentication, authorization, privacy, data integrity*, and *non-repudiation* be enforced. Authentication establishes a user's identity, while authorization establishes what they can do. Privacy ensures that data cannot be seen and understood by unauthorized parties, while data integrity ensures that data cannot be maliciously altered even though it may be seen (regardless of whether it is understood). Non-repudiation between two partners to a transaction means that neither partner can later deny that the transaction took place.

These capabilities are commonly provided by the Grid Security Infrastructure (GSI) [40]. GSI uses public/private key cryptography rather than simply passwords. User A's public key can be widely distributed. Any User B can use this public key to encrypt a message to User A. Only User A can decrypt the message using the private key. In essence, public/private keys make it possible to digitally 'sign' information. GSI uses this concept to build a *certificate* that establishes a *user's identity*. A GSI certificate has a subject name (user or object), the public key associated with that subject, the name of the Certificate Authority (CA) that signed the certificate certifying that the public key and identity belong to the subject, and the digital signature of the named CA. GSI also provides the capability to *delegate trust* to a *proxy* using a *proxy certificate*, thus allowing multiple entities to act on the user's behalf. This, in turn, enables the capability of *single sign-on* where a user only has to 'login once' to be authenticated to all resources that are in use. Using these capabilities, we note that it is possible to securely build *virtual organizations* across physical organizations by establishing one's *grid identity* and *role* within the VO.

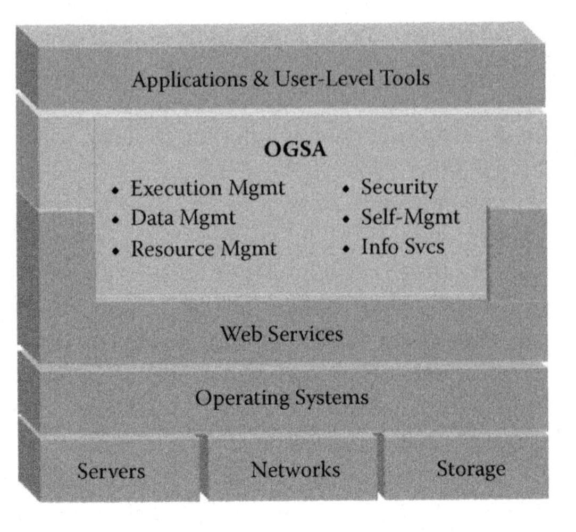

Figure 9.2 The OGSA framework.

The *GridShib* project is extending GSI with work from the Internet2's *Shibboleth* project [24]. Shibboleth is based on the Security Assertion Markup Language (SAML) to exchange attributes between trusted organizations. To use a remote resource, a user authenticates to their home institution, which, in turn, authenticates them to the remote institution based on the user's attributes. GridShib introduces both *push* and *pull modes* for managing the exchange of attributes and certificates.

Up to this point, we have presented and discussed basic Web services and the additional fundamental capabilities that essentially extend these into grid services. We now discuss how these capabilities can be combined into a coherent *service architecture*. The key example here is the emerging standard of the *Open Grid Services Architecture (OGSA)* [10, 39]. Figure 9.2 gives a very high-level view of how OGSA interprets basic Web services to present a uniform user interface and service architecture for the management of servers, storage, and networks. OGSA provides the following broad categories of services:

- Execution Management Services
  - Execution Planning
  - Resource Selection and Scheduling
  - Workload Management
- Data Services
  - Storage Management
  - Transport
  - Replica Management
- Resource Management Services
  - Resource Virtualization
  - Reservations
  - Monitoring and Control

- Security Services
  - Authentication and Authorization
  - Virtual Organizations
  - Policy Implementation and Enforcement
- Self-Management Services
  - Service-Level Agreements and Managers
  - Autonomic Behaviors
  - Configuration Management
- Information Services
  - Resource Discovery
  - Status Monitoring
  - Event Notification

The Globus Toolkit 4 (GT4) [38] is now based on OGSA, which, in turn, is based on WSRF. Standardization of OGSA is underway in the Global Grid Forum [36]. WSRF is also undergoing standardization. We note that while WSRF came out of the GGF OGSA Working Group, it was sent through the standardization process of the Organization for the Advancement of Structured Information Standards (OASIS). This was done with significant overlap in the technical personnel serving in both working groups to get 'buy-in' from the core Web services community.

Since there is so much interest and activity in Web service standards, it has become a difficult topic since there are so many proposed standards across multiple organizations to address different, narrow technical issues. To alleviate the confusion and bring some coordination to the table, the Standards Development Organizations Collaboration on Networked Resources Management (SCRM) Working Group [11] was started at GGF. SCRM has the charter to develop a 'landscape' document and taxonomy of Web service standards, with the goal of identifying areas of synergy and redundancy, and ultimately promoting consensus in the technical marketplace.

As a case in point, even though Globus is felt by many to be a *de facto* grid standard since it is used by so many projects and organizations, competing camps have emerged motivated by the desire to develop simple standards that can even more quickly promote stability and the widespread adoption of grid technology in the marketplace. This has been characterized as the *minimalist* approach as opposed to the *advanced* approach [31]. *Web Services Interoperability Plus* (WS-I+) is the WS-I Basic Profile (including SOAP and WSDL) [17] plus UDDI, BPEL, WS-ReliableMessaging, and WS-Addressing to provide the most basic capabilities for building grids. The fact that these standards are considered to be more stable can be a significant issue for grid projects that must be operational in the near term.

Besides near-term versus long-term issues, there has also been fragmentation along corporate boundaries. Fortunately, however, in March 2006, Hewlett-Packard, IBM, Intel, and Microsoft announced their intent to converge some of the competing Web standards [18]. At the risk of over-simplification, this effort will attempt to merge WS-Management and WS-Distributed Management, WS-Notification and WS-Eventing, and refactor services for managing resources and metadata. While no time table was announced for this convergence, the need for a commonly accepted infrastructure is clear to all.

This convergence is actually indicative of the larger convergence between Web and grid services. While the Web and grid service concepts had different motivations, there is clearly extensive overlap in the fundamental capabilities they are addressing. Hence, it is natural that there should be convergence. Whether the terms 'Web' and 'grid' persist as distinct concepts is less important since there will always be support for the fundamental capabilities of information discovery, remote job management, data movement, etc. Any distinctions will be based on the style of application and how it uses these fundamental capabilities. That is to say, service architectures will enable a wider range of peer-to-peer computing, Internet computing, high-throughput computing, distributed supercomputing, or ubiquitous computing.

9.4.3 End-User Tools

Thus far, we have described the service-oriented architecture concept and current approaches to building grids. While these concepts and standards are intended to enable the construction of enterprise-scale grids, they are generally not intended for use by end-users. Depending on the scope of deployment, grids will be installed by system builders and maintained by administrators. The real benefits of grids, however, will be seen when end-users have simple tools that hide most of the complexity of the infrastructure while enabling the flexible integration of distributed resources [20]. To this end, much work is being done on making the fundamental operations of *file transfer* and *job submission* as easy as possible.

GridFTP. GridFTP [37] is an extension of the established File Transfer Protocol (FTP) that is integrated with the Globus Toolkit. GridFTP can use GSI to manage authentication across different administrative domains. GridFTP can also manage striped transfers where a data file is striped across multiple disks, thereby achieving higher aggregate disk bandwidth. GridFTP can also stripe across multiple network connections, thereby utilizing higher aggregate network bandwidth. Most importantly, the integration with GSI enables GridFTP to manage *secure, third-party transfers*, i.e., Client A can have data transferred between Servers B and C.

GridRPC. GridRPC [22] defines a basic API for Remote Procedure Calls (RPC). Calls are provided to do lookups for desired remote services that are identified by *function handles*. Function handles are then used to make service invocations using data passed essentially as arguments in a remote procedure call. In this context, GridRPC manages the data transfer from the client to server and the return of results. The GridRPC standard was motivated by *network-enabled service* systems, such as NetSolve, Ninf-G, and DIET.

SAGA. The Simple API for Grid Applications (SAGA) Working Group at GGF is endeavoring to define an API that is simple yet will serve the widest segment of grid applications [13, 35]. Data movement is managed using namespaces, physical files and directories, and logical files and replicas. Service execution is managed as synchronous jobs and asynchronous tasks. Interprocess communication is supported using a stream construct. Security is managed as part of a *session* with a *session* *handle*. Different language bindings are possible with C++ and Java implementations underway.

Workflow Engines. With the flexible management of data movement and service invocation, many grid applications are essentially instances of a *workflow*. Hence, grid *workflow engines* have been developed to make it easier to specify and manage these service aggregations. DAGMan (Directed Acyclic Graph Manager) [34] expresses workflows as a DAG and follows dependencies between the tasks to determine the set of ready tasks. GridAnt [4] extends the Ant Java build tool, which manages dependencies in the project build process, to manage task dependencies in a distributed environment. Other systems, such as Triana [32], Kepler [3], and Taverna [25], enable users to graphically construct workflows and then execute them in a variety of environments such as peer-to-peer networks and grids.

Portals. Another approach to making grids easier to use is portals. A Web-based interface can be used that hides as much complexity as possible while providing simple, domain-specific functionality to the end-user. The Grid Portal Toolkit (GridPort) [23] provides client interface tools that enable custom portal development without requiring the user to have any specialized knowledge of the underlying portal technology. GridSphere [1] is an open-source, JSR-168 compliant portal framework that provides the GridPortlets package to support building grid portals. The Open Grid Computing Environments (OGCE) Portal Toolkit [2] integrates portlets from several different projects including some third-party software, such as GridSphere.

9.5 Current Projects

Having introduced the fundamental concepts of Web services and grid architectures and the infrastructure capabilities they will enable, we now illustrate their actual deployment and use in a number of important grid infrastructure projects. More importantly, we then describe just a few science projects relevant to the remote sensing application domain.

9.5.1 Infrastructure Projects

TeraGrid. The Distributed Terascale Facility, or simply *the TeraGrid* [45], is a national grid project funded by the National Science Foundation that started with four core sites: Argonne National Lab, the National Center for Supercomputing Applications, the San Diego Supercomputing Center, and the Caltech Center for Advanced Computing Research. Five other partner sites have since been integrated. Currently an aggregate 102 teraflops and 15 petabytes have been integrated, along with over 100 domain-specific databases. A wide variety of science domains are involved, including high-energy physics, astronomy, computational fluid dynamics, medical therapies, and geology, to name a few. The NSF's new Office of Cyberinfrastructure plans to support even larger peta-scale facilities.

Naregi. The Japanese NaReGI (National Research Grid Initiative) [21] will facilitate collaboration among academia, industry, and government by supporting many virtual organizations with resources across the country. Super-SINET is the NaReGI network infrastructure and provides a 10-Gbps photonic backbone stretching 6000 km. The initial NaReGI testbed will have almost 3000 processors online with an aggregate of nearly 18 teraflops. When fully integrated, NaReGI will have over 7000 processors online. One of the key application areas for these machines is the development of nanoscience simulation applications for the discovery and development of new materials that will lead to the next-generation nano-devices.

UK e-Science Programme. The UK e-Science Programme [27] is supported through a joint collaboration of the UK Research Councils and likewise aims to facilitate joint academic and industrial research. Supported application domains include astronomy, physics, biology, engineering, finance, and health care. The Open Middleware Infrastructure Institute (OMII-UK) maintains an open-source Web service infrastructure and provides comprehensive training for application stakeholders. Besides just building the infrastructure, the e-Science Programme also has a Digital Curation Centre that addresses the preservation of digital documents and data products. In addition to cataloging and making documents available, the DCC captures the provenance of data and guards against the technical obsolescence of storage media.

EGEE. The Enabling Grids for E-SciencE (EGEE) [8] project is primarily funded by the European Union but is comprised of more than 90 institutions in 30 countries, spanning Europe, the Americas, and the Far East. Specific EGEE projects target collaboration with underrepresented regions such as the Baltic nations and Latin America. This wide collaboration is accomplished through an International Grid Trust Federation (IGTF) where EGEE sites trust IGTF root Certificate Authorities. Again, a wide variety of application domains are supported, including astrophysics, biomedicine, chemistry, earth science, fusion, finance, and multimedia, with an Industry Task Force and Industry Forum to facilitate technology transfer. Grid education is also supported for both students and educators with the development of curricula and educational events that will produce future generations of grid professionals.

9.5.2 Scientific Grid Projects

While these large grid projects will provide national and international grid infrastructures and facilitate science discovery and engineering advancement, the immediate goal now is to connect them to remote sensing applications.

DAME. As part of the UK e-Science Programme, the Distributed Aircraft Maintenance Environment (DAME) project endeavored to build a grid-based diagnosis and prognosis system for jet engines [5]. Vibration sensor data from in-service Rolls-Royce jet engines was captured and later downloaded when the aircraft was on the ground. This time series data was processed against historical data by a suite of analysis codes that would identify anomalies and determine failure mode probabilities. A portal-based Signal Data Explorer could be used to interactively examine and process data that were managed and archived using the Storage Resource Broker. The ultimate goal of the project was to process data from multiple aircraft whenever they arrived at an airport and have enough throughput to do an analysis within the aircraft's turn-around time. Clearly, though, with the advent of satellite-based, in-flight Internet access, such sensor data could be streamed in real-time to analysis facilities.

IVOA. The goal of the International Virtual Observatory Alliance (IVOA) [16] is to develop a grid infrastructure for the utilization of international astronomical archives as 'integrated and interoperating virtual observatory.' At least fifteen funded, national virtual observatory projects are collaborating in eight working groups to address issues such as space-time coordinate metadata, uniform access standards for spectra and images, and virtual observatory resource metadata. This collaboration is motivated by the tremendous amount of digital astronomy data being generated by new observatories and sky surveys with instruments that produce two giga-pixel images every few seconds. Using the type of remote sensing, 31 new supermassive black holes were discovered.

SERVOGrid. SERVOGrid is the computational infrastructure for the Solid Earth Research Virtual Observatory (SERVO) project [7] whose goal is the study and prediction of earthquakes. SERVO will integrate many distributed data sources with high-performance computational resources to facilitate discovery in this area. Data will be assimilated from GPS-located sensors in the field that measure 3-D earth surface displacements in the California region. Space-based, interferometric, synthetic aperture radar (InSAR) will also be used to collect geodetic data characterizing geodynamic crustal deformations. Once the data are acquired, hidden Markov models, pattern informatics, finite element models, and other techniques will be used to make and evaluate predictions.

CEOS and GEOSS. The Committee on Earth Observation Satellites (CEOS) (www.ceos.org) is an international coordinating body for civil spaceborne missions for the study of the planet Earth. In support of this goal, CEOS maintains a Working Group on Information Systems and Services (WGISS) with the responsibility to promote the development of interoperable systems for the management of Earth Observation data internationally. Hence, the CEOS/WGISS Grid Task Team is coordinating efforts such as the European Space Agency (ESA) Global Satellite O_3 Measurements project (also supported by EGEE), the National Oceanic and Atmospheric Administration's (NOAA) Operational Model Archive and Distribution System (NOMADS), and the NASA LandSat Data Continuity Mission Grid Prototype. In a closely related effort, the Group on Earth Observations plans to build a Global Earth Observation System of Systems (GEOSS) in 10 years [14]. GEOSS's 10 Year Implementation Plan, which targets 2015 for completion, calls for the development of a global, interoperable geospatial services architecture.

9.6 Future Directions

Grids have come a long way from initial research efforts to build distributed infrastructure prototypes to the current efforts in Web/grid service architectures. Grids still, however, have a tremendous amount of unrealized potential.

Seamless Integration of Resources in a Ubiquitous Service Architecture: Grids will not begin to reach their full potential until (1) there is a core set of grid services, with (2) sufficient reliability, that are (3) widely deployed enough to be usable. At this point, it will be useful to speak of a seamless integration of resources in a service architecture that is just expected to be there. In much the same way that networking research produced what is now known as the Internet and the World Wide Web once the technology became sufficiently reliable and widespread, grid computing and grid services will become more commonplace. Besides various national grid projects, enterprise-scale grids are also being deployed. At some point, some grids will start peering to one another and begin to form a ubiquitous infrastructure. The ability to do resource and data discovery along with resource scheduling and management in a secure, scalable, open-ended environment based on well-known and widely adopted services will enable a wide variety of application domains and styles of computation. Likewise, a wide variety of machines, from wireless PDA/smartphones, to big-iron supercomputers, to embedded devices will be combinable into a flexible spectrum of virtual organizations driven by the needs and requirements of the participants.

Autonomic Behaviors: As soon as we begin to speak about computing infrastructures with this inherent scale, heterogeneity, dynamism, and non-determinism, current system management paradigms begin to break down, making both the infrastructure and applications brittle and insecure. The notion of a system administrator that has complete, or even 'good enough,' information about the system status to be effective is just not reasonable anymore. Hence, to manage large distributed infrastructures, system components will have to become more *autonomous* and require a minimum of human intervention. In general terms, autonomic systems are considered to be *self-configuring, self-managing, self-optimizing*, and *self-healing*. To express this a little more concretely, autonomic systems must be able to follow the autonomic control cycle of *monitor, analyze, plan*, and *act*. This means that systems must be able to monitor their environment and events, analyze and understand their meaning, plan some feasible response to achieve a goal state, and finally act on that plan. Besides requiring the integration of event notification and workflow management, this requires well-known semantics and effective planning (inference) engines. These are fundamental aspects of artificial intelligence that have been worked on for many years.

Non-Technical Barriers to Acceptance: Finally, we wish to emphasize that the *non-technical* issues facing grid computing will be just as important, if not more so, than any technical issues, such as scalability, interoperability, reliability, and security. In many aspects, grid computing is about the *managed sharing of resources* while the 'corporate culture' of many organizations is diametrically opposed to this. Organizational units may jealously guard their machines or data out of a perceived economic or security threat. On a legal level, grid computing may require the redefinition of ownership, copyrights, and licensing. Clearly, as grid computing progresses, such cultural, economic, and legal issues will have to be resolved by adjusting our cultural and economic expectations and our legal statutes to integrate what the technology will provide.

References

[1] The GridSphere Portal Framework. http://www.gridsphere.org.

[2] The OGCE Portal Toolkit. http://www.collab-ogce.org/ogce2.

[3] I. Altintas, C. Berkley, E. Jaeger, M. Jones, B. Lud%oscher, and S. Mock. Kepler: An Extensible System for Design and Execution of Scientific Workflows. In *16th International Conference on Scientific and Statistical Database Management*, 2004.

[4] K. Amin et al. GridAnt: A Client-Controllable Grid Workflow System. In *37th Annual Hawaii International Conference on System Sciences*, January 2004.

[5] J. Austin et al. DAME: Searching Large Data Sets Within a Grid- Enabled Engineering Application. *Proceedings of the IEEE*, pages 496–509, March 2005. Special issue on Grid Computing, M. Parashar and C. Lee, guest editors.

[6] E. Deelman et al. Mapping Abstract Complex Workflows onto Grid Environments. *Journal of Grid Computing, 2003.* 1(1), 2003.

[7] A. Donnellan, J. Parker, G. Fox, M. Pierce, J. Rundle, and D. McLeod. Complexity Computational Environment: Data Assimilation SERVOGrid. In *Proceedings of the 2004 NASA ESTO Conference*, June 2004.

[8] Enabling Grids for E-SciencE. http://www.eu-egee.org.

[9] Entropia, Inc. PC Grid Computing. *http://www.entropia.com.*

[10] I. Foster, C. Kesselman, J. Nick, and S. Tuecke. The Physiology of the Grid: An Open Grid Services Architecture for Distributed Systems Integration. http://www.globus.org/ogsa, January 2002.

[11] Global Grid Forum. Standards Development Organizations Collaboration on Networked Resources Management (SCRM) Working Group. http://forge.gridforum.org/projects/scrm.

[12] Globus Alliance and IBM and HP. Web Services Resource Framework. http://www.globus.org/wsrf, 2004.

[13] Goodale, Jha, Kaiser, Kielmann, Kleijer, von Laszewski, Lee, Merzky, Rajic, and Shalf. SAGA: A Simple API for Grid Applications – High-Level Application Programming on the Grid. *Computational Methods in Science and Technology*, 2006. To appear.

[14] Group on Earth Observations. http://www.earthobservations.org.

[15] The Math Works, Inc. *MATLAB Reference Guide.* 1992.

[16] International Virtual Observatory Alliance. http://www.ivoa.net.

[17] K. Ballinger et al. Basic Profile, v1.0. http://www.ws-i.org/Profiles/BasicProfile-1.0-2004-04-16.html.

[18] K. Cline and others. Towards Converging Web Service Standards for Resources, Events, and Management. http://devresource.hp.com/drc/specifications/wsm/index.jsp.

[19] C. Lee, B. S. Michel, E. Deelman, and J. Blythe. From Event-Driven Workflows Towards a posteriori Computing. In Reinefeld, Laforenza, and Getov, editors, *Future Generation Grids*, pages 3–28. Springer-Verlag, 2006.

[20] C. Lee and D. Talia. Grid Programming Models: Current Tools, Issues and Directions. In Berman, Fox, and Hey, editors, *Grid Computing: Making the Global Infrastructure a Reality*, pages 555–578. Wiley, 2003.

[21] S. Matsuoka et al. Japanese Computational Grid Research Project: NAREGI. *Proceedings of the IEEE*, pages 522–533, March 2005. Special issue on Grid Computing, M. Parashar and C. Lee, guest editors.

[22] Nakada, Matsuoka, Seymour, Dongarra, Lee, and Casanova. A GridRPC Model and API for End-User Applications. *Global Grid Forum Proposed Recommendation GFD.52*, September 26 2005.

[23] National Partnership for Advanced Computational Infrastructure. The Grid Portal Toolkit. http://gridport.npaci.edu.

[24] NCSA and The Globus Project. GridShib. http://gridshib.globus.org.

[25] T. Oinn et al. Taverna: A Tool for the Composition and Enactment of Bioinformatics Workflows. *Bioinformatics Journal*, 20(17):3045–3054, 2004.

[26] M. Parashar and C. Lee. Grid Computing: Introduction and Overview. *Proceedings of the IEEE*, March 2005. Special issue on Grid Computing, M. Parashar and C. Lee, guest editors.

[27] Research Councils UK. The UK e-Science Programme. http://www.rcuk.ac.uk/escience.

[28] S. Graham and others. Publish-Subscribe Notification for Web Services. http://www-106.ibm.com/developerworks/library/ws-pubsub, 2004.

[29] SETI at Home. Search for Extraterrestrial Intelligence. http://setiathome.berkeley.edu.

[30] D. Skillicorn and D. Talia. Models and Languages for Parallel Computation. *ACM Computing Surveys*, 30(2), June 1998.

[31] Steven Newhouse. UK Contributions to OGSA. dedalus.ecs.soton.ac.uk/OGSA-UK-24Apr05.pdf.

[32] I. Taylor, I. Wang, M. Shields, and S. Majithia. Distributed Computing with Triana on the Grid. *Concurrency and Computation:Practice and Experience*, 17(1–18), 2005.

[33] The Condor Project. http://www.cs.wisc.edu/condor.

[34] The Condor Project. DAGMan (Directed Acyclic Graph Manager. http://www.cs.wisc.edu/condor/dagman.

[35] The GGF SAGA Working Group. Simple API for Grid Applications. http://wiki.cct.lsu.edu/saga/space/start.

[36] The Global Grid Forum. www.ggf.org.

[37] The Globus Alliance. GridFTP. http://www.globus.org/grid_software/data/gridftp.php.

[38] The Globus Alliance. The Globus Toolkit. http://www.globus.org.

[39] The Globus Alliance. Towards Open Grid Services Architecture. http://www.globus.org/ogsa.

[40] The Globus Project. Grid Security Infrastructure. www.globus.org/security.

[41] The Legion Project. Legion: Worldwide Virtual Computer. http://legion.virginia.edu.

[42] The Object Management Group. Internet Inter-ORB Protocol. http://www.omg.org/technology/documents/ formal/corba_iiop.htm.

[43] The Object Management Group. CORBA 3 Release Information. http://www.omg.org/technology/corba/ corba3releaseinfo.htm, 2000.

[44] The Open Group. Distributed Computing Environment. http://www.opengroup.org/dce.

[45] The TeraGrid Project. http://www.teragrid.org.

[46] The UNICORE Forum. http://www.unicore.org/forum.htm.

[47] The World Wide Web Consortium. http://www.w3c.org.

[48] The WS-Context Specification. xml.coverpages.org/WS-ContextCD-9904.pdf.

[49] R. Wolski. Dynamically Forecasting Network Performance to Support Dynamic Scheduling Using the Network Weather Service. In *6th High-Performance Distributed Computing Conference*, August 1997.

Chapter 10

Remote Sensing Grids: Architecture and Implementation

Samuel D. Gasster,
The Aerospace Corporation

Craig A. Lee,
The Aerospace Corporation

James W. Palko,
The Aerospace Corporation

Contents

10.1 Introduction

In the last 40 years, remote sensing of the Earth has seen a continuous growth in the capabilities of the instrumentation (satellites, airborne, and ground-based sensors that monitor and measure the environment) that provides the fundamental data sets and an increase in the complexity of the data analyses and modeling that these data sets support. The rate of increase in the remote sensing data volume continues to grow. Additionally, the number of organizations and users is also expanding and is now a worldwide community struggling to share data and resources. These trends necessitate a shift in the way in which remote sensing systems are designed and implemented in order to manage and process these massive data sets and support users worldwide.

Grid computing, as originally described by Foster et al. [1, 2], provides a new and rich paradigm with which to describe and implement various distributed computing system architectures. The promise of grid computing for science users is a shared environment that will facilitate their scientific research and provide them access to an unprecedented range of resources: instrumentation, data, high-performance compute engines, models, and software tools. Ultimately this access will be two-way, one in which the remote sensing scientist will not only receive and process data from remote sensors and instrumentation, but will be able to reconfigure them as well.

In a 2002 *GridToday* [3] article, Ian Foster further clarifies the definition of a *grid* as a system with these three elements: (1) coordination of resources not subject to centralized control; (2) utilizes standard, open, and general-purpose protocols and interfaces; (3) delivery of non-trivial qualities of service. We will show that these elements map nicely into the remote sensing domain.

The goal of this chapter is to apply the grid computing paradigm to the domain of Earth Remote Sensing Systems. These systems involve the collection, processing, and distribution of large amounts of data and often require massive computing resources to generate the data products of interest to users. Current systems, such as the NASA EOS mission, generate hundreds of gigabytes of raw sensor data per day that must be ingested and processed to produce the mission data products. The trend in both scientific and operational[1] weather forecasting is a steady increase in the amount, and types, of data necessary to support all the applications in this problem domain.

10.1.1 Remote Sensing and Sensor Webs

There are many definitions of remote sensing in the literature [4, 5, 6]. The common theme in all of these definitions is the measurement of some physical property associated with the object under observation without being in physical contact with the object

[1]The term *operational* as used here is intended to convey the fact that the capability is required on a continuous and reliable basis and that the performance meets some minimum quality of service requirements. In the case of weather forecasting, the capability to continuously generate weather forecasts with specific latency and accuracy is the key requirement.

(hence the term *remote*). This form of remote sensing generally involves the detection of some form of electromagnetic energy either reflected by, or emitted from, the object (e.g., visible light reflected from the ocean surface or infrared radiation emitted from a cloud top), which is then fed to an inversion algorithm to retrieve specific geophysical parameters such as sea surface temperature or atmospheric vertical moisture profile.

For the sake of the present discussion, we generalize this definition to include not only the traditional concept of *remote sensing* but also that of *in-situ* sensing, where the measurements are made by instrumentation in contact with, or close proximity to, the object of interest. In either case, the results of these measurements are sent to a location *remote* from the source for further processing and distribution. Thus we include a wide range of sensors and instrumentation that perform their measurements either remotely or *in-situ* and then transmit those observations to remote data collection, processing, and distribution sites. We also consider the concept of a *sensor web* as defined by Higgins, et al., in their report to the NASA Earth Science Technology Office (ESTO) [7]:

> *A sensor web is a distributed, organized system of nodes, interconnected by a communications fabric that behaves as a single, coherent instrument. Through the exchange of measurement data and other information, produced and consumed by its sensing and non-sensing nodes, the sensor web dynamically reacts causing subsequent sensor measurements and node information processing states to be appropriately modified to continually ensure optimal science return.*

The sensor web concept considers a highly distributed system that includes feedback between various nodes. Such a concept is consistent with the fundamental notions of grid computing. By combining the sensor web and grid computing paradigms we create what we term a *remote sensing grid* (RSG). This is a highly distributed system that includes resources that support the collection, processing, and utilization of the remote sensing data. Many of the resources are not under a single central control, yet we have the ability to coordinate the activities of any of these resources. It is possible to construct a remote sensing grid using standard, open, protocols and interfaces. Finally, many of the operational remote sensing systems are required to support highly non-trivial quality of service requirements, such as availability.

10.1.2 Remote Sensing System Architectures and Grid Computing

The goal of this chapter is to provide a description of remote sensing grids by combining the concepts of remote sensing or sensor web systems with those of grid computing. In order to do this one needs to understand how remote sensing systems are described and specified, and similarly how grids are described and specified, and how these two approaches are merged to describe and specify remote sensing grids.

We draw on the approach traditionally employed by systems engineering to specify a system as consisting of various elements and subsystems [8]. The remote sensing system is designed to be an organized assembly of resources and procedures united and

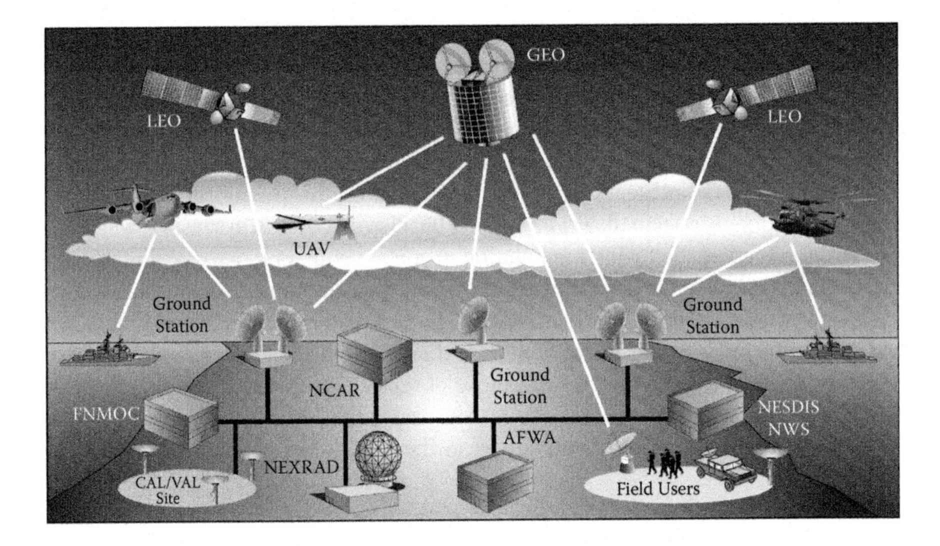

Figure 10.1 High-level architectural view of a remote sensing system.

regulated by interaction or interdependence to accomplish a set of specific functions, which are performed by the various elements and subsystems. The system, as composed of these elements and subsystems, is described by specifying, at various levels of detail, the system architecture. The term *architecture* refers to a formal description of a system, defining its purpose, functions, externally visible properties, and interfaces. It also includes the description of the system's internal components and their relationships, along with the principles governing its design, operation, and evolution.

In order to provide a specific example and context for discussing remote sensing grids, we consider the design of a *notional* weather forecasting and climate science (WFCS) grid. This notional system is motivated by several projects including LEAD [13] and NOMADS [17] (discussed in Section 10.4.1), and The Aerospace High Resolution Forecast Prototype [20]. This system is intended to support weather forecasting and climate studies and is described in more detail in Section 10.2.1.

A high-level view of the WFCS system architecture is shown in Figure 10.1. This figure describes the architecture primarily in terms of the various elements and subsystems, and in later sections of this chapter we will work towards a description in terms of grid computing concepts.

We envision the WFCS system as being made up of resources from a variety of organizations (the notional system is illustrated with examples that include FNMOC, AFWA, NCAR, NESDIS/NWS, and others). The organizations provide specific capabilities necessary to implement such a system, but there may also be additional capabilities required for the realization of a grid architecture. Figure 10.1 illustrates the following elements of the WFCS architecture:

- Observing Elements: in-situ sensors, ground based, airborne, and spaceborne instruments that collect the basic environmental measurements

- Data Management Elements: data transport, storage, archive discovery, and distribution
- Data Processing and Utilization Elements: user applications, modeling and assimilation, forecasting, etc.
- Communications, Command, and Control Elements: the resources that allow all elements to work together; includes interaction and feedback between any of the sensor web elements
- Core Infrastructure: underlying resources needed to tie all elements together, including networks, communication links, etc.

In order to map the system engineering view of the WFCS system into grid computing terms, we need to understand how grid architectures are described. The grid computing community has developed an approach to describing grids known as the Open Grid Services Architecture (OGSA) [9], using the concepts of a service-oriented architecture (SOA). It is therefore prudent to define some of the terminology used in OGSA and SOA. The Global Grid Forum (GGF) and others [10, 11] provide the following definitions:

Service is a software component that can be accessed via a network to provide functionality to a service requester.

Service Oriented Architecture refers to a style of building reliable distributed systems that deliver functionality as services, with the additional emphasis on loose coupling between interacting services.

Workflow is the structured organization of a set of activities, their relative ordering and synchronization, and data needs necessary to accomplish a specific task or goal. The workflow may also specify any necessary participants.

Given these definitions, we see that SOA refers to the design of a system and not how it is implemented. It is also possible to describe, at least in part, various workflows using a service-oriented description. Following Srinivasan and Treadwell [11], we employ SOA as an architectural style that utilizes components as modular services (generally considered to be atomic in that they provide one service) that can be discovered and used to build workflows by clients. We further assume that services have the following characteristics [11]:

Composition — may be used alone or combined with other services.

Communication via Messages — communicates with clients or other services by exchanging messages.

Workflow Participation — services may be aggregated to participate in a specified workflow.

Interaction — services may perform their key functions as stand-alone entities or require interactions with other services to fulfill their functions.

Advertise — services *advertise* their capabilities, interfaces, polices, etc., to clients.

It should also be remarked that SOA reinforces the software engineering principles of encapsulation, modularization, and separation between implementation and interface. Grid computing concepts and frameworks may be used to not only describe the architecture of a remote sensing system, but also to build the specific services required to implement such an architecture.

There are many advantages for remote sensing systems to employ a grid computing approach to system architecture and implementation:

- Cost savings through the sharing of resources, the ability to *grow as you go,* and avoid cost impact and technology obsolescence common with over-provisioning;
- Scalability to meet variations in resource demands and balance work loads;
- Shorter time to results, which allows for the provisioning of extra time and resources to solve problems that were previously unsolvable; this is critical for operational systems that have stringent latency requirements;
- Increased flexibility and resilient operational infrastructures (allows for improved fault tolerance);
- Enable collaboration across organizations and among widely distributed users, by sharing resources (data, software, and hardware);
- More efficient use of available resources; ability to combine resources for systems that are naturally distributed and whose user communities are naturally distributed;
- Increased productivity through standardization and access to additional resources.

We hope that after reading this book the reader will have a better appreciation of these benefits.

The remaining sections of this chapter describe the notional WFCS architecture and implementation, addressing issues relevant to constructing such a grid, providing examples of similar systems in production or research and finally ending with a discussion of various issues related to the adoption of grid computing. Section 10.2 discusses the architecture of our notional example of a remote sensing grid, the WFCS grid. This discussion is framed using the grid computing architectural principles expressed by the OGSA. This allows us to define the architecture and describe the system using a common set of grid services. In Section 10.3 we discuss the implementation of the WFCS using grid technologies that are consistent with the OGSA approach. We discuss the relevant technological, managerial, and policy issues and identify specific current technologies one might consider in building the system described in Section 10.2. Issues such as workflow management and problem solving environments are briefly discussed. Section 10.4 provides a brief discussion of several examples of remote sensing grid projects. Section 10.5 discusses the paths to adoption of grid computing technology for scientific computing and remote sensing applications.

10.2 Remote Sensing Grids: Architecture

In order to discuss remote sensing system architectures, we first define the key requirements that drive these architectures and implementation, then express these architectures using the principles of grid computing (as discussed in Chapter 9). We will use the specifications and standards provided by the OGSA as our basis for describing the *notional* WFCS architecture. It is important to note that OGSA builds upon Web services and the Web Services Resource Framework (WSRF). Many of these underlying technologies, as well as OGSA, are still evolving. That is not to say that they are not sufficiently mature for deployment; rather, one should be aware that the standards and technologies are still evolving.

To define the architecture of the WFCS grid we follow a top-down approach. We present a high-level summary of the mission goals for our notional system, followed by some operational requirements that are derived from the high priority needs of the operational users, and then discuss the use of existing architectural elements, as this is one of the mission goals. We also discuss the needs of the different users of the WFCS grid and provide example workflow scenarios from which we can derive additional requirements. These requirements are mapped into the functional requirements and resource needs of the WFCS grid. From these functional requirements we then define the grid services (following OGSA) necessary to meet these requirements, and, finally, we present the notional WFCS grid architecture using these services.

10.2.1 Weather Forecasting and Climate Science Grid

To define the *notional* WFCS grid architecture we first start with an elucidation of the mission and system requirements for this notional system. We envision building this as a sensor web using a grid service architecture that links a wide array of instrumentation with data storage and computation facilities that enable the operational and scientific workflows necessary to accomplish the WFCS mission. We next translate those requirements into specific functionality and resources necessary and then into the specific grid services that allow construction of the the WFCS grid using an SOA. Given that a key requirement for the WFCS is operational weather forecasting, we also assume the use of existing capabilities at many of the national weather forecasting sites within the United States as elements within the architecture.

In defining the requirements we follow the approach used by many projects, and similar to that employed by the GGF, we use case examples to present a description of the WFCS grid requirements [21].

Summary Description of the WFCS Grid. The WFCS grid is envisioned as a notional sensor web created by combining existing resources from government forecast centers, national laboratories, university research centers, and a wide array of instrumentation, and adding new resources as required, to support operational weather forecasting and climate science research. The *glue* that binds all the elements together and allows for interoperability is the SOA implementation using grid services.

The mission-level requirements (high-level system goals) of the WFCS grid may be stated as:

- Provide operational synoptic and regional/mesoscale weather forecasting[2];
- Provide research scientists an environment to analyze meteorological and climate data in support of their climate science studies;
- Provide an environment for research in the improvement of forecast and climate models;
- Insure that proper security is maintained throughout the system regarding access to all systems, data, and information; all users require authentication and authorization for resource utilization and data access;
- Create a virtual organization to achieve these goals from existing organizations that currently perform the necessary functions, augmenting them as necessary.

The WFCS grid will also be required to meet several operational-level requirements that are primarily intended to address quality of service (QoS) requirements and entail service-level agreements (SLA) between the organizations participating in the WFCS grid. These requirements would include latency and data refresh requirements for the production of weather forecasts; ability to maintain long-term data archives and retrieval; ability to reprocess specific volumes of data faster than real time to support climate research; and automation of as much of the workflow as possible. To achieve the required QoS, it is necessary to define a high-level policy model for the WFCS grid that addresses the high priority and SLA requirements of the operational users and the needs of the research community. These would include:

- Operational weather forecasting workflow has priority
- Climate studies and science users have lower priority
- Some resources are only shared among forecasting services
- Some resources may be shared by both forecasting and climate workflow

In order to meet these high-level requirements, the WFCS will require the types of elements shown in Figure 10.1, which may be categorized as:

Observational Elements — includes all ground, air, and space assets that perform basic measurements of the environmental and geophysical parameters; satellites in LEO or GEO orbits, research aircraft or UAVs, ground sensor networks or webs, etc. These components provide the basic observational data necessary to support both operational weather forecasting and climate research;

[2]Synoptic generally refers to weather phenomena with spatial scales $\geq$ 2000 km and temporal scales of days, and Mesoscale refers to spatial scales less than 2000 km, down to 2 km, and temporal scales from days down to minutes.

Data Management and Distribution Elements —provides data transport and storage, including long-term archives (LTA); provides metadata search engines and tools to map logical data pointers to physical data locations;

Data Processing Elements — includes CPU subsystems (HPC, clusters, etc.), models, and software tools;

Infrastructure Elements — networks (wired and wireless), communication links (satellite-ground, aircraft-ground), system monitoring tools, and security

From these basic requirements we must also derive additional requirements for interfaces and interoperability of all of these elements.

Customers. Since the intent of the WFCS grid is to provide the various user communities with the services and data products they require, it is important to capture their specific needs. These are of course captured at a high level in the overall mission requirements previously mentioned. As this is not intended to be a complete requirements analysis, we focus on the driving requirements for the system architecture and implementation. These users include operational weather forecast centers (e.g., FNMOC, AFWA, NOAA/NWS in the USA or ECMWF in Europe) and a set of science users distributed worldwide (interested in weather forecast model development and climate modeling and research).

We have identified three categories of users:

- Operational Users — These are users from government organizations (NOAA, DoD, NASA, etc.) that have a sustained requirement for timely weather forecasts to support the needs of various civilian agencies and military operations.
- Science Users — These are research scientists from a variety of organizations (NOAA, NASA, ESA, universities, private industry) interested in studying some aspect of the Earth system or weather and climate modeling.
- Commercial Users — These include users that apply the data products and model results for business or commercial purposes. These might include the news media (Web, TV, and radio), agri-business, logging, and shipping (land, air, sea).

Each category of users has specific workflow and data utilization that they need to implement to exploit the remote sensing data collected by the observational resources and models. Operational users are interested in performing timely synoptic and mesoscale weather forecasting, or measuring the tracks of hurricanes in order to predict location and time of landfall; science users may be interested in studying the Earth's climate variation on decadal time scales; commercial users maybe interested in resource monitoring or techniques to provide improved crop yield estimates.

It is not possible to list all user requirements in this chapter, but rather we wish to provide the reader with examples of the types of user requirements that drive the need for grid computing technology to best achieve the desired goals of these users. To do this we draw upon typical examples of user requirements that also provide examples of the needs that different users have now and in the future. We draw upon examples from operational weather forecasting and from various climate research activities. These are presented below in the form of a few representative scenarios.

Scenarios. We now present several examples of remote sensing data processing and analysis workflows for the WFCS grid. These workflows are presented as typical scenarios that describe *what* is done but not *how* it is implemented. The implementation is discussed in Section 10.3 when we consider the actual implementation using grid technology. These examples are intended to be illustrative of the kinds of workflows that we envision for the WFCS grid and have their basis in several existing or proposed projects [22, 23, 24]. The scenarios include:

- Weather Forecast Model (WFM) Runs: Operational and Research
- Climate Data Reprocessing
- Calibration and Validation (Cal/Val) and Data Ingest

Scenario 1: Weather Forecast Model Runs: Operational and Research. This scenario involves running a high-resolution, near-real-time, mesoscale weather forecast model (such as MM5 or Weather Research and Forecasting (WRF)) [25, 26, 27] to produce routine operational weather forecasts for specific geographical regions on a fixed time line. For example, the forecast might be for a region covering southern California, with forecast outputs at (0, 3, 6, 9, 12, 15, 18, 21) *Z* hours.[3] This scenario is motivated by a prototype system developed at The Aerospace Corporation [20] and the NOMADS project [17]. Various organizations currently run various numerical weather prediction (NWP) models: AFWA runs MM5 and WRF, FNMOC runs COAMPS, and both NASA and NOAA/NWS run a variety of synoptic and mesoscale models (and many of the research organizations within NOAA also run forecast models). NASA and NOAA also run a wide range of climate forecast and study models.

This scenario illustrates an example of workflows that are similar but have different constraints. The first, operational weather forecasts, is configured such that the users are only consumers of the output results and have no ability to adjust the processing, and in the second, research weather forecast model runs, the user is given complete control over the parameter space needed to fully specify the computation.

Specifically, this scenario requires the following capabilities:

- NWP Models — Run configurable NWPs covering both synoptic and mesoscale spatial and temporal ranges; various users may require the ability to tailor the model runs to their specific needs, for example, optimizing performance and forecast skill for a given geographical region or types of weather conditions. In addition to the operational NWP runs, we also consider the case where a researcher would like the ability to specify more details regarding the model runs and parameters.

[3] *Z* refers to the reference time zone, which is the same as GMT or currently referred to as UTC.

- Data Assimilation — Includes a variety of data sources (satellite observations, aircraft observations, ground networks, ocean buoys, etc.) that are ingested by the model assimilation system to prepare these diverse inputs for the forecast model; define the boundary and initial conditions for the model run.
- Post-Processing — Provides model output in a variety of standard formats and includes a set of standard analysis and visualization tools.
- Data Archives — The output products are stored in near-line storage for a specified period of time, after which they are purged (these operational forecasts are not stored in LTA as they can be recreated since all input data and models are archived).
- Software Management — Ensure configuration management of all NWP and related software across the participating organizations; maintain both operational and research code branches. Due to the nature of the operational weather forecasts, it is critical to maintain detailed provenance of all software. This requires the creation of a software management working group to review and approve all revisions and version releases. The group would include members from all operational and research organizations involved in developing and running these models.
- Verification and Validation (V&V) — Ability to perform software and model V&V for all NWP models and software tools; perform automated regression testing and operational testing of model quality and performance. This may require the capability to archive operational forecasts as workflow instructions for retrospect analyses.
- Ensemble Forecasts — Most NWPs runs are deterministic, however, it is often useful to run a set of multiple NWP runs that vary the initial and boundary conditions and the model physics to generate a set of predictions. This set of predictions is then analyzed to assess, for example, the ensemble variance and evaluate the forecast uncertainties.
- Tools for the Human Forecaster — Ultimately, most of the forecasts will still require a human-in-the-loop to make the final forecast and issue warnings or hazard assessments. These forecasters will require visualization and analysis tools to support these tasks.

As a specific example, consider that running NWP models such as MM5 and WRF typically requires HPC clusters with on the order of 200 processors, 500 GB of memory per run, 10–100 GB of near-line storage per run, throughput on the order of 1 GB/s, and latencies of 1 hour or less.

Scenario 2: Climate Data Reprocessing. A very important aspect of climate data analysis is the ability to reprocess various data sets as algorithms and knowledge improve. This is necessary since the goal is to observe small, long-term trends in various environmental parameters (such as sea surface temperature (SST)) that may be easily masked by systematic errors in the observing system or data processing algorithms.

The key requirement is to be able to reprocess possibly years' worth of satellite and related data at a rate much higher than the original real-time data acquisition rate. To illustrate this we consider the following example: 5 years of raw sensor data (Level 0) are collected and archived. An improvement to the algorithm to produce calibrated data is developed and a scientist wishes to reprocess all 5 years' worth of data from Level 0 to Level 1. However, she doesn't want to do this at the original data rate, but rather wants it reprocessed in 3 months (90 days). As a comparison we assume the original data rate was 10 Mbps, and the full 5 years' worth of Level 0 data would be $\simeq$ 200 pB. The effective reprocessing data rate would be about $10 \times (365 \times 5)/90 = 200$ Mbps. This is a tremendous throughput requirement and has significant impact on how the data stored, retrieved, and reprocessed.

To put this into perspective, consider that NASA has currently archived massive amounts of satellite data stored in robotic tape drives. These archives have limited throughput ($\simeq$ 25 TB/month) and are the main bottleneck in climate data analysis [28]. In order to facilitate timely climate data studies and other data mining activities, much higher data access rates are desired by the climate researchers. As a goal they would like to be able to analyze 1 pB/month. Thus the throughput rate requires about a factor of 40 increase. The WFCS grid architecture must provide a mechanism to address this data access bottleneck.

Scenario 3: Cal/Val and Data Ingest. An important remote sensing workflow involves the calibration and validation of new instruments as they are deployed and their associated data retrieval algorithms (most commonly applied for spaceborne sensors). As new instrumentation is deployed, it is necessary to verify the calibration and validate the derived data products against other measurement systems and standards. This effort usually involves focused measurement campaigns over specific *Cal/Val sites,* such as the DOE ARM sites [29]. The Cal/Val activities involve not only the collection of data from the new instrument(s), but also include a wide range of ground-based and airborne validation sensors deployed at the *Cal/Val sites.* All of there data must be collected, quality controlled, archived, and made available to the Cal/Val science teams for timely analysis in conjunction with the satellite data. A common set of data preprocessing is typically involved that includes subsetting the satellite data to the spatial and temporal coordinates of the Cal/Val activities and performing *match-ups,* between the Cal/Val instrumentation and the satellite data.

10.2.2 Derived Functional Requirements and Resources

We now translate the scenarios and mission requirements discussed above into the functional and resource requirements for the WFCS grid. These requirements are then mapped into the definition of specific grid services that will be used to construct the WFCS as an SOA.

Workflow Management: The management of workflow is probably one of the most critical requirements for the WFCS grid. Support is required at the user level in a very high level form, using domain-specific terminology to allow users to create their workflows. Support is also required down to much lower levels to create workflows that will be invisible at the user level. Users must be able to create workflows to execute their applications and models with specified data sets and perform post-processing and visualization of the results.

Data Management and Processing: The various WFCS grid workflows all involve the management of massive amounts of data, from raw sensor data to high-level data products created by the system. The WFCS grid will have to support various metadata standards for all data, data formats (HDF, NetCDF, BUFR, etc.), and tools to support reading and writing these formats. All WFCS grid clients (human users and applications) will require support for data discovery, replica location, and data movement. In many cases it will be necessary to provide data preprocessing tools collocated with the data to facilitate various filtering or data mining operations locally, so that user data requests don't unnecessarily utilize bandwidth. For example, there maybe cases where a user only requires a specific spatial/temporal subset of a larger data set, so it should be possible to specify the subsetting operation as part of the data request. Federation of various types of databases will also be required to support ease of access to data and resources, as well as provide system performance and reliability.

Resource Management: Management of grid resources involves resource discovery, brokering, and scheduling in order to support the QoS requirements. The WFCS grid must provide resource registries so that clients may automatically discover what resources are available, what access they have to these resources, and the specific attributes these resources provide. This is critical to the execution of workflows. The automation of WFCS grid workflow execution requires sophisticated tools for the scheduling, reservation, and brokering necessary to carry out the workflow execution. Implementing these functions relies on the ability to perform *load balancing*, *notification/messaging*, and *logging*.

Core Functions: These include the base level functions that will be pervasive across the whole WFCS grid and include security, grid monitoring, and resource fault and recovery management.

Given the complex nature of the WFCS and the fact that this system is created by combining services and resources from many different service providers, each with their own local policies, there will be a significant amount of coordination to develop *grid-wide* security policies and practices. These policies will address client authentication and authorization; however, to facilitate automated and arbitrary user workflows, a single sign-on process will be required. Role-based authorization will also be an important approach. This will also need to be extended to all the data that flow within the WFCS grid, regarding all aspects of data life cycle (creation, edit, read/write, delete).

The large simulations must be monitored constantly to make sure they have the compute resources to continue. The entire grid of instruments and the compute/data grid must be constantly monitored.

The WFCS grid must be highly fault tolerant and have robust disaster recovery mechanisms in place. Operational forecasts are required 24×7 so that the systems and software must have robust fault detection and handling mechanisms. In the event of a system failure, there must be a very fast mechanism for fail-over. This maybe accomplished using local backups or having mirrored resources that provide duplication services for all operational forecasts.

Policy Specification: Given the high-level priorities specified above and the fact that the WFCS grid requires utilization of resources that cross different administrative domains, there are specific policy requirements that will need to be agreed upon by all participating organizations and users and enforced using the appropriate software and hardware mechanisms:

- Workflow Prioritization and Conflict Resolution: operational weather forecasting workflow has priority over climate studies and science users;
- User Authentication and Authorization: single sign-on and proxying not only for hardware resources but data resources as well;
- Resource Usage: may require defining roles and categories that specify resource access, utilization, and priorities.

There are a very large number of policies that must be defined in a system such as the WFCS grid. This is discussed further in Section 10.3.

Performance Considerations: The performance of the WFCS grid is one of the key requirements. Performance may be specified by the various Quality of Service requirements that are either explicitly stated or derived. Performance may be broken down into a potentially large number of metrics. Here we will just discuss a few of the more critical performance metrics. For example, since operational weather forecasting has priority over all other applications, and has specific latency requirements, this application will receive all required computing and network bandwidth resources necessary to meet this requirement.

Performance Parameters: these quantify the behavior of the system using well-defined and measurable quantities. Some of these parameters will have a first order impact on the user experience, while others will have a higher order impact and are utilized by the underlying infrastructure to keep the system running to maintain system performance requirements. Also, many of these parameters are not completely independent (for example, throughput and latency are often key parameters from a user perspective as they can interact to determine the user's time to solution). We are interested in quantifying the performance of algorithms, software, and systems at various levels of abstraction.

An important aspect of performance measurement is that many of the grid services may need to utilize performance parameter measurements to optimally perform their functions. One example might be a scheduler that requires resource utilization and loading in order to perform job scheduling and planning. The term often applied is *performance aware* services.

It is not possible within this chapter to provide a complete list, however, some important examples include:

Throughput: generally measured as the amount of information (bits/seconds) that may be transfered between two points over a specified channel or connection.

Latency: the time (seconds) taken between when a request is made and a response is received.

Resource Utilization: the definition will depend on the specific resource under consideration, CPU, node, cluster, disk storage space, etc.

Given this overview of the WFCS grid system requirements, we next discuss how this may be mapped into a grid services architectural description.

10.2.3 WFCS Grid Services Architecture

The high-level system architecture and requirements analysis for the WFCS system addressed the system elements and subsystems required to achieve the necessary functionality for the WFCS. Given this view of the WFCS system architecture, we now wish to cast it in terms of a service-oriented architecture using the framework provided by the OGSA. The OGSA defines a set of basic capabilities with which to express service-oriented functionality (see [9] for complete details): Execution Management Services, Data Services, Resource Management Services, Security Services, Self-Management Services, and Information Services. These services allow one to achieve the necessary resource virtualization, common management capabilities, resource discovery, and standardization (protocols and schemas) throughout the WFCS grid. These OGSA defined services are built upon the physical resources that include subsystems and interconnections (CPUs, storage systems, networks, etc.) as well as instrumentation (e.g., spaceborne sensors).

Using these principles as a starting point we define an SOA view of the WFCS Grid as shown in Figure 10.2. This figure illustrates the standard layered view (similar to that shown in the overview to Section III, Introduction to Grids). The layers present the physical resources at the lowest level that are distributed throughout the different organizations that would constitute the WFCS grid, as well as the elements that provide connectivity of the compute and sensor resources. The physical resource layer is virtualized from the perspective of the user or application layer by the grid services layer. This layer includes web services as a fundamental building material upon which all services are built. At the user or application layer, the scientists have

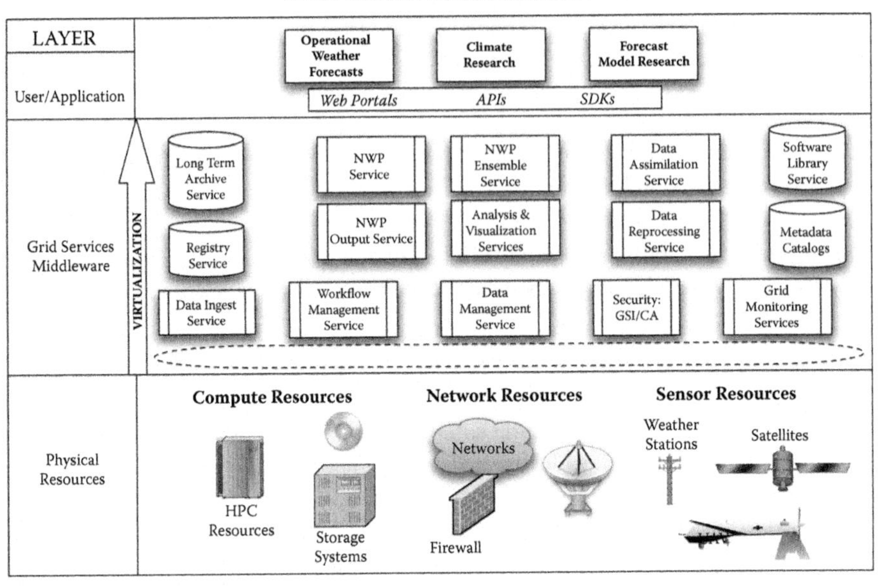

Figure 10.2 WFCS grid services architecture.

access through a variety of high-level applications, APIs, or SDK. Some examples have been discussed in the Grid Overview, including Web-based portals and specific end-user grid applications.

These services are summarized below:

Workflow Management Services — general set of tools to allow clients to configure and execute their operational and scientific research workflows within the WFCS grid. These provide access to various underlying resources (such as HPC systems) and are configured to enforce the prioritization policies. Following the work of Deelman and others [31, 32], we envision workflow management that allows users to express abstract workflows at the application domain level, and these workflows are then translated into concrete workflows that are automatically built from application services and optimally executed. By employing Artificial Intelligence-based planning and scheduling techniques, it may be possible to optimize workflow execution. This automation has several advantages, as discussed later in this chapter:

- Workflow Composition Services – allowing users to define specific workflows to be executed within the WFCS grid. This service would operate in the application domain, providing users a familiar set of terms with which to define their workflow.
- Workflow Scheduling, Brokering, and Execution Service – handles the actual execution of the workflow on behalf of the user. Ideally, these services

would employ a high level of automation and optimization capabilities that free the operational and science user from the low-level job execution details. Given the QoS requirements for operational forecasting, these workflows would require a high degree of reliability in their execution. This service should allow for the execution of very complex workflows, such as those that might be built by a climate scientist analyzing several decades worth of climate record data from various sources.

- Execution Monitoring Service – allows clients to monitor the progress and status of their workflow.

Data Management Services — this covers a range of services, including:

- Data Ingest Services – ingest data into the WFCS grid from all sources (ground, air and space), apply Quality Control/Quality Assurance analysis, format and reformat and apply specific metadata schemas, and store the data in various storage systems, including an LTA.
- Metadata Catalog, Replica Location, and Retrieval Services – allow users to find data using high-level domain language queries and retrieve or move physical replicas to where the data are needed. Also performs replica management (data revision control). The retrieval services include tools such as GridFTP.
- Long-Term Archive Service – long-term archival storage of all raw instrument data as well as other data products generated within the WFCS grid.

Weather Forecast and Climate Model Services — services to support both operational and research forecast generation and analysis.

- Synoptic Model Services – provides access to operational or research synoptic weather forecast models.
- Mesoscale Model Services – provides access to operational or research mesoscale weather forecast models.
- Ensemble Forecast Service – interfaces with the synoptic or mesoscale services to allow users to compose ensemble forecasts.
- Data Assimilation (DA) Services – a set of DA tools that can be used with a variety of models for data assimilation.
- Post-Processing Tools — basically a library of tools that the user can have applied to model output; include output format selection, statistical analysis tools, and plot generation. This would include tools to analyze ensemble forecasts.
- NWP Verification and Validation Service – a set of services that support the verification and validation of NWP models; most likely constructed from a set of lower level services and components (a generalized testing framework that is implemented in such a manner as to allow grid-enabled testing).
- Forecaster Data Analysis and Visualization Service – a suite of tools that allows human forecasters to visualize and analyze the model outputs on their local workstations.

- Climate Model Services – services to set up and execute specified climate models.

Core Grid Services — underlying services that support all the higher level grid services.

- Security and Policy Management – provides authentication and authorization for all users across the entire WFCS grid. The goal would be to have single sign-on as well as proxy capabilities to allow services to perform their functions on behalf of the authenticated user.
- Grid Monitoring Services – general performance monitoring as well as monitoring to support *performance-aware* grid services the require these data to achieve the necessary QoS.
- Software Configuration Management Services – grid-wide configuration management and software repository/registry services.
- Fault Handling and Disaster Recovery Services – due to the high QoS for operational workflows, the underlying resources required to maintain the necessary reliability would be handled by these services, catching system faults and switching to backup systems and resuming workflows.

Next we consider some of the issues one would face in the implementation of the WFCS grid architecture.

10.3 Remote Sensing Grids: Implementation

Given the requirements and architecture discussed in the previous section, we now turn our attention to the practical implementation of a remote sensing grid such as the WFCS. When combining heterogeneous and distributed resources to build the WFCS grid one must address issues related not only to the grid middleware technology but also the issues of grid management and policies. We first summarize the management, policy, and staffing issues, and then discuss the necessary grid services to support specific WFCS grid architectures, concentrating on data management, job management, and monitoring.

10.3.1 Management and Policy Considerations

One of the key aspects of grid technology is the construction of virtual organizations that constitute remote sensing grids or sensor webs. These virtual organizations are constructed using the physical resources provided by different organizations, each with their own administrative domains. The importance of understanding and addressing the policy aspects of building remote sensing grids or sensor webs cannot be overstated and often is more of a bottleneck to implementation than any aspect of the underlying technologies. In the early planning stages it is important to build working groups that include the relevant information technology (IT) staff to address issues related to network management, information security, and resource allocation.

The fundamental considerations for security deal with authentication and authorization. Authentication addresses the ability to verify that a user or client (other process or agent) is indeed who they claim to be. Authorization addresses the issue of whether that user has rights or permissions to access and utilize grid resources as part of the current request. Another real security issue for the construction of a grid, such as the WFCS, is firewall policy. Local organization firewall and security administrators have developed specific policies that do not necessarily include the needs and requirements of the grid users. In the past, many grid tools required a wide range of ports to be open in order to properly operate, and this was often in conflict with existing firewall rules. The transition to a Web services-based implementation has helped to alleviate this to some extent.

Beyond authentication and authorization is the general issue of identity management. This is not only managing user accounts across multiple organizations but also managing virtual organizations that rely on multiple certificate authorities. Another important issue, driven by the creation of VOs, is that of the timely propagation of certificate revocation lists among multiple certificate authorities.

An often overlooked area when setting up a grid environment is that of the types and levels of expertise required by the support staff who maintain the grid resources at each site and across the entire VO. The types of support staff and their functions that may be required to support the grid are summarized below:

- Systems Administration — provides administration of all computer systems at a particular grid site. A key function is keeping the system up-to-date with respect to software patches, especially security patches. Those resources that provide operational QoS to the WFCS grid would need full-time systems administration support.
- Database Administration — provides administration of the various databases that exist at any given grid site. A key function of the database administrator (DBA) position is the initial setup and optimization of a given DB. An operational system may require a full-time DBA with on-call support.
- Grid Services Administration and Application Support — provides specific support to particular grid service applications. This does not necessarily need to be a person separate from the systems administrator, but could be a sysadmin who is familiar with the grid software. They also provide support to scientific applications running on the WFCS grid as a service. Depending on the application, it may require small teams that maintain the specific applications.
- Network Support — provides administration and maintenance of network infrastructure at a particular grid site. The key function here is to interface with network support personnel at each local site to guarantee QoS and security. An operational system may require full-time network support including on-call support. The VO will also require support for intra-grid network management

between sites and interfaces with various service providers. Another often overlooked issue is that of IP address space management. Grid implementation often requires large blocks of IP addresses or mechanisms to dynamically manage a range of addresses.

- Programmers — provide specialized tools based on various grid tool APIs or SDKs. For applications with stressing QoS requirements these will have to be programmers very familiar with how to develop optimized code on a range of platforms, or may be a team at a particular organization within the WFCS grid VO.
- Training — provides training to grid users. The level of effort required will vary depending on the user base of the system.

10.3.2 Data Management

Clearly the WFCS will need to manage data, from ingest to long-term archiving. At the lowest level, data will reside in local file systems. This could be in simple flat files, various types of databases, or structured file formats for scientific data such as HDF [41] and NetCDF [42]. These file formats are sometimes referred to as *self-describing* since the files contain metadata describing the data, such as the number, dimensionality, size, and type of arrays.

Of course, WFCS users will be distributed and want to access data that are also distributed. This means that files must be accessible from remote locations. One way of doing this is with a *distributed file system*. The Network File System (NFS) is essentially a distributed file system since it allows files to be accessed over a local area network. Systems such as the Andrew File System (AFS), however, enable multiple, distributed file servers to be federated under the same file namespace [43]. This is a conceptually simple approach where remote files can be read and written, assuming that the path name to the physical file is explicitly known. Systems such as the Storage Resource Broker (SRB) [44] and the Metadata Catalogue Service (MCS) [45], however, allow data discovery using metadata catalogs. Data are then accessed by *attribute* or *logical file name*, such that the user does not have to know the physical location of the data. Replica management services may be invoked to determine an optimal physical location for data access based on the user and specific data destination. It should be pointed out that often the data transfer is not from the physical storage location to the user proper, but rather to a third location for data processing. This allows applications, such as the SRB, to transparently manage multiple storage facilities and archives, resulting in a virtualization of the data storage from a user perspective.

However the desired data are identified, they may have to be moved from one grid location to another for processing or any other subsequent use. For basic data transfers GridFTP was developed that extends the traditional File Transfer Protocol by using the Grid Security Infrastructure. Not only does this enable strong authentication, it also enables secure *third-party* transfers. This is very important in grid environments where a client may want to use multiple resources in different locations.

10.3.3 Job Management and Workflow

Even with an installed grid infrastructure, each user application will need specific services built to provide the capabilities to achieve the user's goals. While domain-specific applications can be built using grid middleware and Web services (e.g., Globus toolkit) directly, many scientific users feel that this interface is too low-level and detailed and desire higher level access to the underlying capabilities.

The development of grid enabled applications, either new applications or grid-enabling legacy applications, has proved to be somewhat of an impediment to the implementation of grid technology. Tools are also being developed, however, that are allowing for easier application development.

At the highest, most specific level, grids may be used to provide services to users in a manner that is essentially indistinguishable from desktop implementations. One approach to this is via network-enabled services (NES). This follows the traditional paradigm of servers providing services to clients over networks, but in the case of network-enabled services, there is a resource manager that brokers the client-server interaction to take advantage of the distributed and dynamic nature of the available resources. This formalism is implemented in the GridRPC protocol, which extends the remote procedure call (RPC) protocol, a well-established client-server interface standard.

GridRPC has been implemented in various network-enabled services that aim to provide high level functionality to the end user, such as the NetSolve/GridSolve project that provides a client interface in the Matlab and Mathematica computing environments, Ninf and DIET which provide a framework for writing GridRPC enabled applications in high level programming languages.

There have been several attempts to produce more general purpose application programming interfaces (API) to facilitate grid application authoring, resulting in the creation of the Simple API for Grid Applications (SAGA), within the GGF [39]. SAGA is an attempt to provide a basic interface to the most commonly used grid functionalies. It does not attempt to encompass the full capabilities of the underlying middleware, but allows for rapid development of software. It attempts to capture 80 percent of the functionality with perhaps 20 percent of the effort as compared to directly interfacing at the lowest level. The SAGA working group of the GGF is basing the requirements for the API on submitted use cases for a variety of grid applications. A *strawman* API specification is being developed using the object-oriented Scientific Interface Description Language (SIDL) with object abstractions for complex entities such as a handle to a remote procedure call. The API includes functionalies for Security via a session paradigm; Data Management including remote file access and replica cataloging; Job Management including remote job startup and asynchronous operations; and Inter-Process Communications via a stream mechanism. Work is in progress on specific bindings to C++ and Java. Commodity Grid Kits (CoG Kits) also provide a direct interface to the functionality of grid back-ends via high level languages such as Java and Python.

Although stand-alone services may be offered with grid back-ends, more often the desired product requires considerable flexibility that is best accomplished by combining and chaining grid applications and services into workflows. From the user perspective, one would like to focus their attention on their specific workflow and not the underlying grid infrastructure and computing technologies. Thus grid technology should enable the remote sensing scientists to build and execute their workflows with little or no understanding of the underlying technologies (they don't need to be computer scientists). They would like to be free from the low-level details involved with the execution of these workflows; this is the virtualization of scientific workflow, one of the visions for the grid.

Automation of workflow has several advantages. It improves usability from the perspective of the scientific user, who is now freed from the complexity of the grid and can focus on their application. The users interested in very complex workflows (such as the climate scientist wishing to analyze and data mine massive amounts of satellite data) are freed from the details of how their application level workflow specifications are mapped to the concrete workflow and executed on the grid. It is also possible to address the policy issues of user authorization and task priorities (recall that operational tasks have priority over research tasks in the WFCS grid). Additional automation helps to also address issues with cost to solution, reliability (fault detection and handling), and performance issues.

The development of sophisticated workflow managers that allow users to employ high level, domain-specific commands is a critical need. There are several ongoing projects attempting to address this problem. Pegasus (Planning for Execution in Grids) [30, 31, 32] is a workflow construction tool that allows the mapping of an abstract workflow to a specific grid implementation. The abstract workflow is portable in the sense that the operations are not tied to particular grid resources. Pegasus uses artificial intelligence (AI) methods to heuristically optimize the resulting concrete workflow. Pegasus guarantees that the necessary data are transferred to the appropriate execution nodes, and also attempts to minimize redundancy by removing processes from the workflow that produce intermediate data that are already available. The input to Pegasus is in the form of abstract Directed Acyclic Graphs (DAGs), which may be created by the Chimera package, and it produces concrete DAGs compatible with the Condor DAG Manager (DAGMan) for execution [46].

XWorkflow [47] is a graphical tool for producing general workflows from Web services. It provides a GUI that allows the developer to chain outputs from one Web service to another's input and generates Jython [48] scripts that execute the workflows.

10.3.4 Grid Monitoring and Testing

The purpose of a grid performance monitoring (GPM) system is to reliably provide timely and accurate measurements regarding various grid services and resources, without perturbing the grid operation or performance. Monitoring is important for understanding the overall operation of the WFCS grid, detection of failures, and optimization of performance. Information on grid health and status is used by some services to optimize and schedule resource utilization.

GPM provides the data and analysis to identify performance problems and issues so that performance may be optimized for specific applications or conditions. In the

WFCS grid we are interested in not only monitoring the underlying grid infrastructure, resources, and services, but the applications (e.g., weather forecast model execution) as well. The GPM system must provide real-time information about the state of the WFCS, as well as measurements of specific events (e.g., failures). GPM also provides fault tolerance required by operational systems and those that have service level agreements to provide minimum quality of service. GPM also provides the data required by performance prediction models that may assist with various resource scheduling and load balancing services. The GPM system must itself be fault tolerate in order to provide continuous and reliable observations. Typical parameters monitored as part of a GPM system include:

- Network bandwidth utilization (past, present, and future)
- Storage utilization and availability
- Storage archive transfer rates
- Processor utilization
- Data transfer latency
- Metadata query rates

Performance monitoring in a grid environment is different than that typically performed within a local administrative domain. Grid performance monitoring systems must be scalable over a wide-area network and will involve a large number of heterogeneous resources, many within different administrative domains.

The GGF has defined a baseline architecture for grid performance monitoring, the Grid Monitoring Architecture (GMA) [33]. The GMA consists of three components: producers, consumers, and a registry. The producers are sources of grid performance measurements and are registered with the registry service so that consumers may discover their existence. Once a consumer discovers the existence of a particular producer, the consumer and producer may communicate directly. The producers are the source of grid measurements, coupled with either software or hardware sensors that perform the basic measurements. Producers and consumers may be composed to build complex producers that provide higher level measurement information (for example, by ingesting network bandwidth measurements, one could build a bandwidth prediction producer).

Ignoring for the moment the details of instrumentation and data collection, we observe that the WFCS grid performance data resembles, in many respects, the sensor data that the WFCS grid is designed to catalog, store, and distribute. In other words, from a sufficiently abstract perspective, the WFCS grid performance data are produced by sensors (performance probes implemented in hardware or software) on platforms (hosts, CPUs, disk arrays, network routers, and the like). Consequently, many of the generic attributes of the metadata schemata such as producer, platform, sensor, and time of collection (to name a few) can also be applied to data sets of WFCS grid performance measurements. In addition, these performance data sets can be stored, archived, and replicated using the same infrastructure that the WFCS grid uses for its science data. Finally, the same execution and scheduling mechanisms that are used by the science codes executing on the WFCS grid can also be used by performance analysis tools for postmortem analysis of the measurement data sets generated by the performance monitors. In other words, WFCS grid performance data are essentially no different, in principle, from any other science data set, and the WFCS grid infrastructure can support performance monitoring and analysis just as easily as it supports the collection, generation, and analysis of earth sciences data products.

Testing is an area that is often only given consideration late in system development. However, we would argue that testing should be designed into the grid architecture and implementation from the very beginning. The GPM architecture and tools may also be used for testing the grid.

In addition to the core services there will be the need to develop services unique to grids such as the WFCS that provide capabilities unachievable without the integration of the components. One such example is the steering of instruments based on in-depth data analysis and model predictions. Such feedback is extremely difficult currently because of the barriers separating the observation and computational elements. This capability would greatly enhance observational efficiency and is an important goal of LEAD (Linked Environments for Atmospheric Discovery), a remote sensing/weather forecasting grid currently under development and discussed in the next section.

Finally, we mention a few examples of some of the grid performance monitoring tools that have been employed by other grid projects. NetLogger (short for Networked Application Logger) is a system for performing detailed end-to-end analysis of distributed applications [35]. It includes tools for instrumenting applications, host systems, and networks and has a powerful visualization tool for correlating monitoring data from all components. NetLogger is extremely useful for debugging and tuning distributed applications, and for bottleneck detection. The Network Weather Service (NWS) is a distributed system that periodically monitors and dynamically forecasts the performance that various network and computational resources can deliver over a given time interval [36]. There is also an open source host, service, and network monitoring application called Nagios [37]. Nagios uses a plug-in architecture and provides current status information, historical logs, and various reports all accessible from a Web browser.

10.4 Remote Sensing Grids: Examples

10.4.1 Example 1: Linked Environments for Atmospheric Discovery (LEAD)

Linked Environments for Atmospheric Discovery (LEAD) is a National Science Foundation (NSF) Large Information Technology Research (ITR) project that is creating an integrated, scalable cyber-infrastructure for mesoscale meteorology research and education [12]. It is designed to allow weather forecasts and atmospheric analysis that can adapt to rapidly changing conditions and take advantage of large amounts of computing resources that may only be needed for a relatively short amount of time but are required on-demand. These applications are also characterized by large amounts of streaming data from sensors that would ideally interact with the analysis. The general concept of LEAD is a group of services that may be linked together in workflows to accomplish the task at hand, but whose workflows may be dynamically and automatically modified based on feedback from the sensor observations [13].

A grid architecture is an ideal solution to this sort of problem. The architecture consists of a sequence of layers spanning the distributed resources to the user, but there are several 'cross-cutting' services that are available to all layers, encompassing authorization, authentication, monitoring, notification, and a personal metadata catalog service, known as MyLEAD. LEAD stores metadata using an XML schema that is an extension of the Federal Geographic Data Committee (FGDC) standard. These services, along with the portal service, are persistent and form a robust core upon which workflows can be built with specific instances of the layered services.

The base software layer, the resource access layer, is built on Globus grid services, Unidata's Local Data Manager (LDM), Open-Source Project for a Network Data Access Protocol (OPeNDAP), the Open Grid Service Architecture Data Access and Integration (OGSA-DAI) service, and other data access services. On top of this is a LEAD-specific middle layer that contains five basic functional blocks: a resource broker; application and configuration services that provide the facilities to launch instances of the needed forecasting and analysis applications (WRF, ADaM, ADAS, etc.); a catalog service with a virtual organization catalog allowing access to public-domain data sources and services; data services that allow for access, query, and transformation of data products; and workflow services to build and monitor workflows.

LEAD specifies workflows using the Business Process Execution Language for Web Services (WS-BPEL), which has facilities allowing for event detection and dynamic modification of workflows. This is a key requirement for LEAD's goal of adaptability. Currently, only static workflows are being supported, but dynamic user-initiated modification capabilities are being developed with automatic modification to follow. Workflow status is monitored via a system that utilizes the Autopilot and SvPablo performance monitoring toolkits as well as a newly built Health Application Monitoring Interface and a workflow annotation system to report on currently running workflow elements.

Figure 10.3 shows a high-level view of the LEAD architecture. This is similar to the WFCS Grid architecture illustrated in Figure 10.1, with the resource, middleware, and application layers. In the case of the LEAD system, the middleware services are accessed by various *portlets* that provide access to the various high-level functions, such as workflow design and data visualization. Workflows are graphically defined, then converted into WS-BPEL for execution by the workflow service. The user interacts with the portlets through a Web-based portal or using desktop *serviceware*. Also note that the cross-cutting services in Figure 10.3 are the core grid services mentioned earlier in this chapter.

The current instance of LEAD is being prototyped on a grid based at several member institutions with hardware varying from single-CPU servers to large clusters. Later instances will be expanded to other grids such as the TeraGRID.

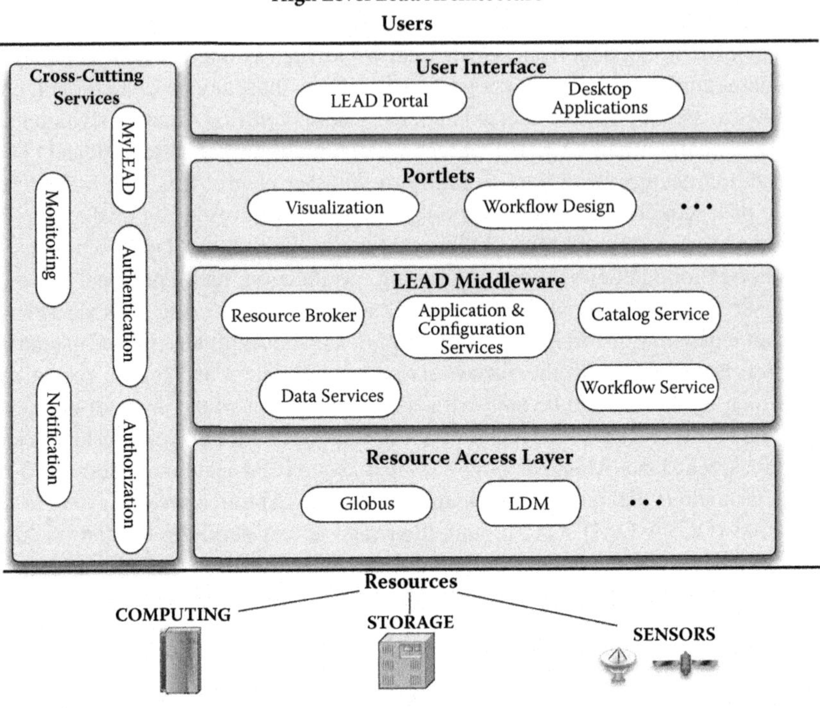

Figure 10.3 LEAD software architecture.

10.4.2 Example 2: Landsat Data Continuity Mission (LDCM) Grid Prototype (LGP) Project

The Landsat Data Continuity Mission (LDCM) Grid Prototype (LGP) offers a specific example of distributed processing of remotely sensed data. The LGP is a proof-of-concept system using distributed computing to generate single, cloud and shadow free Landsat-like scenes from the composite of multiple input scenes, the data for which may be physically distributed [14, 15, 16]. The system ingests multiple scenes from the Landsat Enhanced Thematic Mapper+ (ETM+), calibrates the intensities, applies cloud and shadow masks, calculates surface reflectance (based on ancillary data for atmospheric ozone and water vapor profiles), registers the images with respect to their geographic location, and forms a single composite scene.

These computational tasks are performed within a virtual organization with computing resources consisting of Linux servers located at USGS offices in South Dakota, USA, and the NASA Goddard Space Flight Center in Maryland, USA. The Globus toolkit (version 2.4.3) provides middleware services with the Java Commodity Grid Kit (CoG) overlayer. Workflows are initiated via Perl scripts that form a *command line* user interface and generate XML files that are interpreted by the Karajan workflow engine. Specifically, two elements of Globus are used in this project: GridFTP for file transfer, and the Globus Resource Allocation Manager (GRAM) for resource allocation and job execution. Authentication is accomplished via the Grid Security Infrastructure (GSI) using host certificates issued by the Committee on Earth Observation Satellites (CEOS) certificate authority.

The user initiates a workflow by creating a file that specifies the geographic areas and times of interest. (Additionally, the network location of initial data is currently specified manually.) XML files generated from this product specification direct the transfer of input data files to the compute machines, which are allocated in a round-robin fashion. Processing of an input image takes place serially on the compute node to which it is assigned. The intermediate results are then gathered for final compositing and transferred back to the initiating computer.

10.4.3 Example 3: NOAA National Operational Model Archive and Distribution System (NOMADS)

The National Oceanic and Atmospheric Administration (NOAA) National Operational Model Archive and Distribution System (NOMADS) is a pilot project allowing access to historical and real-time climate and weather data both from observations and models [17, 18, 19]. It is built as a collection of Web services that provide data spanning multiple government agencies and academic institutions in a format-neutral manner. The data available include input and output from the Numerical Weather Prediction (NWP) models from National Centers for Environmental Prediction (NCEP); Global Climate Models (GCM); simulations from Geophysical Fluid Dynamics Laboratory (GFDL); global and regional reanalysis from NCEP and the National Center for Atmospheric Research (NCAR); and limited surface, upper-air and satellite observational data sets from NCDC, NOAA's National Ocean Data Center (NODC), and NOAA's Forecast System Laboratory (FSL).

While NOMADS is not strictly implemented as a grid service, it is based on Web services and as such fits into the architectural approach outlined for the WFCS grid. NOMADS consists of a variety of services, but it is primarily based on the Open Source Project for a Network Data Access Protocol (OPeNDAP). OPeNDAP is a system to allow abstract access to differently formatted, physically distributed data. It handles a wide variety of formats for translation (e.g. HDF, netCDF, GRIB, etc.) and allows for subsetting and variable isolation on the server side. There are multiple NOMADS servers located at NCEP, GFDL, and NCDC. The NCDC instance provides 20 TB of online storage where near-term data are archived. The remainder of the *deep* archives are accessed via the High-Performance Storage System (HPSS).

The data are served to the user via HTTP or FTP using the Grid Analysis and Display System (GrADS) Data Server (GDS) and the Live Access Server (LAS). There are also lower level access capabilities that utilize http transfers of data served by Perl scripts. These capabilities allow high-volume access to experienced users with more specific knowledge of the data.

10.5 Remote Sensing Grids: Adoption

Building a system such as the WCFS grid will actually require *paths to adoption* for all the major stakeholders. Besides defining and building desired new capabilities, many legacy systems will have to be incorporated into the grid architecture. Grid enabling the legacy systems requires creating the necessary services based on these capabilities. One of the primary advantages offered by the service-oriented architecture approach is the ability to provide continuity to the user during system evolution. Only the client-service interface need remain fixed to guarantee compatibility. Generally, the implementation of a grid such as the WFCS would be undertaken with the aim of minimally disrupting the existing structure of the individual organizations forming the VO. Given this constraint, there is a wide range of accessibility that the individual organizations can offer. An operational weather forecasting organization will still be expected to generate an accurate and timely product without disruption and consequently is unlikely to offer low-level resource access to outside users. Such an entity is likely to join the VO as a highly vertically integrated module with a service *cap* that provides an interface to the final product. Conversely, a university or other research organization may be able to offer access at all levels providing finer granularity modules for tighter horizontal integration of the VO. The ability to handle this heterogeneity (of organizational structure as well as system architecture) is a key aspect of grid computing. Furthermore, as the system evolves, the grid allows for gradual blurring of the lines of the vertically integrated organizations to share resources at lower levels. As such, the system can evolve in capabilities and efficiency while still providing continuity to the user.

Ultimately we want to enable the complete non-computer specialist to be a routine grid user. This means that we must build easy-to-use tools using widely accepted user interfaces. Perhaps the most popular mechanism to provide these end-user tools is via Web interfaces or grid portals. Tools are available to speed the development of these interfaces, such as GridPort, which allows for the generation of customizable Web interfaces. The concepts discussed here have been brought together in grid-based problem solving environments (PSE). PSEs provide all the functionality and resources necessary to solve a specific domain of problems at a high level. The syntax of these environments is that of the target application domain removing the user from any of the details of the underlying computational system, here the grid computing paradigm. Research is progressing in the use of PSEs to provide remote sensing functionality, e.g., processing of synthetic aperture radar (SAR) images to the science user [40].

Grids have come a long way and the technology is maturing with standards emerging in the marketplace and major vendors staking out positions. However, we still do not quite have turnkey grid *solutions*. Designing and deploying a grid infrastructure to support a particular project or community still requires a core IT staff that is knowledgeable about grids. Nonetheless, grid middleware has matured over the past few years, and the development of grid services by leveraging Web services has improved adoptability. As the grid standards become more mature and stable, it is expected that the rate of adoption and implementation of remote sensing grids will increase.

10.6 Acronyms

Term	Description
3DVar	3-Dimensional Variational (data assimilation)
AdaM	Algorithm Development and Mining System
ADAS	Atmospheric Data Assimilation System
AFS	Andrew File System
AFWA	Air Force Weather Agency
AI	Artificial Intelligence
API	Application Programming Interface
ARM	Atmospheric Radiation Measurement
BUFR	Binary Universal Form for Representing meteorological data
CA	Certificate Authority
CEOS	Committee on Earth Observation Satellites
CPU	Central Processing Unit
COAMPS	Coupled Ocean/Atmosphere Mesoscale Prediction System
DB	Database
DBA	Database Administrator
DAG	Directed Acyclic Graph
DoD	Department of Defense
DOE	Department of Energy
ECMWF	European Center for Medium-range Weather Forecasting
EOS	Earth Observing System
ESTO	Earth Science Technology Office
FGDC	Federal Geographic Data Committee
FSL	Forecast Systems Laboratory
FNMOC	Fleet Numerical Meteorology and Oceanography Center
FTP	File Transfer Protocol
GCM	Global Climate Models
GDS	GrADS Data Server
GFDL	Geophysical Fluid Dynamics Laboratory
GGF	Global Grid Forum
GPM	Grid Performance Monitoring
GrADS	Grid Analysis and Display System
GRAM	Globus Resource Allocation Manager
HDF	Hierarchical Data Format
HPC	High Performance Computing
HTTP	Hypertext Transfer Protocol
IDV	Integrated Data Viewer
LAS	Live Access Server
LEAD	Linked Environments for Atmospheric Discovery
LEO	Low Earth Orbit
LDCM	LandSat Data Continuity Mission
LGP	LDCM Grid Prototype
LTA	Long Term Archive
MM5	PSU/NCAR mesoscale model

MCS	Metadata Catalog Service
NASA	National Aeronautics and Space Administration
NCAR	National Center for Atmospheric Research
NCDC	National Climatic Data Center
NCEP	National Centers for Environmental Prediction
NES	Network Enabled Services
NetCDF	Network Common Data Format
NOAA	National Oceanic and Atmospheric Administration
NODC	National Ocean Data Center
NOMADS	National Operational Model Archive & Distribution System
NWP	Numerical Weather Prediction
NWS	National Weather Service
OGSA	Open Grid Services Architecture
OGSA-DAI	Open Grid Services Architecture Data Access & Integration
QoS	Quality of Service
PSE	Problem Solving Environment
RPC	Remote Procedure Call
RSG	Remote Sensing Grid
SAGA	Simple API for Grid Applications
SDK	Software Development Kit
SIDL	Scientific Interface Description Language
SLA	Service Level Agreement
SOA	Service Oriented Architecture
SOAP	Simple Object Access Protocol
SRB	Storage Resource Broker
USGS	United States Geological Survey
VO	Virtual Organization
V&V	Verification and Validation
WFCS	Weather Forecasting and Climate Science
WRF	Weather Research and Forecasting Model
WS	Web Services
WS-BPEL	Business Process Execution Language for Web Services
WSDL	Web Services Description Language
XML	eXentsible Markup Language

References

[1] I. Foster, C. Kesselman, S. Tuecke. *The Anatomy of the Grid: Enabling Scalable Virtual Organizations,* International J. Supercomputer Applications, 15(3), 2001.

[2] I. Foster, C. Kesselman, J. Nick, S. Tuecke. *The Physiology of the Grid: An Open Gird Services Architecture for Distributed Systems Integration. Global Grid Forum,* http://www.ggf.org, June 22, 2002.

[3] I. Foster. *What is the Grid? A Three Point Checklist,* GridToday (July 20, 2002). http://www.gridtoday.com/02/0722/100136.html

[4] Several definitions may be found at: http://amsglossary.allenpress.com/glossary, and http://en.wikipedia.org/wiki/Remote_sensing.

[5] James B. Campbell, *Introduction to Remote Sensing,* Third Edition The Guilford Press, NY (2002).

[6] W. G. Rees, *Physical Principles of Remote Sensing (Topics in Remote Sensing),* Cambridge University Press, Cambridge (2001).

[7] G. Higgins et al., *Advanced Weather Prediction Technologies: Two-way Interactive Sensor Web & Modeling System,* Phase II Vision Architecture Study, NASA ESTO, November 2003.

[8] *Systems Engineering Handbook,* International Council on Systems Engineering (INCOSE), INCOSE-TP-2003-016-02, Version 2a, Seattle (1 June 2004).

[9] Global Grid Forum, Open Grid Services Architecture (OGSA-WG): http://www.ggf.org/ggf_areas_architecture.htm.

[10] Global Grid Forum, OGSA Glossary of Terms v1.0: http://www.gridforum.org/documents/GFD.44.pdf.

[11] *An Overview of Service-oriented Architecture, Web Services and Grid Computing,* Latha Srinivasan and Jem Treadwell, HP Software Global Business Unit, http://devresource.hp.com/drc/technical_papers/grid_soa/index.jsp?jumpid=reg_R1002_USEN (November 2005).

[12] *Service-Oriented Environments for Dynamically Interacting with Mesoscale Weather,* Droegemeier KK, Brewster K, Xue M, Weber D, Gannon D, Plale B, Reed D, Ramakrishnan L, Alameda J, Wilhelmson R, Baltzer T, Domenico B, Murray D, Ramamurthy M, Wilson A, Clark R, Yalda S, Graves S, Ramachandran R, Rushing J, Joseph E, Morris V. Computing in Science & Engineering Vol. 7, No. 6, pp. 12–29 (Nov-Dec 2005).

[13] Linked Environments for Atmospheric Discovery (LEAD) http://lead.ou.edu/.

[14] Jeff Lubelczyk and Beth Weinstein, *An Introduction to the LDCM Grid Prototype,* available at http://isd.gsfc.nasa.gov/technology/TechWS/Lubelczykt.pdf (January 14, 2005).

[15] Beth Weinstein, *NASA GSFC Landsat Data Continuity Mission (LDCM) Grid Prototype (LGP),* CEOS WGISS 21, Budapest, (8-12 May 2006), available at wgiss.ceos.org/meetings/wgiss21/Tech-and-Servs/GRID/LGP_CEOS_21_May2006.ppt.

[16] Ananth Rao, Beth Weinstein, *Landsat Data Continuity Mission (LDCM) Grid Prototype (LGP) Design Report,* (September 30, 2005, available at sgt.sgt-inc.com/ arao/lgp-delivery-2005/LGP_SDP_9_2005.doc.

[17] NOAA National Operational Model Archive and Distribution System (NOMADS) http://nomads.ncdc.noaa.gov/.

[18] Rutledge, G.K., J. Alpert, R. J. Stouffer, B. Lawrence, *The NOAA Operational Archive and Distribution System (NOMADS)*, 2002, Proceedings of the Tenth ECMWF Workshop on the Use of High Performance Computing in Meteorology — Realizing TeraComputing Ed., W. Zwieflhofer and N. Kreitz, World Scientific, pp 106-129 2003, Reading, UK.

[19] Glenn K. Rutledge, Danny Brinegar, Andrea Fey, Jordan Alpert, Dan Swank, Michael Seablom, *The NOAA National Operational Model Archive and Distribution System (NOMADS): Growing Pains and a Look to the Future* (2006) available at http://nomads.ncdc.noaa.gov/publications/rutledge-ams-atlanta-2006.pdf

[20] http://www.aerospaceweather.com/.

[21] Global Grid Forum, Use Cases: http://www.gridforum.org/.

[22] *Advanced Data Grid (ADG) Prototype System Description Document (SDD),* Release Version V1.0, NASA Goddard Space Flight Center, Greenbelt, Maryland (April 25, 2003).

[23] NPP Web Site: http://jointmission.gsfc.nasa.gov/.

[24] Space Studies Board, National Research Council (2000). Ensuring the Climate Record from the NPP and NPOESS Meteorological Satellites, Committee on Earth Studies, Space Studies Board, Division on Engineering and Physical Sciences, Washington, D.C., available at http://www.nas.edu/ssb/cdmenu.htm, and Space Studies Board, National Research Council (2002). Assessment of the Usefulness and Availability of NASAs Earth and Space Science Mission Data, National Research Council, National Academy Press, Washington, D.C.

[25] Fifth-Generation NCAR/Penn State Mesoscale Model (MM5), http://www.mmm.ucar.edu/mm5/.

[26] Weather Research and Forecasting (WRF) Model, http://wrf-model.org/.

[27] List of commonly used operational NWP models is available from: http://www.meted.ucar.edu/nwp/pcu2/index.htm.

[28] B. R. Barkstrom, T. H. Hinke, S. Gavali, W. J. Seufzer, *Enhanced Product Generation at NASA Data Centers Through Grid Technology,* presented at the Workshop on Grid Applications and Programming Tools June 25, 2003, Seattle, WA. [Online]. Available at: http://www.cs.vu.nl/ggf/apps-rg/meetings/ggf8/barkstrom.pdf.

[29] U.S. Department of Energy (DOE), Atmospheric Radiation Measurement (ARM) Program, [Online]. Available: http://www.arm.gov/.

[30] http://pegasus.isi.edu/.

[31] *Mapping Abstract Workflows onto Grid Environments,* Ewa Deelman, Jim Blythe, Yolanda Gil, Carl Kesselman, Gaurang Mehta, Karan Vahi, Kent

Blackburn, Albert Lazzarini, Adam Arbree, Richard Cavanaugh, and Scott Koranda. Journal of Grid Computing, Vol. 1, No. 1, 2003.

[32] *Artificial Intelligence and Grids: Workflow Planning and Beyond,* Yolanda Gil, Ewa Deelman, Jim Blythe, Carl Kesselman, and Hongsuda Tangmurarunkit. IEEE Intelligent Systems, January 2004.

[33] *A Grid Monitoring Architecture,* (August 2002). http://www.ggf.org

[34] http://ganglia.sourceforge.net.

[35] http://www-didc.lbl.gov/NetLogger.

[36] http://nws.cs.ucsb.edu/.

[37] http://www.nagios.org/.

[38] Global Grid Forum, WS-Agreement: http://www.gridforum.org/ .

[39] GGF. (2004) Simple API for Grid Applications Research Group. [Online] Available: http://forge.gridforum.org/projects/saga-rg/.

[40] Aloisio, G., Cafaro, M., Epicoco, I., Quarta, G., *A Problem Solving Environment for Remote Sensing Data Processing,* Information Technology: Coding and Computing, 2004. Proceedings. ITCC 2004. International Conference on Publication, Vol.2, pp. 56–61 (2004).

[41] http://hdf.ncsa.uiuc.edu/.

[42] http://www.unidata.ucar.edu/software/netcdf/.

[43] http://www.openafs.org/.

[44] http://www.sdsc.edu/srb/index.php/Main_Page.

[45] http://www.isi.edu/~deelman/MCS/.

[46] http://www.cs.wisc.edu/condor/.

[47] http://www.myxaml.com/wiki/ow.asp?Workflows.

[48] http://www.jython.org/Project/index.html.

Chapter 11

Open Grid Services for Envisat and Earth Observation Applications

Luigi Fusco,
European Space Agency

Roberto Cossu,
European Space Agency

Christian Retscher,
European Space Agency

Contents

The ESA Science and Application Department of Earth Observation Programmes Directorate at ESRIN has focused on the development of a dedicated Earth Science grid infrastructure, under the name Earth Observation Grid Processing On-Demand (G-POD). This environment provides an example of transparent, fast, and easy access to data and computing resources. Using a dedicated Web interface, each application has access to the ESA operational catalogue via the ESA Multi-Mission User Interface System (MUIS) and to storage elements. It furthermore communicates with the underlying grid middleware, which coordinates all the necessary steps to retrieve, process, and display the requested products selected from the large database of ESA and third-party missions. This makes G-POD ideal for processing large amounts of data, developing services that require fast production and delivery of results, comparing scientist approaches to data processing, and permitting easy algorithm validation.

11.1 Introduction

Following the participation of the European Space Research Institute (ESRIN) at ESA in DataGrid, the first large European Commission funded grid project [1], the ESA Science and Application Department of Earth Observation Programmes Directorate has focused on the development of a dedicated Earth Science grid infrastructure, under the name Earth Observation Grid Processing on-Demand [2]. This generic grid-based environment (G-POD) ensures that specific Earth Observation (EO) data handling and processing applications can be seamlessly plugged into the system. Coupled with high performance and sizeable computing resources managed by grid technologies, G-POD provides the necessary flexibility for building a virtual environment that gives applications quick access to data, computing resources, and results. Using a dedicated Web interface, each application has access to a catalogue like the ESA Multi-Mission User Interface System (MUIS) and storage elements. It furthermore communicates with the underlying grid middleware, which coordinates all the necessary steps to retrieve, process, and display the requested products selected from the large database of ESA and third-party missions.

Grid On-Demand provides an example of transparent, fast, and easy access to data and computing resources. This makes G-POD an ideal environment for processing large amounts of data, developing services that require fast production and delivery of results, comparing approaches, and fully validating algorithms. Many other grid-based systems are being proposed by various research groups using similar and alternative approaches, although sharing the same ambition for improved integration of the emerging Information and Communication Technologies (ICT) technologies exploitable by the Earth Science community.

In the Sections 11.2 and 11.3 we give an overview of selected ESA Earth Observation missions and related software tools that ESA provides for facilitating data handling and analysis. In Section 11.4 we describe how the EO community can benefit from grid technology for data access and sharing. In this context, some examples of ESA and EU projects are described. Section 11.5 describes in detail the G-POD environment, its infrastructure, the intermediary layer developed to interface with the application, and the grid computer and storage resources, the Web portals. Different examples of EO applications integrated in G-POD are described in Section 11.6. Section 11.7 briefly documents the use of grid technology in Earth Science Knowledge Infrastructures. Conclusions are drawn in Section 11.8.

11.2 ESA Satellites, Instruments, and Products

This section briefly overviews the ESA European Remote Sensing satellite (ERS) and Envisat missions and the sensors on-board these satellites, with special attention to the data used in the context of ESA's activities on grids.

11.2.1 ERS-2

The ERS-2 Earth Observation mission [3] has been operating since 1995. The ERS-2 satellite carries a suite of instruments to provide data for scientific and commercial applications. ERS-1, the ERS-2 predecessor, was launched in July 1991 and was ESA's first sun-synchronous polar-orbiting remote sensing mission, operated until March 2000. It continued to provide excellent data, far exceeding its nominal lifetime. ERS-2 is nearly identical to ERS-1. The platform is based on the design developed for the French SPOT satellite. Payload electronics are accommodated in a box-shaped housing on the platform; antennas are fitted to a bearing structure. On-board ERS-2 there are seven instruments to support remote sensing applications: RA, ATSR, GOME, MWR, SAR, WS, and PRARE. In particular we wish to refer to:

- SAR: Synthetic Aperture Radar (SAR) wave mode provides two-dimensional spectra of ocean surface waves. For this function the SAR records regularly spaced samples within the image swath. The images are transformed into directional spectra providing information about wavelength and the direction of the wave systems. Automatic measurements of dominant wavelengths and directions will improve sea forecast models. However, the images can also show the effects of other phenomena, such as internal waves, slicks, small-scale variations in wind, and modulations due to surface currents and the presence of sea ice.
- GOME: The GOME instrument, which stands for Global Ozone Monitoring Experiment, is a newly developed passive instrument that monitors the ozone content of the atmosphere to a degree of precision hitherto unobtainable from space. This highly sophisticated spectrometer was developed by ESA in the record time of five years. GOME is a nadir-scanning ultraviolet and visible spectrometer for global monitoring of atmospheric ozone. It was launched on-board ERS-2 in April 1995. Since the summer of 1996, ESA has been delivering to users three-day GOME global observations of total ozone, nitrogen dioxide, and related cloud information, via CD-ROM and the Internet. A key feature of GOME is its ability to detect other chemically active atmospheric trace gases as well as the aerosol distribution.
- ATSR: The Along-Track Scanning Radiometer consists of an InfraRed Radiometer (IRR) and a Microwave Sounder (MWS). On-board ERS-1, the IRR is a four-channel infrared radiometer used for measuring sea-surface temperatures (SST) and cloud-top temperatures, whereas on-board ERS-2 the IRR is equipped with additional visible channels for vegetation monitoring.

11.2.2 Envisat

The Environmental Satellite (Envisat) [4] is an advanced polar-orbiting Earth Observation satellite that provides measurements of the atmosphere, ocean, land, and ice. The Envisat satellite has an ambitious and innovative payload that ensures the continuity of the data measurements of the ERS satellites. The Envisat data support Earth Science research and allow monitoring of the evolution of environmental and climatic changes. Furthermore, they facilitate the development of operational and commercial applications. On-board Envisat there are ten instruments: ASAR, MERIS, AATSR, GOMOS, MIPAS, SCIAMACHY, RA-2 (Radar Altimeter 2), MWR (Microwave Radiometer), DORIS (Doppler Orbitography and Radio-positioning), LRR (Laser Retro-Reflector). In particular we wish to refer to:

- ASAR: ASAR is the Advanced Synthetic Aperture Radar. Operating at C-band, it ensures continuity with the image mode (SAR) and the wave mode of the ERS-1/2 AMI (Active Microwave Instrument). It features enhanced capability in terms of coverage, range of incidence angles, polarization, and modes of operation. This enhanced capability is provided by significant differences in the instrument design: a full active array antenna equipped with distributed transmit/receive modules that provide distinct transmit and receive beams, a digital waveform generation for pulse 'chirp' generation, a block adaptive quantization scheme, and a ScanSAR mode of operation by beam scanning in elevation.
- MERIS: MERIS is a programmable, medium-spectral resolution imaging spectrometer operating in the solar reflective spectral range. Fifteen spectral bands can be selected by ground command, each of which has a programmable width and a programmable location in the 390 nm to 1040 nm spectral range. The instrument scans the Earth's surface by the so-called push-broom method. Linear CCD arrays provide spatial sampling in the across-track direction, while the satellite's motion provides scanning in the along-track direction. MERIS is designed so that it can acquire data over the Earth whenever illumination conditions are suitable. The instrument's 68.5° field of view around nadir covers a swath width of 1150 km. This wide field of view is shared between five identical optical modules arranged in a fan-shape configuration.
- AATSR: The Advanced Along-Track Scanning Radiometer (AATSR) is one of the Announcement of Opportunity (AO) instruments on-board Envisat. It is the most recent in a series of instruments designed primarily to measure Sea Surface Temperature (SST), following on from ATSR-1 and ATSR-2 on-board ERS-1 and ERS-2. AATSR data have a resolution of 1 km at nadir and are derived from measurements of reflected and emitted radiation taken at the following wavelengths: 0.55 μm, 0.66 μm, 0.87 μm, 1.6 μm, 3.7 μm, 11 μm, and 12 μm. Special features of the AATSR instrument include its use of a conical scan to give a dual view of the Earth's surface, on-board calibration targets, and use of mechanical coolers to maintain the thermal environment necessary for optimal operation of the infrared detectors.
- GOMOS: The Global Ozone Monitoring by Occultation of Stars instrument is a medium-resolution spectrometer covering the wavelength range from 250 nm

to 950 nm. The high sensitivity down to 250 nm required the design of an all-reflective optical system for the UVVIS part of the spectrum and the functional pupil separation between the UVVIS and the NIR spectral regions. Due to the requirement of operating on very dim stars (magnitudes ≤ 5), the sensitivity requirement for the instrument is very high. Consequently, a large telescope with 30 cm $\times$ 20 cm aperture had to be used in order to collect sufficient signals. Detectors with high quantum efficiency and very low noise had to be developed to achieve the required signal to noise ratios (SNR).

- MIPAS: The Michelson Interferometer for Passive Atmospheric Sounding is a Fourier transform spectrometer for the detection of limb emission spectra in the middle and upper atmosphere. It observes a wide spectral interval throughout the mid infrared with high spectral resolution. Operating in a wavelength range from 4.15 μm to 14.6 μm, MIPAS detects and spectrally resolves a large number of emission features of atmospheric trace gas constituents playing a major role in atmospheric chemistry. Due to its spectral resolution capabilities and low-noise performance, the detected features can be spectroscopically identified and used as input to suitable algorithms for extracting atmospheric concentration profiles of a number of target species.
- SCIAMACHY: The Scanning Imaging Absorption Spectrometer for Atmospheric Cartography instrument is an imaging spectrometer whose primary mission objective is to perform global measurements of trace gases in the troposphere and in the stratosphere. The solar radiation transmitted, backscattered, and reflected from the atmosphere is recorded at high resolution (0.2 μm to 0.5 μm) over the range 240 nm to 1700 nm, and in selected regions between 2.0 μm and 2.4 μm. The high resolution and the wide wavelength range make it possible to detect many different trace gases despite low concentrations. The large wavelength range is also ideally suited for the detection of clouds and aerosols. SCIAMACHY has three different viewing geometries: nadir, limb, and sun/moon occultations, which yield total column values as well as distribution profiles in the stratosphere and even the troposphere for trace gases and aerosols.

11.3 Example of Specialized User Tools for Handling ESA Satellite Data

To facilitate users in accessing ERS and Envisat instrument's data products, ESA has developed a set of software utilities with the contribution and validation of key instrument scientists. All these tools can be downloaded for free at [5].

Among these tools, some of them have been integrated in the ESA grid environment, and for this reason we briefly describe them in the following. Greater details can be obtained from the aforementioned Website.

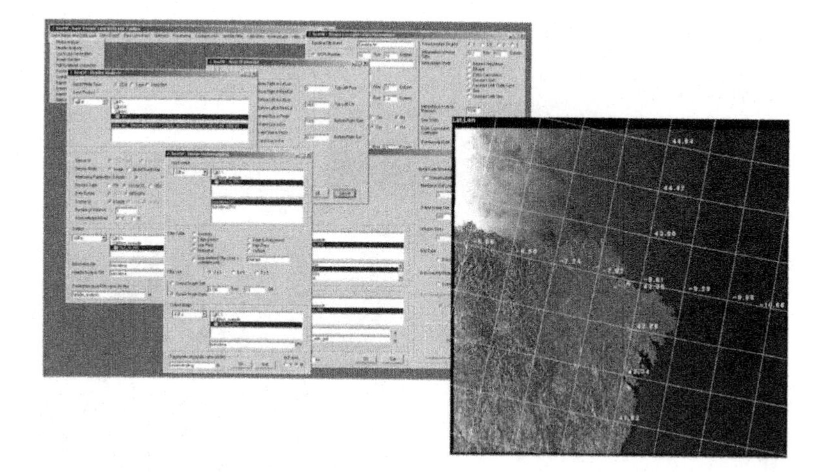

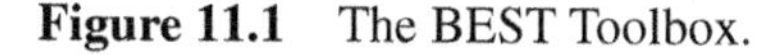

Figure 11.1 The BEST Toolbox.

11.3.1 BEST

The Basic Envisat SAR Toolbox (BEST) is a collection of executable software tools that has been designed to facilitate the use of ESA SAR data. The purpose of the Toolbox is not to duplicate existing commercial packages, but to complement them with functions dedicated to the handling of SAR products obtained from ASAR and AMI on-board Envisat, ERS-1, and ERS-2, respectively. BEST has evolved from the ERS SAR Toolbox (see Figure 11.1).

The Toolbox operates according to user-generated parameter files. The interface does not include a display function. However, it includes a facility to convert images to TIFF or GeoTIFF format so that they can be read by many commonly available visualization tools. Data may also be exported in the BIL format for ingestion into other image processing software.

The tools are designed to achieve the following functions: data import and quick look, data export, data conversion, statistical analysis, resampling, co-registration, basic support for interferometry, speckle filtering, and calibration.

11.3.2 BEAM

The Basic ERS & Envisat (A)ATSR and MERIS Toolbox is a collection of executable tools and APIs (Application Programming Interfaces) that have been developed to facilitate the utilization, viewing, and processing of ERS and Envisat MERIS, (A)ATSR, and (A)SAR data. The purpose of BEAM is to complement existing commercial packages with functions dedicated to the handling of MERIS and AATSR products. The main components of BEAM are:

- A visualization, analyzing, and processing software (VISAT).
- A set of scientific data processors running either from the command line or invoked by VISAT.

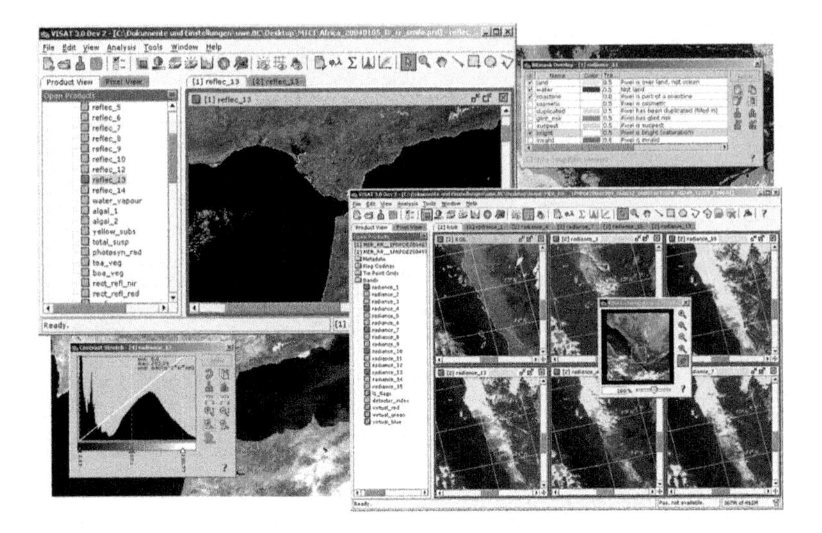

Figure 11.2 The BEAM toolbox with VISAT visualization.

- A data product converter tool allowing a user to convert raw data products to RGB images, HDF-5, or the BEAM-DIMAP standard format.
- A Java API that provides ready-to-use components for remote sensing related application development and plug-in points for new BEAM extensions.
- MERIS/(A)ATSR/(A)SAR product reader API for ANSI C and IDL, allowing read access to these data products using a simple programming model.

VISAT (see Figure 11.2) and the scientific data processors use a simple data input/ output format, which makes it easy to import ERS and Envisat data in other imaging applications. The format is called DIMAP and has been developed by SPOT-Image in France. The BEAM software uses a special DIMAP profile called BEAM-DIMAP, which has the following characteristics:

- A single product header (XML) containing the product metadata.
- An associated data directory containing ENVI-compatible images for each band.

Each image in the directory is composed of a header file (ASCII text) and an image data file (flat binary) source code. The complete BEAM software has been developed under the GNU public license and comes with full source code (Java and ANSI C). All main components of the toolbox are programmed in pure Java for maximum portability. The product reader API for C has been developed exclusively with the ANSI-compatible subset of the C programming language. The BEAM software has been successfully tested under MS Windows 9X, NT4, 2000, and XP, as well as under Linux and Solaris operating systems. BEAM is intended to also run on other Java-enabled UNIX derivates, e.g., Mac OS X.

11.3.3 BEAT

The Basic ERS and Envisat Atmospheric Toolbox aims to provide scientists with tools for ingesting, processing, and analyzing atmospheric remote sensing data. The project consists of several software packages, with the main packages being BEAT and VISAN. The BEAT package contains a set of libraries, command line tools, and interfaces to IDL, MATLAB, FORTRAN, and Python for accessing data from a range of atmospheric instrument product files. The VISAN package contains an application that can be used to visualize and analyze data retrieved using the BEAT interface. The primary instruments supported by BEAT are GOMOS, MIPAS, SCIAMACHY (Envisat), GOME (ERS-2), OMI, TES, and MLS (Aura), as well as GOME-2 and IASI (MetOp). BEAT, VISAN, and an MIPAS processor called GeoFit are provided as Open Source Software, enabling the user community to participate in further development and quality improvements.

The core part of the toolbox is the BEAT package itself. This package provides data ingestion functionalities for each of the supported instruments. The data access functionality is provided via two different layers, called BEAT-I and BEAT-II:

- BEAT-I: The first layer of BEAT provides direct access to data inside each file that is supported by BEAT. The supported instruments include GOMOS, MIPAS, SCIAMACHY, GOME, OMI, TES, and MLS. All product data files are accessible via the BEAT-I C library. On top of this C library there are several interfaces available to directly ingest product data using, e.g., FORTRAN, IDL, MATLAB, and Python. Furthermore, BEAT also comes with a set of command line tools (beatcheck, beatdump, and beatfind).
- BEAT-II: The second layer of BEAT provides an abstraction to the product data to make it easier for the user to get the most important information extracted. Using only a single command you will be able to ingest product data into a set of flexible data types. These predefined data types make it easier to compare similar data coming from different instruments and also simplify the creation of general visualization routines. Furthermore, the BEAT-II layer provides some additional functions to manipulate and import/export these special data types. The layer 2 interface is built on top of the BEAT-I C library, but BEAT-II also supports reading of additional products that are stored in, e.g., ASCII, HDF4, or HDF5 format. As for BEAT-I, all BEAT-II functionality is accessible via the BEAT-II C. Moreover, BEAT contains interfaces of BEAT-II for FORTRAN, IDL, MATLAB, and Python, and a command line tool.
- VISAN: VISAN (see Figure 11.3) is a cross-platform visualization and analysis application for atmospheric data, where the user can pass commands in Python language. VISAN provides powerful visualization functionality for two-dimensional plots and worldplots. The Python interfaces for BEAT-I and BEAT-II are included so one can directly ingest product data from within VISAN. By using the Python language and some additional included mathematical packages it is possible to perform an analysis on selected data.

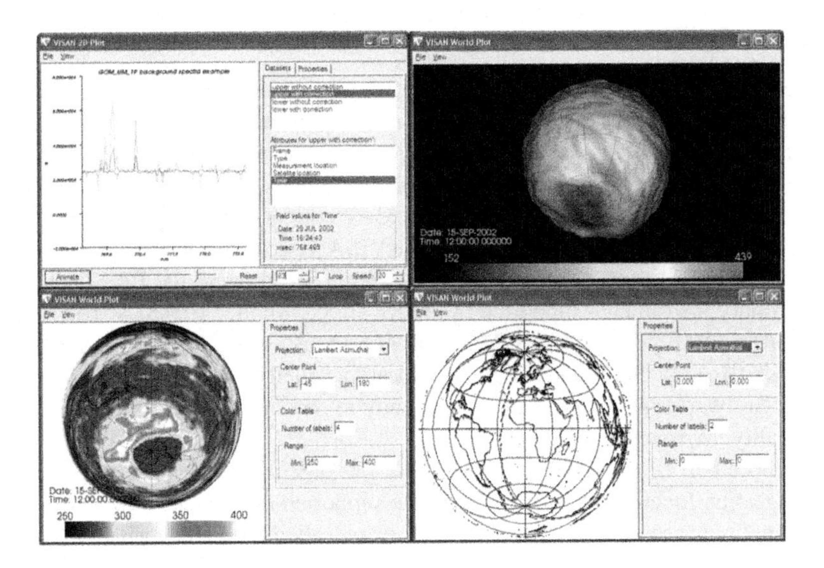

Figure 11.3 The BEAT toolbox with VISAN visualization.

- GeoFit: BEAT also contains the GeoFit software package, which is used to process MIPAS special mode measurements.

11.4 Grid-Based Infrastructures for EO Data Access and Utilization

While conducting their research, Earth scientists are often hindered by difficulties locating and accessing the right data, products, and other information needed to turn data into knowledge, e.g., interpretation of the available data. Data provision services are far from optimal for reasons related both to science and infrastructure capabilities. The process of identifying and accessing data typically takes up the most time and money. Of the different base causes of this, those most frequently reencountered relate to:

- *The physical discontinuity of data.* Data are often dispersed over different data centers and local archives distributed all over Europe and abroad and, inherent to this, the different policies applied (e.g., access and costs), the variety of interoperability, confidentiality, and search protocols as well as the diversity of data storage formats. To access a multitude of data storage systems, users need to know how and where to find them and need a good technical/system background to interface with the individual systems. Furthermore, often only the metadata catalogues can be accessed online, while the data themselves have to be retrieved offline.

- *The diversity of (meta)data formats.* New data formats are being introduced daily, not only due to the individual needs of a multitude of data centers, but also due to advances in science and instrumentation (satellites and sensors) creating entirely new types of data for research.
- *The large volume of data.* The total quantity of information produced, exchanged, and requested is enormous and is expected to grow exponentially during the next decades, even faster than it did before. This is partly the result of the revolution in computational capacity and connectivity and advances in hardware and software, which, combined together, are expanding the quality and quantity of research data and are providing scientists with a much greater capacity for data gathering, analysis, and dissemination [6]. For example, the ESA Envisat satellite [4] launched in early 2002, with ten sensors on-board, increases the total quantity of data available each year by some 500 Terabytes, while the ESA ERS satellites produced roughly five to ten times less data per year. Moreover, large volume data access is a continuous challenge for the Earth Science community. The validation of Earth remote sensing satellite instrument data and the development of algorithms for performing the necessary calibration and geophysical parameters extraction often require a large amount of processing resources and highly interactive access to large amounts of data to improve the statistical significance of the process. The same is true when users need to perform data mining or fusion for specific applications. As an alternative to the traditional approach of transferring data products from the acquisition/storage facilities to the user's site, ad-hoc user-specified data processing modules could be moved in real-time to available processing facilities situated more optimally for accessing the data, in order to improve the performance of the end-to-end EO data exploitation process.
- *The unavailability of historic data.* Scientists do not only work with 'fresh' data, they also use historic data, e.g., global change research, over multiple time periods. Here, different problems can be distinguished. First, it is evident that often no metadata are defined, or no common metadata standards are being used, and auxiliary knowledge needed by scientists to understand and use the data is missing, e.g., associated support information in science and technical reports. Although the problem also exists for fresh data, it is exacerbated when using historic data. Metadata will be at the heart of every effort to preserve digital data in the next few decades. It will be used to create maintenance and migration programs and will provide information on collections for the purpose of orienting long-term preservation strategies and systems [7]. Second, there are insufficient preservation policies in place for accessing historical data. After longer periods of time, new technologies may have been introduced, hardware and software upgraded, formats may have changed, and systems replaced. For example, it is almost impossible today to read files stored on 8-inch floppy disks that were popular just 25 years ago. Vast amounts of digital information from just 25 years ago are lost for all practical purposes [8].

- *The many different actors involved.* Science is becoming increasingly international and interdisciplinary, resulting in an increased total number of different actors involved (not only human). For example, ESA currently serves approximately 6000 users in the Earth Observation domain, many of whom need to exchange data, information, and knowledge.

The International Council for Science, for example, deals with data access issues on a global scale [6]. In Europe, different initiatives are supported by the European Commission (EC), e.g., as part of their specific action on research infrastructures (part of the 6th Framework Programme), which aims to promote the development of a fabric of research infrastructures of highest quality and performance, and their optimum use on a European scale to ensure that researchers have access to the data, tools, and models they need.

ESA is participating in different initiatives focusing, in particular, on the use of emerging technologies for data access, exploitation, user information services and long-term preservation. For example, [9] provides an overview of the use of grid, Web services, and Digital Library technology for long-term data preservation. The same technologies can be used for accessing data in general. Moreover, emerging technologies can support data access, e.g., via infrastructures based on high-speed networks that could drastically speed up the transfer of the enormous quantities of data; the use of grids for managing distributed heterogeneous resources including storage, processing power, and communication, offering the possibility to significantly improve data access and processing times; and digital libraries that can help users locate data via advanced data mining techniques and user profiling. A shared distributed infrastructure integrating data dissemination with generic processing facilities shall be considered a very valuable and cost-effective approach to support Earth Science data access and utilization.

Of the specific technologies that have had an important role in the ES community, Web services in particular have played a key role for a long time. Web services technologies have emerged as a de facto standard for integrating disparate applications and systems using open standards. One example of a very specialized ES Web service is the Web mapping implementation specification proposed by the OpenGIS Consortium [10]. Thanks to Web services, the Internet has become a platform for delivering not only data but also, most importantly, services. After a Web service is deployed on a Web server and made discoverable in an online registry of services, other applications can discover and invoke the deployed service to build larger, more comprehensive services, which in turn deliver an application and a solution to a user. Web-based technologies also provide an efficient approach for distributing scientific data, extending the distribution of scientific data from a traditional centralized approach to a more distributed model. Some Web services address catalogue services to help users to locate data sets they need or at least narrow the number of data sets of interest from a large collection. The catalogue contains metadata records describing the datasets.

As discussed in Chapter 9 of the present volume, Web services provide the fundamental mechanism for discovery and client-server interaction and have become a widely accepted, standardized infrastructure on which to build simple interactions. On the other hand, grids were originally motivated by the need to manage groups of machines for scientific computation. For these reasons, Web services and grids are somehow complementary and their combination results in grid services (e.g. Open Grid Services Architecture).

In the following subsections we briefly describe some specific European experiences involving Earth Science users at various levels for data access, sharing, and handling as well as service provisions based on interfacing grid infrastructures.

11.4.1 Service Support Environment

The Service Support Environment (SSE) can be considered as a market place that interconnects users (e.g. customers) and Earth observation providers (data, value-adding industry, and service industry), and allows them to register and provide their services via the SSE portal [11]. Depending on their profiles, SSE users gain access to a set of services on the SSE portal via an Internet connection.

The SSE is aimed at providing an opportunity for improving the market expansion and penetration of existing or prototyped Earth observation products and services, as well as into the Geographic Information Systems (GIS) world, through an enabling, open environment for service providers and potential users. The SSE will also offer the European development and service industry the opportunity to take a leading role in the installation, maintenance, and operation on request of personalized systems and services related to the future EO related business-to-business (B2B) market.

The SSE service directory provides access to a continuously expanding set of basic and complex Earth observation and GIS services, and also a large variety of services from a diverse set of contributors such as space agencies, data processing centers, data providers, educational establishments, private companies, and research centers.

11.4.2 GeoNetwork

The United Nations (UN) Food and Agriculture Organization (FAO) has developed a standardized and decentralized spatial information management environment called GeoNetwork [12]. The GeoNetwork Open Source system implements and extends the ISO 19115 geographic metadata standard. It facilitates sharing of geographically referenced thematic information between different FAO Units, UN agencies, NGOs, and other institutions. GeoNetwork is designed to enable access to georeferenced databases, cartographic products, and related metadata from a variety of sources, enhancing the spatial information exchange and sharing between organizations and their audience, by using the capacities of the Internet. This approach of geographic information management aims to give a wide community of spatial information users easy and timely access to available spatial data and existing thematic maps to support informed decision making. ESA/ESRIN hosts a GeoNetwork node.

GeoNetwork has improved the accessibility of a wide variety of data, together with the associated information/metadata, at different scales and from multidisciplinary sources, organized and documented in a standard and consistent way. This has enhanced the data exchange and sharing between the organizations, avoiding duplication, and has increased the cooperation and coordination of efforts in collecting data. The data are made available to benefit everyone, saving resources and at the same time preserving data and information ownership.

FAO, the World Food Programme (WFP), and the United Nations Environment Programme (UNEP) have combined the strategy to effectively share their spatial databases including digital maps, satellite images, and related statistics. The three agencies make extensive use of computer-based data visualization tools, based on Open Source, proprietary Geographic Information System, and Remote Sensing (RS) software, used mostly to create maps that combine various layers of information. GeoNetwork offers a single entry point for accessing a wide selection of maps and other spatial information stored in different databases worldwide.

11.4.3 CCLRC DataPortal and Scientific Metadata Model

The Central Laboratory of the Research Councils (CCLRC), on behalf of the UK research community, operates on a multitude of next-generation of powerful scientific facilities and recognizes the vital role that e-Science will have for their successful exploitation. These facilities (synchrotrons, satellites, telescopes, and lasers) will collectively generate many Terabytes of data every day. Their users will require efficient access to geographically distributed leading-edge data storage, computational and network resources in order to manage and analyze these data in a timely and cost-effective way. Convenient access to secure and affordable medium- to long-term storage of scientific data is important to all areas of CCLRC's work and to all users of CCLRC's facilities. It will help to facilitate future cross-disciplinary activities and will constitute a major resource within the UK e-Science grid. CCLRC is exploring the opportunities within this context for developing a collaborative approach to large-scale data storage spanning the scientific program of CCLRC and the other Research Councils. To support data description and facilitate data reuse, CCLRC has developed the scientific metadata model and the CCLRC DataPortal [13]. In addition, CCLRC is collaborating with the San Diego Super Computing Centre (SDSC) on the development and deployment of the Storage Resource Broker (SRB) for large-scale, cross-institutional data management and sharing, bringing secure long-term data storage to the scientist's desktop and supporting secure international data sharing amongst peers. In collaboration with the Universities of Reading and Manchester, CCLRC will be investigating the state of the art in long-term metadata management and the usage of Data Description Languages for data curation.

ESA and CCLRC cooperate in many Earth Science related technologies and application domains. In particular it is worthwhile to mention the cooperation for long-term scientific data and knowledge preservation via the CASPAR project [14] (cf. Section 11.7.4).

11.4.4 Projects@ReSC

The Reading e-Science Center (ReSC) [15] is very active in promoting e-Science methods in the environmental science community. As for other EO domains, modern computer simulations of the oceans and atmosphere produce large amounts of data on the Terabyte scale. Consequently, data providers need a manageable system for storing these data sets whilst enabling the data consumer to access them in a convenient and secure manner. The matter is complicated by the plethora of file formats (e.g. NetCDF, HDF, and GRIB) that are used for holding environmental data. For this reason ReSC has set up database management systems for storing and manipulating gridded data. Among operational and demonstration projects, the following examples are worth introducing here:

- Grid Access Data Service (GADS), a Web service that provides access to distributed climatological data in an intuitive and flexible manner. Users do not need to know any details about how, where, or in what format the data are stored. Data can be downloaded in a variety of formats (e.g., netCDF and GRIB) and the service is readily extensible to accommodate new formats.
- GODIVA (Grid for Ocean Diagnostics, Interactive Visualization and Analysis) allows users to interactively select data from a file access server for download and for creating movies on the fly. Recent features include the visualization of environmental data via the Google Maps and Google Earth clients [16].

11.4.5 OPeNDAP

An Open Source Project for a Network Data Access Protocol [17] is a data transport architecture and protocol widely used by Earth scientists. The protocol is based on HTTP, and the current specification includes standards for encapsulating structured data, annotating the data with attributes, and adding semantics that describe the data.

An OPeNDAP server can handle an arbitrarily large collection of data in any format including a user-defined format. OPeNDAP offers the possibility to retrieve subsets of files, and to aggregate data from several files in one transfer operation. OPeNDAP is widely used by governmental agencies such as the National Aeronautics and Space Administration (NASA) and the National Oceanic & Atmospheric Administration (NOAA) to serve satellite, weather, and other observed Earth Science data.

11.4.6 DataGrid and Follow-up

DataGrid was the first large-scale international grid project and the first aiming to deliver a grid infrastructure to several different Virtual Organizations (High Energy Physics, Biology, and Earth Observation) simultaneously. The objective was to build a next-generation computing infrastructure, providing intensive computation and analysis of shared large-scale databases, from hundreds of Terabytes to Petabytes, across widely distributed scientific communities. After a very successful final review by the European Commission, the DataGrid project was completed at the end of March 2004.

Many of the products (e.g., technologies and infrastructure) of the DataGrid project have been included in the follow-up EU grid project called Enabling Grids for E-sciencE (EGEE) [18], already introduced in Chapter 10 of this book. EGEE, funded by the EC Framework Programme (FP), aims to develop a European-wide service grid infrastructure available to scientists 24 hours a day. The EGEE project also focuses on attracting a wide range of new users to the grid. The second 2-year phase of the project started 1 April 2006 and includes:

- More than 90 partners in 32 countries, organized in 13 Federations.
- A grid infrastructure spanning almost 200 sites across 39 countries.
- An infrastructure of over 20000 CPUs available to users 24 hours a day, 7 days a week.
- About 5 Petabytes of storage.
- Sustained and regular workloads of 20000 jobs/day.
- Massive data transfers > 1.5 Gigabytes/s.

A few companion DataGrid and EGEE projects have been focusing on Earth science applications, responding to Earth science key requirements, such as handling spatial and temporal metadata, near-real-time (NRT) features, dedicated data modeling, and data assimilation. ESA has been involved in various workshops and publications organized specifically and jointly by the grid and the Earth Science community, for example:

- EOGEO: It exists to deliver sustainable Earth Observation and Geospatial Information and Communication Technologies (EOGEO ICTs), which are vital to the operation of the Civil Society Organization and to the well-being of individual citizens [19].
- CEOS: The purpose of the Committee on Earth Observation Satellites (CEOS) Task Team is to investigate the applicability of grid technologies for CEOS needs, to share experience gained from the effective use of these technologies, and to make recommendations for their application [20].
- ESA grid and e-collaboration workshops: ESA periodically organizes workshops dedicated to reviewing the status of grid and e-collaboration projects for the Earth science community [21].

11.4.7 CrossGrid

CrossGrid [22] is an example of other EC Information Society Technologies (IST) FP5 funded projects that are focusing on key functionalities dedicated to the Earth science community. This R&D project aimed at developing techniques for real-time, large-scale grid-enabled simulations and visualizations. The issues addressed include:

- Distribution of source data.
- Simulation and visualization.
- Virtual time management.
- Interactive simulation.
- Platform-independent virtual reality.

The application domains addressed by the CrossGrid project include environmental protection, flood prediction, meteorology, and air pollution modeling.

With regard to floods, the usefulness of grid technology for supporting crisis teams is being studied. The challenges in this task are the acquisition of significant resources at short notice, NRT response, the combination of distributed data management and distributed computing, the computational requirements for the combination of hydrological (snowmelt-rainfall-runoff) and hydraulic (water surface elevation, velocity, dam breaking, damage assessment etc.) models, and, eventually, mobile access under adverse conditions.

The interactive use and scalability of grid technology is being investigated, in order to meet atmospheric research and application user community requirements. A complete application involves grid tools that enable remote, coordinated feedback from atmospheric models and wave models, based on local coastal data and forced by wind fields generated by atmospheric components of the system.

11.4.8 DEGREE

DEGREE (Dissemination and Exploitation of GRids in Earth sciencE) [23] is a coordinated action, funded within the last grid call of EC FP6. It is proposed by a consortium of Earth Science (ES) partners that integrates research institutes, European organizations, and industries, complementary in activity and covering a wide geo-cultural dimension, including Western Europe, Russia, and Slovakia. The project aims to promote the grid culture within the different areas of ES and to widen the use of grid infrastructures as platforms for e-collaboration in the science and industrial sectors and for select thematic areas that may immediately benefit from it.

DEGREE aims to achieve this by showing how grid services can be integrated within key selected ES applications, approaching the operational environment and shared within thematic community areas. The DEGREE project will also tackle certain aspects presently considered as barriers to the widespread uptake of the technology, such as the perceived complexity of the middleware and insufficient support for certain required functionality. The ES grid expertise, application tools, and services developed so far will be promoted within the DEGREE consortium and throughout the ES community. Collective grid expertise gathered across various ES application domains will be exchanged and shared in order to improve and standardize application-specific services. The use of worldwide grid infrastructures for cooperation in the extended ES international community will also be promoted.

In particular, the following objectives are to be achieved:

- Disseminate, promote uptake of grid in a wider ES community, and integrate newcomers.
- Reduce the gap between ES users and grid technology.
- Explain and convince ES users of grid benefits and capability to tackle new and complex problems.

11.5 ESA Grid Infrastructure for Earth Science Applications

In previous sections we analyzed how Web services and grid technologies can complement each other forming so-called grid services.

The ESA-developed Grid on-Demand Service Infrastructure allows for autonomous discovery and retrieval of information about data sets for any area of interest, exchange of large amounts of EO data products, and triggering concurrent processes to carry out data processing and analysis on-the-fly.

Access to grid computing resources is handled transparently by the EO grid interfaces that are based on Web services technologies (HTTP, HTTPS, and SOAP with XML) and developed by ESA within the DataGrid project. As a typical application, the generation of a 10-day composite (e.g., Normalized Difference Vegetation Index (NDVI)) over Europe derived from Envisat/MERIS data involves the reading of some 10–20 Gigabytes of Level 2 MERIS data for generation of a final Level 3 product of some 10–20 Megabytes, with a great saving of data circulation and network bandwidth consumption.

In the following, we analyze in detail the Grid on-Demand Service Infrastructure.

11.5.1 Infrastructure and Services

Following the successful experience in the EU DataGrid project (2001–2004) [1], in which the focus was to demonstrate how Earth Observation could take benefit from the large infrastructure deployed by the High Energy Physics community in Europe, the Grid on-Demand Infrastructure and Services project was initiated. Since then it has demonstrated how internal and external users can benefit from a very articulated organization of applications that can interface locally and remotely accessible computing resources, in a way that is completely transparent to the Earth Science end user.

Using an ubiquitous Web interface, each application has access to the ESA catalogue and storage facilities, enabling the definition of a new range of Earth Observation services.

The underlying grid middleware coordinates all the necessary steps to retrieve, process, and display the requested products selected from a vast catalogue of remote sensing data products and third-party databases. The integration of Web mapping and EO data services using a new generation of distributed Web applications and the OpenGIS [10] specification provided a powerful new capability to request and display Earth Observation data products in a given geotemporal coverage area.

The ESA Grid on-Demand Web portal [2] is a demonstration of a generic, flexible, secure, re-usable, distributed component architecture using grid and Web services to manage distributed data and computing resources. Specific and additional data handling and application services can be seamlessly plugged into the system. Coupled with the high-performance data processing capability of the grid, it provides the necessary flexibility for building an application for virtual communities with quick accessibility to data, computing resources, and results.

At present, the ESRIN-controlled infrastructure has a computing element (CE) of more than 150 PCs, mainly part of four clusters with storage elements of about 100 Terabytes, all part of the same grid LAN in ESRIN, partially interfaced to other grid elements in other ESA facilities such as the European Space Research and Technology Centre (ESTEC), the European Space Astronomy Centre (ESAC), and EGEE.

The key feature of this grid environment is the layered approach based on the GRID-ENGINE, which interconnects the application layer with different grid middleware (at present interfaced with three different brand/releases of middleware: Globus Toolkit 4.0 [24], LCG 2.6 [25], and gLite 3.0 [26]). This characteristic enables the clear separation and development path between the Earth Observation applications and the middleware being used.

11.5.2 The GRID-ENGINE

The GRID-ENGINE is an intermediary layer developed to interface the application and the grid computer and storage resources. In computational terms, the GRID-ENGINE is an application server accessed by SOAP Web services that enables the instantiation of different services. These services are the responsibility of an application manager that defines and implements all the application-specific requirements and interfaces, thus enabling their direct parameterization by the users.

The services are made of script templates that define three major operations: the preparation phase, the wrapper execution, and the completion phase.

In the preparation phase the template scripts allow the application developer to define the execution of auxiliary application templates that will enable the correct parameterization of the application. This might involve requests to the storage catalogue, elaborations to define specific parameters, and the description of all the necessary application input and auxiliary files.

After this preparatory phase, the wrapper execution module will evaluate the degree of parallelism supported by the application. Currently, only two main factors will be taken into consideration. These are the required data files and their spatial (in geographical terms) distribution. The first case is for services that elaborate outputs directly and independently based on the inputs (n inputs to n outputs approach). An automatic splitter algorithm was implemented based on the application computational and data weight, and the user permissions. On the other hand, for applications that require n input files for the elaboration of one or more files, a spatial or geo-splitter method was defined that will try to minimize the computational time required based on the resources available. Although of limited usefulness for other domains, this method was born for and its usefulness has been demonstrated in the Earth Observation and Geosciences domains, where the data are spatially distributed in nature and the spatial integration methods are common (e.g., elaboration of global maps of environmental variables such as vegetation, chlorophyll, or water vapor from independent measures stored in different files). By dividing the spatial domain (e.g., continents or latitude/longitude boxes), a straightforward division of the corresponding process is achieved.

The applications are then submitted to the computing elements and their state is automatically monitored by the system until their completion (successful or not). In the case of a job failure, the user can retrieve directly from the Web portal the standard error and standard out of the application and report the error to the system administrator or the application manager.

The completion phase terminates the service instantiation. As in the preparation phase, the application manager is allowed to define auxiliary applications that might analyze, register, or store the results obtained. All the resulting data resources, not specifically stored as such by the application manager, will be automatically cleaned and deleted by the system.

On top of this, the GRID-ENGINE allows the definition of simple service chaining (more in the line of information flow) where the services can be stitched together with their results being automatically defined as input parameters for the subsequent services. This capability allows the definition of generic services that can be reused in diverse domains (e.g., image and charts creation, image analysis, and geographical data re-projection).

The parameters necessary to execute all the templates of the three phases and the job chaining definition are sent directly from the Grid on-Demand Web portal using SOAP through a secure channel. With the necessary variables requested by the user and the parameters defined by the application manager for the actual service, the Web portal will send to the GRID-ENGINE all the necessary information for the instantiation of all templates defining the service.

All necessary grid operations performed in all phases, such as applications and data files transfer, grid job status, exception, and error management, are virtualized in order to enable the development and integration of the different grid concepts and implementations (e.g., Globus, LCG, and gLite). Because of the operational nature of the infrastructure, in terms of quality of service and maintenance requirements, the supported grid middleware is restrained to Globus Toolkit 4.0 and LCG 2.6 (with gLite in testing phase). Even though the Web Services Resource Framework (WSRF) actually demonstrates an enormous potential, its current use in this infrastructure is being limited to proof-of-concepts experiments and for test trials in the development environment. The current framework implementations tested so far (in Java, C++, and .NET) have shown new application development paths, but together with old shortcomings and instabilities that are unsuitable for an environment that needs to guarantee a near-real-time production level. As new developments and more stable and mature specifications arise, its integration will be performed.

11.5.3 The Application Portals

While the grid middleware provides low-level services and tools, the EO applications need to access the available grid resources and services through user-friendly application portals connected to back-end servers. The back-end servers then access the grid using the low-level grid middleware toolkits.

The ESA Grid on-Demand portal demonstrates the integration of several technologies and distributed services to provide an end-to-end application process, capable of being driven by the end user. The portal integrates:

- User authentication services.
- Web mapping services for map image retrieval and data geolocation.
- Access to metadata catalogues such as the ESA Multi-Mission User Interface System to identify the data sets of interest and access the ESA Archive Management System (AMS) to retrieve the data.
- Access to grid FTP transfer protocols to stage the data to the grid.
- Access to the grid computing elements and storage elements to process the data and retrieve the results in real time.

The architectural design of the Grid on-Demand portal application includes a distinct application-grid interfacing layer (see Figure 11.4). The core of the interface layer is implemented by the EO GRID-ENGINE, which receives Web service requests from grid client applications and organizes their execution using the available services provided by several different grids.

The underlying grid infrastructure coordinates all of the steps necessary to retrieve process and display the relevant images, selected from a vast range of available satellite-based EO data products. Using a new generation of distributed Web applications and OpenGIS specifications, the integration of Web mapping and EO data services provides a powerful capability to request and display Earth Observation

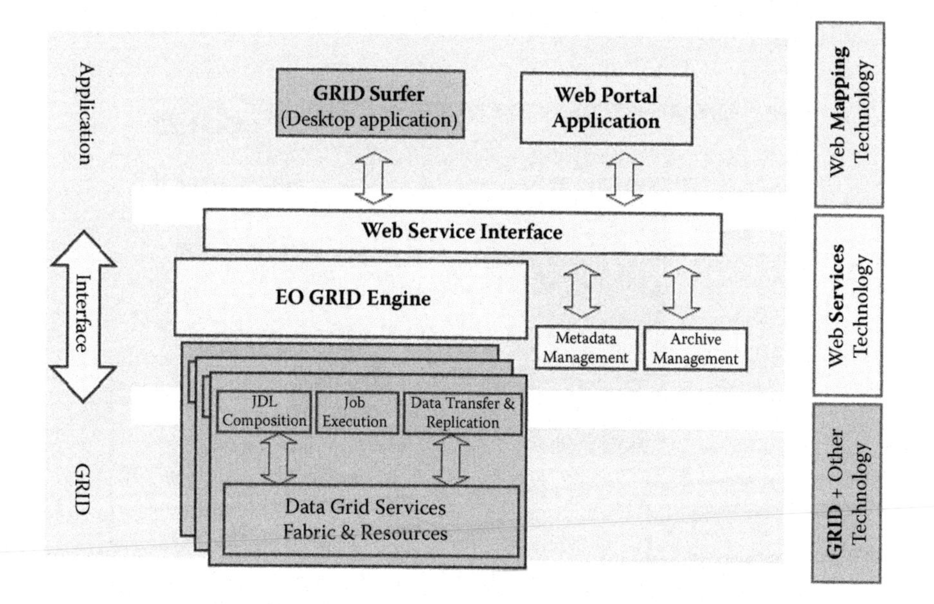

Figure 11.4 The architecture model for EO Grid on-Demand Services.

information in any given time range and geographic coverage area. The main functionality offered by the Grid on-Demand environment can be summarized as follows:

- It supports science users with a common accessible platform for focused e-collaborations, e.g., as needed for calibration and validation, development of new algorithms, or generation of high-level and global products.
- It acts as a unique and single access point to various metadata and data holdings for data discovery, access, and sharing.
- It provides the reference environment for the generation of systematic application products coupled with direct archives and NRT data access.

11.5.3.1 An Example of an Application Portal: Computation and Validation of Ozone Profile Calculation Using the GOME NNO Algorithm

To demonstrate the Web portal, in the following we refer to a specific application, which calculates the ozone profiles using the GOME NNO algorithm and performs validation using ground-based observation data. The user selects the algorithm, geographic area, and time interval, and the Web portal retrieves the corresponding Level 1 data orbit numbers by querying MUIS, the ESA EO product catalogue. Using the orbit numbers, it is then possible to query a Level 2 metadata catalogue to retrieve the current status of the requested orbits. The Level 2 orbits may be already processed, not yet processed, or currently being processed.

In the first case, the Service Layer Broker searches the grid replica catalogue to obtain the Level 2 data logical file names, and then retrieves the data from the physical grid locations. The processed orbits are then visualized by the Web portal (see Figure 11.5).

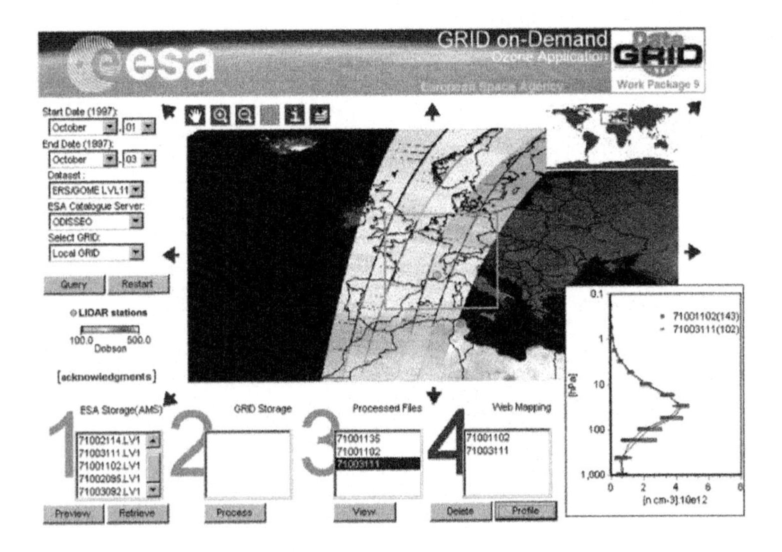

Figure 11.5 Web portal Ozone Profile Result Visualization.

In the second case, the EO product catalogue also provides the necessary information to retrieve the Level 1 orbit data from EO archives. After the Level 1 data have been transferred to grid storage, jobs are submitted to the grid in order to process the orbits. Once the processing has terminated, the resulting Level 2 products are also transferred to grid storage (from the WNs) and the logical file names are registered in the replica catalogue. A Level 2 metadata catalogue is also updated.

In the third case (currently orbits are being processed), the request ID is appended to the current job ID and awaits the job conclusion as in the second case.

For the validation application, the Web portal has a dedicated graphical user interface (GUI) where the user accesses the Lidar catalogue of L'Institut Pierre-Simon Laplace (IPSL) and cross checks that information with the ESA catalogue. It returns the orbit information, file names for the Light Detection and Ranging instrument (Lidar), and calculates the necessary geographical parameters for input to the validation job. The input parameters are translated into grid job parameters, generating several jobs for each of the corresponding Lidar files. The status of the different jobs can be viewed using the portal, and when all jobs are terminated the Web portal is used to retrieve and view the results.

11.6 EO Applications Integrated on G-POD

The first significant example of the ESRIN G-POD system is described in [27]. A GOME Web portal was set up, which constitutes a prototype integration of grid and Web services and allows the users to select a given geographical area and time period, retrieve the corresponding GOME Level 1 products, and process them into Level 2 products. The processing load is automatically distributed across several available grid resources, in a completely transparent way to the user.

Following the success of this test bed, other EO applications (and related Web portals) were developed. Some of these applications are now fully operational and available through the ESA EO grid portal. In the following, we describe some services that have been obtained by integrating EO processing toolboxes on the grid and by setting up ubiquitous user-friendly Web portals.

11.6.1 Application Based on MERIS and AATSR Data and BEAM Tools

11.6.1.1 MERIS Mosaic as Displayed at EO Summit in Brussels, February 2005

Using spectral bands 2, 3, 5, and 7 [28] from the entire May to December 2004 data set of Envisat/MERIS Reduced Resolution Level 2 products (1561 satellite orbit passes), the Grid on-Demand Services and Infrastructure produced a 1.3 km resolution TIF image (see Figure 11.6) that maximizes the sun light in both hemispheres using the MERIS PR/COM processor available on Grid on-Demand.

This service, motivated mostly by public relations teams, aims at the on-demand generation of mosaics using MERIS Level 2 products. These products, which are

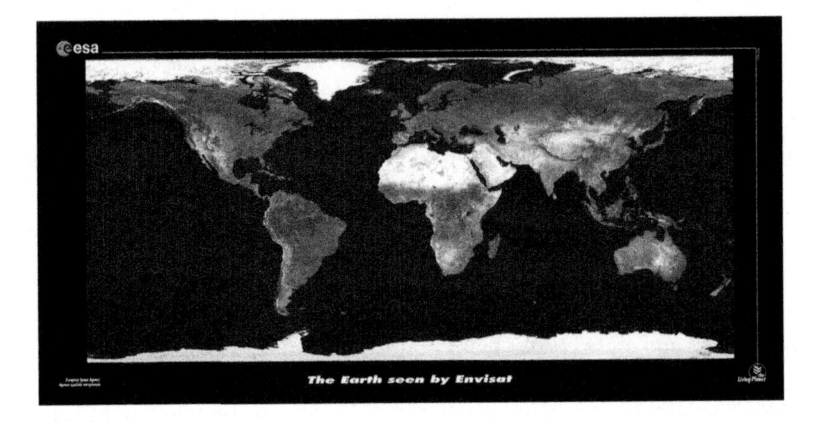

Figure 11.6 MERIS mosaic at 1.3 km resolution obtained in G-POD from the entire May to December 2004 data set.

automatically updated and registered each day from the ground segment, can be selected over user-defined areas and temporal coverage for producing public-relations material. The final image is a mosaic made up of true color images using four out of 15 MERIS spectral bands (bands 2, 3, 5, and 7) with data combined from the selected separate orbital segments, with the intention of minimizing cloud cover as much as possible by using the corresponding data flags. The output file can be downloaded in TIFF format, a JPEG scale-pyramid, or used directly as a Web map service to be combined with other geographical information.

The mosaic was donated by ESA to the United Nations in Geneva, as a testimony to the current state of our planet, to be handed down to future generations. The image will be exhibited permanently in the new access building by the Pregny gate in the Palais des Nations compound [29].

11.6.1.2 MERIS Global Vegetation Index

Vegetation indexes are a measure of the amount and vigor of vegetation at the surface. The Envisat/MERIS vegetation index called MERIS Global Vegetation Index (MGVI) uses information from the blue part of the recorded spectrum from Earth, providing a major improvement to vegetation monitoring. The information in the blue wavelengths improves the correction of the atmospheric noise and the precision of the vegetation index. This service generates maps of geophysical products at monthly and in 10-day intervals. Each individual value represents the actual measurement or product for the day considered the most representative of that period. The geometry of illumination and observation for the particular day selected is saved as a part of the final product.

11.6.1.3 MERIS Level 3 Algal 1

This service comprises a binning of Level 2 for the creation of an Algal Level 3 map. In addition to the original algorithm, this implementation gives the user the ability to select one of three possible binning algorithms (maximum likelihood, arithmetic mean, and minimum/maximum) and predefine a subset region as minimum/maximum latitude/longitude. All pixels outside this region are rejected. Define the bin size without restrictions and finally update long-term means step by step as the input data become available. Another possible method is the selection of the most representative value as the sample that is the closest to the temporal average value estimated over the compositing period. The output file can be downloaded in TIFF format and in a JPEG scale-pyramid.

11.6.1.4 Volcano Monitoring by AATSR

The Volcano Monitoring by InfraRed (VoMIR) service allows the user to extract in a short time and over the large AATSR product archive the thermal radiances at different wavelengths measured by AATSR during night-time. Envisat passes over a user-defined selection of volcanoes. For the selected volcanoes, the output is presented in the form of a spreadsheet, gathering all time-stamped measures, statistics, and quick-look images and summarizing the volcano thermal activity along time, enabling the analysis of the activity trends and patterns in the long-term. In addition, the user may tailor the VoMIR algorithm and customize its pre-defined rules and parameter settings driving the elaboration of the analysis.

11.6.2 Application Based on SAR/ASAR Data and BEST Tools

11.6.2.1 A Generic Environment for SAR/ASAR Processing

Synthetic Aperture Radar sensors are becoming more and more important thanks to their ability to acquire measures that are almost completely independent of atmospheric conditions and illuminations. For these reasons SAR data can play an important role in several applications including risk assessment and management (e.g., landslides and floods) and environmental monitoring (e.g., monitoring of coastal erosion, wetland surveying, and forest surveying). Every day approximately 10 Gigabytes of ASAR Wide Swath medium resolution (WSM) products and 1.5 Gigabytes of ASAR Image Mode medium resolution (IMM) products are acquired by the ASAR sensor on-board the Envisat satellite and stored at the ESRIN archiving center. Unfortunately, the use of SAR data is still limited in comparison to their potentialities and availability.

For the above considerations, it was decided to create a generic SAR processing environment on a grid [30]. Different applications are now available for internal use through user-friendly Web portals that allow transparent access to grids. Different SAR toolboxes have been integrated on EO grids allowing fully automatic SAR image despeckling, backscattering computation, image co-registration, flat ellipsoid projection for medium resolution images, terrain correction using Shuttle Radar Topography Mission (SRTM) Digital Elevation Model (DEM) v3 [31], and mosaicking. These capabilities are obtained by using different toolboxes such as BEST and in-house developed software.

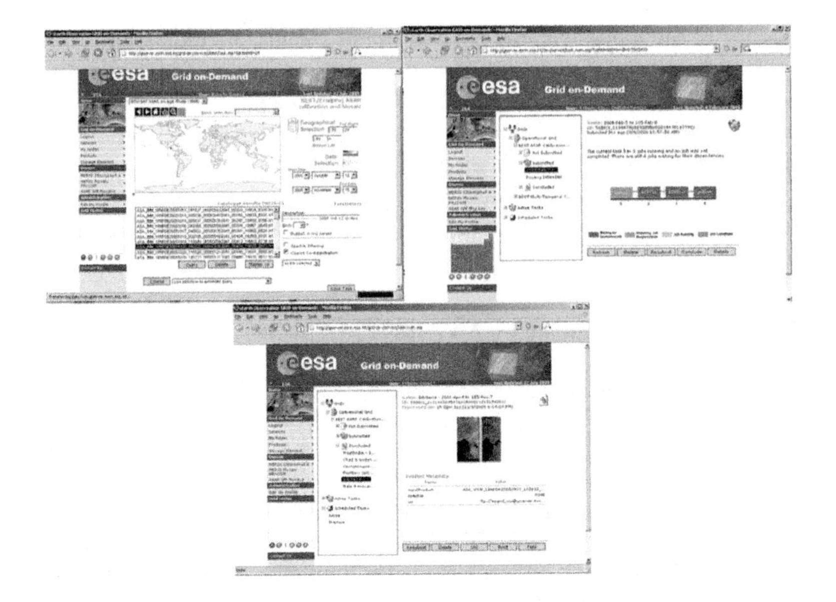

Figure 11.7 The ASAR G-POD environment. The user browses for and selects products of interest (upper left panel). The system automatically identifies the subtasks required by the application and distributes them to the different computing elements in the grid (upper right panel). Results are presented to the user (lower panel).

Based on these tools, higher level functionalities aimed at the analysis of multitemporal images have been developed. Users can browse for the required products specifying the geographical area of interest as well as the acquisition time, and, if required, limiting the search for a given mode or pass. Afterwards they can specify the service of interest (e.g., co-registration of multitemporal images and mosaics). The system automatically retrieves data stored on different storage elements (e.g., distributed archive), identifies the jobs needed for accomplishing the task required by the user, and distributes them on different computing nodes of the grid (see Figure 11.7).

This environment can be used to produce high resolution orthorectified mosaics on a continental/global scale. As an example, the G-POD SAR processing application produced a 3 arcsec ($\sim$ 90 m) pixel size orthorectified Envisat ASAR mosaic over Europe (see Figure 11.8). This mosaic was obtained by using ASAR WSM data acquired between January and May 2006, the DEM at 3 arcsec derived from the global SRTM, and the GTOPO30 DEM (for latitudes above 60° N). The whole process was achieved in a few hours. To cover the whole of Europe, 143 ASAR WSM stripline products were automatically selected and retrieved from G-POD storage together with required SRTM and GTOPO30 DEM tiles. Products were orthorectified and normalized for near-range/far-range effects over sea before being aggregated in a mosaic. The so-obtained result was used to produce the SAR European mosaic poster

Figure 11.8 Three arcsec (~ 90 m) pixel size orthorectified Envisat ASAR mosaic obtained using G-POD. Political boundaries have been manually overlaid. The full resolution result can be seen at [34].

distributed at the Envisat Symposium 2007 the [33] and published as the ESA Image of the week on the March 16th ESA Web portal [34].

11.6.2.2 EnviProj – Antarctica ASAR GM Mapping System

The ESA Antarctica Mapping System processes any ASAR Global Monitoring Mode product over Antarctica to obtain radar mosaics of the continent at 400 m resolution.

A mosaic is a composite image of several Global Monitoring Mode (GMM) stripes. GMM products are very useful to rapidly cover wide areas. Since the sensor is carried on ESA's polar-orbiting satellite Envisat, 14 images a day are acquired over Antarctica. Thanks to the 8000 km long and 400 km wide swaths, Antarctica is mapped in a few days at 400 m resolution (see Figure 11.9).

11.6.2.3 ASAR Products Handling and Analysis for a Quasi Systematic Flood Monitoring Service

Earth Observation is becoming a recognized source of information for disaster management [35], in response to natural and man-made hazards, in Europe and in the rest of the world. EO-based crisis mapping services are generally delivered via projects like the Global Monitoring for Environment and Security (GMES) Services Element (GSE), such as GSE Flood and Fire and RESPOND, the GMES Services Supporting Humanitarian Relief, and the Disaster Reduction Reconstruction, alongside the International Charter Space and Major Disasters, which enables timely access to crisis data from a variety of EO missions.

The all-weather capability of high resolution SAR observations provides useful input to crisis and damage mapping. This is particularly relevant for flood monitoring, and SAR is considered a useful information source for river plain flooding events, a

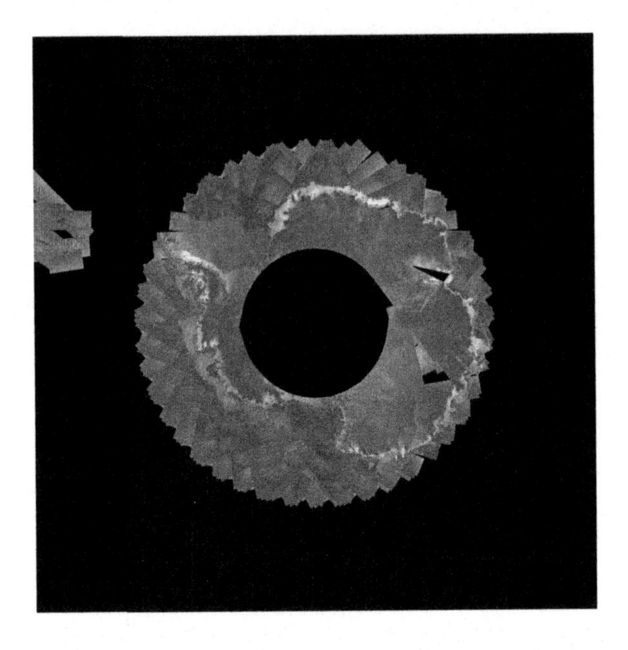

Figure 11.9 ASAR mosaic obtained using G-POD considering GM products acquired from March 8 to 14, 2006 (400 m resolution).

frequent and important type of hazard in both Europe and the rest of the world. In this context, the access and exploitation of Envisat/ASAR data can benefit from grid-based processing to enable accurate, rapid, and large coverage observations of flooded features. Such a capability would facilitate the provision of crisis mapping products combining ASAR-based observations with other EO crisis data. This investigation is based on the analysis of the requirements of RESPOND users, who come from both the humanitarian aid and the disaster management communities.

The capabilities of G-POD to calibrate and co-register images related to a flood affected area in a fast and accurate way were investigated. A series of Envisat data sets of the Chinese Poyang lake were selected with advice from the RESPOND partner Service Régional de Traitement d'Image et de Télédétection (SERTIT), which runs flood monitoring tests in the framework of the ESA Dragon Programme [36].

In all tests, images were selected and the co-registration task was run. The system calibrated and co-registered image pairs in a fully automatic and unsupervised way. BIL floating point calibrated images, compressed JPEG images for visualization, and KMZ (zipped Keyhole Markup Language files) files to be imported in Google Earth were produced as a result.

Based on the results obtained and discussions with specialists, it is expected that G-POD can provide a significant contribution to develop an enhanced flood monitoring capability. Users can take advantage of the underlying grid technology that results in both transparent access to the huge distributed data archive and a significant reduction of the time required for data processing.

11.6.3 Atmospheric Applications Including BEAT Tools

11.6.3.1 GOME Processing

GOME is one of several instruments on-board ESA's ERS-2 remote sensing satellite, which has been orbiting the Earth since 1995. Every day, some thousand GOME measurements of atmospheric ozone are transmitted to the ERS ground stations. The raw readings are sent to the dedicated Processing and Archiving Facility in Germany (D-PAF), which produces the standard data products and distributes them to scientific investigators.

In recent years, a research activity has started to derive special higher-quality data products, so-called GOME Level 2 products, which include the Earth's ozone profile and total ozone column, which give the precise gas concentrations at different altitudes above the Earth's surface at any location. By analyzing the global GOME data set over the whole period together with ground-based measurements, it is possible to obtain an accurate picture of the speed with which the ozone concentrations in our atmosphere are changing. This allows scientists to improve the forecasting models for future ozone concentrations in the near- and long-term. Two different ozone-profiling algorithms [37, 38, 39], one developed by KNMI and the other in ESRIN together with Università Tor Vergata, have been selected for this purpose.

11.6.3.2 3D-Var Data Assimilation with CHAMP Radio Occultation (RO) Data

Assimilation techniques such as 3D-Var, which are state of the art within Numerical Weather Prediction (NWP) systems, are seldom used within climate analysis frameworks, partly because of the enormous numerical processing cost. The Grid on-Demand high-performance computing environment offers the opportunity to compute many time layers in parallel, significantly reducing the computing time.

In order to conduct the climate study we are using Global Navigation Satellite System (GNSS) based radio occultation observations. A remote sensing technique provides a new kind of precise atmospheric observation that will supplement the database used for climate research, numerical weather prediction, and atmospheric process studies. Since the launch of the Challenging Mini Satellite Project (CHAMP) in summer 2000 and the start of the RO experiment in February 2001, a comprehensive data set for experiments and impact studies is available. The potential of this novel type of observation has already been demonstrated. First impact trials at the UK Met Office (UKMO) and the European Centre for Medium Range Forecast (ECMWF) show that RO data comprise additional information content not present in other observations, proving to be nonredundant. This can be seen as a remarkable result considering the limited amount of RO observations that entered the NWP systems to conduct the impact studies.

The global coverage, all-weather capability, long-term stability, and accuracy make the observations an ideal supplement to the extensive data set assimilated into numerical weather systems. Furthermore, the inherent properties of RO observations and the long-term perspective offered by the Meteorological Operational Satellite (MetOp) program make these types of data ideal for also studying long-term atmospheric and climate variability, providing an ideal candidate to build global climatologies (a vast amount of data are already available today due to CHAMP and the recently launched COSMIC constellation). This contribution to atmospheric applications using Grid on-Demand investigates the application of the 3D-Var methodology within a global climate monitoring framework. It studies the assimilation of radio occultation derived refractivity profiles into first guess fields, derived from 21 years of ECMWF's ERA40 re-analysis data set on a monthly mean basis, divided in the four synoptic time layers to take the diurnal cycle into account.

The system is tuned for high vertical and moderate horizontal resolution, best suited to the spatial characteristics of these satellite-based measurements. Analyses are performed using a General (Global) Circulation Model (GCM) compliant Gaussian grid, comprising 60 model levels up to a height of $\sim$ 60 km and a horizontal resolution corresponding to a Gaussian grid N48 (e.g., 192×96). The control variables used are temperature, specific humidity, and surface pressure; the background is compared with the observations in refractivity space. During the optimization procedure the control variables are updated. Results indicate a significant analysis increment that is partly systematic, emphasizing the ability of RO data to add independent information to ECMWF analysis fields, with the potential to correct biases. RO data sets offer a new, accurate kind of atmospheric observation comprising long-term stability stemming from the measurement principle (the RO technique needs no calibration; the basis of the observation is a measurement of time, thus a direct compilation of observation series from different RO instruments is possible), with global coverage, high vertical resolution, and due to the used wavelengths the observations are not disturbed by rain or clouds.

Global atmospheric fields (see Figure 11.10) are derived from the 3D-Var implementation within the Grid on-Demand framework, comprising the summer seasons (June, July August (JJA)) from 2002 to 2005, demonstrating the potential of assimilation techniques in combination with high-performance computing grids [40]. CHAMP observations used within these assimilation experiments have been processed by GeoForschungsZentrum Potsdam (GFZ) to Level 2 and by the Wegener Center for Climate and Global Change, University of Graz (WegCenter) to refractivity (Level 2a).

11.6.3.3 YAGOP: GOMOS Non-operational Processing

The beat2grid [41] service embraces the exploitation of data of several atmospheric instruments, such as GOMOS, MIPAS, SCIAMACHY, and GOME. Here we present the service corresponding to GOMOS data products.

The provided GOMOS metadata definition consists of information about the geolocation (latitude, longitude, and orbit height) of the satellite, the geographical location of the measurement (latitude, longitude, and tangent height), and the occultation date and time. Once the data, generated by a preliminary processing step, are stored in the database (described in Section 11.6.2), the results are immediately available for users by logging onto the portal.

The beat2grid GOMOS service offers a graphical selection of the occultation measurement's site (see Figure 11.7). Predefined lists help the user to select the search criteria, and the direct selection of latitude and longitude position is also possible. The

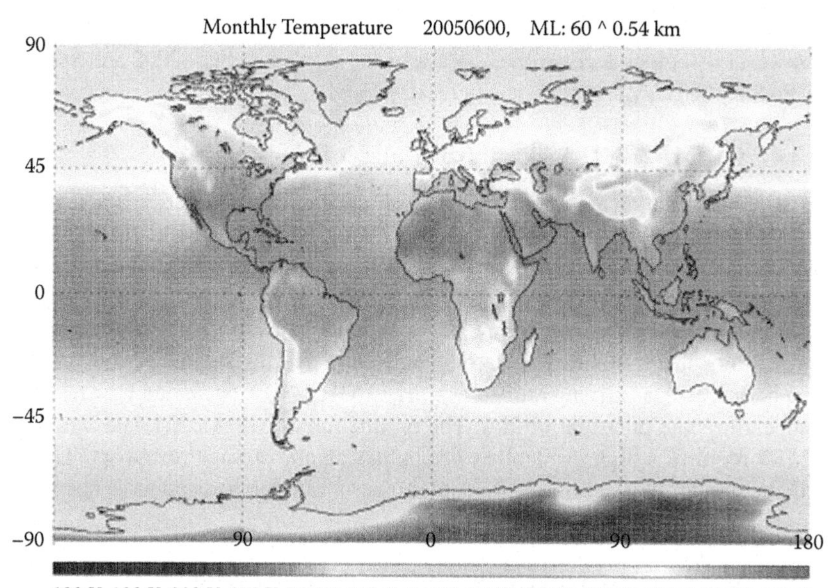

Figure 11.10 Global monthly mean near surface temperature profile for June 2005, time layer 0 h.

latter helps expert users to select a distinct small area (e.g., 300 km $\simeq 2.5°$) in order to derive the diurnal, monthly, seasonal, and even yearly variations of atmospheric products. This also requires the exact selection of date and time of the occultation measurements. Once date and time and the geographical positions are chosen, the user can decide which atmospheric products should be retrieved. In the actual setup of the service, we offer ozone, NO_2, NO_3, O_2, and H_2O profiles, as well as air and aerosol density. The aerosol product is not always available in good quality in the original GOMOS data set. There is a new effort to acquire a better quality aerosol density from stellar occultation measurements. When this is available, the Grid on-Demand service will be enhanced using this information.

In addition to the atmospheric constituent profiles, as discussed for the current situation on Grid on-Demand, the GOMOS product allows the exploitation of Steering Front Assembly and Star Acquisition and Tracking Unit (SAF/SATU) data. These profiles, giving the instruments mirror position and accuracy, are inverted to produce density, pressure, and temperature profiles. In cooperation with the University of Graz (Wegener Center for Climate and Global Change) the Yet Another GOMOS Processor (YAGOP) is running on Grid on-Demand [42]. This processor not only enables the derivation of temperature profiles; the calculation of non-operational GOMOS Level 2 ozone products was also implemented, which is a processing option based on an optimal estimation technique, inverting ozone and NO_2 profiles simultaneously.

The retrieved ozone profiles are compared with operational GOMOS ozone data (see Figure 11.11). Temperature profiles for a selected data set in Sep. 2002 are compared to MIPAS and CHAMP (not shown here) data.

11.6.3.4 GRIMI-2: MIPAS Prototype Dataset Processing

The GRIMI-2 initiative embraces the grid technology, data production, and exploitation for a prototype processor of an ongoing ESA mission. GRIMI-2 stands for Grid MIPAS Level 2 prototype processing and refers to the production of test data sets for research and verification purposes for the operational MIPAS Level 2 product. MIPAS is the Michelson Interferometer for Passive Atmospheric Sounding and one of three atmospheric chemistry instruments on-board Envisat. It measures Earth's atmospheric trace gas mixing ratios of O_3, H_2O, CH_4, N_2O, NO_2, and HNO_3, as well as temperature and pressure profiles. The production of MIPAS test data sets within GRIMI-2 requires a high availability of computing and storage resources. This service demonstrates the usefulness of a grid infrastructure for the support of operational processing for Earth Observation purposes.

11.6.3.5 SCIA-SODIUM: SCIAMACHY Sodium Retrieval

Mesospheric sodium is believed to be generated by the ablation of meteors in the upper atmosphere. Despite its importance, mesospheric sodium distribution is poorly known, due to the limited number of ground-based instruments (e.g., Lidars) able to detect the Na presence at around 90 km of altitude, and to the lack of satellite instruments probing the upper mesosphere.

In this context, upper atmospheric SCIAMACHY limb measurements in the visible part of the spectrum provide the unique opportunity to estimate some key parameters of the upper mesospheric sodium chemistry; the input of atomic sodium to the Earth atmosphere via meteorite ablation; and the concentration profiles of O_3, H_2O, and CO_2 driving sodium chemistry in this region.

Detailed analysis of the SCIAMACHY limb spectra in the 585–595 nm wavelength range and for tangent altitudes above 70 km revealed detectable radiance emission signals in correspondence of the sodium D1 and D2 lines. This signal is due to resonant scattering of atomic sodium and is triggered by the solar irradiance, which is measured daily by SCIAMACHY by directly looking at the sun. The measured solar irradiance spectrum is used as a normalization factor to the earthshine limb spectra in order to account for possible changes in the radiative flux forcing (as solar activity changes in time) and to minimize radiometric/instrumental calibration effects. Thus, the sun-normalized sodium radiances are the input spectra for two independent algorithms (DOAS-like and a two-line differential scheme) that have been developed to estimate the slant distribution of atomic sodium. The SCIA-SODIUM processor running on Grid on-Demand analyzes the SCIAMACHY limb and nadir Level 1 product to estimate global scale vertical profiles and total amounts of atmospheric H_2O, O_2, and Na. The Limb spectra are used to estimate the vertical profiles of water vapor and oxygen from Earth's surface to 30 km altitude, and Na vertical profiles in the 70–105 km altitude range. From Nadir viewing measurements the vertical column density of H_2O

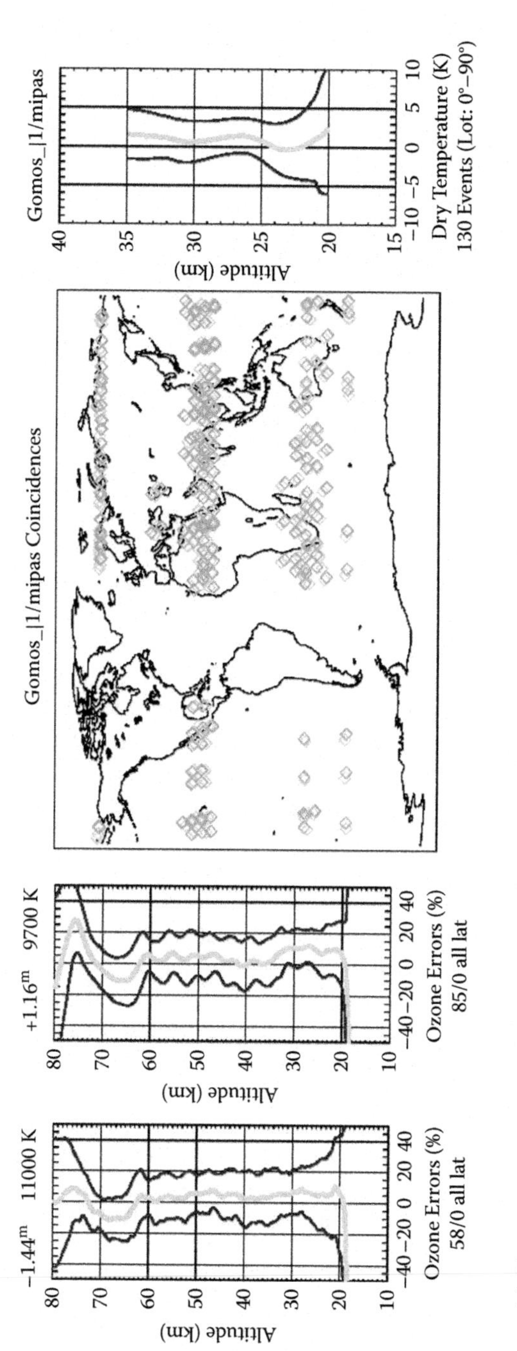

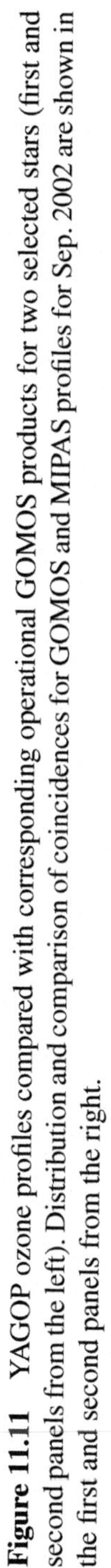

Figure 11.11 YAGOP ozone profiles compared with corresponding operational GOMOS products for two selected stars (first and second panels from the left). Distribution and comparison of coincidences for GOMOS and MIPAS profiles for Sep. 2002 are shown in the first and second panels from the right.

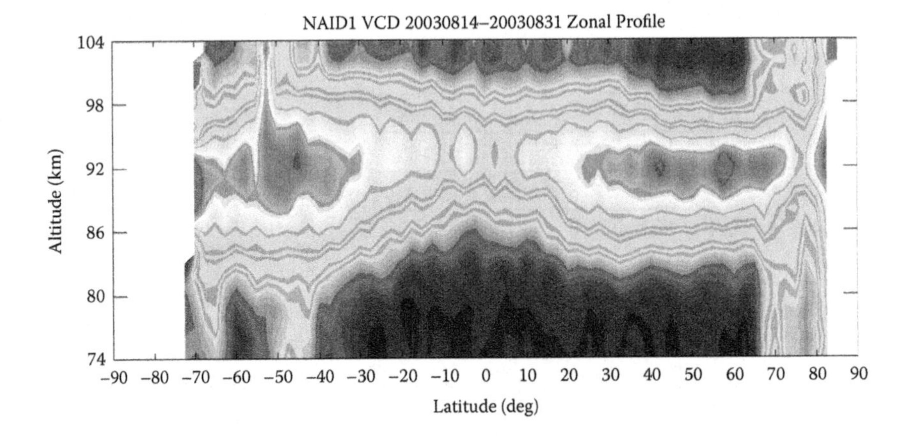

Figure 11.12 Zonal mean of Na profiles of 14–31 August 2003.

and O_2 are estimated with very good accuracy. The quality of the H_2O and O_2 vertical column densities estimated using this technique is well known [43, 44, 45] while the Na vertical profiles from satellite daytime measurements are estimated for the first time here, and extensive product validation is ongoing. Nevertheless, preliminary validation results using sodium Lidar data from the University of Colorado (NCAR-CEDAR database) and from the Instituto Nacional de Pesquisas Espaciais (Brazil) showed very good agreement between these data and SCIA-SODIUM products.

In Figure 11.12 we show a 15-day zonal average (14–31 August 2003) of the sodium profiles estimated by the SCIA-SODIUM processor. The Na layer is very well characterized, with its maximum located at around 92 km and a layer width of about 8 km.

The 2002–2006 SCIAMACHY products were analyzed using the grid infrastructures allowing for a global and long-term analysis of the sodium layer. Given the large amount of data to be processed per orbit (300 Megabytes) and the complex calculations implemented in the SCIA-SODIUM processor, the use of a grid infrastructure increases the processing speed by a large extent. The development of a dedicated service for the SCIA-SODIUM processor allows Grid on-Demand users to retrieve mesospheric sodium profiles in a very efficient and user-friendly way.

11.7 Grid Integration in an Earth Science Knowledge Infrastructure

Knowledge Infrastructure (KI) is an ICT-based organized place open to defined communities and members, according to agreed policies. Actors (not only people) collaborating using a KI can bring, offer, and share data, information, and resources. In order to exploit data and information and derive knowledge for use in its context, a KI provides and integrates, dynamically, different resources (physical resources, network, software, etc.)

The objective of a KI is to generate and support community shared interests and collaboration (e.g., in multidisciplinary research). The key functionalities of a KI are:

- Produce, control, share, preserve, retrieve, access, and consumer knowledge.
- Organize and enrich the 'chaotic' environment in the user community. The enriching process consists, for example, of adding context with relevance to user communities.
- Support multidisciplinary applications (needed for wide common standards, framework, approach, etc.).
- Interoperability to allow cross-KI interaction with other distributed KIs.
- Respond to actors with defined quality services (e.g., time response).

KIs are owned by communities (e.g., the responsible bodies), in order to ensure the functionalities last for a long time. This implies development and use of organization elements, policies, and standards.

ESA/ESRIN is working to build an open Earth Science KI. The main contribution is in integrating a common infrastructure. In this context, ESA is active in demonstrating extensive use of applications in stable grid environments. In the summer of 2006, ESA opened an 'Announcement of Opportunity' to provide online access to large ESA Earth Observation archives, computing and storage elements, and user tools for handling data. Finally, the e-collaboration projects described in Section 11.6 also significantly contribute to building an ESA open Earth Science KI. In the following subsections, we describe some ESA contributions in KI.

11.7.1 Earth Science Collaborative Environment Platform and Applications – THE VOICE

THE VOICE, short for Thematic Vertical Organizations and Implementation of Collaborative Environments, is a two-phase, ESA General Studies Programme (GSP) financed study [46] started in early 2004, looking at how e-collaboration technologies can support the Earth Science community. During its first phase, a survey of e-collaboration technologies was performed that was matched with the results of an analysis of Earth Science e-collaboration service requirements to define a service-oriented architecture and derive a so-called generic collaborative environment node (GCEN) to serve as a basis for the implementation of selected prototypes, including atmospheric instruments calibration and validation, agricultural production support and decision planning, forest management, ocean monitoring, and urban area monitoring during the second phase of the study that started in December 2004. The first phase has demonstrated the principal need to relate to seamless (and getting the delivery in a relatively short time) access to and/or use of data, information, and knowledge without having to worry about where they are, their format, their size, security issues, multiple logins, etc. After a careful analysis of prototype requirements, essential and additional services have been derived, and technologies and tools have been selected for implementation as given in the tables below. Besides the mentioned technologies, wireless technologies are also used [47].

The study has already implemented the essential services as part of the GCEN and will complete the prototypes before the end of 2005. At the end of the project it will demonstrate near-real-life scenarios with distributed actors, resources, data, and other relevant items. Besides the mentioned technologies and tools, it is also looking into the use of standards like the ones defined by the Open Geospatial Consortium (OGC) and the World Wide Web Consortium (W3C) to facilitate data access.

11.7.2 Earth Science Digital Libraries on Grid

Digital Libraries (DL) are seen as an essential element for communication and collaboration among scientists and represent the meeting point of a large number of disciplines and fields including data management, information retrieval, library sciences, document management, information systems, the Web, image processing, artificial intelligence, human-computer interaction, etc.

ESA/ESRIN is leading the Implementation of Environmental Conventions (ImpECt) scenario as part of the EC project Diligent, short for A Digital Library Infrastructure on Grid Enabled Technology [48], which focuses on integrating grid and digital library technologies towards building a powerful infrastructure that allows globally spread researchers to collaborate by publishing and accessing data and knowledge in a secure, coordinated, dynamic, and cost-effective manner.

The main ImpECt requirements concern retrieval of Earth Sciences related information based on spatial, topic, and temporal selection criteria and the accessibility of services and applications able to process this information. Existing Earth Sciences related digital library systems cannot handle such queries in a sufficient manner and do not host any similar services such as those required by the ImpECt scenario.

A first ImpECt implementation uses well-known data sources and services, including Envisat and other satellite products as well as services capable of generating and elaborating them. Grid on-Demand has a strategic role as service and data provider. The core feature is the automatic interaction between separated entities as the test digital library and external services able to accept queries from ImpECt users process the information on the ESA grid and publish the results on the DL. The test DL is based on the Digital Library Management System (DLMS) OpenDLib [49], while the grid infrastructure relies on the gLite middleware [50]. The specific information provided in the test DL concerns ocean chemistry, in particular ocean color, being an Earth Sciences consolidated topic with many and different types of information (e.g., environmental reports, images, videos, specific applications, data sets, scientific publications, and thesaurus).

This activity is intended to allow users to annotate available contents and services, to arrange contents in user-defined collections, to submit advanced search queries for retrieving geo-referenced information, to build user-defined compound services to run specific processing, and to maintain heavy documents as environmental conventions reports by an automatic refresh of the information they hold.

Future work will allow virtual organizations to create on-demand ad-hoc defined DLs, to get newly generated information processed on the grid in a totally transparent way, and to navigate the information with the support of domain-specific and top-level ontologies.

11.7.3 Earth Science Data and Knowledge Preservation

Earth Observation data are large in volume and range from local to global in context. Within initiatives like the GMES [51] and ESA Oxygen [52], large amounts of digital data, which are potentially of great value to research scientists in the Earth Observation community and beyond, need to be acquired, processed, distributed, used, and archived in various facilities around the world. For example, Envisat [4], the advanced European polar-orbiting Earth Observation satellite, carries a payload of ten instruments that provide huge amounts of measurements of the atmosphere, ocean, land, and ice. These data are processed, corrected, and/or elaborated using auxiliary data, models, and other relevant information in support of Earth Science researche (e.g., monitoring the evolution of environmental and climatic changes). These need to be preserved for future generations to better understand the evolution of our Earth. That is, preservation does not relate to the data only, but implies also the maintenance of information regarding when and how such data were processed, the reason(s) why this or that way was used. It may include knowledge, documented results, scientific publications, and any other information, needed to support scientists in their research. Currently, however, there is no clear mandate to preserve Earth Observation mission data, relevant information, and knowledge at the European level and the responsibility falls under the remit of the individual mission owners and/or national archive holders. Coordinating efforts on standards and approaches to preserve the most valuable European EO products will be required in order to guarantee the accessibility and reusability of these frequently distributed data.

It should be noted that the availability and accessibility to most of the above-mentioned information is considered relevant for efficient and adequate exploitation of Earth Observation data and derived products. As such, their preservation is also considered important for optimal short-term archive and data reuse by different actors, including multidisciplinary exchange of experiences on the same data set. In other words, problems encountered in long-term preservation certainly include those encountered in short-term exploitation, and as such the chapter's focus is on some aspects of data preservation. Solid infrastructures are needed to enable timely access, now and in the future. Much more digital content is available and worth preserving. Researchers increasingly depend on digital resources and assume that they will be preserved [53].

The workshop on 'Research Challenges in Digital Archiving and Long-term Preservation' [54], which was held in 2002, identified many challenges in long term data preservation. The overall challenge is guaranteeing access to and usability of data, independent from underlying hardware and software systems; systems evolve along time, technologies are being renewed and replaced, data formats may change, and so may as well related information and scientific knowledge.

Over the past few years different communities have tackled this problem, experimenting with different technologies. These include Semantic Web focusing on semantic Web access, data grid technology focusing on management of distributed data, digital library technology focusing on publication, and persistent archive technology focusing on management of technology evolution [55, 56].

In Earth Science, accessing historical data, information, and related knowledge may be nowadays quite complex and sometimes impossible, due to the lack of descriptive information (metadata) that could provide the context in which they fit as well as the lack of the information and knowledge. Different initiatives are focusing on these issues. One of them is the Electronic Resource Preservation and Access Network, short ERPANET/CODATA [57], but also the Persistent Archive Research Group and the Data Format Description Language research group, both part of the Global Grid Forum, are looking at similar questions [58]. Within the EO community, the Committee of Earth Observation Satellites (CEOS) is looking at the use of the Extended Markup Language XML for Science Data Access [59]. Based on results achieved in ongoing projects and expected results in planned projects, a possible technical solution to approach long-term data preservation may consider technologies such as digital libraries and grids.

In summary, long-term data preservation has to be based on a distributed environment capable of handling multiple copies of the same information. Grid and DL technology could help in performing long-term data preservation since, as said above, the preservation task of migrating from old to new technology is really similar to managing access to data distributed across multiple sites, while the organization of data and metadata in information collections requires discovery and access techniques as provided within DLs.

11.7.4 CASPAR

CASPAR (Cultural, Artistic and Scientific knowledge for Preservation, Access and Retrieval) [14] is an Integrated Project co-financed by the European Union within the Sixth Framework Programme. It intends to design services supporting long-term digital resource preservation, despite changes in the underlying computing (hardware and software) and storage systems.

In this context, CASPAR will:

- Enhance the techniques for capturing Representation Information and other preservation-related information for content objects.
- Integrate digital rights management, authentication, and accreditation as standard features of CASPAR.
- Research more sophisticated access to, and use of, preserved digital resources, including intuitive query and browsing mechanisms.

Different case studies will be developed to validate the CASPAR approach to digital resource preservation across different user communities and assess the conditions for a successful replication.

11.7.5 Living Labs (Collaboration@Rural)

Collaboration@Rural aims to boost the introduction of Collaborative Working Environments (CWE) as key enablers for catalyzing rural development. To achieve this priority objective, Collaboration@Rural will advance the specification, development, testing, and validation of a powerful and flexible worker-centric collaborative platform that will significantly enhance the capabilities of rural inhabitants both at work and at life, thus leading to a better quality of life and a revaluation of rural settings.

From the technical standpoint, Collaboration@Rural will organize the work in three layers: Collaborative Core Services – CCS (layer 1), Software Collaborative Tools – SCT (layer 2), and Rural Living Labs – RLL (layer 3). Layer 1 will encapsulate all core services and resources (networks, sensors, devices, software modules, localization sources, etc.) in reusable software components.

A key piece of Collaboration@Rural's framework is the upper-layer service architecture, or C@RA, which combines in a synergetic manner the layer 1 components according to orchestration of high-level capabilities resulting in a set of high-level software tools, at layer 2. C@RA will be highly customizable in the sense of providing mechanisms to incorporate any proprietary or open solutions, and any standard. This approach will permit Collaboration@Rural to substantially contribute to the definition of a user-centric Open Collaborative Architecture (OCA).

Collaboration@Rural layer 3 will articulate Rural Living Labs as innovative research instruments involving rural users. The RLL user-oriented methodology will guarantee to meet the highly specific rural users' expectations and will provide mechanisms to gather technical requirements for the C@RA.

Several innovative scenarios with an expected high impact on rural development have been selected to enable later validation of the C@RA.

11.8 Summary and Conclusions

In this chapter we described some of the ESA activities related to the use of grid technology for Earth Observation. The ESA Grid Processing on-Demand environment, a generic, flexible, secure, re-usable, distributed component architecture using grid and Web services to manage distributed data and computing resources, has been described. Several EO applications have been plugged into G-POD, benefiting from easy and transparent access to EO data and computing resources. This opens the possibility to address new applications that would be unfeasible without the data access and computing resource distribution offered by grid technology. In the future the collocation of a Grid on-Demand node with the EO facilities performing data acquisition or data archiving can minimize and optimize the need and availability of high-speed networks.

Following the interest and expectations raised as a result of several presentations made at workshops and conferences, and of some significant scientific results obtained using the grid for analysis of satellite-based data, we expect more EO science teams (and also other communities) will begin to deploy their applications on the grid. For instance, applications in the fields of seismology and climatology need to handle large, regularly updated databases. Data may come from various types of observatories, e.g., ground-based, airplane, and balloon measurements, or from simulation and modeling.

Recently, ESA offered the opportunity to Principal Investigators of exploring the EO data archives with their own algorithms/processors, using the ESA available grid computing and dynamic storage resources (please consult ESA Earth Observation Principal Investigator Web portal [60]). This represented a new and unique opportunity for scientists to perform bulk processing and/or validation of their own algorithms over the large ESA Earth Observation archive at very limited associated cost and effort. We have presented some examples of EO applications of Grid on-Demand, such as applications for SAR, MERIS, AATSR, GOME, GOMOS, MIPAS, and SCIAMACHY data. More of these applications will come and enrich the selection of valuable services for the EO community.

ESA also sees the grid as a powerful means to improve the integration of data and measurements coming from very different sources to form a Collaborative Environment.

The need for coordinated e-collaboration within the Earth Science community has been clearly confirmed in the ESA GSP project THE VOICE. Detailed needs have been confronted with outcomes of a survey of current e-collaboration technologies including grid and Web services, workflow management technology, and Semantic Web, to provide a basis for the implementation of a series of prototypes. A service-oriented architecture has been adapted as a basis for the prototype generic collaboration platform offering a certain number of essential and additional services. Web services have proven to be the most flexible and powerful instrument to build such an architecture. Completed with other technologies and the use of selected standards, such a platform will ease the interaction between the different actors involved.

Digital Libraries are seen as an essential element for communication and collaboration among scientists and represent the meeting point of a large number of disciplines and fields including data management, information retrieval, library sciences, document management, information systems, the Web, image processing, artificial intelligence, human-computer interaction etc.

The use of grid, Web services, and Digital Library technologies can help in easing the accessibility of data and related information needed to interpret the data, as is being demonstrated in different initiatives at ESA. However, there are still many technological challenges to be explored further and ESA intends to follow these closely because of their interest in guaranteeing users' transparent access to the ever-growing amounts of Earth Science data, products, and related information from ESA and third-party missions.

For the Earth Science community, it is important to continue and invest in activities focusing on grid and e-collaboration. Initiatives like the mentioned ESA G-POD and projects like DILIGENT and THE VOICE are demonstrating their relevance. Moreover, these projects demonstrate that an emerging technologies-based underlying infrastructure is a real asset needed by this community; it improves significantly the accessibility and usability of Earth Science data, information, and knowledge, and the way Earth Science users collaborate.

Finally, the means and extent to which grid technology can be exploited in the future are highly dependent on the adoption of common standards to enable different grids to collaborate and interoperate.

11.9 Acknowledgments

We are very grateful to all members of the ESA EO grid team who in one way or another have directly contributed and impacted to the design, implementation, validation, and operation of the Grid on-Demand services. Same thanks to the colleagues having influenced the generation, editing, and proof reading of this chapter. We also thank all external to ESA partners, scientists, and PIs for their excellent input, creative ideas, and for running their applications on the ESA Grid on-Demand infrastructure, thus extending the EO data exploitation and analysis and at the same time enlarging the grid culture in our community.

References

[1] European Data Grid Project Website: http://www.eu-datagrid.org.

[2] ESA Grid Processing on-Demand Web portal: http://eogrid.esrin.esa.int.

[3] ERS Website: http://earth.esa.int/ers.

[4] Envisat Website: http://envisat.esa.int.

[5] ESA software tools Website: http://www.envisat.esa.int/resources/softwaretools.

[6] International Council for Science, *Report of the CSPR Assessment Panel on scientific Data and Information*, December 2004, ISBN 0-930357-60-4, 2004.

[7] P. Gauthier et al.: Ensuring the Sustainability of Online Cultural and Heritage Content: From an Economic Model to an Adapted Strategy, M2W FIAM, 2003.

[8] H. Besser, Digital Longevity, in Maxine Sitts (ed.), *Handbook for Digital Projects: A Management Tool for Preservation and Access*, Andover MA: Northeast Document Conservation Center, pp. 155–166, 1999.

[9] L. Fusco and J. van Bemmelen, Earth Observation Archives in Digital Library and Grid Infrastructures, *Data Science Journal*, Vol. 3, pp. 222–226, 2004.

[10] OpenGIS Website: http://www.opengis.org.

[11] ESA Service Support Environment Website: http://services.eoportal.org.

[12] FAO Geonetwork Website: http://www.fao.org/geonetwork.

[13] Central Laboratory of the Research Councils Website: http://www.e-science.clrc.ac.uk/web/projects/dataportal.

[14] CASPAR Website: http://www.casparpreserves.eu.

[15] Reading e-Science Center Website: http://www.resc.rdg.ac.uk.

[16] GODIVA Project Website: http://www.nerc-essc.ac.uk/godiva.

[17] OPeNDAP Website: http://www.opendap.org.

[18] Enabling Grid for E-sciEnce Website: http://www.eu-egee.org.

[19] Website: http://www.eogeo.org/Workshops/EOGEO2004.

[20] WGISS GRID Task Team Website: http://grid-tech.ceos.org/new.

[21] Website: http://directory.eoportal.org/info_GRIDeCollaborationfortheSpaceCommunity.html.

[22] CrossGrid Website: http://www.crossgrid.org/main.html.

[23] EC DEGREE Project Website: http://www.degree-project.eu.

[24] Website: http://www.globus.org.

[25] Website: http://lcg.web.cern.ch/LCG.

[26] Website: http://glite.web.cern.ch/glite.

[27] L. Fusco, P. Goncalves, J. Linford, M. Fulcoli, A. Terracina, and G. D'Acunzo, Putting Earth Observation Applications on the Grid, *ESA Bulletin*, Vol. 114, 2003.

[28] Envisat-1 Products Specifications Vol. 11: *MERIS Products Specifications*, ESA Publication.

[29] Website: http://www.esa.int/esaEO/SEM2CXEFWOE_index_0.html.

[30] L. Fusco, P. Goncalves, F. Brito, R. Cossu, and C. Retscher, A New Grid-Based System to Assist Users in ASAR Handling and Analysis, European Geosciences Union General Assembly, Vienna, Austria, 2006.

[31] Shuttle Radar Topography Mission (SRTM). The mission to map the world. Website: http://www2.jpl.nasa.gov/srtm.

[32] Zoomify Website: www.zoomify.com.

[33] Envisat Symposium 2007 Website: http://www.envisat07.org.

[34] ESA Portal Website: http://www.esa.int/esaEO/SEM9QLQ08ZE_index_0.html.

[35] International Charter "Space and Major Disasters" Website: http://www.disasterscharter.org.

[36] Dragon Cooperation Programme Website: http://earth.esa.int/dragon.

[37] F. Del Frate, M.F. Iapaolo, and S. Casadio, Intercomparison Between GOME Ozone Profiles Retrieved by Neural Network Inversion Schemes and ILAS Products, *International Journal of Atmosphere Oceanic Technology, 22*, No. 9, pp. 1433–1440, 2005.

[38] F. Del Frate, M. Iapaolo, S. Casadio, S. Godin-Beekmann, and M. Petitdidier, Neural Networks for the Dimensionality Reduction of GOME Measurement Vector in the Estimation of Ozone Profiles, *Journal of Quantitative Spectroscopy and Radiative Transfer*, pp. 275–291, 2005.

[39] F. Del Frate, A. Ortenzi, S. Casadio, and C. Zehner, Application of Neural Algorithm for Real Time Estimation of Ozone Profiles from GOME Measurements, *IEEE Transactions on Geoscience and Remote Sensing, 40*, pp. 2263–2270, 2002.

[40] A. Loescher, C. Retscher, L. Fusco, P. Goncalves, F. Brito, and G. Kirchengast, Assimilation of Radio Occultation Data into ERA40 Derived Fields for Global Climate Analysis on ESA's High Performance Computing Grid on-Demand, *Remote Sensing of Environment: Special Issue*, revised, 2007.

[41] C. Retscher, P. Goncalves, F. Brito, and L. Fusco, Grid on-Demand Enabling Earth Observation Applications: The Atmosphere, HPDC GELA Workshop, 2006.

[42] C. Retscher, G. Kirchengast, and A. Gobiet, Stratospheric Temperature Sounding with Envisat/GOMOS by Exploitation of SFA/SATU data, in G. Kirchengast and U. Foelsche and A.K. Steiner eds., *Atmosphere and Climate* Springer, 2006.

[43] S. Casadio, G. Pisacane, C. Zehner, and E. Putz, Empirical Retrieval of the Atmospheric Air Mass Factor (ERA) for the Measurement of Water Vapour Vertical Content Using GOME Data, *Geophys. Res. Lett., 27*, pp. 1423–1426, 2000.

[44] S. Casadio and C. Zehner, GOME-MERIS Water Vapour Total Column Inter-Comparison on Global Scale: January–June 2003, MERIS-AATSR Workshop, September 26–30, ESRIN, Italy, 2005.

[45] S. Casadio, C. Zehner, E. Putz, and G. Pisacane, Empirical Retrieval of the Atmospheric Air Mass Factor (ERA) for the Measurement of Water Vapour Vertical Content Using GOME Data. Comparisons with Independent Measurements, ERS ENVISAT Symposium, October 16–20, Goteborg, Sweden, 2000.

[46] THE VOICE, Phase 1 Executive summary, TVO-SYS-DAT-TN-025-1.0.PDF, ESRIN Contract No. 18104/04/I-OL, April 2005; see also http://www.esa-thevoice.org.

[47] P. Betti, C. Camporeale, K. Charvat, L. Fusco, and J. van Bemmelen, Use of Wireless and Multimodality in a Collaborative Environment for

the Provision of Open Agricultural Services. XIth Year of International Conference–Information Systems in Agriculture and Forestry—on the Topic e-Collaboration, Prague, Czech Republic, 2005.

[48] Diligent Project Website: http://www.diligentproject.org.

[49] OpenDLib DLMS Website: http://www.opendlib.com.

[50] gLite Website: http://glite.web.cern.ch/glite/default.asp.

[51] Global Monitoring for Environment and Security (GMES) Website: http://www.gmes.info.

[52] Oxygen, A New Strategy for Earth Observation, *Earth Observation Quarterly*, June 2003, EOQ, N° 71, http://www.esa.int/publications, http://esapub.esrin.esa.it/eoq/eoq71/chap1.pdf.

[53] R. W. Moore, Preservation of Data, *SDSC Technical Report, 2003-06*, San Diego Supercomputer Center, University of California, San Diego, 2003.

[54] Report on the NSF Workshop on Research Challenges in Digital Archiving: Towards a National Infrastructure for Long-Term Preservation of Digital Information, *Workshop Report—Draft 2.0 (Pre-Publication Draft)*, August 12, 2002.

[55] C. Schwarts, Digital Libraries—Preservation and Digitization, Digital Libraries course, Simmons College, Boston, http://web.simmons.edu/˜schwartz/dl-preserve.html.

[56] C. Schwarts, Digital Libraries—LIS 462 Definition, Digital Libraries course, Simmons College, Boston, http://web.simmons.edu/˜schwartz/462-defs.html.

[57] Final Report, The Selection, Appraisal and Retention of Digital Science Data, ERPANET/CODATA Workshop, Biblioteca Nacional, Lisbon, December 15–17, 2003.

[58] Global Grid Forum Website: http://www.gridforum.org.

[59] R. Suresh and K. McDonald, XML for Science Data Access, CEOS Joint Sub-Group Meeting, Frascati, Italy, 2002.

[60] Earth Observation Principal Investigator (EOPI) Web portal: http://eopi.esa.int.

Chapter 12

Design and Implementation of a Grid Computing Environment for Remote Sensing

Massimo Cafaro,
Euromediterranean Center for Climate Change & University of Salento, Italy

Italo Epicoco,
Euromediterranean Center for Climate Change & University of Salento, Italy

Gianvito Quarta,
Institute of Atmospheric Sciences and Climate, National Research Council, Italy

Sandro Fiore,
Euromediterranean Center for Climate Change & University of Salento, Italy

Giovanni Aloisio,
Euromediterranean Center for Climate Change & University of Salento, Italy

Contents

This chapter presents an overview of a Grid Computing Environment designed for remote sensing. Combining recent grid computing technologies, concepts related to problem-solving environments, and high-performance computing, we show how a dynamic Earth Observation system can be designed and implemented, with the goal of management of huge quantities of data coming from space missions and for their on-demand processing and delivering to final users.

12.1 Introduction

The term remote sensing was first used in the United States in the 1950s by Evelyn Pruitt of the U.S. Office of Naval Research, and is now commonly used to describe the science of identifying, observing, and measuring an object without coming into direct contact with it. This process involves the detection and measurement of different wavelength radiations, reflected or emitted from distant objects or materials, by which they may be identified and categorized by class, type, substance, and spatial distribution. Remote Sensing Systems are thus made of:

- sensors mounted on an aircraft or a spacecraft that gather information from the Earth's surface;
- acquisition and archiving facilities that store the acquired raw data;
- computing resources that process and store them as images into distributed databases;
- on-line systems provided by space agencies for their distribution to final users.

Information can be achieved by means of the so-called passive sensors, which detect the radiation emitted by the sun and the one spontaneously emitted by the ground. It should be noted that passive sensors do not work during the night and their efficiency is strongly influenced by atmospheric conditions. Active sensors instead detect the backscattered radiation emitted by a radar installed on-board the spacecraft and, unlike passive sensors, can be used for monitoring both day and night, whatever the weather conditions. The *Synthetic Aperture Radar* (SAR) is such an active sensor, and it is widely used in remote sensing missions to achieve high-resolution Earth images. What we are interested in are the system components belonging to the ground segment and devoted to the archiving, processing, and delivering of remote sensing images to the final user. The scenario is thus characterized by big distributed archives in which information is stored as raw data, that is, in the original format acquired on-board the spacecraft, or as images derived from processing of the raw data. As a matter of fact, further post-processing is usually required to generate standard products to be delivered to final users.

To access the huge quantity of remote sensing data stored in the archives, some user interfaces have been developed. An example of a Web interface is the Intelligent Satellite Data Information System (ISIS) at the German Remote Sensing Data Center

(DFD). This interface provides catalogue search and visualization of digital quick-looks and electronic order placement. It is interesting to remark that links to external Earth Observation Systems (EOS) archives are also provided. Using the ISIS interface, the user is allowed to search for the information through a clickable map of the world to set the geographic region of interest and other parameters like the campaign, the selected data center, the time range, and the processing level. The ISIS interface is similar to that provided by the European Space Agency Earthnet On Line.

These on-line systems are in our opinion good examples of Web interfaces to static Earth Observation Systems: The user is just allowed to access a static catalogue (i.e., only images previously processed and stored in the data base can be retrieved) and no on-demand processing is permitted. Moreover, a limited number of post-processing facilities is provided (no true real-time services) and the level of transparency in the data access is very low, i.e., the user must know in advance where the data are stored, how to access them, etc. Through a nice Web interface, the user is well guided on a clickable map of the world, where she can select the region of interest, the data source, the acquisition time range, the data centre, but she needs to know too many details and no high-level queries are allowed, due to a lack of intelligence capable to translate a high-level user's request to low-level actions on the system. The level of integration of multi-sources data is also very low, so that these systems are not capable of fusing multiple data sources together to infer new knowledge. For these reasons, although the EOS data are so useful, the number of real users is very limited compared to the big investments of the International Space Agencies.

In this chapter, we present an overview of a Grid Computing Environment designed for remote sensing. Combining recent grid computing technologies, concepts related to problem solving environments, and high-performance computing, we show how a dynamic Earth Observation system can be designed and implemented, with the goal of management of huge quantities of data coming from space missions and for their on-demand processing and delivering to final users. The rest of the chapter is organized as follows. Section 12.2 introduces Grid Computing Environments (GCEs), and Section 12.3 describes common grid components of GCEs. We describe the design of our GCE for remote sensing highlighting implementation details in Section 12.4 and report about the implementation best practices in Section 12.5. We discuss the GCE approach and compare it against classic approaches in Section 12.6. Finally, we draw our conclusions in Section 12.7.

12.2 Grid Computing Environments

Due to the rapid evolution of grid computing technologies with respect to both concepts and implementations, people increasingly think about grids clearly separating the user and the system sides. Indeed, a useful distinction in a grid system is made by considering separately how the users access computing services and how these services interface themselves with back-end resources. Usually, the user side is called the Grid Computing Environment whereas the underlying distributed system is called the Core Grid. A GCE therefore embraces the tools provided by a grid system that allow users accessing grid resources and applications; it includes graphical interfaces (for authentication/authorization, job composition and submission, file management, job monitoring, resource management and monitoring, etc.), run-time support, result visualization, knowledge sharing modules, and so on [1].

Often, GCEs are implemented as Web applications (so-called grid portals) that provide the users with a friendly interface to grid applications and thus, to a set of resources and applications. Usually, grid portals reside on the top level of a multi-tier application development stack; the major reasons for using the Web as a transfer protocol are ubiquity, portability, reliability, and trust. It is a comfortable, low-tech delivery system that is available at the lab or at home, or from a laptop in a hotel room: The only requirement that users must satisfy to access a Web-based GCE is the availability of a Web browser.

GCEs have also been called grid-based Problem Solving Environments (PSEs) in order to stress the natural tendency of these systems to provide services and functionalities to solve users' problems. Nevertheless, GCEs and PSEs are often confused and misunderstood: A PSE represents a complete, integrated environment for composing, compiling, and executing applications belonging to a specific area (or areas). It includes advanced features in order to assist users in creating and solving a problem on a specific platform. With this in mind, a useful distinction between GCEs and PSEs is related to their focus: While a GCE is meant to control a grid environment that can be potentially composed of thousands of resources, a PSE is specialized to one (or more) applicative domain and, thanks to its specialization, it can provide high-level services (such as assistance) to users in composing their applications. A PSE may access virtual libraries, knowledge containers, and sophisticated control systems related to execution and result visualization. PSEs are able to provide rapid prototyping systems and a high scientific production rate without worrying about hardware or software details. In this chapter we argue that the design of a grid environment for remote sensing must take into account all of the aspects related to the specific applicative domain, not just the nature of data involved in the processing (with regard to format heterogeneity, file size, etc.), the kind of application really used by scientists, etc., but especially and more than anything else the users. Consequently, we will describe the GCE design process and its specialization to this applicative field.

Focusing now on general GCEs features, we can describe them taking into account the functional point of view and considering the technologies used for the implementation. With regard to the former, we define a GCE as a grid subsystem able to satisfy at least the following two requirements [2]:

- programming the users' side of the grid by means of best-suited programming technologies;
- controlling users' interactions in order to implement functionalities such as authentication, jobs composition, and submission.

These two requirements of a generic GCE can be analyzed in depth taking into account the functionalities that must be provided: users' authentication, data management, job management, and resource management. Indeed, a computing environment in which the resources are geographically distributed and owned by different organizations, as in a grid system, requires proper authentication, authorization, and accounting to address the key aspects related to access and resource usage control. Moreover, we refer to all of these issues using the general term security, recalling that confidentiality and integrity of data must also be provided. Job submission and monitoring, and file transfers are two other services essential to any computing environment in which users can perform some kind of processing. Finally, in grid environments, the number of involved resources can rapidly change, and thus an efficient mechanism to manage them is required in order to react rapidly to changes in the environment due to such a dynamic behavior. It is worth noting here that a resource can be a computing node, software, an instrument, and so on, consequently the heterogeneity of these resources must by properly considered. All of these aspects are faced by the core grid services available as grid middleware, but again, the mechanisms provided by the middleware itself are not immediately accessible to end users. A GCE must implement, using the underlying service functionalities, a set of higher-level interfaces.

Let us consider now the second point of view related to the characterization of GCEs: technologies and programming languages. While the GCE functionalities can be rapidly summarized, the way to implement a GCE is more variegated and requires some care. Indeed, GCEs usually exploit the core grid services provided by the underlying software modules (Grid middleware). For instance, the Globus Toolkit [3] is the most widely used grid middleware, but it does not provide direct support for building GCEs: It makes available to developers a set of modules that face all of the aspects involved in the development of grid systems. The modules provided can be accessed directly harnessing their clients and servers, but, besides offering limited functionalities, this approach is rather inconvenient for those scientists who are not computer experts and requires learning the gory details of the Globus software. Instead, scientists would rather employ their time computing and producing useful scientific results. Nevertheless, GCEs can be built using the Globus APIs with the mission to hide all of the low-level details in order to provide a friendly interface. With respect to the programming languages and technologies suited for GCE development, we note here that many different GCE implementations do exist based on C/C++, Java, and related technologies (JavaBeans, Java Server Pages, Portlets, CORBA, XML, Web services, etc.). Often, several architectural layers have been defined in order to de-couple raw resources and low-level services from user interfaces (see Figure 12.1). Examples of toolkits for development of layered GCEs include: the Grid Portal Development Toolkit [4] (GPDK, a suite of JavaBeans components suitable for Java based GCEs), the Java [5], CORBA [6], Python [7], and Perl [8] Commodity Grid interfaces to the Globus Toolkit (COG kits) [9]. In general, distributed object technology is often used due to the distributed nature of both the software and hardware objects.

It is clear that GCEs can be classified with regard to the technologies and languages used. In this case, considerations about the performances and their relation with the programming language can be drawn in order to support the design phase [10].

We conclude this section by introducing OGSA and WSRF. Open Grid Services Architecture (OGSA) and Web Services Resource Framework (WSRF) specifications

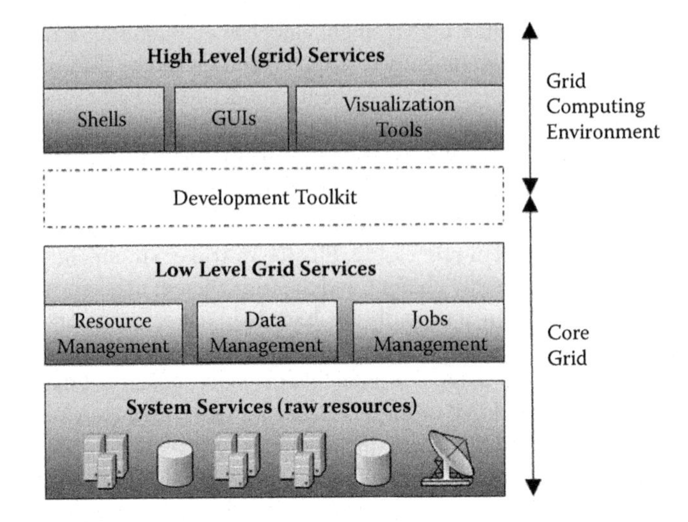

Figure 12.1 Multi-tier grid system architecture.

were recently published respectively by The Global Grid Forum and several vendors in order to define a common standard and an open architecture for grid-based applications including a technical specification of what a Grid Service is.

12.3 Common Components of GCE

GCEs must allow users to access grid resources and applications and thus must include several graphical interfaces and functionalities exploiting the underlying core grid services. In order to satisfy this requirement we can identify common GCEs components related to security (in order to allow authentication and authorization), job management (for task composition, job submission and monitoring, file staging, etc.), data management (to allow users searching and retrieving data from distributed data repositories, storing simulation results, etc.), and information management. In the following subsections we detail these modules referring to the Globus Toolkit.

12.3.1 Security

Security is concerned with the following problems: establishing the identity (authentication) of users or services, determining who (principal) has the rights to perform what actions (authorization), and protecting communications, as well as with supporting functions, e.g., managing user credentials and maintaining group membership information. The Globus Toolkit GSI (Grid Security Infrastructure) leverages X.509 end entity certificates and provides support for proxy certificates, which are used to identify persistent entities such as users and servers and to allow the temporary delegation of privileges to other entities. GSI has been devised to deal with authenticated and confidential secure communication among clients and grid services in a distributed setting that avoids the use of a centrally managed security system and provides single sign-on access to grid resources and delegation of credentials. GSI is based on public key cryptography and provides an implementation of the GSS APIs (Generic Security Services) layered on top of TLS (Transport Layer Security), credential management through the MyProxy service, which is an online credential repository, and the SimpleCA package, which is a basic Certification Authority. Group membership is addressed by the Community Authorization Service (CAS), and a GSI-enabled version of the SSH protocol is also available.

12.3.2 Job Management

Job management is provided in Globus by the Grid Resource Allocation and Management (GRAM). Submitting a job, while conceptually simple, may involve complex operations, including job initiation, monitoring, management, scheduling, and/or coordination of remote computations. In GCEs, a typical job may also require additional work to stage the executable and/or input files to the remote machine prior to execution, and to coordinate staging output files following the execution. On machines where a local scheduler is available, GRAM provides a uniform interface allowing job submission independently of the actual local resource scheduler. A job submitted may be interactive, a classic batch job, and even parallel. Job monitoring is easy, since GRAM can notify asynchronously job state changes. Security leverages GSI, and file stage-in and stage-out takes advantage of GridFTP, detailed in the next subsection.

12.3.3 Data Management

The Globus Toolkit provides GridFTP tools to handle file transfers. GridFTP is a protocol extending the classic FTP with a number of extensions especially designed for grid environments. The primary aim of the protocol and related tools is the provision of a grid service for secure, robust, fast, and efficient transfer of bulk data. GridFTP indeed leverages GSI and provides support for the automatic restart of failed transfers, for high-performance file transfer exploiting multiple parallel streams, automatic negotiation of TCP buffer size, third-party and partial file transfer.

Besides GridFTP, other tools are also available including the Reliable File Transfer Service (RFT), which leverages GridFTP, the Replica Location Service (RLS), the Data Replication Service (DRS), and the Data Access and Integration (OGSA-DAI).

Currently in the same area, there are other grid data management toolkits, which address the same issues providing extreme performance and enhanced functionalities. One of them is the Grid Relational Catalog (GRelC) toolkit [15], which is composed of the following data grid services: GRelC Service (data access to interact with databases), GRelC Storage (to manage workspace areas), and GRelC Gather (to provide a data federation service). The three services are entirely developed in C as GSI-enabled Web services to address efficiency, security, and robustness.

12.3.4 Information Services

Grid computing emerged as a new paradigm distinguished from traditional distributed computing because of its focus on large-scale resource sharing and innovative high-performance applications. The grid infrastructure ties together a number of Virtual Organizations (VOs) that reflect dynamic collections of individuals, institutions, and computational resources. A Grid Information Service in Globus-based grids aims at providing an information-rich environment to support service/resource discovery and decision-making processes. The main goal of Globus-based grids is indeed the provision of flexible, secure, and coordinated resource sharing among virtual organizations to tackle large-scale scientific problems, which in turn requires addressing, besides other challenging issues like authentication/authorization, access to remote data, service/resource discovery, and management for scheduling and/or co-scheduling of resources. Information thus plays a key role, allowing, if exploited, high-performance execution in grid environments: The use of manual or default/static configurations hinders application performance, whereas the availability of information regarding the execution environment fosters the design and implementation of so-called grid-aware applications. Obviously, applications can react to changes in their execution environment only if these changes are somehow advertised. Therefore, self-adjusting, adaptive applications are natural consumers of information produced in grid environments where distributed computational resources and services are sources and/or potential sinks of information, and the data produced can be static, semi-dynamic, or fully dynamic.

Nevertheless, providing consumer applications with relevant information on-demand is difficult, since information can be (i) diverse in scope, (ii) dynamic, and (iii) distributed across many VOs. Moreover, obtaining information regarding the structure and state of grid resources, services, networks, etc., can be challenging in large-scale grid environments. In this context, the usefulness of a Grid Information Service providing timely access to accurate and up-to-date information related to distributed resources and services is evident. This is why the Globus Toolkit provided the grid community with a Grid Information Service since its inception, as one of the fundamental middleware services, the Monitoring and Discovery System (MDS). However, due to MDS-inherent performance problems, in our work we adopted the iGrid Information Service, which we have been developing in the context of the European GridLab project [23][24].

12.4 The Design and Implementation of GCEs for Remote Sensing

Earth Observation (EO) products provided by different EO facilities differ in terms of format, geographic reference system, projection, and so on. Usually, domain specialists are increasingly willing to integrate, compare, and fuse data coming from different EO sensors. EO systems are thus characterized by the management of heterogeneous data. Moreover, several processing algorithms are generally composed by many steps, and some intermediate partial results are used as input to other algorithms defining a workflow. In order to design a GCE tuned to this specific context it is extremely important to take into account these issues. This section describes the GCE requirements and architecture focusing on each aspect related to the considered applicative domain. More in depth, a GCE for EO must allow at least:

- sharing of computational resources and sensors among different organizations;
- sharing of EO and geospatial data;
- sharing of software resources (with particular emphasis on EO and geospatial data processing applications);
- transparent access to computing resources through a graphical interface;
- efficient usage of the resources (this implies handling job scheduling, resource management, and fault tolerance);
- efficient usage of the network links (indeed, the amount of data to be transferred from end users to nodes and vice versa is quite large. The data movement, when considering that it can be a bottleneck, impacts on the global performance of the system);
- composing and compiling new processing applications based on existing ones; the system should allow users building complex applications by means of a user interface that will hide low-level details related to grid infrastructure and resources;
- supporting different, heterogeneous data formats;
- knowledge management (gathering and sharing user's practices).

In order to meet the previous requirements, the designed architecture includes the following modules, as depicted in Figure 12.2:

- **Web interface**: This component allows accessing services and resources. It permits one to search data over a distributed catalogue, to use the available services, and to administer the resources. The Web interface consists of three main sub-components: the data search engine interface, the workflow interface, and the system management interface.
- **Knowledge container**: It is a component that gathers and provides information related to the user's behavior and contains an ontology related to the software tools available in the environment.
- **Distributed data management system**: Based on a common metadata schema for describing EO and geospatial products, it uses some modules belonging to the Grid Relational Catalog (GRelC) Project, which provides efficient, transparent, and secure data access, integration, and management services in a grid environment.
- **Workflow management system**: An integrated workflow management system for EO data processing that includes a resource manager optimized for EO applications.

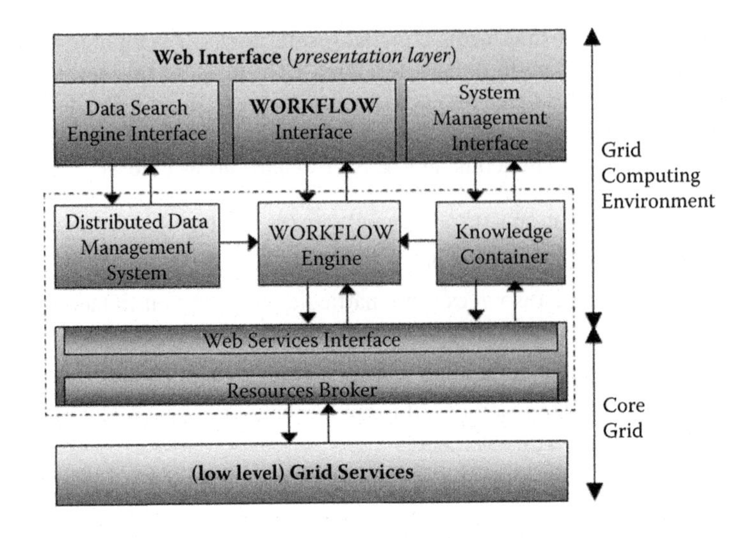

Figure 12.2 System architecture.

- **Core grid services**: The high-level and low-level grid services that provide all of the functionalities needed to access grid resources (authentication, authorization and accounting services, job submission, file transfer, monitoring, and so on). This layer has been developed using specific libraries belonging to the Grid Resource Broker (GRB) [16][17] that exploits the Globus toolkit v4.x as grid middleware. In the following subsections we describe in depth the design and implementation of a GCE for remote sensing.

12.4.1 System Overview

The presentation layer of the GCE consists of a set of interfaces (implemented using classic Web technologies such as dynamic Web pages developed using Java server pages, Java applets, servlets, CGI, and so on) that exploit the logic provided by the underlying business logic modules. This layer contains the following interfaces:

- **Data search engine interface**: Leverages fast CGI programs (for performance reasons) to retrieve data and present them (via the HTTP protocol) within properly formatted HTML pages. This sub-component is layered on top of the distributed data management system and allows users submitting geographic or metadata-based queries (the users can define a bounding box over an Earth map or define as search criteria a time interval or an image type).
- **Workflow interface**: Allows building complex applications by means of a graphical user interface (a Java applet) that hides low-level details related to grid infrastructure and resources. This component relieves the user from the burden of taking into account low-level details and implements mechanisms to assist and guide the user during the specification of her workflow.

- **System management interface**: It is the component that permits configuration of grid resources. Through this component, the administrator user has the possibility to properly configure the resources while contributor users can configure services and applications; allowed operations include addition, modification, and deletion of resources and services. This component interacts directly with the knowledge container through the exchange of XML messages.

12.4.2 Knowledge Container

In grid computing the description of resources is crucial, and several researches have been focused on this aspect. In our approach we have considered separately the information specifically related to the EO domain from the other ones that are common to each grid environment. Indeed, the knowledge container described in this section represents the component that allows the contextualization of our system to a specific domain. Other information related to resource monitoring and discovery is provided by Core Grid Services.

A well-known approach for managing information can be found in the Grid Information Service (GSI) developed within the Globus project. The Globus Monitoring and Discovery Service (MDS-2) provides a large distributed collection of generic information providers that extract information from local resources. The gathered information is structured in term of a standard data model based on LDAP. Moreover, MDS-2 provides a set of higher level services that collect, manage, and index information provided by one or more information providers. The MDS-2 includes a configurable information provider framework (Grid Resources Information Service or GRIS) and a configurable aggregate directory component (Grid Index Information Service or GIIS). An information provider for a computing resource might provide information about the number of nodes, amount of memory, operating system, load average, and so on. An information provider for a running application might provide information about its configuration and status. The MDS schema can be easily extended and additional information providers can be developed in order to manage and publish an extended set of information.

The DataGrid project took a relational approach to GIS, called Relational Grid Monitoring Architecture (R-GMA). This solution uses a relational approach to structure the information about grid resources and is composed by three main components: Consumer, Producer, and a directory service, called Registry (as specified in the Global Grid Forum Grid Monitoring Architecture). In the R-GMA, Producers register themselves with the Registry and describe the type and structure of information they want to make available to the grid. Consumers can query the Registry to find out what type of information is available and to locate Producers that provide such information. When this information is known, the Consumer contacts the Producer directly to obtain the data. In this implementation, the information provided by Producers is available to Consumers as relations are used to handle the registration of Producers and Consumers. R-GMA considers Virtual Organization (VO) information to be organized logically as one relational database in which the implementation is based on a number of loosely coupled components. The database is partitioned and the description of the partitioning is held in the Registry. The Consumer can access information made available by several Producers through a single query expressed as an SQL select statement.

iGrid is a novel Grid Information Service we have been developing within the European GridLab project. Among iGrid requirements there are performance, scalability, security, decentralized control, support for dynamic data, and the possibility to handle users' supplied information. The iGrid Information Service has been specified and carefully designed to meet these requirements. Core iGrid features include:

- *Web service interface* built using the open source gSOAP Toolkit;
- *Distributed architecture* based on two kind of nodes, the iServe and the iStore Web services;
- *Based on relational DBMS* we are currently using PostgreSQL as the relational back-end for both iServe and iStore nodes;
- *Support for Globus GSI* through the GSI plug-in for gSOAP that we have developed;
- *Support for TLS (Transport Layer Security) binding to DBMS* to secure database communications;
- *Support for Administrator-defined Access Control List* to allow for local decisions;
- *Support for heterogeneous relational DBMS* through the GRelC library;
- *Support for multiple platforms* iGrid builds on Linux, Mac OS X, FreeBSD, and HP-UX;
- *Extensible* by adding new information providers (this requires modifying the iGrid relational schema);
- *Extreme performances* as needed to support grid-aware applications;
- *Fault tolerance*.

In a distributed environment, information related to deployed software can play an important role in allowing a scheduling algorithm to perform efficiently. Moreover, the software objects deployed on a computing element can require specific libraries, environment variables, and so on. With regard to software resources, there are several relevant approaches meant to characterize applications. A formal specification of the software objects in a grid setting has been derived from BIDM standard (Basic Interoperability Data Model, IEEE standard 1420.1) by expanding the classes of objects defined in the standard itself [11]. Another approach is the Open Software Description Format (OSD) [12]. OSD is an application of XML that provides a vocabulary used for describing software packages and their dependencies.

We have designed an ad-hoc knowledge container specialized to describe applications and tools for remote sensing data processing and management. In our approach, we have considered separately three kinds of resources: hardware, software, and data. For each component, we have derived and implemented an information model that describes the component.

In particular, in order to characterize the application domain, we have analyzed the EO systems from two points of view: the remotely sensed products and the software tools able to process them. For both, we have derived and implemented an information model that fully describes them. The information schemas are developed in XML and are composed by:

- The MetaSoftware schema, in which we have collected a set of relevant information able to describe the software tools. The collected information related to an application includes: (a) input and output data formats; (b) application capabilities (we distinguish applications with pre-processing capabilities, post-processing or data format conversion capabilities); and (c) information needed to launch the application, i.e., hostname, pathname, shared libraries on which the application depends, environment variables, application arguments, and so on.
- The MetaData schema: In order to realize a MetaData schema that involves the most important information about remote sensing data, we have considered the ISO TC/211-19115 standard [13]. From this standard we have derived a set of metadata. The most important metadata we have considered address product identification and distribution, data quality, platform and mission, spectral properties, maintenance, generic information, spatial representation, reference system, and other information related to the file data format. This metadata set is structured utilizing a relational schema.

This component is also able to store the processing patterns that the users define during their work sessions. For example, if a user defines a new processing algorithm starting from the services available on the environment (possibly by means of the Workflow interface), it will be showed to the other users as a new service. This way the environment can learn form the users' experience and can provide users with a rich set of tools.

12.4.3 Distributed Data Management System

Remote sensing data, acquired by ground-segments or processing facilities, comes from different sensors and different space missions. Often the data format produced for each mission and each sensor differs, and if we consider that the ground-segment can store the acquired data using internal formats, it is immediately evident how the management and format conversion is an important issue that must be taken into account. An environment like a PSE must support several data formats and must allow converting to/from different formats. In order to simply handle acquired raw data or post-processed data, it is fundamental to associate a set of metadata to each remote sensing product. Moreover, through these metadata it is possible to perform advanced operations like thematic searches. The Committee on Earth Observation Satellites (CEOS) is an international organization aimed at coordinating international civil spaceborne missions with the purpose to observe and study the Earth planet. CEOS comprises 41 space agencies and other national and international organizations, and is recognized as the major international forum for the coordination of Earth observation satellite programs and for interaction of these programs with users of satellite data worldwide. One of the activities of CEOS is to coordinate the Earth observation data exchange, through the publication of a set of principles and the definition of a standard data format. The CEOS format is, indeed, the standard format adopted by several agencies to distribute remote sensing data. Another relevant standard to consider is ISO TC/211, by the International Organization for Standardization (ISO). The aim of ISO TC/211 is the standardization in the field of digital geographical information or to establish a structured set of standards for information concerning objects or phenomena that are directly or indirectly associated with a location relative to the Earth. These standards specify methods, tools, and services for data management, acquiring, processing, analyzing, accessing, presenting, and transferring such data in digital/electronic form among different users, systems, and locations. The work will link appropriate standards for information technology and data where possible, and will provide a framework for the development of sector-specific applications using geographical data. The information model we propose aims at describing and modeling the remote sensing products. It is based on the CEOS data format and on ISO TC/211-19115 specification. ISO 19115:2003 defines the schema required for describing geographical information and services. It provides information about the identification, extent, quality, spatial and temporal schema, spatial reference, and distribution of digital geographical data.

The distributed data management system [14] (see Figure 12.3) consists of several components such as the GRelC Service (GS), the Enhanced GRelC Gather Service (EGGS), and the GRelC Data Storage (GDS). A Web application is available to easily access data through a Web interface.

The GS module provides an efficient and standard data access interface to the EOS metadata repositories. It performs a first level of data virtualization in a grid environment, providing both DBMS and location transparency. It offers (i) a wide set of APIs related to the interaction with relational and not relational data sources; (ii) efficient delivery mechanisms leveraging both gridFTP and HTTPG (HTTP over GSI) protocols, compression, and streaming mechanisms to speed up performance; (iii) fine-grained access control policies; (iv) transactional support; (v) access to a wide range of relational DBMSs (Postgresql, Mysql, UnixODBC data sources, etc.), and (vi) advanced APIs exploiting interaction with storage servers to upload resultsets or database dumps into a common workspace area, etc. Moreover, a user-friendly graphical interface allows users interacting with EOS metadata repositories without requiring detailed knowledge about underlying technologies or technical issues.

Security is provided through GSI and leverages (i) mutual authentication between the GS-contacting-user (user that tries to bind to the GS) and the GS, (ii) authorization based on the Access Control List, (iii) data encryption (to assure confidentiality), and (iv) delegation to allow the GS user acting (exploiting proxy credentials) on behalf of the GS-contacting-user with respect to interaction with other services (i.e., the storage service).

The EGGS module provides an efficient data integration interface to the GS modules and thus to the EOS metadata repositories. It performs a second level of data

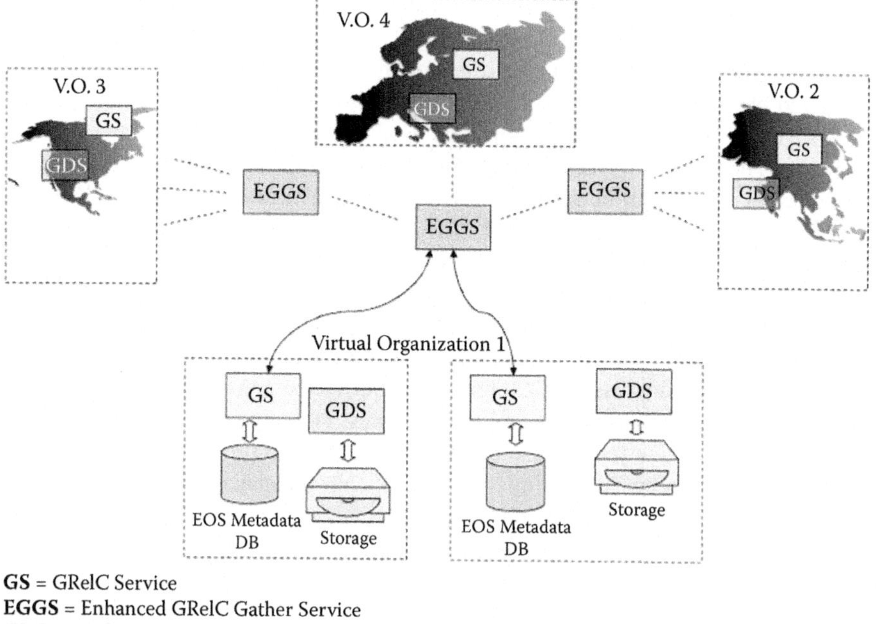

Figure 12.3 Distributed data management architecture.

virtualization in a grid environment providing data source location and distribution transparency. It exploits a peer-to-peer data federated approach in order to join transparently information coming from different EOS metadata repositories. The federated approach versus the centralized one provides local autonomy and scalability and represents a key element within the proposed data management architecture.

Within an effective computational grid scenario, several EGGS can be linked together in a connected graph, originating a federation of multiple EGGSs. Each EGGS has full control of a set of GSs (local data sources) belonging to its own administrative domain (the same virtual organization/institution). Moreover, the EGGS is able to (i) submit queries to the local GSs, (ii) route queries to the EGGS neighbors, (iii) reject queries already checked, (iv) collect partial results (transit node) and deliver the entire resultset (agent node), (v) manage time to live and hops to live, and (vi) support both synchronous and asynchronous queries.

It is worth nothing here that each EGGS manages its own domain independently from the other ones and performs local access control policies to meet administrative issues (these represent strong requirements related to an EOS scenario). Even in this case security is performed by means of the GSI and delegation is also supported to allow fine-grained access control to remote EOS data sources.

The query language used to query the GS and the EGGS is the Standard Query Language (SQL) containing GIS extensions in order to satisfy complex user requests and meet specific needs.

The GDS module provides an efficient data access interface to the storage resources located within the EOS system. It is able to manage efficiently, securely, and transparently collections of data (files) on the grid (allowing interaction between users and physical storage systems), fully exploiting the grid-workspace concept. It allows sharing data among groups of users, leveraging different data access policies and roles. The GDS also manages metadata related to files, workspaces, etc., providing different data transport protocols for file transfer. Regarding static workspaces, the storage service provides functionalities (i) to define and configure permanent workspaces; (ii) to define user profile, privileges, and roles within the Grid Storage and workspaces; (iii) to transfer data using GridFTP, HTTPG, and other protocols; (iv) to manage group of users; (v) to remotely access files using Posix-like functions, and (vi) to perform a coherence check of the system and provide report notification to the workspace administrator. Moreover, owing to the nature of a computational grid (heterogeneous, distributed, prone to failures, etc.) along with static services, a more dynamic framework is also needed. The GDS also provides a basic support for dynamic workspaces leveraging two key concepts: the management of workspace lifetime and quota. A preliminary disk-cache management of the data storage is based on the Storage Resource Manager specification.

It is worth nothing here that the proposed peer-to-peer approach based on EGGS is able to federate both metadata stored within the GS (EOS Metadata Catalogue, as previously described) and metadata stored within the GDS (a distributed workspace area).

12.4.4 Workflow Management System

Workflow management is emerging as one of the most important grid services of GCEs and captures the linkage of constituent services together to build larger composite services.

The reasons that induced us to consider the Earth observation field are the following. First, several processing algorithms are composed by many steps and some intermediate partial results are used as input to other algorithms or to choose, under some conditions, the next task. Another important reason is that, according to the problem-solving environment approach, new complex processing algorithms composed using simpler ones can be defined, saved, and then reused in the next work session or shared among different users.

More in depth, from the architectural point of view, the main components of the Workflow Management System are: i) the workflow interface (WI) included in the presentation layer, ii) a service discovery module (SDM) in charge of discovering services or applications on the basis of some criteria, and iii) a workflow verification and reduction module (WFVRM) that implements the workflow verification and reduction functionality. In Figure 12.4 is depicted the interaction among the workflow management system modules and descriptions of the functions performed by each one. The SDM provides service and application lookup functionalities by accessing the knowledge container. When a user selects the EO products to be processed, the SDM queries the knowledge container using as filter criteria the list of the product

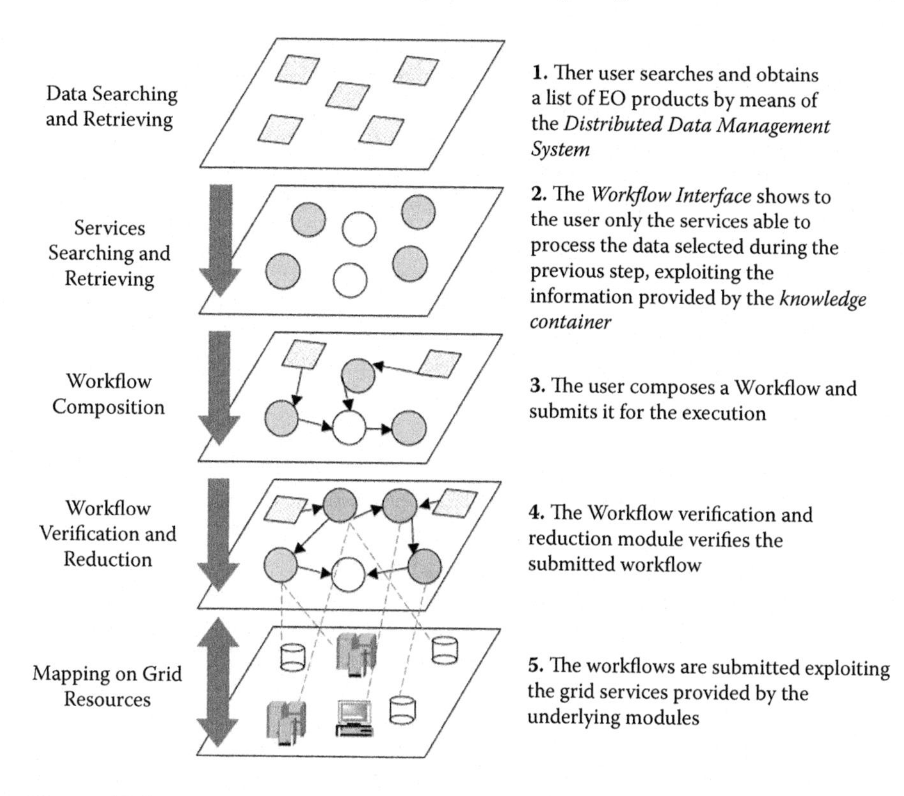

Data Searching and Retrieving
1. Ther user searches and obtains a list of EO products by means of the *Distributed Data Management System*
Services Searching and Retrieving
2. The *Workflow Interface* shows to the user only the services able to process the data selected during the previous step, exploiting the information provided by the *knowledge container*
Workflow Composition
3. The user composes a Workflow and submits it for the execution
Workflow Verification and Reduction
4. The Workflow verification and reduction module verifies the submitted workflow
Mapping on Grid Resources
5. The workflows are submitted exploiting the grid services provided by the underlying modules

Figure 12.4 Workflow management stack.

formats, and it retrieves all of the applications or services able to process these data. SDM also uses the user's profile as an additional information source. The SDM provides a mechanism to use previously defined workflows as a building block for composing recursively new applications. Moreover, SDM also accesses the grid information service to verify the availability of the resources needed by the workflow. The SDM module has been developed in Java.

For the WFVRM, an optimized algorithm has been developed in order to verify the graph in terms of data flow consistency. Indeed, this module is in charge of checking the compatibility between data formats produced by a task and data formats admitted by the next one. Whenever two linked applications have different data formats, WFVRM will automatically look up an available application with data format conversion capability that will be used to convert the data. The user's workflow is preliminarily checked to verify if a direct connected acyclic graph has been properly designed. Using an adjacency matrix representation of the graph, the WFVRM module applies a basic algorithm for cycle detection and to define the graph connectivity. Exploiting the information related to the ontology of the EO applications, WFVRM is also able to define the consistency of a workflow; for instance, a workflow designed to have a post-processing application as an ancestor of a pre-processing application does not make sense.

This module is also in charge of extending or reducing the workflow definition in order to make two nodes compliant with each other, whenever this does not happen. Let us suppose that node u in the graph is directly connected to node v. Let us also suppose that the application instantiated on node u produces as outcome data with format f_1 and the application instantiated on node v admits data with format f_2; WFVRM will find an application or service able to convert data format from f_1 to f_2, and it will extend the workflow definition instantiating a new node between u and v. Finally, redundant data paths will be detected and removed, optimizing the workflow. As an example of a redundant definition, consider multiple instantiation of the same application with the same input parameters; the WFVRM module detects such situations and optimizes the workflow definition removing redundant paths.

12.4.5 Resource Broker

The GRB Scheduler [16] [17] acts as a meta-scheduler among the available grid resources. It has been designed to be fully compliant with respect to the JSDL (Job Submission Description Language) specification [18]. The JSDL language has been extended in order to provide better support for batch job definition, parameter sweep job definition, and workflow graphs. Moreover, the extended JSDL allows defining multiple VOs, managing multiple user's credentials, and defining a set of candidate hosts for resource brokering. The GRB scheduler has also been designed to meet the following requirements:

- Independence from a specific and predefined Virtual Organization. The GRB scheduler can act on behalf of the user among all of the specified computational resources; this means that if a user gains access to resources belonging to different VOs, the GRB will be able to use all of the user's credentials matching the remote resources security policies.
- Support for multiple and heterogeneous grid services. Owing to the GRB libraries, the GRB scheduler can contact different remote resource managers such as Globus Toolkit GRAM, and batch systems such as PBS and LSF.
- Modularity. The GRB scheduler has been designed to support different scheduling algorithms; moreover, new algorithms can be easily plugged in.
- Security. The GRB scheduler supports the Globus Toolkit GSI security infrastructure and exploits the user's delegated credentials to act on grid resources on behalf of the user.

The scheduling process uses simple heuristics such as *MinMin, MaxMin* [19][20], *Workqueue* [21], and *Sufferage*. These iteratively assign tasks to hosts by considering tasks not yet scheduled and computing predicted Completion Times (CT_{ij}) for each task i and host j couple. The task with the best metric is assigned to the host that lets it achieve its Completion Time. For each algorithm the best metric is defined as follows: the minimum of CT_{ij} for MinMin; the maximum of CT_{ij} for MaxMin; let *diff* be the difference between the second minimum CT_{ij} and the minimum CT_{ij}. Then the best is defined as the maximum *diff* over all i and j subscripts.

The GRB scheduler front-end presents a Web service interface developed using the gSOAP Toolkit with GSI plug-in [22]. In its back-end the GRB scheduler makes use of the GRB libraries to access grid resources and services provided by the underlying grid middleware. GRB also automatically refreshes a job credential, allowing long-running jobs, and notifies users by email each time a change of state (e.g., from pending to running, etc.) related to submitted jobs occurs. The GRB scheduler Web service currently advertises the following methods:

- **grb_schedule_job**: This method gets an extended JSDL job description related to the job to be scheduled and returns a GRB job identifier. The *GRB-job-id* is a job handle to be used for cancelling or checking job status. On user invocation, the scheduler authenticates the user and authorizes her on the basis of its access policy. The scheduler then parses the provided JSDL, acquires the credential(s) from the specified MyProxy server, and, if needed, acts as a broker on behalf of the user before submitting the job. All of the scheduler actions are logged to stable storage for subsequent accounting and auditing, and transactions are used when storing job data using an embedded hash table provided by Berkeley DB. The GRB scheduler can submit to different resource managers and is able to contact them directly through specialized driver libraries that interface with Globus GRAM, LSF, PBS, etc. Execution of graphical applications is also allowed: The GRB Scheduler automatically redirects the remote X-Window display, so that users can steer graphical applications. Of course, this feature may require configuring as needed firewalls.
- **grb_check_job**: The method can be used to check a job status. Only the job owner is authorized to check the status of a given job; the owner can also retrieve job output.
- **grb_cancel_job**: The method cancels the execution of a given job.

12.5 The Implementation of GCE, Best Practices

The architecture described in the previous subsections includes several modules. In this section three main modules will be described: front-end interface, data management, and resource management.

12.5.1 Front-End Interface

The user interface is based on Web technologies and has been implemented using CGIs developed in C. A user can access the system after successful authentication and authorization; the authentication mechanism is based on log-in and password, protected by means of the HTTPS protocol. When the user enters the grid portal, she needs a valid proxy stored on a given MyProxy Server; the authentication mechanism used for job submission and in general to access grid resources and services is based on mutual authentication with X509V3 certificates. Secure, ephemeral cookies are exploited to establish secure sessions between the users and the GRB portal.

The Workflow editor has been developed as a Java Applet component that includes tools for composing a generic acyclic graph; each vertex represents one task to be executed on the grid and the edges describe the interaction and dependencies between tasks. The Workflow editor also includes the ontology related to applications for Earth Observation. The ontology actively guides the user during the definition of the workflow; whenever the user connects the output produced by a task with a task that is not compliant with the data format specified as input, the Workflow editor warns the user and automatically tries to supply a new mediator task able to convert the data format.

12.5.2 Information and Data Management

In order to extract the appropriate set of information used to characterize remote sensing applications, a set of software packages has been analyzed. The purpose of this analysis is to obtain a set of common properties and parameters that allow the design of a complete MetaSoftware schema that takes into account all needed information for running applications in a distributed environment. These properties can include required libraries, the list of supported operating systems, the execution parameters, the supported input data formats, the produced output data format, and so on.

The applications available for remote sensing data processing can be classified into three categories:

- Pre-processing applications: Belonging to this class are all of those applications used to process raw data coming directly from sensors installed on remote sensing satellites and acquired by ground-segments. The products obtained in this initial phase represent semi-finished products, and the end user have to apply further processing in order to extract relevant information.
- Post-processing applications: In this class are all of those applications used to extract relevant information from semi-finished products produced by the pre-processing application. These applications usually perform advanced processing like filtering, classification, data analysis, and so on.
- Utility applications: Another kind of software tool we have considered are those applications that perform simple data manipulations, e.g., data format conversion, data header analysis and extraction, image cropping, and so on.

The analysis of these applications has allowed us to extract all relevant information needed and to design the MetaSoftware schema. We have collected a set of relevant information able to completely describe a software object. The collected information mainly belongs to two classes: information characterizing the applications from the functional point of view and information about the performance. The former is useful for resource discovery; the latter can be used by the scheduler to define a submission schedule that minimizes, for example, the completion time. This information has been structured into a relational schema, and it includes:

- application definition: information about the general properties of the application like name, required processor speed, required amount of memory, required disk space, and interface type (which gives an indication about the mechanism to be used for remotely starting and monitoring jobs);
- application class: information about the application typology, which, as shown in the previous section, can be for instance pre-processing, post-processing, and utility;
- data formats: information about the supported input data formats and information about data formats that the application is able to produce;
- execution command: information to be used to remotely execute the application, thus here we include information like the executable pathname, list of accepted arguments and their default values, and list of environment variables and their default values;
- required libraries: list of required libraries, the version number, a short description, and so on;
- operating system: list of the operating systems and related versions, on which the application can be executed;
- performance: the ensemble of information that describes the running time as a function of the type and size of input data obtained from a series of experimental executions.

In order to realize a MetaData schema that involves the most important information about remote sensing data, we have considered the ISO TC/211-19115 standard. From this standard, we have derived a set of raw metadata. This set is mainly composed of 13 groups of information. The most important considered metadata concerns are product identification and distribution, data quality, platform and mission, spectral properties, maintenance, generic information, spatial representation, reference system, and other information related to the TIFF data format.

This set of raw metadata is mapped into a uniform metadata set derived from the ISO standard itself in order to have a homogeneous set related to the following missions: ENVI, ERS1, ERS2, RadaraSat1, SLR1, SLR2, and SRTM. We have obtained a set made of about 200 metadata that describes all of the mentioned Earth Observation products with sufficient thoroughness. This metadata set is structured in a relational schema.

Moreover, the CEOS data format specification is considered in order to achieve a good description of input and output data format for remote sensing applications. We have considered, for each product, the files associated to the product.

The MetaSoftware and MetaData schemas have been structured into relational databases accessible on the grid through the GS interfaces. We also used Postgresql as a back-end database management system. The low-level access functionalities, needed for managing these catalogues, have been implemented both in Java using

JDBC and in C using UnixODBC; moreover, high-level data access interfaces are provided as Web services methods for users who want to access the EOS metadata repositories within a computational grid. We have also developed the MetaSoftware and MetaData modules using an XML schema definition because of two main reasons: (a) the data belonging to the catalogues can be easily presented as dynamic html pages; (b) the XML language is well suited to implement data exchange among heterogeneous components belonging to a grid. In the MetaSoftware catalogue, the user's management functionality has been realized through Java Servlets that use the JAXP package for parsing XML documents and XSL for processing documents. Moreover, with regard to these two catalogues, a set of additional functionalities has been realized using some Java modules and Java Servlets. These functionalities allow manual and automatic ingestion into the system of software objects and data. We have adopted our iGrid information service developed in the GridLab project to handle all other general information that is not related to remote sensing aspects.

12.5.3 Use Case

In order to give an idea about how our GCE can be usefully used, in this section we describe a use case in which our components have been deployed. This environment, as already described in previous sections, provides the following functionalities: sharing of computational resources, sharing of software resources, management of resources, secure access to services available on the grid according to local access policies, transparent access to computing resources trough a graphical interface, efficient usage of the resources, efficient usage of the network links, composition of new processing applications based on existing ones, and so on. The knowledge container module allows storing all of the information about the resources and can be used by several architectural components to perform service discovery, according to the user's request. A fundamental component of a PSE is the Graphical User Interface (GUI), which allows the composition of complex applications built from single application components. The GUI initially presents to the user the services available on the system, querying the knowledge container. Once the user selects a set of applications and combines them defining a workflow and specifying the input data, the system queries the knowledge container in order to obtain the metadata attributes related to each involved application. By carefully checking the retrieved metadata attributes, the system can verify by inspection the compatibility between the data produced during each processing step and use them to map the high-level request, specified through the workflow, into a set of low-level actions that perform the needed processing.

Let us suppose now that the following resources have been registered into the PSE:

- the grid is composed of four computing resources (H1, H2, H3, and H4) and two storage resources (S1 and S2);
- an SAR processor that performs image focalization (A1). Let us suppose that it supports the data formats F1 and F2 as input formats, it produces data in the F3 format, and it is available on hosts H1 and H2;

- a co-registration tool that performs the co-registration of two SAR focused images (A2). Let us suppose that it supports the data formats F3 and F4 as input format, it produces data in the F5 format, and it is available on hosts H1 and H3;
- an interferogram generator, which performs the generation of an interferogram starting from two co-registered SAR images (A3). Let us suppose that it supports the data format F5 as input format, it produces data in the F6 format, and it is available on hosts H2 and H4;
- finally, let us suppose that on the storage resource S1 a raw SAR frame (D1) in the format F1 and a focused SAR frame (D2) in the format F3 are available. On the storage resource S2, the orbital data D3 for the datasets D1 and D2 are available.

Let us now suppose that a potential PSE user wants to produce an interferogram (D4) starting from the two available dataset D1 and D2. She asks the system to discover all of the available applications, composing them through the GUI (see Figure 12.5a). After the workflow submission, the system queries the knowledge container and retrieves all of the metadata attributes related to the data and applications. It can verify if the input datasets are compatible with the applications; if they are not compatible, the system warns the user or activates, whenever possible, an automatic data format

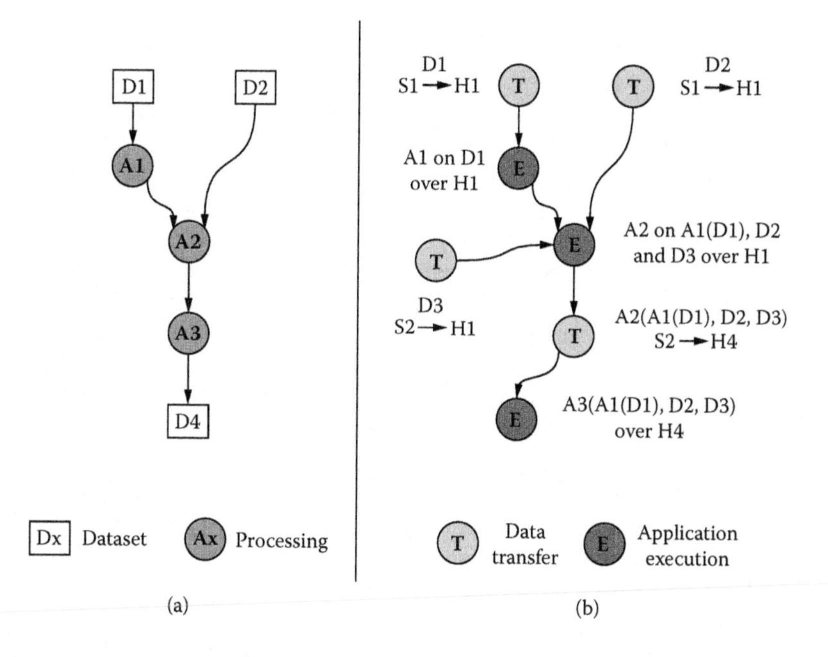

Figure 12.5 a) Task sequence showing interferogram processing; b) task sequence mapped on grid resources.

conversion. The system discovers also that the application A1 requires the orbital dataset D3. Moreover, through the metadata attributes, the system discovers where the applications are available and on the basis of this information it can make an informed decision related to scheduling, leading to better choices as a function of performance parameters, the execution machine performance, and data transfer rate (see Figure 12.5b).

As shown in this example, the knowledge container plays a crucial role for the PSE. Indeed, this component provides all of the needed information both to users (in order to discover resource and available services) and to the system in order to perform job scheduling. Another fundamental aspect that the previous example did not show is the functionality provided to the users by the distributed MetaData catalogue, accessible through the EGGS interface, to make efficiently and transparently advanced queries on the available data sets, without knowing anything about data location and distribution. Indeed, through the metadata attributes related to the available products, the user can query the system to find all of the available products that satisfy her criteria, and thus she can perform the needed processing on retrieved data.

12.6 Comparison Between a GCE for Remote Sensing versus the Classical Approach

Several works are strictly related to the use of distributed and parallel computing for remote sensing data storage, processing, and access. In some of the works the authors show how distributed computing can be used for on-demand processing [25][26][27][29][30][31][32]; in other works it is shown how distributed computers linked together through wide-area networks can be used for data storage and access [25][26][27][28][29][30]. In other works, the authors show how to do high-performance processing on data through clusters of PCs [26][30][31][32][33][34]. This work presents many of the involved technologies; however, all of the presented solutions can be characterized as having a static architecture and the addition of new functionalities requires reengineering and recoding. In scientific environments, people rapidly develop new algorithms and software tools. In particular, in the remote sensing field the greater quantity of information is extracted through post-processing operations that involve new techniques and algorithms. It is immediately evident how a dynamic and easily reconfigurable architecture can be a better choice. For instance, in the Digital Puglia project [25] we emphasize the retrieval of remotely sensed data to the client's workstation, but additionally we customize the processing of the data. It may be that the data are too large for downloading to be practical, or the client may not have the relevant processing software. The processing at the server may be as simple as a change of format, or it may be that a user does compute-intensive image processing such as principal component analysis, supervised classification, or pattern matching. By 'processing on demand' we mean a data archive connected to a powerful compute server at high bandwidth, controlled by a client that may be connected at low bandwidth. Nevertheless, the architecture of current approaches is rather static since the addition or removal of applications or computational resources requires recoding several software components.

12.7 Conclusions

This chapter presented an overview of grid computing environments and discussed their usefulness in the context of remote sensing. We showed how, combining recent grid computing technologies, concepts related to problem-solving environments, and high-performance computing, it is possible to design and implement a GCE based on the Globus Toolkit that allows users seamless access to EO data and grid resources, providing data retrieval services and on-demand processing of remote sensing data, leveraging dynamically acquired resources and services.

References

[1] M. P. Thomas, G. Fox, and D. Gannon, A Summary of Grid Computing Environments. *Concurrency and Computation: Practice and Experience*, vol. 14, pp. 1035–1044, 2002.

[2] G. Fox, M. Pierce, D. Gannon, and M. Thomas, Overview of Grid Computing Environments, *Global Grid Forum*, Tech. Rep., 2002.

[3] *The Globus Grid Project*, http://www.globus.org.

[4] J. Novotny, The Grid Portal Development Kit, *Concurrency and Computation: Practice and Experience*, vol. 14, Grid Computing Environments Special Issue 13–15, pp. 1129–11444, 2002.

[5] G. von Laszewski, J. Gawor, P. Lane, N. Rehn, and M. Russell, Features of the Java Commodity Grid Kit, *Concurrency and Computation: Practice and Experience*, vol. 14, Grid Computing Environments Special Issue 13–15, pp. 1045–1056, 2002.

[6] G. von Laszewski, M. Parashar, S. Verma, J. Gawor, K. Keahey, and N. Rehn, A CORBA Commodity Grid Kit, *Concurrency and Computation: Practice and Experience*, vol. 14, Grid Computing Environments Special Issue 13–15, pp. 1057–1074, 2002.

[7] K. Jackson, pyGlobus: A Python Interface to the Globus Toolkit, *Concurrency and Computation: Practice and Experience*, vol. 14, Grid Computing Environments Special Issue 13–15, pp. 1075–1084, 2002.

[8] S. Mock, M. Dahan, M. Thomas, and G. von Lazewski, The Perl Commodity Grid Toolkit, *Concurrency and Computation: Practice and Experience*. vol. 14, Grid Computing Environments Special Issue 13–15, pp. 1085–1096, 2002.

[9] G. von Laszewski, J. Gawor, S. Krishnan, and K. Jackson, Commodity Grid Kits — Middleware for Building Grid Computing Environments, Chapter 26, pp. 639–658 in *Grid Computing: Making the Global Infrastructure a Reality*, ISBN 0-470-85319-0, John Wiley & Sons Ltd, Chichester, 2003.

[10] F. Berman, G. Fox, and T. Hey, *Grid Computing: Making the Global Infrastructure a Reality*, ISBN 0-470-85319-0, John Wiley & Sons Ltd, Chichester, 2003.

[11] J. Millar, Grid Software Object Specification, Information Service, University of Tennessee, request for comment: GWD-GIS-008.

[12] *The Open Software Description Format*, http://www.w3.org/TR/NOTE-OSD.

[13] ISO TC 211 standard of the International Organization for Standardization. Webpage: http://www.isotc211.org.

[14] G. Aloisio, M. Cafaro, S. Fiore, and G. Quarta, A Grid-Based Architecture for Earth Observation Data Access, *Proceedings of the 2005 ACM Symposium on Applied Computing (SAC 2005)*, ACM Press, ISBN 1-58133-964-0, March 13–17, 2005, Santa Fe, USA, vol I, pp. 701–705.

[15] *The GRelC Project: An Easy Way to Manage Relational Data Sources in the Globus Community*, http://grelc.unile.it/.

[16] G. Aloisio and M. Cafaro, Web-Based Access to the Grid Using the Grid Resource Broker Portal, *Concurrency and Computation: Practice and Experience*, Special Issue on Grid Computing Environments, vol. 14, pp. 1145–1160, 2002.

[17] G. Aloisio, M. Cafaro, E. Blasi, and I. Epicoco, The Grid Resource Broker, a Ubiquitous Grid Computing Framework, *Journal of Scientific Programming*, Special Issue on Grid Computing, IOS Press, Amsterdam, pp. 113–119, 2002.

[18] A. Anjomshoaa, F. Brisard, M. Drescher, D. Fellows, A. Ly, S. McGough, D. Pulsipher, and A. Savva, GFD-R-P.056, Job Submission Description Language (JSDL) Specification v1.0 (2005), http://forge.gridforum.org/projects/jsdl-wg.

[19] M. Maheswaran, S. Ali, H. J. Siegel, D. Hensgen, and R. Freud, Dynamic Matching and Scheduling of a Class of Independent Tasks onto Heterogeneus Computing Systems, *Journal of Parallel and Distributed Computing*, Special Issue on Software Support for Distributed Computing, vol. 59, pp. 107–131, 1999.

[20] O. H. Ibarra and C. E. Kim, Heuristic Algorithms for Scheduling Independent Tasks on Nonidentical Processors, *Journal of the ACM*, pp. 280–289, 1977.

[21] T. Hagerup, Allocating Independent Tasks to Parallel Processors: An Experimental Study, *Journal of Parallel and Distributed Computing*, pp. 185–197, 1997.

[22] G. Aloisio, M. Cafaro, I. Epicoco, and D. Lezzi, The GSI Plug-In for gSOAP: Enhanced Security, Performance, and Reliability, *Proceedings of Information Technology Coding and Computing*, IEEE Press, vol. I, pp. 304–309, 2005.

[23] *GridLab Project*, http://www.gridlab.org.

[24] G. Aloisio, M. Cafaro, I. Epicoco, S. Fiore, D. Lezzi, M. Mirto, and S. Mocavero, iGrid, a Novel Grid Information Service, *Proceedings of Advances in Grid Computing—EGC 2005*, European Grid Conference, Amsterdam, The Netherlands, February 14–16, 2005, Revised Selected Papers, Lecture Notes in Computer Science, vol. 3470, pp. 506–515, 2005.

[25] G. Aloisio, M. Cafaro, and R. Williams, The Digital Puglia Project: An Active Digital Library of Remote Sensing Data, *Proceedings of the 7th International Conference on High Performance Computing and Networking Europe*, vol. 1593, Lecture Notes in Computer Science, pp. 563–572, 1999.

[26] S. C. Taylor, B. Armour, W. H. Hughes, A. Kult, and C. Nizman, Operational Interferometric SAR Data Processing for RADARSAT Using Distributed Computing Environment, *Proceedings of the IV International Conference on GeoComputation*, Fredericksburg, VA, USA, 1999.

[27] H. A. James and K. A. Hawick, A Web-based Interface for On-Demand Processing of Satellite Imagery Archives, *Australian Computer Science Communications*, vol. 20, Springer-Verlag Pte Ltd, 1998.

[28] P. D. Coddington, K. A. Hawick, K. E. Kerry, J. A. Mathew, A. J. Silis, D. L. Webb, P. J. Whitbread, C. G. Irving, M. W. Grigg, R. Jana, and K. Tang, Implementation of a Geospatial Imagery Digital Library using Java and CORBA, *Proceedings of Technologies of Object-Oriented Languages and Systems (TOOLS) Asia'98*, pp. 280–289, 1998.

[29] G. Aloisio, M. Cafaro, R. Williams and P. Messina, A Distributed Web-Based Metacomputing Environment, *Proceedings of HPCN97*, eds. L.O. Herzberger and P.M.A. Sloot, Lecture Notes in Computer Science, 1997.

[30] K. A. Hawick, H. A. James, K. J. Maciunas, F. A. Vaughan, A. L. Wendelborn, M. Buchhorn, M. Rezny, S. R. Taylor, and M. D. Wilson, Geostationary-Satellite Imagery Applications on Distributed High Performance Computing, *Proceedings HPCAsia*, Seoul, Korea, 1997.

[31] W. Walcher, F. Niederl, and A. Goller, A WWW-Based Distributed Satellite Data Processing System, *Proceedings of ISPRS Joint Workshop Form Producer to User*, 1997.

[32] K. A. Hawick and H. A. James, Distributed High-Performance Computation for Remote Sensing, *Proceedings of Supercomputing '97*, San Jose, November 1997.

[33] G. Petrie, C. Dippold, G. Fann, D. Jones, E. Jurrus, B. Moon, and K. Perrine, Distributed Computing Approach for Remote Sensing Data, *Proceedings of the 34th Symposium on the Interface*, Montreal, Quebec, Canada 2002.

[34] K. T. Rudahl and S. E. Goldin, PC Clusters as a Platform for Image Processing: Promise and Pitfalls, *Proceedings of the 23rd Asian Conference on Remote Sensing*, Birendra, Nepal, November 2002.

[35] C. Yang and C. Hung, Parallel Computing in Remote Sensing Data Processing, *Proceedings of the 21st Asian Conference on Remote Sensing*, Taipei, Taiwan December 2000.

[36] C. A. Lee, C. Kesselman, and S. Schwab, Near-Real-Time Satellite Image Processing: Metacomputing in C++, *IEEE Computer Graphics & Applications*, vol. 16, pp. 74–84, 1996.

Chapter 13

A Solutionware for Hyperspectral Image Processing and Analysis

Miguel Vélez-Reyes,
University of Puerto Rico at Mayaguez

Wilson Rivera-Gallego,
University of Puerto Rico at Mayaguez

Luis O. Jiménez-Rodríguez,
University of Puerto Rico at Mayaguez

Contents

This chapter describes the concept of a solutionware for hyperspectral image analysis. Solutionware is a set of catalogued tools and toolsets that will provide for the rapid construction of a range of hyperspectral image processing algorithms and applications. The proposed hyperspectral solutionware will span toolboxes, visualization toolsets, and application-specific software systems at different computational resolution levels. A MATLAB hyperspectral image analysis toolbox (HIAT) provides the lowest resolution level but the friendliest interface where scientists and engineers in hyperspectral image processing can try different combinations of hyperspectral image processing algorithms in a simple fashion and add their own algorithms via the MATLAB programming language. As applications require the processing of large data sets in a timely fashion, the solutionware will provide grid, parallel, and hardware computational platforms to provide the user with computational alternatives that can be used to optimize performance and take full advantage of the data. In this chapter, the MATLAB hyperspectral toolbox is presented and parallel processing implementations of some of its components in the Itaniun architecture are described. A prototype version of the hyperspectral image processing toolbox over the grid, Grid-HSI, which extends the hyperspectral image processing environment developed in HIAT to take advantage of computational resources that can be distributed over the network, is depicted.

13.1 Introduction

Hyperspectral image analysis usually consists of performing a series of highly computational intensive operations on large data sets. The analysis extracts information of interest from the data contained in a region for the application scientist. This information of interest may include the extraction of features, the classification of a region in an image, or simply the detection of some specific object. However, two of the main constraints in obtaining analysis results in a timely manner are the efficient computation of the operation itself and the efficient management of large volumes of data. The massive volumes of data involved in hyperspectral imagery is the main limitation in testing different varieties of algorithms on the data as well as in the extraction of features in a timely fashion.

When describing the methods to solve a computational problem in hyperspectral imaging, the levels of abstractions in the architecture are of primary importance on the performance observed in the computation. Figure 13.1 illustrates the different levels of abstractions where a computing problem may be solved. It is interesting to note that the higher the level of abstraction, the design description syntax the to the 'language' spoken by the application closer is the programmer. By programming at a high level of abstraction in an environment such as MATLAB^{TM} [1] or IDL^{TM} [2], the programmer can quickly construct a set of algorithms to solve a problem. Also, these environments are capable of providing a framework for proper software engineering practices to be followed. However, this implies that the application's performance might decrease as the abstraction level increases. In theory, working at a

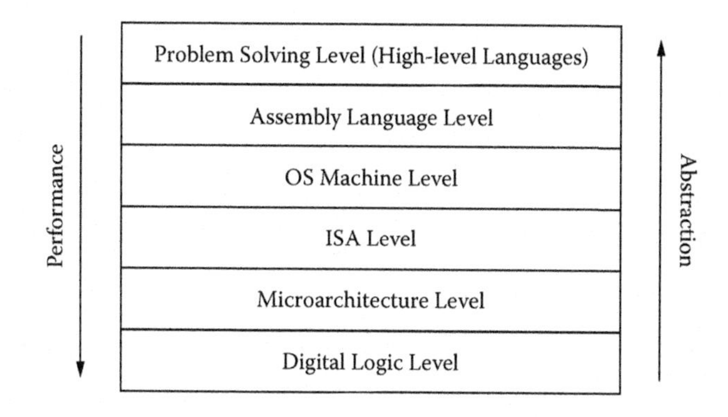

Figure 13.1 Levels in solving a computing problem.

lower level of abstraction can result in better system performance since the developer has more control over the computational structures. However, as the level of abstraction decreases, the complexity in the design process increases, making it harder for the developer to have a complete grasp of the whole design process. The objective of our work is to develop a hyperspectral solutionware or a set of catalogued tools and toolsets that will provide for the rapid construction of a range of hyperspectral image processing algorithms and applications. Solutionware tools will span toolboxes, visualization toolsets, and application-specific software systems that have been developed at the Center for Subsurface Sensing and Imaging Systems[1] (CenSSIS). The chapter is organized as follows. First, an overview of the MATLAB hyperspectral image analysis toolbox is given. Second, parallel and grid implementations of some of the algorithms in the toolbox are described. Future directions of the work are summarized at the end.

13.2 Hyperspectral Image Analysis Toolbox

The Hyperspectral Image Analysis Toolbox (HIAT) is a collection of algorithms that extend the capability of the MATLAB numerical computing environment for the processing of hyperspectral and multispectral imagery. The purpose of HIAT is to provide information extraction algorithms to users of hyperspectral and multispectral imagery in different application domains. HIAT has been developed as part of the NSF Center for Subsurface Sensing and Imaging (CenSSIS) Solutionware that seeks to develop a repository of reliable and reusable software tools that can be shared by researchers across research domains. HIAT provides easy access to supervised and unsupervised classification algorithms, unmixing algorithms, and visualization tools

[1] http://www.censsis.neu.edu.

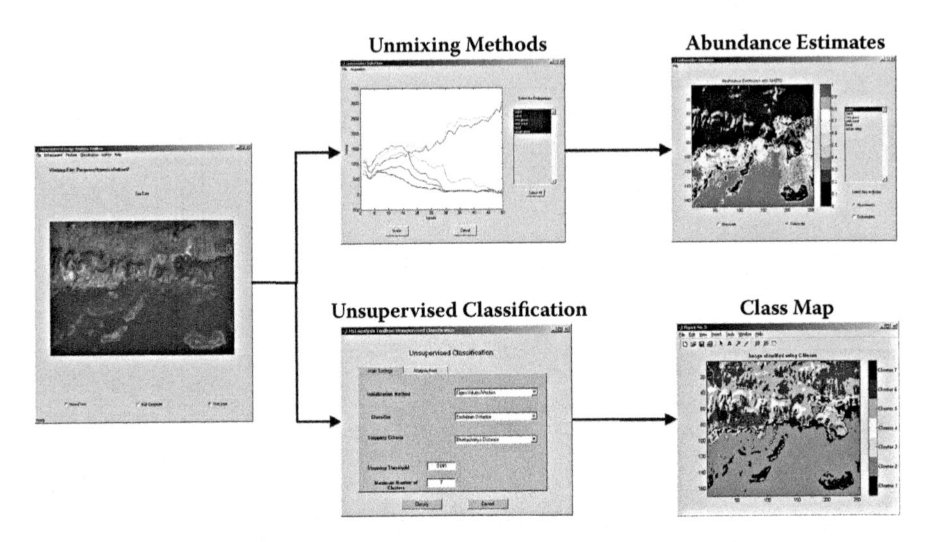

Figure 13.2 HIAT graphical user interface.

developed at UPRM Laboratory for Applied Remote Sensing and Image Processing (LARSIP) over the last 8 years.

HIAT is implemented within an optimized MATLAB environment. It provides useful image analysis techniques for educational and research purposes, allowing the interaction and development of new algorithms, data management, results comparisons, and post-processing. It is an easy-to-use and powerful tool for researchers involved in hyperspectral/multispectral image processing. In addition, MATLAB provides portability of the code to the different platforms in which MATLAB runs: Windows family, Mac OS, and UNIX systems.

13.2.1 HIAT Functionality

The GUI of the toolbox is shown in Figure 13.2. MATLAB version 6.5 was used for the implementation of the HIAT. Tests are currently being conducted using MATLAB version 7.2 (MATLAB2006a) to ensure the toolbox is upward compatible. Figure 15.3 shows the processing schema implemented in HIAT. The processing phases of HIAT are divided into four groups: Feature Selection/Extraction, Classification, Unmixing, and Post-Processing. As Figure 13.3 shows, HSI data could be processed with feature selection/extraction algorithms (or not) before the classification or unmixing and to enhance the classification map post-processing algorithms that are used. Users can combine different processing algorithms to generate different data products.

13.2.1.1 Pre-Processing

In the toolbox, it is assumed that the image has undergone any sensor specific pre-processing, de-glinting, or geometric and atmospheric correction before processing.

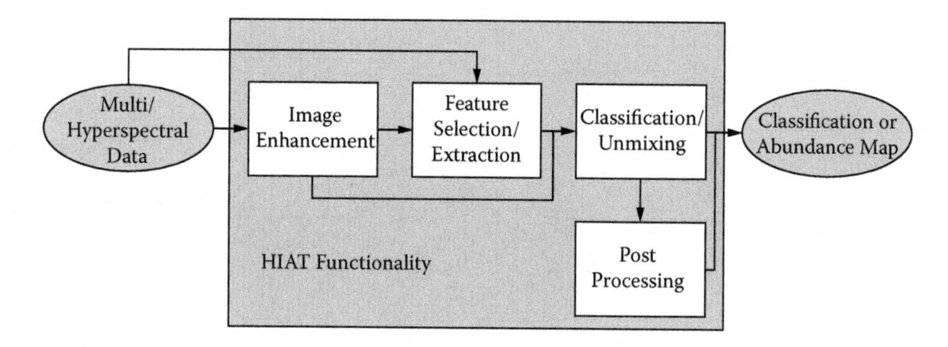

Figure 13.3 Data processing schema for hyperspectral image analysis toolbox.

Pre-processing in the toolbox is limited to image enhancement including noise reduction.

One of the most widely used noise reduction algorithms for hyperspectral imagery is Reduced Rank Filtering (RRF). In this type of filtering, a principal component decomposition is performed using the singular value decomposition (SVD); the small singular values are set to zero, which produces a reduced noise (low rank) approximation of the original image. In addition to this algorithm, a noise reduction method based on oversampling filtering, developed by the authors [3], is available in the toolbox.

The oversampling filtering technique takes advantage that hyperspectral imagers typically collect 100-300 contiguous spectral bands at a high spectral resolution (10nm in most sensors), which results in more samples than are needed to represent the spectra of many objects. Having more samples than are needed is known as oversampling. Oversampling is defined as sampling a signal higher than its Nyquist rate. Specifically, the oversampling rate can be written as

$$OSR = \frac{f_s}{2f_m} \tag{13.1}$$

where f_m is the maximum frequency in the signal and f_s is the sampling frequency. The maximum frequency of the sampled signal power spectral density (PSD) is directly proportional to the maximum frequency of the original signal and inversely proportional to the sampling rate. This means that for a fixed maximum frequency in a signal, the higher the sampling rate, the lower the maximum frequency of the sampled signal PSD. This is illustrated in Figure 13.4. Figure 13.4(a) shows the sampled signal PSD when the signal is sampled at the Nyquist rate while Figure 13.4(b) shows the sampled signal spectra when the signal is sampled at twice the Nyquist rate.

The usefulness of oversampling for noise reduction is that if the signal has been oversampled and there is noise (anything other than the signal of interest) in the frequency range not occupied by the signal, it can be lowpass filtered without changing the signal. The reduction in noise means an increase in the signal-to-noise ratio. It has been shown that typical reflectance spectra are oversampled by a factor of 4 when they

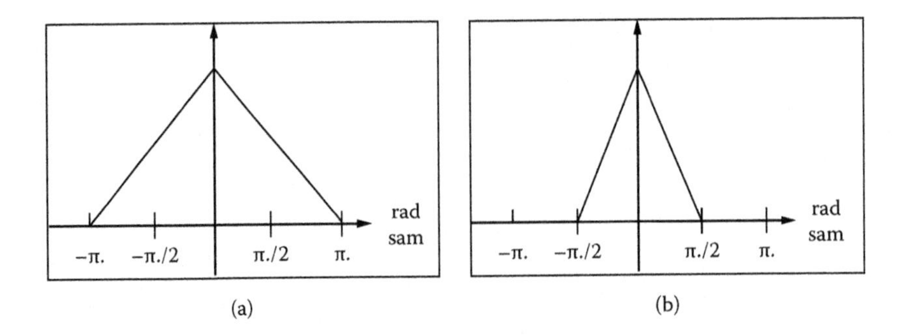

Figure 13.4 Spectrum of a signal sampled at (a) its Nyquist frequency, and (b) twice its Nyquist frequency.

are sampled at 10 nm, which is the spectral sampling rate used in the AVIRIS sensor [3]. This implies that the signal PSD only occupies 1/4 of the available bandwidth, and the other 3/4 can be filtered to reduce noise. The reflectance spectrum of grass is shown in Figure 13.5, along with its power spectral density. As can be seen, most of the power of this signal is concentrated in the lower frequencies.

The only parameter specified in the Oversampling Filtering (OF) algorithm is the cutoff frequency of the lowpass filter. In the case of supervised classification, the program uses the user supplied classes to determine the cutoff frequency. It first calculates the spectra of each of the classes, and then determines the bandwidth of that class. Since these are finite signals, they cannot be bandlimited and the bandwidth used in the algorithm is defined as the frequency below which 95% of the power of that spectra lies. The cutoff frequency of the lowpass filter is the highest of the different class bandwidths. With unsupervised classification, the cutoff frequency is specified by the user. A default of $\pi/4$ is used.

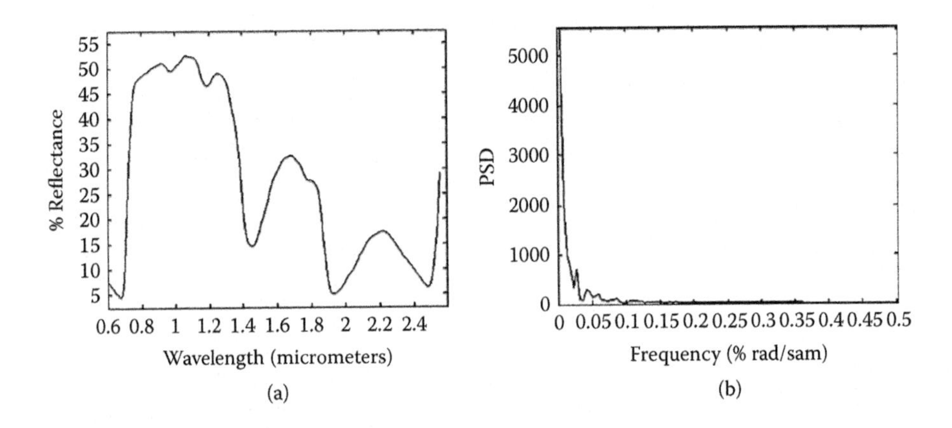

Figure 13.5 (a) The spectrum of grass and (b) its power spectral density.

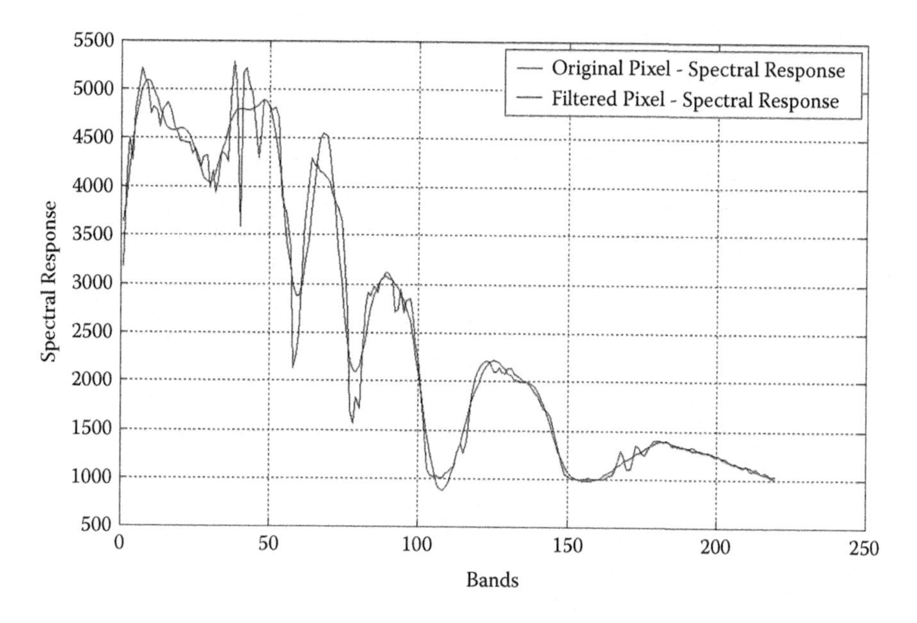

Figure 13.6 Sample spectra before and after lowpass filtering.

Figure 13.6 shows typical spectra of one pixel before and after filtering. Figure 13.7 shows the effect of the OF on one band of a HYPERION image of the Enrique Reef in Lajas, Puerto Rico. The HYPERION sensor was designed as a proof of concept prototype with low signal-to-noise ratio, which is problematic for remote sensing of coastal environments. Any increase in the signal-to-noise ratio is particularly useful in this application.

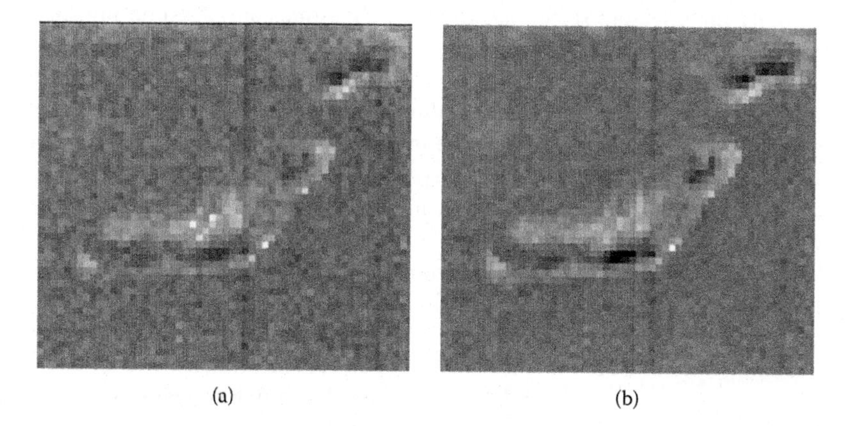

Figure 13.7 HYPERION data of Enrique Reef (band 8 at 427 nm) before (a) and after (b) oversampling filtering.

In [4], the Oversampling Filtering (OF) method was compared to RRF in terms of complexity and classification improving over non-processed images. The RRF algorithm uses the SVD, so its computational complexity is fairly high. The OF algorithm is based on a linear lowpass filter that is implemented using the FFT, so its complexity is much lower. Specifically, let m be the number of pixels in the image and n the number of bands. The computational complexity of the SVD transform is $O(mn2)$ [5], and the computational complexity of the OF algorithm [4], based on linear filtering, is $O(mn \log(n))$. It is clear that resolution enhancement is significantly faster for typical values of m and n. The filtering algorithms were included in the enhancement stage of the HIAT classification system and their effect in classification performance was studied and compared to classification results using non-filtered data. Using a variety of hyperspectral imagery, it was shown that both methods increase the percentage of correct classification on testing data over the non-filtered data and that the increase caused by the use of OF was larger than the increase due to the use of RRF.

13.2.1.2 Feature Extraction and Band Subset Selection

Most classification algorithms are based on statistical methods. As the number of bands grows, the number of labeled data samples needed for training can have a linear, or even an exponential, growth depending on the classifier. Hence, in many applications, classification is not done at the full spectral dimensionality. Feature extraction and selection are used to reduce the dimensionality of the feature vector for classification while trying to maintain class separability.

Two widely known feature extraction algorithms, Principal Component (PC) analysis [6] and Fisher's Discriminant analysis [7], are implemented in the feature extraction/selection stage, along with four unsupervised algorithms developed at LARSIP. The first of these is the Singular Value Decomposition Band Subset Selection (SVDSS) [8], which selects a subset of bands that have the smallest canonical correlation with the first few principal components.

The other three feature selection/extraction algorithms are based on Information Divergence and Projection Pursuit. One of these selects a subset of bands, and the others are linear transformations of the data. The basic idea behind both algorithms is to search for projections onto a lower dimensional space that preserves class separability without any a priori information in terms of labeled samples. The measure of class separability is how different the probability density function (PDF) of the feature is from a Gaussian distribution. This is based on the observation that two or more classes that are separate will have a multi-modal density function, with one mode for each class. A good projection onto a lower dimensional space will preserve multi-modality, while a bad projection will not.

Based on this, a measure of how different the feature is from having a Gaussian distribution is used as a performance measure. Let $f(y)$ be the density function of the feature and $g(y)$ be the Gaussian density function. The difference between $f(y)$ and $g(y)$ is measured using the information divergence given by

$$I(f, g) = d(f \| g) + d(g \| f) \tag{13.2}$$

where

$$d(f\|g) = \int_{-\infty}^{\infty} f(y) \ln\left(\frac{f(y)}{g(y)}\right) dy \tag{13.3}$$

is the relative entropy of $f(y)$ with respect to $g(y)$.

The first of these methods, Information Divergence Band Subset Selection (IDSS) [9], calculates the divergence of the PDF of each band from a Gaussian distribution, and then selects those bands that have the greatest divergence. The second method, motivated by the one proposed in [10], is called Information Divergence Projection Pursuit (IDPP). In [10], the data are first whitened, and then each pixel is tested as a possible candidate for projections of the whole data set. This exhaustive search is time consuming and does not guarantee an optimal projection. A modification of the IDPP algorithm, called the Optimized Information Divergence Projection Pursuit (OIDPP) [9], is also available in the toolbox. This algorithm uses a numerical optimization routine as an alternative to the exhaustive search in IDPP. The reduced feature set is related to the original spectral signature by a projection matrix **A** as follows:

$$\mathbf{y} = \mathbf{A}^T\mathbf{x} \tag{13.4}$$

where **y** is the reduced order feature set, **x** is the measures spectral signature, and **A** is the projection matrix. The matrix **A** is found as follows. The first column of **A** is found using

$$\mathbf{a}_{opt} = \arg\left(\max_i I\left(f\left(\mathbf{a}_i^T, \mathbf{Z}\right), \mathbf{q}\right)\right) \tag{13.5}$$

where **q** is the Gaussian distribution. The columns of **A** are found one at a time, where the present column is the optimal projection vector that is orthogonal to the previous columns. A detailed comparison and examples for these approaches are described in [9].

13.2.1.3 Classification

Once the pre-processing and feature selection is completed, the next step is to classify the data. There are five algorithms implemented in the toolbox for this. These are a Euclidean distance, a Mahalanobis distance, Fisher's Linear Discriminant, Maximum Likelihood, and Angle detection classifiers. All are well-known algorithms in the literature and have been widely applied to multispectral and hyperspectral imagery; see, for instance, [12].

13.2.1.4 Unsupervised Spatial-Spectral Post-Processing

Typical hyperspectral/multispectral classification algorithms perform a pixel by pixel classification. The post-processing algorithms implemented in the toolbox use spatial information to improve the overall classification. The algorithm in the toolbox is an unsupervised version of the well-known supervised ECHO classifier [6],[12] called

UnECHO [13]. UnECHO is divided into two stages. The first stage is a conventional C-means classification algorithm. The post-processing part uses the spectra of the pixels along with the class map to improve classification results.

During the classification stage, spectra that are similar may be assigned to different classes. The basic idea behind the UnECHO algorithm is to look at a neighborhood of pixels, and if they are similar to force them to be assigned to the same class. The image is sectioned into non-overlapping square neighborhoods. Let $\hat{\mathbf{X}}^{(k)}$ be the set of pixels that are members of the kth neighborhood. The degree of similarity or homogeneity of the group is defined as

$$Q_j\left(\hat{\mathbf{X}}^{(k)}\right) = \frac{1}{L}\sum_{m=1}^{L} q_j\left(\mathbf{X}_m\right) \tag{13.6}$$

where $\mathbf{X}_m$ is the mth pixel in the neighborhood, L is the number of pixels in the neighborhood, and $q_j\left(\mathbf{X}_m\right)$ is the distance between the pixel and the cluster mean. The algorithm first tests whether the pixels are similar (homogeneity) using

$$O_j\left(\hat{\mathbf{X}}^{(k)}\right) < T_i \tag{13.7}$$

where T_i is a constant threshold value. If $Q_j\left(\hat{\mathbf{X}}^{(k)}\right)$ is smaller than the threshold, then the spectra of all the pixels in the neighborhood are similar, and the pixels will be assigned to the same class. The next step assigns all the pixels in the neighborhood to the ith cluster if

$$Q_j\left(\hat{\mathbf{X}}^{(k)}\right) < O_j\left(\hat{\mathbf{X}}^{(k)}\right), \forall j \neq i \tag{13.8}$$

If the cluster does not pass the homogeneity test, the original pixel-by-pixel classification is kept. Three different neighborhood sizes: 2×2, 3×3, or 4×4, are available in the toolbox. A detailed description and evaluation of UNECHO can be found in [13].

13.2.1.5 Covariance Estimators

As mentioned previously, estimating statistical parameters in high-dimensional space is difficult because of the large amount of training data needed. One method that can be used to improve covariance estimates for use in classifying hyperspectral data is based on regularization. This method includes a priori knowledge to improve the estimate.

With quadratic classification algorithms, the covariance matrix must be estimated. The effect of having too few training samples with which to compute the covariance estimate results in a rank-deficient or ill-conditioned matrix. The approach used in [14] is to include a priori information as a regularization technique to reduce the rank-deficiency. This approach is a variation of regularized discriminant analysis (RDA) [15], where a regularization technique based on biasing the maximum likelihood covariance matrix is used. The regularized covariance matrix estimate is given by

$$\hat{\sum_{REG}} = (1-\gamma)\hat{\sum_{ML}} + \gamma \cdot c \cdot \mathbf{I} \tag{13.9}$$

where $\hat{\sum}_{ML}$ is the maximum likelihood estimate of the covariance matrix, $\hat{\sum}_{REG}$ is the regularized covariance matrix, c is a constant, and γ is the variable that controls the relative levels of the covariance matrix and a priori knowledge. If $\gamma = 0$, we get $\hat{\sum}_{REG} = \hat{\sum}_{ML}$, and if $\gamma = 1$, we get $\hat{\sum}_{REG} = c \cdot \mathbf{I}$. Here, the a priori covariance matrix is a scaled identity matrix. The optimal value of it is selected in to minimize the probability error (P_E) and the total number of outliers as follows. An element X is an outlier with respect to the ith class if

$$(X - \hat{m}_i)\hat{\sum}^{-1}(X - \hat{m}_i)^T > T_\theta \tag{13.10}$$

where $\hat{m}_i$ is the class mean and T_θ is a constant. The total number of outliers for all classes divided by the total number of samples will be called the probability of missing (P_M). Defining $P_{mix}(\gamma) = \sqrt{P_E^2 + P_M^2}$, the optimum γ is then

$$\gamma_{opt} = \arg_\gamma \min[P_{mix}(\gamma)] \tag{13.11}$$

The reader is referred to [14] for further reading.

13.2.1.6 Unmixing

The algorithms described above classify a pixel as belonging to one single class. However, in most applications, the reflected radiation from a pixel as observed in remote sensing imagery has rarely interacted with a volume composed of a single homogeneous material. The high spectral resolution in HSI enables the detection and classification of subpixel objects from their contribution to the measured spectral signal. The problem of interest is to decompose the measured reflectance (or radiance) into these different spectral responses. This process, called spectral unmixing [16], is the procedure by which the measured spectrum of a pixel is decomposed into a collection of constituent spectra, or endmembers, and a set of corresponding fractions or abundances.

The solution is generally separated into two parts; First the endmembers are determined using one of several methods [16], and then the fractional contribution of each end member to each pixel is determined. The mathematical model of how the spectra interact to form a pixel is called the mixing model. The algorithms implemented in the toolbox are based on a linear mixing model, where the surface is portrayed as a checkerboard mixture. The spectrum of the received light is a linear mixture of the spectra of the different elements in the pixel, proportional to the area of each element and its reflectivity. The linear mixing model is given by

$$\mathbf{b} = \sum_{i=1}^{n} x_i \bar{\mathbf{a}}_i + \mathbf{w} = \mathbf{A}\mathbf{x} + \mathbf{w} \tag{13.12}$$

where $\mathbf{b} \in \Re_{+}^{m}$ is the measured pixel response, $\bar{\mathbf{a}}_i$ is the spectral signature of the ith endmember, x_i is the corresponding fractional abundance, $\mathbf{w}$ is the measurement noise, m is the number of spectral channel, and n is the number of endmembers. The matrix $\mathbf{A} \in \Re_{+}^{mxn}$ is the matrix of endmembers and is the vector of spectral abundances we are trying to estimate. Therefore, $\mathbf{x} \in \Re_{+}^{n}$ is no endmember estimation routine implemented in the toolbox. Endmembers can be selected by the user directly from the image or supplied in a separate file. Once the endmembers are determined, we need to estimate their abundances. Several abundance estimation algorithms are available in the toolbox.

If w in 13.12 is assumed to be independent and identically distributed white Gaussian noise, then the maximum likelihood estimate of $\mathbf{x}$ based on $\mathbf{b}$ is the pseudoinverse given by $\hat{\mathbf{x}} = (\mathbf{A}'\mathbf{A})^{-1}\mathbf{A}'\mathbf{b}$ [17]. This is called the unconstrained solution (ULS) in the toolbox. The advantage is that it is fast; the disadvantages are that the solution can contain negative abundances, and the sum of the fractional parts does not have to equal one. For real spectra measurements, $\mathbf{b}$, $\mathbf{A}$, and $\mathbf{x}$ are constrained to be positive and $\sum_{i=1}^{n} x_i = 1$. If only the sum to one constraint on the abundances is enforced [17], it has a direct solution, too. If only the positivity constraint is enforced in the abundances, this is called in the literature the non-negative least squares (NNLS) problem and a solution can be obtained iteratively [17], [18]. Including positivity and sum to one constraints result in the following constrained linear least squares problem:

$$\hat{\mathbf{x}} = \arg \min_{\mathbf{x}} \|\mathbf{A}\mathbf{x} - \mathbf{b}\|_2^2 \tag{13.13}$$

subject to:

$$\mathbf{x} \geq 0 \tag{13.14}$$

and

$$\sum_{i=1}^{n} x_i = 1 \tag{13.15}$$

where $\hat{\mathbf{x}}$ is the estimate of $\mathbf{x}$ and $\| \|_2$ is the Euclidean norm. We will refer to the above expressions as the fully constrained abundance estimation problem. Only iterative methods can be used to solve this problem. In the toolbox, versions solving the sum to one and the sum to less than or equal to one are available which solve the abundance estimation problem as a least distance minimization problem are implemented in the toolbox and described in [19], [20].

13.2.2 Other Components

In addition to the processing routines, HIAT provides input/output routines to handle MATLAB *.mat, and common remote sensing image formats and heading files such as:

- Remote Sensing file (*.bil, *.bis, *.bsq)
- Remote Sensing file with Header file

- Remote Sensing data in HDF format
- JPG file
- TIFF file

Also, different image visualization methods are available in HIAT:

- Band-by-band on Gray Scale
- Three-band Color Composite
  - Manual Selection
  - Automated band selection algorithms
- True Color (need channel waevelength information).

13.2.3 Toolbox Availability

HIAT can be downloaded from CenSSIS homepage at www.censsis.neu.edu. The toolbox has been downloaded by over 500 users in the past 2 years, covering users from different domains such as remote sensing, agricultural, biomedical, military, and food processing who are interested in applications of hyperspectral and multispectral imagery.

13.3 Implementing Components of HIAT in Parallel Computing Using the Itanium Architecture

HIAT usefulness is limited by MATLAB itself, which in many applications its memory management and data representation schemes do not allow us to manage large hyperspectral images in a reasonable amount of time. In our research work, we are looking at different alternatives where some of the HIAT functionality is implemented in high-performance computing platforms. This section describes the experiences and results on implementing the principal component, minimum distance, and maximum likelihood classifier algorithms from HIAT using the Itanium Processor Family.

On the Itanium architecture, all instructions are transformed into bundles of instructions and these bundles are processed in a parallel fashion by the different functional units. Experimental results show that exploiting implicit parallelism and linking HP Mathematical LIBrary optimized for Itanium yield significant improvement in performance.

For algorithm implementation, we have used two different libraries: ATLAS+ CLAPACK for IA32 and HP MLIB for Itanium. Automatically Tuned Linear Algebra Software (ATLAS)[2] focuses on applying empirical techniques in order to provide

[2]http://math-atlas.sourceforge.net/.

TABLE 13.1 Function Replace Using BLAS Routines

Original Function	Description	BLAS Function
jacobi()	Calculates all Eigen vectors and Eigen values of a symetric matrix	sytrd() stevx()
matmat()	Performs a matrix-matrix multiplication	gemm()
vecmatmul()	Performs a vector-matrix multiplication	gemv()

portable performance. Currently, it provides C and Fortran77 interfaces to a portably efficient BLAS[3] implementation, as well as a few routines from LAPACK.[4] ATLAS was compiled on IA32 systems using gcc compiler version 3.3-3 on an Intel Xeon 2.2GHz machine. The ATLAS version used was 3.7.3. Most of ATLAS routines only provide a subset of LAPACK routines, so for the proper use of these routines CLAPACK[5] should be installed. CLAPACK is the same Fortran LAPACK library built using a FORTRAN to C conversion utility called f2c. The entire Fortran 77 LAPACK library is run thought f2c to obtain C code, and then modified to improve readability. The HP Mathematical LIBrary (HP MLIB)[6] is an HP-based high-performance numerical package optimized for the IA64 and PA-RISC architectures. This package consists of three packages: LAPACK, VECLIB[7] and SCILIB.[8] VECLIB contains the complete set of BLAS routines.

The codes were run on an Intel Xeon 2.2GHz machine running Red Hat 8.0 with 1GB of RAM on the IA32 side, and for IA64 we use an HP rx4640 machine with one IA64 Madison processor 1.5Ghz and 6MB of cache running Red Hat Advanced Server 2.1 with 1GB of RAM. On IA64 we used the Intel 8.0 non-commercial compiler and on IA32 we used the gcc 3.3-3 compiler. For code profiling, we used the gprof tool [21].

When performing code profiling, not surprisingly the routines with several performance penalties were the matrix to matrix multiplication, eigenvector, and eigenvalue calculations and vector to matrix multiplication. In Table 13.1, we summarize the functions and their counterparts on the BLAS library. People familiar with the BLAS library know that there are a lot of routines available that could replace our original functions. These BLAS routines were selected basically because matrices used to represent HSI are in general real matrices and most of the calculations involve symmetric matrices.

Table 13.2 shows the execution times obtained for different versions of the algorithms. The first two columns show their execution times using standard mathematical functions. The two columns on the left show the algorithm after the optimized BLAS routines were used. For the Principal Components Algorithm (PCA), we can see

[3]http://www.netlib.org/blas/.

[4]http://www.netlib.org/lapack/.

[5]http://www.netlib.org/clapack/.

[6]http://www.hp.com/go/mlib.

[7]http://www.nasoftware.co.uk/libraries/veclib.html.

[8]http://www.netlib.org/scilib/.

TABLE 13.2 Algorithms Benchmarks Before and After BLAS Library Replacements

Algorithm	IA32	IA64	IA32 (Optimized)	IA64(Optimized)
PCA	1m 39s	1m 9s	4.13s	3.18s
EDC	5m 3s	2m 17s	1m 5.74s	11.66s
ML	29m 8s	1h 7m	16m 22s	4m 22s

a breakthrough of nearly 23 times faster on both architectures with the optimized libraries versus the standard implementation. For the Euclidean Distance Classifier (EDC), we can see an improvement of twice the speed by just compiling the application on IA64. We also get a boost on performance of more than 4.5 times on IA32 by using BLAS routines and of 11.7 times on IA64. In the Maximum Likelihood (ML) classifier, we encountered a penalty of 2.3 times in the IA64 execution. When the BLAS code is integrated, we then see a speed improvement of 43.8% on IA32 and 93.4% on IA64.

For PCA, there is very little opportunity for parallelization. Since it is based on the covariance matrix of the image data, it requires that all data be on a same node. After the covariance matrix is obtained, its eigenvalues and eigenvectors also should be computed locally on a single node. Since PCA calculation involves a lot of linear algebra calls and there is no obvious parallelization for the algorithm, we use PLAPACK to handle all linear algebra calls and data distribution.

The Euclidean distance and the maximum likelihood classifiers are good algorithms where parallelism can be exploited, since each pixel independently calculates its membership to a class. The problem arrives at calculating the new means and the covariance for each class using the pixel membership. Since a class will have member pixels distributed across nodes, there should be a way to calculate mean and covariance in an efficient parallel way. A master delegate nodes approach was developed for this purpose. The amount of data transferred is minimized as much as possible. It was decided to transfer the local classification vector to the master node at each iteration and to propagate the final vector back to the nodes. In this way, the transfer is for an integer vector of the size of the pixels' resolution. Once each node has its own copy of the global vector, covariances and means are calculated locally. No additional transfers are done until the next iteration. With this approach each node has all the data necessary to calculate pixel memberships without the need to transfer huge amounts of data for each node

Next we will discuss in more detail the performance of the algorithms as well as their parallel implementations

13.3.1 Principal Components Algorithm

From the previous results, there is no apparent performance benefit on the Itanium architecture. We need to further analyze the algorithms to fully exploit the architecture. As shown in Figure 13.8, there are four major components in the PCA algorithm.

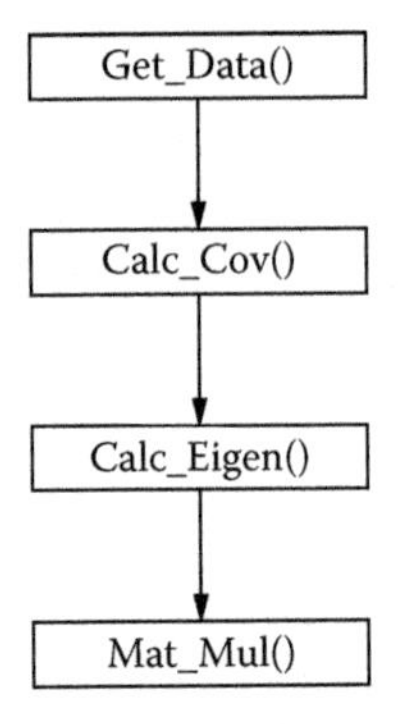

Figure 13.8 Principal component algorithm block components.

The first block is gathering the spectral image from a file, since this task is mostly dependant on the disk I/O and is not covered. The next block is the covariance calculation. The most computing-intensive task of this block is a matrix multiplication, which accounts for more than 80% of the processing time in that block. It explains the performance gains when the BLAS routines were used. The last block is also a matrix multiplication. Therefore, only using the BLAS gemm() routine, the biggest hotspot on block 2 was eliminated and the hotspot on block 4 was minimized. On both BLAS implementations, similar performance benefits are achieved, so both libraries optimized the architectural calls. Using the same algorithm, we can see an improvement on Itanium that, we believe, is mostly due the highest clock speed and some average usage of the Itanium cache.

13.3.2 Euclidean Distance Classifier

The Euclidean distance classifier exhibits the best performance gains because it exploits the Itanium cache benefit. On IA32, the Euclidean() function, which is the one in charge of the main distance calculation, accounts for 73% of the execution time with 47 seconds. On the IA64, the Euclidean() accounts for only 31% of the processing with a self call of 6.64 seconds. The main calculation is an accumulative value: a multiplication followed by an addition to the previous value. The Itanium architecture is optimized for these types of operations. With its three levels of cache ranging from 3 to 9MB, Itanium can outperform all other architectures in this type of sequential read. Moreover, all the data for the whole algorithm can be available on cache for the whole execution. New distances are calculated using the same pixel vectors of the original image, so there should be a small amount of cache misses requesting data. Is is assumed that most of the image data are stored on the nearest cache and then all the operations are executed, but the procedure is mostly the same for the whole algorithm.

Execution results for the EDC shown in Figure 13.9 are very promising. On both architectures, we see a highly scalable curve. It suggests that the algorithm could be used with images with more spectral bands and higher resolutions. Here we found another interesting behavior. On IA32, the parallel execution times were better than

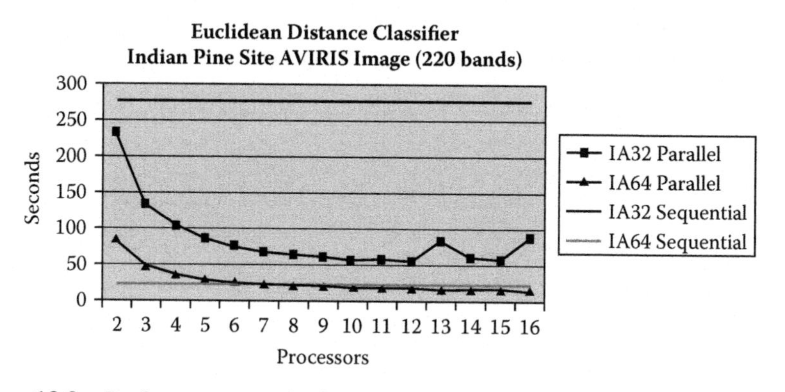

Figure 13.9 Performance results for Euclidean distance classifier.

the sequential approach. This could be an expected result; however, what is surprising is that the sequential execution times on IA64 are very low compared to IA32 and that the parallel approach on IA32 does not compare to IA64. Therefore, for this image, it is faster to run the algorithm sequentially on IA64 than the parallelized version on IA32. On higher resolution images, the IA64 sequential implementation has better execution times than the IA32 parallel version. In this case, it should be better to distribute the means than the classification vector. The following equations provide a guide to know when it will be better to broadcast the classification vector or the means:

$$ClassVec = Height \times Width \times Sizeof(int) \quad (13.16)$$

$$Mean = C \times N \times Sizeof(double) \quad (13.17)$$

These equations calculate the size in bytes for the classification vector and for the means. If the classification vector is greater than the means, then the means should be broadcast, otherwise the vector. Here C is the number of classes and N is the number of the image spectral bands.

13.3.3 Maximum Likelihood Classifier

On the Maximum Likelihood classifier, although we have the same accumulative effect as in the Euclidean distance, it adds a series of matrix manipulations. Moreover, we need to compute the covariance for each class. On the original implementation, the main function hotspot was the matrix multiplication on the covariance function.

With the integration of the gemm() BLAS routine, the performance benefits were tremendous. Results on ML are very similar to those of the EDC. Figure 13.10 shows huge execution time improvement on IA32, but at the same time, the IA32 parallel version, which seems highly scalable, is in the same range as the IA64 sequential approach.

Our experience with these and other algorithms ported to Itanium is that IA64 should provide a boost in performance in the order of 1.3 to 1.5% with just a compilation. If an algorithm ported to IA64 is not on that boundary, then more aggressive

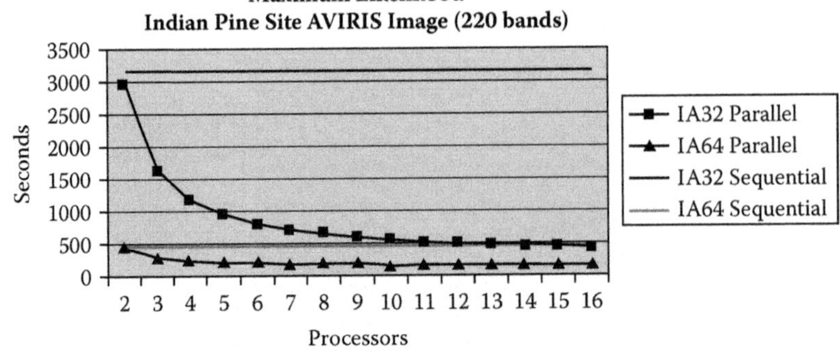

Figure 13.10 Performance results for maximum likelihood.

optimization is needed. Compiler flags were used to help in this optimization at first, but all efforts were unsuccessful. The most significant improvement came when the mathematical functions were linked with the HP MLIB tools. The linking process was straightforward and no major issues were found. The algorithms that took advantage of the new functions saw huge improvements in their performance.

The Itanium architecture relies for most of its performance on the compiler interpretation of the code. All instructions are transformed into bundles of instructions, and these bundles are processed in a parallel fashion between the four different functional units. The idea is that all functional units will be executing instructions simultaneously. But sometimes the compiler cannot generate successful bundles of instructions causing 'split issues,' meaning that functional units are stalled waiting for instructions. This issue can seriously impact the program performance, and it causes programs to run slower on IA64 than on IA32. Also we need to clarify that these are very demanding computing-intensive applications that require specific architectural knowledge to fully exploit the processor capabilities.

13.4 A Grid Service-Based Tool for Hyperspectral Image Analysis

In the previous section, it was shown that parallel computing has been successfully used to significantly reduce the runtime of some of the HIAT components. It is then expected that grid-level resources can play a significant role in improving performance while increasing the pervasity of the image processing algorithms.

This section presents the architecture, design, and implementation of Grid-HSI, which seeks to immerse HIAT analysis capability into a grid platform that supports remote analysis and visualization. The system is based on the Open Grid Service Architecture (OGSA) and implemented on the top of Globus Toolkit 3.0 (GT3). Grid-HSI provides users with a transparent interface to access computational resources and perform remotely hyperspectral imaging analysis through a set of grid services. The Grid-HSI prototype presented here is composed by a Portal Grid Interface, a Data Broker, and a number of specialized grid services to enable HSI analysis. The Grid-HSI architecture and its implementation are described and the suitability of Grid-HSI to perform HSI analysis is presented.

13.4.1 Grid-HSI Architecture

Figure 13.11 depicts the complete Grid-HSI architecture. The Grid Infrastructure includes the local resources, the HSI grid services, and the associated clients' stubs. Each HSI grid service has an associated Data Broker to provide access through a Servlet implemented in the Portal Grid Interface. Figure 13.12 shows the user interface that provides transparent access to resources.

Portal Grid Interface: This interface allows users to enter the required input parameters for executing grid services associated to each of the HSI algorithms implemented in Grid-HSI. Users follow an authentication process to access the resources. This authentication process is based on Grid Security Infrastructure (GSI), which delivers a secure method of accessing remote resources. It enables a secure, single sign-on capability while preserving site control over access control policies and the local security infrastructure. The Portal Grid Interface uses Java Servlets hosted within a Tomcat Servlet container environment. All requests to the Portal Grid Interface go through an Apache server, which forwards requests to the Tomcat using Apache JServ Protocol (AJP).

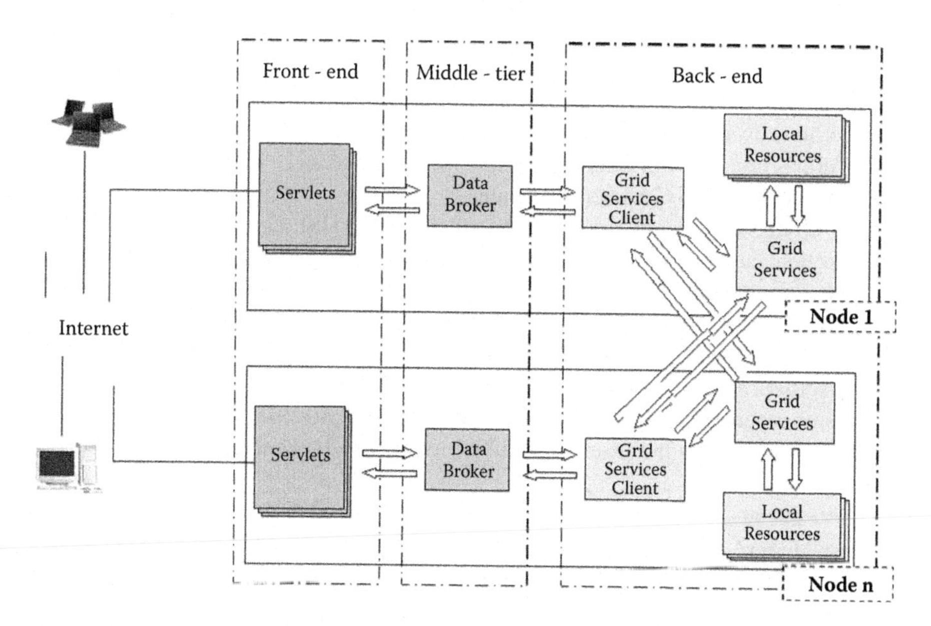

Figure 13.11 Grid-HSI architecture.

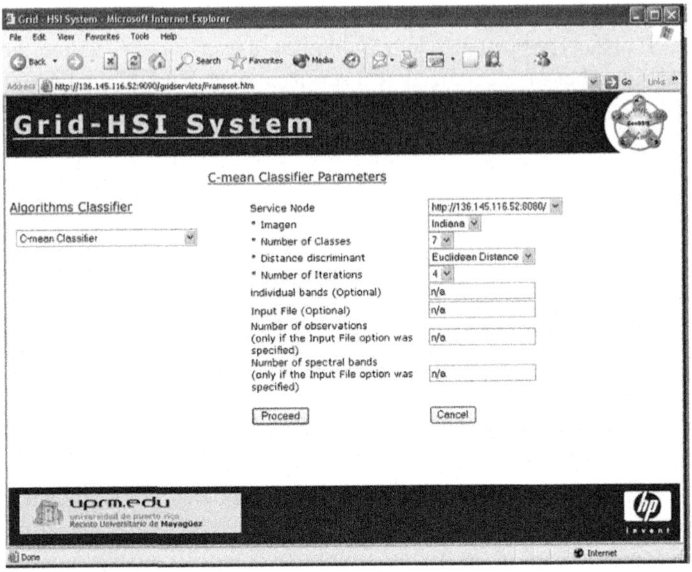

Figure 13.12 Grid-HSI portal.

Data Broker: This component is a link between the Portal Grid Interface and the HSI grid services. For each HSI grid service implemented in Grid-HSI, there is a Data Broker that assures the access from the Portal Grid Interface. The Data Broker manages data related to the grid services available on each resource so a match between local resources and user requests is met. When the Data Broker receives a user request, it seeks node availability and selects the node with the highest performance to respond to the user's request. The Data Broker then sends the request with information about the selected node to a Grid Service Client.

HSI Grid Services: These services implement the HSI algorithms. For each service, a client stub is implemented so continuous access to grid services is performed through the associated client stub. Jobs are submitted by users through the Portal Grid Interface. A Master Manager Job Factory Server (MMJFS), implemented in GT3, executes the task in the remote resources indicated by the Data Broker, examines results of submitted jobs, views information of resources, and so on. After the client stub receives the request from the Data Broker it proceeds to send the request to the node specified by the Data Broker. The main Grid-HSI services are summarized below:

- *CMeansClassifier*. Generates the classification vector (.txt) using the C Means algorithm according selected parameters.
- *PcaReduction*. Generates a new matrix of reduced dimensionality (.txt) using the Principal Component Analysis (PCA) according selected parameters.

- *TxtJpg*. Converts a ClassificationVector.txt file that contains the result membership for each pixel on the image to a jpeg format file.

13.4.2 Functional Process of Grid-HSI

The functional design of Grid-HSI is constituted by a set of components interconnected logically to accomplish the system objective. The functional process is described as follows:

- Initially a user accesses the Pportal Grid and sends a service request with a classifier parameter through an HTML Form.
- The Data Broker Servlet reads these parameters, defines which resource to use, and sends such parameters to the Classifier Grid Client.
- The Classifier Client sends these parameters to the Classifier grid service in the selected resource.
- The Classifier Grid Service invokes its local Classifier Algorithm. This algorithm yields a result. txt File and sends to the Classifier Grid Service an algorithm report.
- The Classifier Grid Service sends back to the Classifier Grid Client the algorithm report and the resulting. txt File ID.
- The Classifier Client receives the algorithm report and the result File ID; it proceeds to send these parameters to the Data Broker.
- The Data Broker sends to the Server Result Displayer the algorithm report and sends to the Txt-Png Converter Grid Client the Result File ID.
- Txt-Png Converter Grid Client sends Result File ID to Txt-Png Converter Grid Service.
- Txt-Png Converter Grid Service invokes its Txt-Png Converter with the Result File ID as a parameter.
- Txt-Png Converter reads from Storage Result Txt File written by the Classifier algorithm and processes it to get it in a Png file.
- The Png File ID is sent back to the Data Broker through Txt-Png Converter Grid Service and Txt-Png Converter Grid Client.
- The Data Broker with this Png File ID invokes a Transfer File Client that, using a TCP Socket, performs the transference from the Resource Storage to Client Storage.
- Finally, the Servlet Data Displayer receives the Png File ID from the Data Broker, reads the Png File from its Storage, and sends the Png file to a Web user.

13.4.3 Experimental Results

In the experiments the resources consisted of a low-cost commodity PC cluster consisting of eight nodes connected using a 100 Mbps Ethernet switch. Each node is

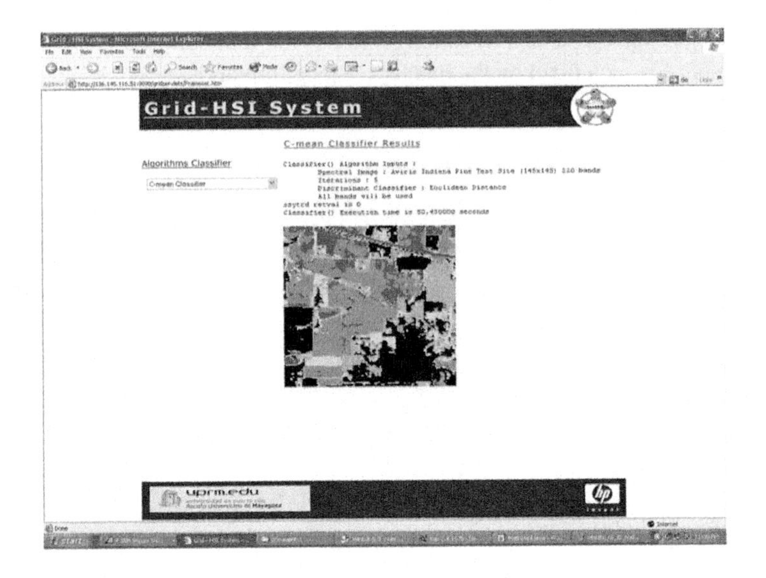

Figure 13.13 Graphical output at node 04.

an Intel P3-651.485Mhz with 256 MB of memory running RedHat Linux 3.2.2-5. A 145 × 145 portion of the June 1992 NW Indian Pines AVIRIS image taken over Indiana [10] is used in the experiments.

Many scenarios were run during the experiments. One case is presented for illustration. More details and cases can be found in [22]. Node 02 receives a classification request via the Internet from node 04. In this case, node 04 does not have the resources needed to serve the user request. After the classifier service classifies the image, a classification vector is returned. This classification vector contains the resulting membership for each pixel in the image. This classification vector is then stored on a directory of the container node as ClassificationVector.txt. The TxtPng service, through JAI API, transforms ClassificationVector.txt to a Portable Network Graphics (PNG) format file. After that the PNG File invokes a TCP Socket and performs the transfer to the Client Browser Hard Disk. The servlet Data Displayer receives the algorithm report and Png File from the Data Broker. Then, it reads the Png file from its hard disk and shows the algorithm report and the png file to the Web browser. The graphical output at node 04 is shown in Figure 13.13.

TABLE 13.3 Results of C-Means Method with Euclidean Distance

# Classes	Iterations	Bands Used	Execution Time (sec)
5	4	220	43.91
5	5	220	53.63
5	6	220	67.92

TABLE 13.4 Results Principal Components Analysis

# Components	Percent amount Energy	Bands Used	Execution Time (sec)
3	90	220	9.52
5	90	220	9.85
7	90	220	10.02

Experiments show that every local resource can accomplish successfully the requests sent by users without regard to the source of the request, and users can submit jobs to several nodes at the same time. Tables 13.3 and 13.4 show the execution timing results of the C-Means clustering with Euclidean distance and the Principal Component Analysis, respectively. As shown in [22], execution time is slightly increased by the overhead of the grid service.

13.5 Conclusions

This chapter described the concept of a solutionware system for the solution of hyperspectral/ multispectral remote sensing image processing problems. We described our experiences and results on implementing a set of hyperspectral image processing algorithms in different platforms. The most comprehensive set of tools is available on the MATLAB HIAT toolbox. It provides users of hyperspectral and multispectral data different processing algorithms that can be combined in different ways to generate data products for image analysis. It also gives users the capability of expanding its functionality by adding their routines via the MATLAB programming language. However, HIAT is limited in its capability to manage large hyperspectral images. To deal with this problem, we are looking into implementating some of the toolbox components in parallel processing using the Itanium architecture. Experimental results showed that exploiting implicit parallelism and linking HP Mathematical LIBrary optimized for Itanium yield significant improvement in performance. To take further advantage of distributed computational resources, grid computing was explored as an alternative for implementing HIAT. We presented Grid-HSI, a Service-Oriented Architecture-Based Grid application to enable hyperspectral image processing. Grid-HSI provides users with a transparent interface to access computational resources and perform remotely hyperspectral image analysis through a set of grid services. The proposed architecture was described and a prototype implementation with few services was presented.

As hyperspectral remote sensing becomes more available, the large data volume will require that users have tools that can be tailored to meet their need of data products in a timely fashion within their available computational resources. The proposed solutionware not only will serve the remote sensing community but also users in other areas where imaging spectroscopy is used.

13.6 Acknowledgment

The work described in this chapter has been supported by the NSF Engineering Research Centers Program under grant EEC-9986821, by the NASA University Research Centers Program under grant NCC5-518, the U.S. Department of Defense under DEP-SCoR grant DAAG55-98-1-0016, and grant NMA2110112014 from the US National Geospatial and Intelligence Agency (formerly US National Imagery and Mapping Agency). Additional support was also provided by the Hewlett-Packard Technology Center of Puerto Rico. We thank the Laboratory for Applications of Remote Sensing (LARS) at Purdue University for the image that is shown in Figure 13.7. We also thank Dr. Nayda Santiago for Figure 13.1 and the interesting discussions about some of the material presented here.

References

[1] http://www.mathworks.com/products/matlab/.

[2] http://www.rsinc.com/envi.

[3] S.D. Hunt and H. Sierra. Spectral Oversampling in Hyperspectral Imagery, *Proccedings of SPIE: Algorithms and Technologies for Multispectral, Hyperspectral and Ultraspectral Imagery IX*, vol. 5093, pp. 643–650, 2003.

[4] S.Morrillo-Contreras, M. Vlez-Reyes and S. Hunt. A Comparison of Noise Reduction Methods for Image Enhancement in Classification of Hyperspectral Imagery, *Proccedings of SPIE: Algorithms and Technologies for Multispectral, Hyperspectral and Ultraspectral Imagery XI*, vol. 5806, pp. 384–392, 2005.

[5] C. Van Loan and G. Golub. *Matrix Computations*, 3rd Edition, Baltimore, MD, John Hopkins University Press, 1997.

[6] J. A. Richards. *Remote Sensing Digital Image Analysis, An Introduction*, 3rd Edition, New York, NY, Springer, 1999.

[7] R. Duda and P. Hart. *Pattern Classification and Scene Analysis*, New York, NY, John Wiley, 1973.

[8] M. Vélez-Reyes, L.O. Jiménez, D.M. Linares and H.T. Velázquez. Comparison of Matrix Factorization Algorithms for Band Selection in Hyperspectral Imagery, *Proccedings of SPIE: Algorithms and Technologies for Multispectral, Hyperspectral and Ultraspectral Imagery VI*, vol. 4049, 2005.

[9] L.O. Jiménez, E. Arzuaga-Cruz and M. Vélez-Reyes. Unsupervised Feature Extraction Techniques for Hyperspectral Data and Its Effects on Supervised and Unsupervised Classification, To appear in *IEEE Transactions on Geosciences and Remote Sensing*, 2007.

[10] A. Ifarragueri and C. Chang. Unsupervised Hyperspectral Image Analysis with Projection Pursuit, *IEEE Transactions on Geoscience and Remote Sensing*, vol. 38, pp. 2529–2538, 2000.

[11] D.A. Landgrebe. *Signal Theory Methods in Multispectral Remote Sensing*, Hoboken, NJ, John Wiley, 2003.

[12] D.A. Landgrebe. The Development of a Spectral-Spatial Classifier for Earth Observational Data, *Pattern Recognition*, vol. 12, pp. 165–175, 1980.

[13] L.O. Jiménez, J. Rivera-Medina, E. Rodríguez-Díaz, E. Arzuaga-Cruz, and M. Ramírez-Vélez. Integration of Spatial and Spectral Information by means of Unsupervised Extraction and Classification for Homogenous Objects Applied to Multispectral and Hyperspectral Data, *IEEE Transactions on Geoscience and Remote Sensing*, vol. 43, pp. 844–851, 2005.

[14] M. Ramírez-Vélez and L. Jiménez-Rodríguez. Regularization Techniques and Parameter Estimation for Object Detection in Hyperspectral Imagery, *Procceddings of SPIE: Algorithms and Technologies For Multispectral, Hyperspectral and Ultraspectral Imagery IX*, vol. 5093, pp. 694–704, 2003.

[15] J.H. Friedman. Regularized Discriminant Analysis, *Journal of the American Statistical Association*, vol. 84, pp. 165–175, 1989.

[16] N. Keshava and J.F. Mustard. Spectral Unmixing, *IEEE Signal Processing Magazine*, pp. 44–57, 2002.

[17] C.L. Lawson and R.J. Hanson. *Solving Least Squares Problems*, Prentice-Hall, Englewood Cliffs, New N.S., 1974.

[18] M. Vélez-Reyes, A. Puetz, R. B. Lockwood, M. Hoke and S. Rosario. Iterative Algorithms for Unmixing of Hyperspectral Imagery, *Proccedings of SPIE: Algorithms and Technologies for Multispectral, Hyperspectral and Ultraspectral Imagery IX*, vol. 5093, pp. 418–419, 2003.

[19] S. Rosario-Torres and M. Vélez-Reyes. An Algorithm for Fully Constrained Abundance Estimation in Hyperspectral Unmixing, *Proccedings of SPIE: Algorithms and Technologies for Multispectral, Hyperspectral and Ultraspectral Imagery XI*, vol. 5806, pp. 711–719, 2005.

[20] M. Vélez-Reyes and S. Rosario. Solving Adundance Estimation in Hyperspectral Unmixing as a Least Distance Problem, *Proceedings 2004 IEEE International Geoscience and Remote Sensing Symposium*, Alaska, vol. 5, pp. 3276–3278, 2004.

[21] S. Graham, P. Kessler and M. McKusick. Gprof: A Call Graph Execution Profiler, *Proceedings of the Symposium on Compiler Construction*, vol. 17, pp. 120–126, 1982.

[22] C.L. Carvajal-Jiménez. Using Grid Computing to Enable Hyperspectral Imaging Analysis, Master Thesis, University of Puerto Rico at Mayagez, Mayagez, 2004.

Chapter 14

AVIRIS and Related 21st Century Imaging Spectrometers for Earth and Space Science

Robert O. Green,
Jet Propulsion Laboratory,
California Institute of Technology

Contents

Imaging spectroscopy (also known as hyperspectral imaging) is a field of scientific investigation based upon the measurement and analysis of spectra measured as images. The human eye qualitatively measures three colors (blue, green, and red) in the visible portion of the electromagnetic spectrum when viewing the environment. The human eye-brain combination is a powerful observing system, however, it generally provides a non-quantitative perspective of the local environment. Imaging spectrometer instruments typically measure hundreds of colors (spectral channels) across a much wider spectral range. These hundreds of spectral channels are recorded quantitatively as spectra for every spatial element in an image. The measured spectra provide the basis for a new approach to understanding the environment from a remote perspective based in the physics, chemistry, and biology revealed by imaging spectroscopy.

The measurement of hundreds of spectral channels for each spatial element of an image consisting of millions of spatial elements creates an important requirement for the use of high-performance computing. First, high-performance computing is required to acquire, store, and manipulate the large data sets collected. Second, to extract the physical, chemical, and biological information recorded in the remotely measured spectra requires the development and use of high-performance computing algorithms and analysis approaches.

This chapter uses the Airborne Visible/Infrared Imaging Spectrometer (AVIRIS) to review the critical characteristics of an imaging spectrometer instrument and the corresponding characteristics of the measured spectra. The wide range of scientific research as well as application objectives pursued with AVIRIS is briefly presented. Roles for the application of high-performance computing methods to AVIRIS data sets are discussed. Next in the chapter a review is given of the characteristics and measurement objectives of the Moon Mineralogy Mapper (M3) imaging spectrometer planned for launch in 2008. This is the first imaging spectrometer designed to acquire high precision and high uniformity spectral measurements of an entire planetary-sized rocky body in our solar system. The size of the expected data set and roles for high performance computing are discussed. Finally, a review is given of one design for an Earth imaging spectrometer focused on investigation of terrestrial and aquatic ecosystem status and composition. This imaging spectrometer has the potential to deliver calibrated spectra for the entire land and coastal regions of the Earth every 19 days. The size of the data sets generated and the sophistication of the algorithms needed for full analysis provide a clear demand for high-performance computing. Imaging spectroscopy and the data sets collected provide an important basis for the use of high-performance computing from data collection to data storage through to data analysis.

14.1 Introduction

Imaging spectroscopy is based in the field of spectroscopy. Sir Isaac Newton first separated the color of white light into the rainbow in the late 1600s. In the 1800s, Joseph von Fraunhofer and others discovered absorption lines in the solar spectrum and light emitted by flames. Through investigation of these absorption lines, the linkage between composition and signatures in a spectrum of light was established. The field of spectroscopy has been pursued by astronomers for more than 100 years to understand the properties of stars as well as planets in our solar system. On Earth, spectroscopy has been used by physicists, chemists, and biologist to investigate the properties of materials relevant to their respective disciplines. In the later half of the

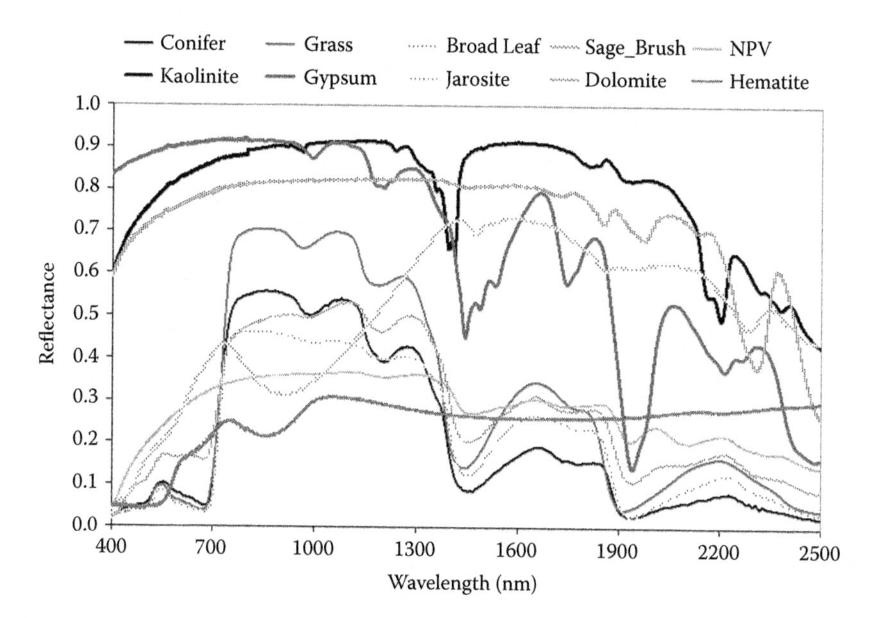

Figure 14.1 A limited set of rock forming minerals and vegetation reflectance spectra measured from 400 to 2500 nm in the solar reflected light spectrum. NPV corresponds to non-photosynthetic vegetation. A wide diversity of composition related absorption and scattering signatures in nature are illustrated by these materials.

20th century Earth scientists developed spaceborne instruments that view the earth in a few spectral bands capturing a portion of the spectral information in reflected light. The AVHRR, LandSat, and SPOT are important examples of this multispectral approach to remote sensing of the Earth. However, the few spectral bands of multispectral satellites fail to capture the complete diversity of the compositional information present in the reflected energy spectrum of the Earth. Figure 14.1 shows a set of measured reflectance spectra from a limited set of rock forming minerals and vegetation spectra. A wide diversity of composition-related absorption and scattering signatures exist for such materials. Figure 14.2 shows these selected reflectance spectra convolved to the band passes of the LandSat Thematic Mapper. When mixtures and illumination factors are included, the 6 multispectral measurements of the multispectral Thematic Mapper are insufficient to unambiguously identify the 10 materials present. In the 1970s, realization of the limitations of the multispectral approach when faced with the diversity and complexity of spectral signatures found on the surface of the Earth lead to the concept of an imaging spectrometer. The use of an imaging spectrometer was also understood to be valid for scientific missions to other planets and objects in our solar system. Only in the late 1970s did the detector array, electronics, computer, and optical technology reach significant maturity to allow design of an imaging spectrometer. With the arrival of these technologies and scientific impetus, the Airborne Imaging Spectrometer (AIS) was proposed and built at the Jet Propulsion Laboratory [1]. The

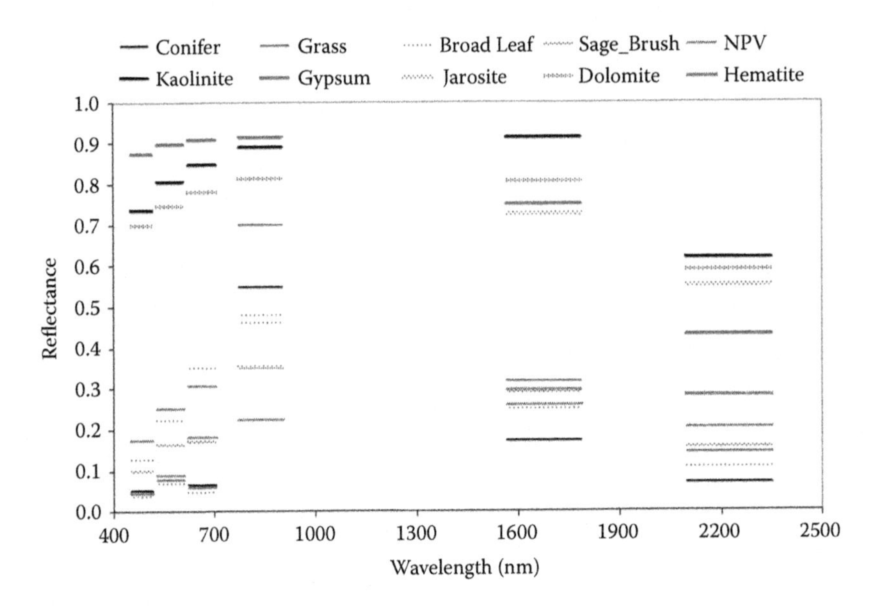

Figure 14.2 The spectral signatures of a limited set of mineral and vegetation spectra convolved to the six solar reflected range band passes of the multispectral LandSat Thematic Mapper. When mixtures and illumination factors are included, the six multispectral measurements are insufficient to unambiguously identify the wide range of possible materials present on the surface of the Earth.

AIS first flew in 1982 as well as in several subsequent years as a technology and science demonstration experiment. Concurrently with the development of the AIS a role for high-performance computing was identified and pursued [2]. The AIS instrument had limited spectral coverage as well as limited spatial coverage. Even as a demonstration experiment, the success of the AIS led to the formulation of the proposal for the Airborne Visible/Infrared Imaging Spectrometer. This next generation instrument was specified to measure the complete solar reflected spectrum from 400 to 2500 nm and to capture a significant spatial image domain. The broader spectral and spatial domain of this full range instrument continued to grow the role for high-performance computing in the field of imaging spectroscopy.

14.2 AVIRIS and the Imaging Spectroscopy Measurement

The Airborne Visible/Infrared Imaging Spectrometer (AVIRIS) [3, 4] measures the total upwelling spectral radiance in the spectral range from 380 to 2510 nm at approximately 10 nm sampling intervals and spectral response function. Figure 14.3 shows a plot of the AVIRIS spectral range in conjunction with an atmospheric transmittance spectrum. Also shown for comparison are the spectral response functions

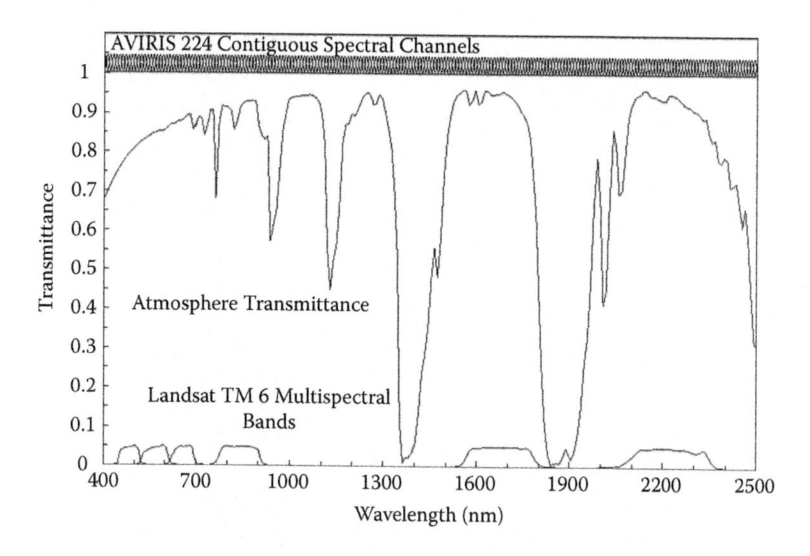

Figure 14.3 AVIRIS spectral range and sampling with a transmittance spectrum of the atmosphere and the six LandSat TM multi-spectral bands in the solar reflected spectrum.

of the multispectral LandSat Thematic Mapper. With AVIRIS a complete spectrum is measured with contiguous spectral channels. Across this spectral range the atmosphere transmits energy reflected from the surface, except in the spectral regions of strong water vapor absorption centered near 1400 and 1900 nm. These strong water vapor absorption regions are used for cirrus cloud detection and compensation. Measurement of this complete spectral range allows AVIRIS to be used for investigations beyond those possible with a multispectral measurement. In addition, measurement of the full spectrum allows use of new, more accurate, computationally intensive algorithms that require high-performance computing.

In the spatial domain, AVIRIS measures spectra as images with a 20 m spatial resolution and an 11 km swath with up to 1000 km image length from NASA's ER-2 aircraft flying at 20 km altitude. On the Twin Otter aircraft flying at 4 km altitude, the spatial resolution is 4 m with a 2 km swath and up to 200 km image length. Figure 14.4 shows an AVIRIS data set collected over the southern San Francisco Bay, California, from the ER-2 platform in image cube representation. The spectrum measured for each spatial element in the data set may be used to pursue specific scientific research questions via the recorded interaction of light with matter.

14.2.1 The AVIRIS Imaging Spectrometer Characteristics

The full set of AVIRIS spectral, radiometric, spatial, temporal, and uniformity characteristics are given in Table 14.1. These characteristics have been refined and improved since the initial development of AVIRIS based upon the requirements from scientists

Figure 14.4 AVIRIS image cube representation of a data set measured of the southern San Francisco Bay, California. The top panel shows the spatial content for a 20 m spatial resolution data set. The vertical panels depict the spectral measurement from 380 to 2510 nm that is recorded for every spatial element.

TABLE 14.1 Spectral, Radiometric, Spatial, Temporal, and Uniformity Specifications of the AVIRIS Instrument

	Spectral properties:
Range	380 to 2510 nm in the solar reflected spectrum
Sampling	10 nm across spectral range
Response	FWHM < 1.1 of sampling
Accuracy	Calibrated to 2% of sampling
Precision	Stable within 1% of sampling
	Radiometric properties:
Range	0 to maximum Lambertian radiance
Sampling	16 bits measured
Response	> 99% linear
Accuracy	> 96% absolute radiometric calibration
Precision (SNR)	As specified at reference radiance
	Spatial properties:
Range	34 degree field-of-view (FOV)
Sampling	0.87 milliradian cross and along track
Response	FWHM of IFOV < 1.2 of sampling
	Temporal properties:
Airborne	As requested 1987 to present
	Uniformity:
Spectral cross-track	> 99% uniformity of position across the FOV
Spectral-IFOV	> 98% IFOVs uniformity over the spectral range

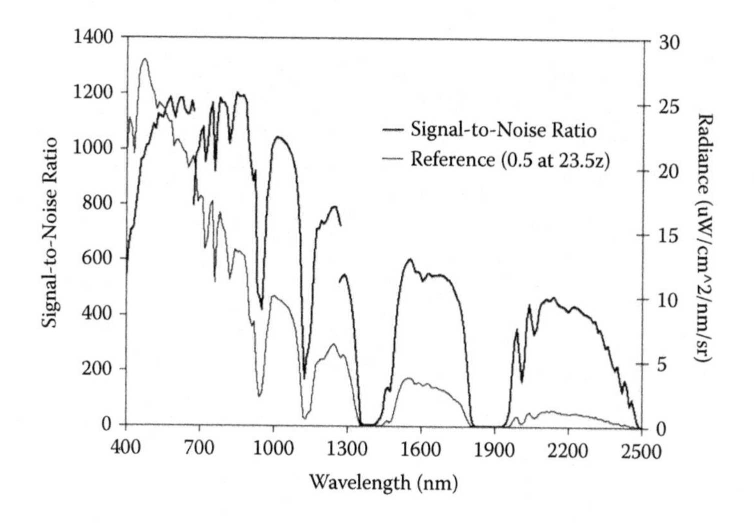

Figure 14.5 The 2006 AVIRIS signal-to-noise ratio and corresponding benchmark reference radiance.

using AVIRIS data. Of particular importance has been the improvement of the signal-to-noise ratio. An increased signal-to-noise ratio has been a critical factor enabling more advanced algorithms and sophisticated analysis approaches. Figure 14.5 gives the 2006 AVIRIS signal-to-noise ratio at the specified AVIRIS reference radiance. The AVIRIS reference radiance was specified in the original AVIRIS proposal as the radiance from a 0.5 reflectance surface illuminated by the sun at a 23.5 degree solar zenith angle through the standard mid-latitude atmospheric model. The current AVIRIS signal-to-noise ratio is 10 to 20 times greater than when the instrument first flew in 1986.

Of special importance for valid physically based imaging spectroscopy science is the uniformity of the imaging spectrometer measurement. Two aspects of uniformity are critical. The first is cross-track spectral uniformity. The spectral cross-track uniformity requirement is that each spectrum in the image have the same spectral calibration to some percentage near 100%. For AVIRIS, the spectral cross-track uniformity exceeds 99% because each spectrum in the image is measured by the same spectrometer. This is inherent in the AVIRIS whiskbroom imaging spectrometer design. For the (approximate) 10 nm spectral sampling of AVIRIS, this 99% uniformity assures that the spectral calibration is the same for all spectra measured in an image to the level of 0.1 nm. Excellent spectral cross-track uniformity is required for all analysis algorithms that are applied directly to all spatial elements in an image. Some of the most powerful algorithms such as spectral dimensional analysis and spectral unmixing require near-perfect spectral cross-track uniformity.

The second critical form of uniformity for an imaging spectrometer is spectral instantaneous-field-of-view (IFOV) uniformity. The IFOV is the sampling area on the surface for a single spatial element. Spectral-IFOV uniformity requires that the IFOV

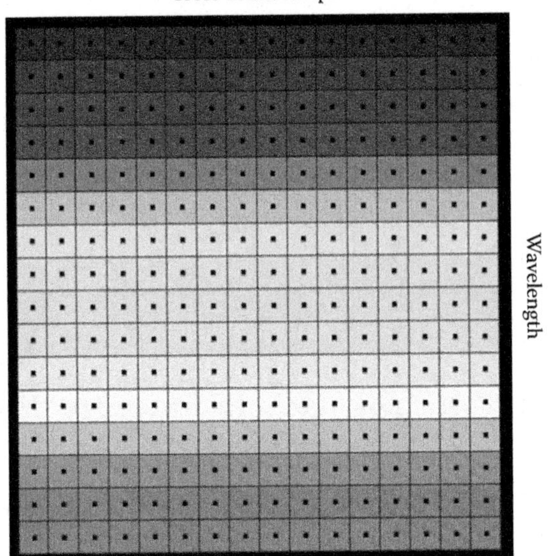

Figure 14.6 Depiction of the spectral cross-track and spectral-IFOV uniformity for a uniform imaging spectrometer. The grids represent the detectors, the gray scale represents the wavelengths, and the dots represent the centers of the IFOVs. This is a uniform imaging spectrometer where each cross-track spectrum has the same calibration and all the wavelengths measured for a given spectrum are from the same IFOV.

for a given spectrum be the same for all wavelengths to some high percentage near 100%. This assures that the same area on the ground is sampled for all wavelengths measured in a spectrum. Again, because AVIRIS is a whiskbroom spectrometer, the spectral IFOV uniformity is high at better than 98%. Figure 14.6 depicts the spectral cross-track and spectral IFOV uniformities for a 100% uniform instrument. Several imaging spectrometers have been constructed with low spectral cross-track and low spectral-IFOV uniformities undermining their potential use.

14.2.2 The AVIRIS Measured Signal

Understanding the detailed nature of the AVIRIS or any imaging spectrometer measurements is essential for appropriate analysis of the data. Figure 14.7 shows the reflectance spectrum of a vegetation canopy. From this reflectance spectrum a wide range of plant composition and status information may be extracted. This information is contained in the molecular absorption and constituent scattering signatures recorded in the vegetation canopy spectrum.

An Earth-looking imaging spectrometer such as AVIRIS does not measure reflectance. AVIRIS measures the total upwelling radiance incident at the instrument

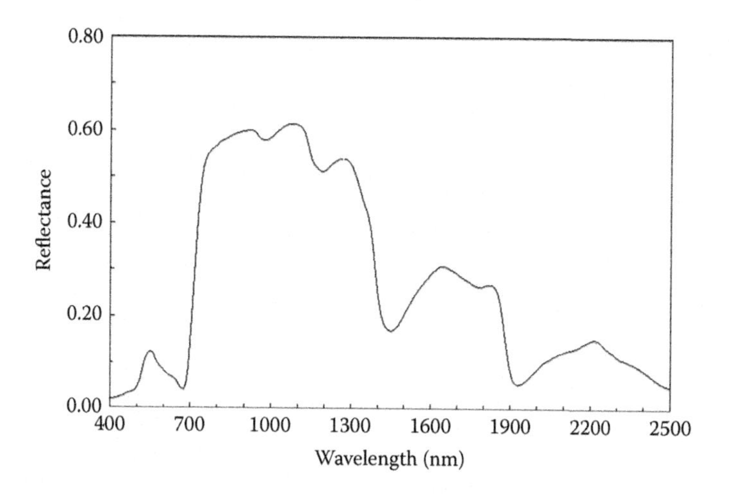

Figure 14.7 Vegetation reflectance spectrum showing the molecular absorption and constituent scattering signatures present across the solar reflected spectral range.

aperture. When flying on the NASA ER-2 aircraft, the aperture is looking down from 20 km. Figure 14.8 show the modeled [5, 6] radiance incident at the AVIRIS aperture for the vegetation canopy reflectance spectrum. This spectrum includes the combined effects of the solar irradiance, two-way transmittance, and scattering of the atmosphere, as well as the reflectance of the vegetated canopy. This is the radiance in

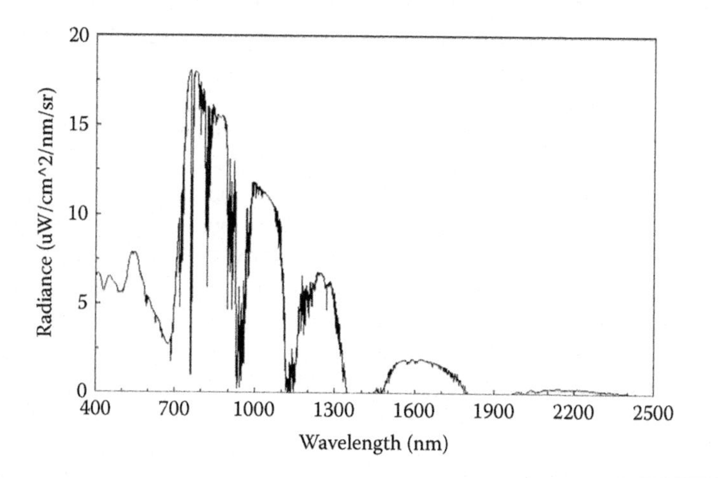

Figure 14.8 Modeled upwelling radiance incident at the AVIRIS aperture from a wel-illuminated vegetation canopy. This spectrum includes the combined effects of the solar irradiance, two-way transmittance, and scattering of the atmosphere, as well as the vegetation canopy reflectance.

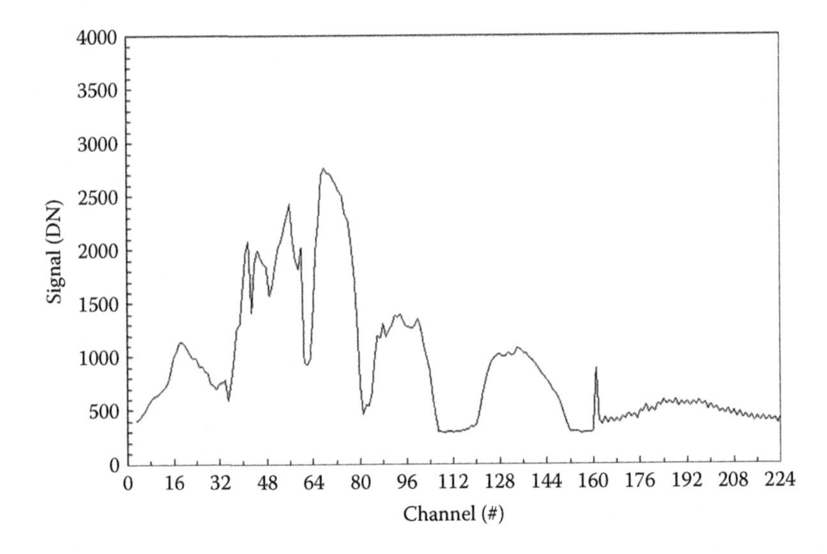

Figure 14.9 AVIRIS measured signal for the upwelling radiance from a vegetation covered surface. The instrument optical and electronic characteristics dominate for recorded signal.

terms of power per area per wavelength per solid angle available to measure for the pursuit of imaging spectroscopy.

As with any radiance measuring instrument, AVIRIS has optical components that collect the incident light and focus it on a detector. At the detector the incident light is converted to measurable signals that are amplified, digitized, and recorded. Figure 14.9 shows the AVIRIS recorded signal for the vegetation canopy upwelling radiance spectrum. The AVIRIS signal has no inherent radiometric or spectral calibration and is recorded as digitized number (DN) versus channel.

The process of spectral and radiometric calibration in the AVIRIS data processing subsystem converts the measured signal to units of spectral radiance. Considerable effort has been expended over the years of AVIRIS' operation to develop spectral, radiometric, and spatial calibration methods in the laboratory [7, 8]. A companion effort has been applied to validate the calibration of AVIRIS in the flight environment [9, 10]. Figure 14.10 shows the AVIRIS calibrated radiance spectrum of the vegetation canopy target. Accurate calibration is an essential requirement for the spectra measured by AVIRIS or any imaging spectrometer to be analyzed quantitatively for science research or application objectives.

If the objective of the investigation is the understanding of surface properties, the calibrated radiance spectra must be corrected for the effects of the atmosphere. Atmospheric correction generally includes compensation for the solar irradiance as well as atmospheric absorbing and scattering effects. Figure 14.11 show the atmospherically corrected spectrum for the vegetation canopy target. The portions of the spectrum located in the strong atmospheric water vapor absorption bands near 1400 and 1900 nm are lost due to the lack of a measurable signal.

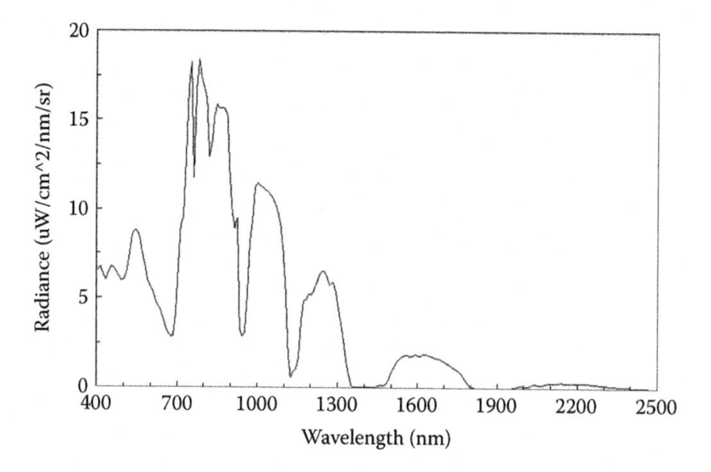

Figure 14.10 Spectrally and radiometrically calibrated spectrum for the vegetation canopy target.

14.2.3 Range of Investigations Pursued with AVIRIS Measurements

The AVIRIS imaging spectrometer was originally proposed to investigate two specific spectral signatures. These were the absorption doublet of the mineral Kaolinite centered near 2200 nm and the red-edge of vegetation in the 700 nm region of the

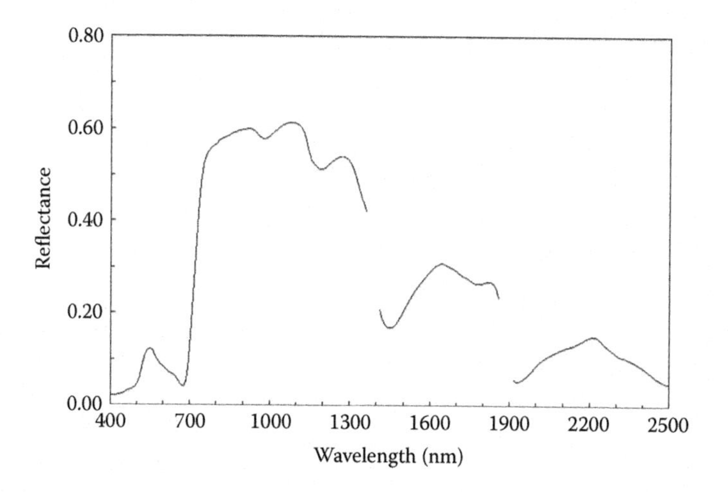

Figure 14.11 Atmospherically corrected spectrum from AVIRIS measurement of a vegetation canopy. The 1400 and 1900 nm spectral regions are ignored due to the strong absorption of atmospheric water vapor. In this reflectance spectrum the molecular absorption and constituent scattering properties of the canopy are clearly expressed and available for spectroscopic analysis.

TABLE 14.2 Diversity of Scientific Research and Applications Pursued with AVIRIS

Atmosphere:	Water Vapor, Clouds Properties, Aerosols, Absorbing Gases
Ecology:	Chlorophyll, leaf water, lignin, cellulose, pigments, structure, nonphotosynthetic constituents
Geology and soils:	Mineralogy, geochemistry, soil type
Coastal and inland waters:	Chlorophyll, plankton, dissolved organics, sediments, bottom composition, bathymetry
Snow and ice hydrology:	Snow cover fraction, grainsize, impurities, melting
Biomass burning:	Subpixel temperatures/extent, smoke, combustion products
Environmental hazards:	Contaminants directly and indirectly, geological substrate
Calibration:	Aircraft and satellite sensors, sensor simulation, validation
Modeling:	Radiative transfer model validation and constraint
Commercial:	Mineral exploration, agriculture, and forest status
Algorithms:	Autonomous atmospheric correction, spectra derivation
Other:	Human infrastructure

spectrum. Fortunately, a full solar reflected energy imaging spectrometer was developed and the AVIRIS instrument has been used to pursue a much broader range of scientific research and application objectives. Table 14.2 summarizes the span of AVIRIS investigations across a range of scientific research and application disciplines. This list illustrates the power of imaging spectroscopy that arises from measurement of the complete solar reflected spectrum from 400 to 2500 nm. Imaging spectroscopy becomes relevant whenever a material absorption or scattering spectral signature can be directly or indirectly linked to the scientific research or application question of interest.

14.2.4 The AVIRIS Data Archive and Selected Imaging Spectroscopy Analysis Algorithms

The AVIRIS archive is maintained at the Jet Propulsion Laboratory. All data in the archive are available for distribution in units of calibrated upwelling spectral radiance. The current volume of data exceeds 10 Terabytes. Geographically the AVIRIS archive includes measurements from northern Alaska to southern Argentina as well as from Hawaii to portions of the eastern Caribbean. From the perspective of surface composition, the archive includes a wide range of vegetation types ranging from tropical to temperate to desert environments. In addition, a wide range of geological surfaces have been measured spanning sedimentary, metamorphic, and igneous rock domains as well as a diversity of soils. Snow and ice data sets have been collected including form the frozen Beaufort Sea in Alaska as well as many snow- and ice-covered mountain regions of the western United States. The AVIRIS archive includes atmospheric conditions from tropical to desert as well as high to low aerosol loading.

A number of active fires have been measured over the years capturing the associated fire spectral signatures. In total, the AVIRIS archive contains the most diverse set of well-calibrated Earth spectral signatures collected to date.

In the years over which the AVIRIS data set has been collected a wide range of new analysis algorithms have been developed and applied. A number of these algorithms have involved forward inversion of the AVIRIS spectra with a physically based model. Algorithms for the inversion of AVIRIS spectra for water vapor [11, 12, 13] as well as simultaneous inversions for water, vapor, liquid water, and frozen water [14] have been developed. Related forward inversion algorithms have been developed for determining the temperature and fractional area of actively burning biomass fires [15, 16]. Inversion of a physically based model with imaging spectrometer measurement is one of the more powerful methods for extracting parameters from well-calibrated spectra. These parameters are then used to pursue the science research or application objectives of interest.

Another important set of algorithms applied to AVIRIS imaging spectroscopy measurements includes spectral mixture analysis. These approaches began with simple unmixing [17, 18] and extended to multiple endmember spectral mixture analysis (MESMA) [19]. An automated Monte Carlo spectral analysis method has also been developed [20, 21], as have techniques for the estimation of spectral endmembers from within imaging spectrometer data sets [22]. Spectral mixture analysis appropriately addressed the fact that every AVIRIS measured spectrum contains a mixture of spectral signatures. The derived component fractions for each spatial element in an image enable pursuit of a wide range of investigations.

Spectral feature fitting is an imaging spectroscopy algorithm approach that has been pursed extensively by scientists of the United States Geological Survey. The result of this multidecade activity is the Tetracorder algorithm [23]. This approach focuses on the direct spectral fitting of measured absorption features with those from a comprehensive spectral library. Results of this and related spectral feature fitting algorithms have been applied successfully to a wide range of geological research investigations as well as other scientific disciplines.

A wide range of supervised and unsupervised classification algorithms have been developed for use with AVIRIS and other imaging spectrometer measurements. The Multispec software suite [24] of the Purdue Research Foundation represents an example containing a wide range of classification algorithms used for the analysis of AVIRIS measurements.

Only a few examples of algorithms employing physical model inversion, spectral mixture analysis, spectral feature analysis, and classification have been briefly described here. In addition, many hybrid algorithms exist that include aspects of two or more of these approaches. Other AVIRIS imaging spectroscopy measurement analysis methods exist as well. To explore the breadth and depth of algorithms that have been successfully applied to AVIRIS measurements, a full search of the refereed and nonrefereed literature is required and is beyond the scope of this chapter. Finally, given the diverse spectral signature content of high precision and high uniformity spectra, there is clear potential for the development of new algorithms and approaches for the extraction of valuable information from existing AVIRIS measurements.

14.3 Objectives and Characteristics of a Spaceborne Imaging Spectrometer for the Moon

The Moon Mineralogy Mapper (M^3) was selected as a NASA Discovery Mission of Opportunity in early 2005. The M^3 instrument is a 21st century high uniformity and high precision imaging spectrometer of the pushbroom type. M^3 measures spectra as images in the solar dominated portion of the electromagnetic spectrum. The basis for the use of imaging spectroscopy for mapping the mineralogy of the moon is found in the diversity of lunar minerals returned to Earth from the Apollo missions of the late 20th century. High precision (signal-to-noise ratio) is required to measure the less pronounced spectral signatures of dust and mineral mixtures as well as to measure high quality spectra near the poles where the solar illumination is reduced. M^3 is planned to be launched as a guest instrument provide by NASA on the Chandrayaan-1 mission of the Indian Space Research Organization (ISRO) in early 2008.

14.3.1 Objectives of the Moon Mineralogy Mapper

The overarching science and exploration objectives of the M^3 instrument and mission are

- Characterize and map the lunar surface composition in the context of its geologic evolution.
- Assess and map the Moon mineral resources at high spatial resolution to support planning for future, targeted missions.

These overarching objectives translate into the following more specific objectives:

- Evaluate the primary components of the crust and their distribution across the highlands.
- Characterize the diversity and extent of different types of basaltic volcanism.
- Explore for, identify, and assess deposits containing volatiles.
- Map fresh craters to assess abundance of small impacts in the recent past.
- Identify and evaluate concentrations of unusual/unexpected minerals.

All of these objectives may be pursued based upon the spectral signatures of the materials on the surface of the moon. Figure 14.12 shows a suite of spectra measured from samples returned to the Earth during the Apollo missions of the 1960s and the 1970s. To pursue these material identification objectives, an imaging spectrometer measuring reflected light in the solar reflected spectrum is required.

14.3.2 Characteristics of the M^3 Imaging Spectrometer

Based upon the scientific objectives and the spectroscopic approach, a detailed specification of the M^3 imaging spectrometer was established [25, 26]. These spectral,

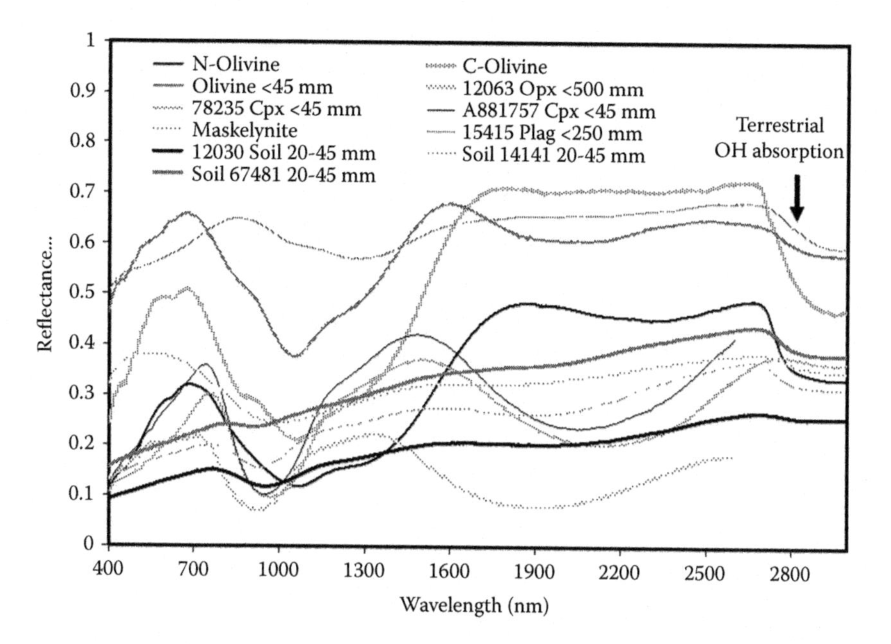

Figure 14.12 Spectra of samples returned by the NASA Apollo missions showing the composition-based spectral diversity of surface materials on the Moon. This spectral diversity provides the basis for pursing the objectives of the M^3 mission with an imaging spectrometer. Upon arrival on Earth the ultradry lunar Samples have adsorbed water, resulting in the absorption feature beyond 2700 nm. These spectra were measured by the NASA RELAB facility at Brown University.

radiometric, spatial, temporal, and uniformity requirements are given in Table 14.3. The reference radiance at which the precision requirement of M^3 is set is the modeled radiance from the returned Apollo 16 soil reflectance (BKR1LR117) illuminated with a 0-degree solar zenith angle.

With the detailed spectral, radiometric, spatial, temporal, and uniformity specifications, a high uniformity and high precision Offner pushbroom imaging spectrometer design was selected [27]. Figure 14.13 shows the mechanical design implementation for M^3. This design uses a three-mirror telescope with fold mirror to feed light through a uniform slit to the Offner spectrometer. The spectrometer consists of one spherical mirror used twice and a custom convex grating. The spectrally dispersed light from the spectrometer passes through an order sorting filter to the detector array that is sensitive from 430 to 3000 nm. This comparatively simple design was enabled by the structured blaze convex grating in the core of the uniform full-range spectrometer.

M^3 is a high uniformity and high precision pushbroom imaging spectrometer. The cross-track swath is 40 km with 70 m spatial sampling in the along-track and cross-track directions. For each 70 m advance of the image swath in the orbit direction around the moon, a full set of 600 cross-track spectra will be read out from the

TABLE 14.3 Spectral, Radiometric, Spatial, Temporal and Uniformity Specifications of The M^3 Imaging Spectrometer for the Moon

	Spectral properties:
Range	430 to 3000 nm in the solar reflected spectrum
Sampling	10 nm across spectral range
Response	FWHM $<$ 1.2 of sampling
Accuracy	Calibrated to 10% of sampling
Precision	Stable within 5% of sampling
	Radiometric properties:
Range	0 to specified saturation radiance (2 $\times$ Apollo 16)
Sampling	12 bits measured
Response	Linear to 1%
Accuracy	$>$ 90% absolute radiometric calibration
Precision (SNR)	$>$ 400 equatorial reference
	$>$ 100 polar reference
	Spatial properties:
Range	24 degree field-of-view (FOV)
Sampling	0.7 milliradian cross and along track
Response	FWHM of IFOV $<$ 1.2 of sampling
	Temporal properties:
Global	$>$ 95% coverage in 2 years (reduced resolution)
Target	$>$ 5% coverage in 2 years (full resolution)
	Uniformity:
Spectral-crosstrack	$>$ 99% uniformity of position across the FOV
Spectral-IFOV	$>$ 90% IFOVs uniformity over the spectral range

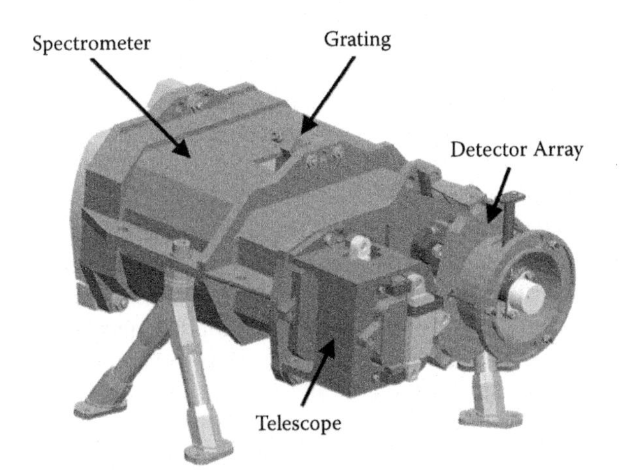

Figure 14.13 Mechanical drawing of the M^3 imaging spectrometer that has been built for mapping the composition of the Moon via spectroscopy. The M^3 instrument has the following mass, power, and volume characteristics: 8 kg, 15 Watts, 25 $\times$ 18 $\times$ 12 cm. The M^3 instrument was built in 24 months.

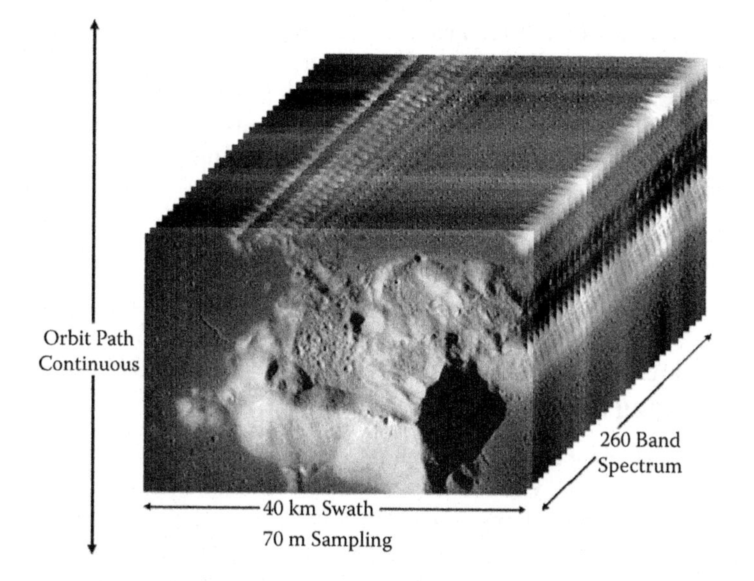

Figure 14.14 Depiction of the spectral spatial and pushbroom imaging approach of the M^3 high uniformity and high precision imaging spectrometer.

detector array simultaneously. Each spectrum will consist of 260 spectral channels from 430 to 3000 nm. A depiction of the type of data to be measured by M^3 is shown in Figure 14.14. To accommodate data rate limitations for transmission to Earth, M^3 will also have a global data acquisition mode with reduced spectral and spatial resolution. In this mode, 260 spectral channels will be selectively summed to give 86 contiguous channels covering the range from 430 to 3000 nm. In the spatial domain the data will by summed to give 140 meter samplings. The balance between coverage of the Moon in global and full resolution modes will depend on the available data downlink at the time of the mission. A critical characteristic of M^3 is uniformity. The cross-track spectral uniformity enables the direct comparison of spectra across the swath and throughout the image and is required for a range of advanced computational analysis techniques. The excellent spectral-IFOV uniformity of M^3 is required for rigorous spectroscopic analysis across the spectral range of M^3. If the area sampled on the surface changes with wavelength, a basic assumption of spectroscopy is violated.

14.3.3 Prospects for the M^3 Imaging Spectrometer Data Set

M^3 is the first imaging spectrometer designed to provide complete coverage of a planetary sized body in our solar system at high spatial resolution. The data set is currently expected to have a volume of at least 5.8 Terabytes at the end of the mission. During the mission, portions of the M^3 data set will be available for analysis through the NASA planetary data system (PDS). After the end of the mission, final calibration and validation the complete data set will become available. Throughout the M^3 mission and following, the imaging spectrometer measurements will be used to characterize and map the lunar surface composition and to assess and map mineral resources at high spatial resolution to support future missions. The volume, quality, and comprehensive nature of the M^3 data set presents a unique opportunity for the application of high-performance computing.

14.4 Objectives and Characteristics of a Future Spaceborne Imaging Spectrometer for the Earth

The diversity and complexity of scientific research questions being pursued for the global Earth system require the use of satellite observations. Due to limits of technology, early satellite observations systems measured only a few spectral bands. With only a few spectral bands the typical algorithms consisted of band ratios and band indices. The normalize vegetation difference index (NDVI) [28] is a prominent example. There are limitations to the interpretation and portability of these simple algorithms using only a few spectral bands, because many more than two components of the Earth's surface and atmosphere system contribute to the measured signals. Realization of these limitations and the advance of technology have led to the current set of Earth observing instruments that in some cases measure tens of spectral bands. However, even with tens of spectral bands, limitations arise for quantitative and portable algorithms due to the undersampling of the available spectral signal. Now in the 21st century, new Earth observing instruments are being designed and proposed with full spectral coverage in the solar reflected spectrum. In this section a description of objectives and characteristics of an Earth imaging spectrometer focused on terrestrial and aquatic ecosystem is discussed.

14.4.1 Objectives of an Earth Imaging Spectrometer for Measuring the State of Terrestrial and Aquatic Ecosystems

The overarching objective of this Earth imaging spectrometer is understanding the health, composition, and productivity of both terrestrial and aquatic ecosystems at a seasonal time scale over the entire globe. The aquatic focus is in the coastal and inland water regions where the ecosystem diversity is greatest. Companion objectives are to understand how these terrestrial and aquatic ecosystems are being altered by human activities and how these changes affect the fundamental processes upon which these ecosystems depend. From these objectives, a set of research area topics arise including: (1) ecosystem function and diversity, (2) biogeochemical cycles, and (3) ecosystem disturbance and response. To address the specific questions in these research areas, a set of measurement products have been identified. These products are given in Table 14.4.

TABLE 14.4 Earth Imaging Spectrometer Products for Terrestrial and Aquatic Ecosystems Understanding

Terrestrial ecosystem science products:
Calibrated full-optical range surface radiance and reflectance
Fractional cover of biotic and abiotic materials
Leaf and canopy water content
Leaf and canopy pigments and nitrogen concentrations
Plant functional types and species dominance
Ecosystem disturbance and response
Plant light-use efficiency and productivity
Aquatic ecosystem science products:
Calibrated full-optical range surface radiance and reflectance
Coastal ocean and inland water optical properties and ecosystem
Component concentrations
Phytoplankton functional groups & species dominance
Benthic and inland water communities
Plant physiological status
Ecosystem productivity
Ecosystem disturbance and response

Based upon the almost 20 years of research results using AVIRIS measurements as well as measurements from other imaging spectrometers and in conjunction with theoretical, laboratory, and field spectroscopic efforts, it has become clear that an Earth spaceborne imaging spectrometer is required to pursue these objectives globally. The use of imaging spectroscopy to pursue these objectives is further supported by the 2007 National Research Council, Committee on Earth Science and Applications from Space: A Community Assessment and Strategy for the Future [29]. Among other important missions, an Earth imaging spectrometer is specified to pursue a set of objectives for understanding and managing ecosystems.

14.4.2 Characteristics of an Ecosystem Focused Earth Imaging Spectrometer

To generate the required science products, a basis for what is measurable from a satellite perspective must be established. The literature now contains numerous publications establishing the physical basis and approach for generating the required science products from spectroscopic measurements. From this work spanning 20 years, a set of spectral, radiometric, spatial, temporal, and uniformity requirements have been defined for such measurements. These represent one set of requirements that are feasible with currently available technology and are given in Table 14.5. One of the most critical measurement requirements is the precision or signal-to-noise ratio. For this imaging spectrometer, the precision is specified at a set of four reference radiances given in Figure 14.15. The corresponding signal-to-noise ratios required are shown in Figure 14.16. The signal-to-noise ratios specified here are high in order to pursue

TABLE 14.5 Nominal Characteristics of an Earth Imaging Spectrometer for Terrestrial and Aquatic Ecosystems' Health, Composition, and Productivity At a Seasonal Time Scale

	Spectral properties:
Range	380 to 2510 nm
Sampling	10 nm (uniform over range)
Response	$< 1.2\times$ sampling (FWHM) (uniform over range)
Accuracy	< 0.5 nm
	Radiometric properties:
Range	0 to specified saturation radiance
Sampling	14 bits measured (possibly 13 down linked)
Response	$> 95\%$ absolute radiometric calibration
Accuracy	$> 90\%$ absolute radiometric calibration
	$> 98\%$ on-orbit reflectance
Stability	$> 99.5\%$
Precision (SNR)	As specified at benchmark radiances
Linearity	$> 99\%$ characterized 0.1%
Polarization	$< 2\%$ sensitivity, characterized to 0.5%
Stray light	$< 1{:}200$ characterized to 0.1%
	Spatial properties (at $\tilde{7}00$ km altitude):
Cross-track samples	2440
Range	146 km FOV (12 degrees)
Sampling	60 m
Response	< 1.2 sampling (FWHM)
	Temporal properties:
Global land coast repeat	19 days at the equator
Rapid response revisit	3 days
	Uniformity:
Spectral cross-track	$> 95\%$ cross-track uniformity
Spectral-IFOV	$> 95\%$ spectral IFOV uniformity

the required science products in the low signal areas of coastal water and dark green vegetation.

14.4.3 Roles for High-Performance Computing

The current specifications for this imaging spectrometer create many opportunities for high-performance computing. The first and perhaps most important is fast lossless data compression. The instantaneous data rate of the imaging spectrometer is 200 megabits per second. Even limiting to only collecting daylight data over the terrestrial and coastal regions of the Earth will overwhelm satellite onboard storage and data downlink capabilities. The science community is also unwilling to accept compression algorithms that lose some of the measured data fidelity. Realization of the data rate and volume challenges of imaging spectrometers have led to new lossless compression approaches [30]. Some recent results are encouraging with fast

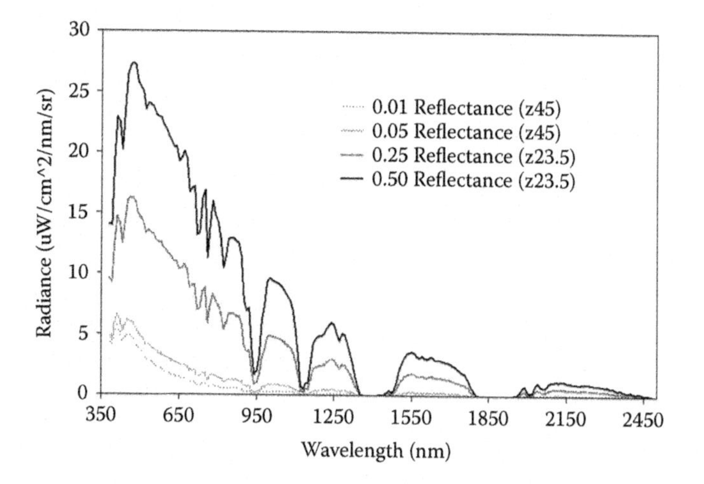

Figure 14.15 Benchmark reference radiance for an Earth imaging spectrometer focused on terrestrial and aquatic ecosystem objectives.

lossless compression ratios of greater than 4 to 1 achieved for data collected by an airborne pushbroom imaging spectrometer that is a close analog to this spaceborne Earth imaging spectrometer. Irrespective of the use of lossless compression on-orbit and for downlink, the size of the uncompressed data set on the ground will be large. Table 14.6 shows the orbit, daily, and yearly data volumes with the spectra held at 16 bit integers. The scale of this data set and the information contained within the measured spectra demand the use of high-performance computing in the areas of storage, algorithms, data I/O, computations, and visualization.

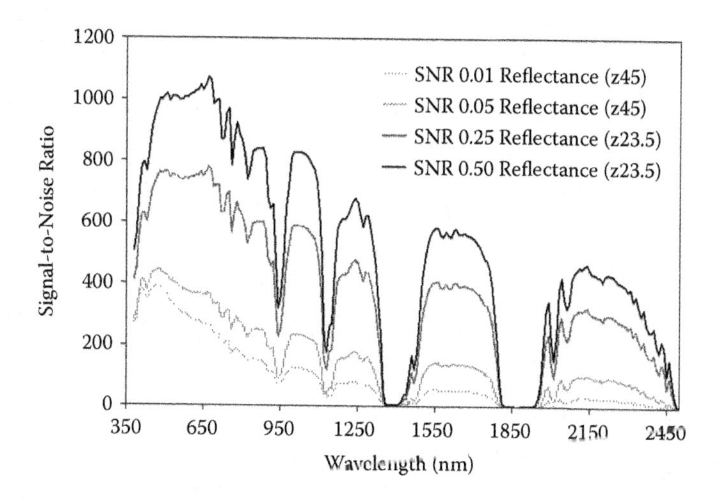

Figure 14.16 The signal-to-noise ratio requirements for each of the bench-mark reference radiances.

TABLE 14.6 Earth Ecosystem Imaging Spectrometer Data Volumes

Period	Volume	Period	Volume	Period	Volume	Period	Volume
Orbit	94 GB	Day	1.37 TB	Year	502 TB	Mission	1.5 PB

14.5 Acknowledgments

This work was carried out at the Jet Propulsion Laboratory of the California Institute of Technology, Pasadena, California, under contract with the National Aeronautics and Space Administration. The broad and deep advances in the field of imaging spectroscopy over the past two decades derive from the contributions of an extensive group of investigators pursuing research questions spanning a wide range of scientific disciplines.

References

[1] G. Vane, A. F. H. Goetz, J. B. Wellman, Airborne imaging spectrometer: A new tool for remote-sensing. *IEEE Transactions on Geoscience and Remote Sensing*, vol. 22, pp. 546–549, 1984.

[2] A. S. Mazer, M. Martin, M. Lee, et al., Image-processing software for imaging spectrometry data-analysis. *Remote Sensing of Environrment*, vol. 24, pp. 101–110, 1988.

[3] G. Vane, R. O. Green, T. G. Chrien, et al., The Airborne Visible Infrared Imaging Spectrometer (AVIRIS). *Remote Sensing of Environment*, vol. 44, pp. 127–143, 1993.

[4] R. O. Green, M. L. Eastwood, C. M. Sarture, et al., Imaging spectroscopy and the Airborne Visible Infrared Imaging Spectrometer (AVIRIS). *Remote Sensing of Environment*, vol. 65, pp. 227–248, 1998.

[5] A. Berk, L. S. Bernstein, D.C. Robertson, MODTRAN: A moderate resolution model for LOWTRAN 7. *Final report*, GL-TR-0122, AFGL, Hanscomb AFB, MA., pp. 42, 1989.

[6] G. P. Anderson, M. L. Hoke, J. H. Chetwynd, Jr., A. J. Ratkowski, L. S. Jeong, A. Berk, L. S. Bernstein, S. M. Adler-Golden, et al., MODTRAN4 radiative transfer modeling for remote sensing. *Proc. SPIE*, vol. 4049, pp. 176–183, 2000.

[7] T. G. Chrien, R. O. Green, M. Eastwood, Laboratory spectral and radiometric calibration of the Airborne Visible/Infrared Imaging Spectrometer (AVIRIS). *Proc. SPIE*, vol. 1298, Imaging spectroscopy of the terrestrial environment, 1990.

[8] T. G. Chrien, R. O. Green, C. Chovit, M. Eastwood, J. Faust, P. Hajek, H. Johnson, H. I. Novack, C. Sarture. New calibration techniques for the Airborne Visible/Infrared Imaging Spectrometer (AVIRIS), *Proc. Fifth Annual Airborne Earth Science Workshop*, JPL Pub. 95–1, 1995.

[9] Green, R. O., J. E. Conel, V. Carrere, C. J. Bruegge, J. S. Margolis, M. Rast, G. Hoover, In-flight validation and calibration of the spectral and radiometric characteristics of the Airborne Visible/Infrared Imaging Spectrometer (AVIRIS). *Proc. SPIE Conference on Aerospace Sensing*, Imaging Spectroscopy of the Terrestrial Environment, Orlando, Florida, 1990.

[10] R. O. Green, J. E. Conel, C. Margolis, C. Chovit, J. Faust, In-flight calibration and validation of the Airborne Visible/Infrared Imaging Spectrometer (AVIRIS). *Proc. Sixth Annual Airborne Earth Science Workshop*, Jet Propulsion Laboratory, JPL Pub. 96–4, vol. 1, March 3–5, 1996.

[11] R. O. Green, V. Carrere, J. E. Conel, Measurement of atmospheric water vapor using the Airborne Visible/Infrared Imaging Spectrometer. *Proc. of the ASPRS conference on Image Processing* '89, pp. 23–26, 1989.

[12] B. C. Gao, A. F. H. Goetz, Column atmospheric water-vapor and vegetation liquid water retrievals from airborne imaging spectrometer data. *Journal of Geophysical Research-Atmospheres*, vol. 95, pp. 3549–3564, 1990.

[13] R. O. Green, J. E. Conel, J. Margolis, C. Bruegge, G. Hoover, An inversion algorithm for retrieval of atmospheric and leaf water absorption from AVIRIS radiance with compensation for atmospheric scattering. *Proc. of the Third AVIRIS Workshop*, JPL Publication 91–28, pp. 51–61, 1991.

[14] R. O. Green, T. H. Painter, D. A. Roberts, et al. Measuring the expressed abundance of the three phases of water with an imaging spectrometer over melting snow. *Water Resources Research*, vol. 42, no. W10402, 2006.

[15] R. O. Green, Estimation of biomass fire temperature and areal extent from calibrated AVIRIS spectra. *Proc. Sixth Annual Airborne Earth Science Workshop*, Jet Propulsion Labratory, JPL Public 96, vol. 1, March 3–5, 1996.

[16] P. E. Dennison, K. Charoensiri, D. A. Roberts, et al. Wildfire temperature and land cover modeling using hyperspectral data. *Remote Sensing of Environment*, vol. 100, pp. 212–222, 2006.

[17] D. A. Roberts, M. O. Smith, J. B. Adams, Green vegetation, nonphotosynthetic vegetation, and soils in AVIRIS data. *Remote Sensing of Environment*, vol. 44, pp. 255–269, 1993.

[18] J. B. Adams, M. O. Smith, A. R. Gillespie, Imaging spectroscopy: interpretation based on spectral mixture analsyis. In: *Remote Geochemical Analysis: Elemental and Mineralogical Composition*, C. M. Pieters and P. Englert, Eds., Cambridge University Press, New York, pp. 145–166, 1993.

[19] D. A. Roberts, M. Gardner, R. Church, et al., Mapping chaparral in the Santa Monica Mountains using multiple endmember spectral mixture models, *Remote Sensing of Environment*, vol. 65, pp. 267–279, 1998.

[20] G. P. Asner, Biophysical and biochemical sources of variability in canopy reflectance. *Remote Sensing of Environment*, vol. 64, pp. 234–253, 1998.

[21] G. P. Asner, D. B. Lobell, A biogeophysical approach to automated SWIR unmixing of soils and vegetation. *Remote Sensing of Environment*, vol. 74, pp. 99–112, 2000.

[22] A. Plaza, P. Martinez, R.M. Perez, J. Plaza. A new approach to mixed pixel classification of hyperspectral imagery based on extended morphological profiles. *Pattern Recognition*, vol. 37, no. 6, pp. 1097–1116, 2004.

[23] R. N. Clark, G. A. Swayze, K. E. Livo, et al., Imaging spectroscopy: Earth and planetary remote sensing with the USGS Tetracorder and expert systems, *Journal of Geophyisical Research-Planets*, vol. 108, no. 5131, 2003.

[24] L. Biehl, D. Landgrebe, MultiSpec—A tool for multispectral-hyperspectral image data analysis. *Computers & Geosciences*, vol. 28, pp. 1153–1159, 2002.

[25] C. Pieters, J. Boardman, B. Buratti, R. Clark, R. O. Green, J. W. Head, T. B. McCord, J. Mustard, C. Ryunyon, M. Staid, J. Sunshine, L. A. Taylor, S. Tompkins, Global mineralogy of the moon: A cornerstone to science and exploration, *Proc. Lunar & Planetary Science Conference*, Houston, 2006.

[26] R. O. Green, C. Pieters, P. Mouroulis, G. Sellars, M. Eastwood, S. Geier, J. Shea, The Moon Mineralogy Mapper: Characteristics and early laboratory calibration results, *Proc. Lunar & Planetary Science Conference*, Houston, 2006.

[27] P. Mouroulis, R. O. Green, T. G. Chrien, Design of pushbroom imaging spectrometers for optimum recovery of spectroscopic and spatial information. *Applied Optics*, vol. 13, pp. 2210–2220, 2000.

[28] J. W. Rouse, R. H. Haas, J. A. Schell, D. W. Deering, Monitoring vegetation systems in the Great Plains with ERTS. *Proc. Third ETRS Symp.*, Washington, DC, NASA NASA SP353, 1973.

[29] *Earth Science and Applications from Space: National Imperatives for the Next Decade and Beyond*, Committee on Earth Science and Applications from Space: A Community Assessment and Strategy for the Future, The National Academies Press, Washington, D.C., ISBN: 978-0-309-10387-9, 400 pages, 2007.

[30] M. Klimesh, Low-complexity adaptive lossless compression of hyperspectral imagery. *Proc. SPIE*, vol. 6300, 9 pages, San Diego, CA, 2006.

Chapter 15

Remote Sensing and High-Performance Reconfigurable Computing Systems

Esam El-Araby,
George Washington University

Mohamed Taher,
George Washington University

Tarek El-Ghazawi,
George Washington University

Jacqueline Le Moigne,
NASA's Goddard Space Flight Center

Contents

The trend for remote sensing satellite missions has always been towards smaller size, lower cost, more flexibility, and higher computational power. On-board processing, as a solution, permits a good utilization of expensive resources. Instead of storing and forwarding all captured images, data processing can be performed on-orbit prior to downlink, resulting in the reduction of communication bandwidth as well as simpler and faster subsequent computations to be performed at ground stations. Reconfigurable computers (RCs) combine the flexibility of traditional microprocessors with the power of Field Programmable Gate Arrays (FPGAs). Therefore, RCs are a promising candidate for on-board preprocessing.

15.1 Introduction

The ability to preprocess and analyze remote sensing data onboard in real time can significantly reduce the amount of bandwidth and storage required in the production of space science products. Consequently, onboard processing can reduce the cost and the complexity of the On-the-Ground/Earth processing systems. Furthermore, it enables autonomous decisions to be taken onboard that can potentially reduce the delay between image capture, analysis, and action. This leads to faster critical decisions, which are crucial for future reconfigurable Web sensors missions as well as planetary exploration missions [1].

The new generation of remote sensing detectors produces enormous data rates. This requires a very high computing power to process the raw data, e.g., the onboard computer has to provide a performance of 3×10^{10} operations/second in space to process and classify hyperspectral raw data to get useful data [2]. Currently there is no space computer with such a performance.

Recently, Field Programmable Gate Array (FPGA) based computing, also known as 'adaptive' or 'reconfigurable computing (RC),' has become a viable target for the implementation of algorithms suited to image processing and computationally intensive applications. These computing systems combine the flexibility of general purpose processors with the speed of application-specific processors. By mapping hardware to FPGAs, the computer designer can optimize the hardware for a specific application resulting in acceleration rates of several orders of magnitude over general-purpose computers. In addition, they are characterized by lower form/wrap factors compared to parallel platforms and higher flexibility than ASIC solutions. RC technology allows new hardware circuits to be uploaded via a radio link for physical upgrade or repair. Therefore, RCs are a promising candidate for onboard data preprocessing. High-speed, radiation-hardened FPGA chips with million gate densities have recently emerged that can support the high throughput requirements for the remote sensing applications [3].

Figure 15.1 shows an example of onboard processing for hyperspectral images using reconfigurable processing. Instead of storing and forwarding all captured images, data processing can be performed on-orbit prior to downlink, resulting in the reduction of communication bandwidth as well as simpler and faster subsequent computations to be performed at ground stations.

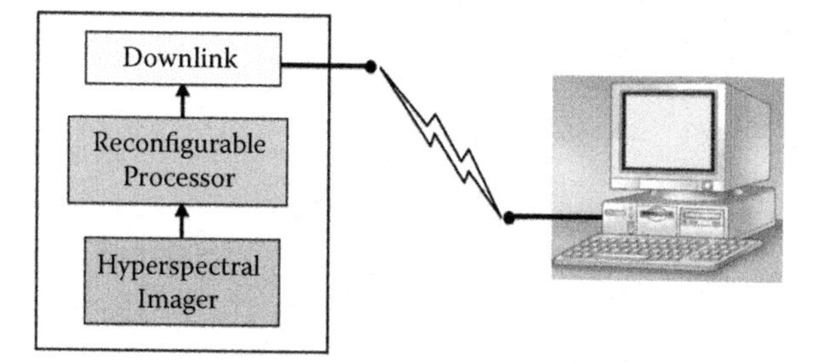

Figure 15.1 Onboard processing example.

15.2 Reconfigurable Computing

Applications have been traditionally implemented either in hardware using, for example, custom VLSI and Application-Specific Integrated Circuits (ASICs), or in software running on processors, such as Digital Signal Processors (DSPs), microcontrollers, and general-purpose microprocessors. These two extremes trade performance with flexibility, and vice versa. For example, ASICs are designed specifically to solve a given problem. Therefore, they are fast and efficient when compared with a microprocessor-based design. However, an ASIC circuit cannot be modified after fabrication. Due to their programmability, microprocessors offer more flexibility, but at the expense of speed.

Reconfigurable hardware [4] introduces a trade-off between traditional hardware and software by achieving hardware-like performance with software-like flexibility. Reconfigurable hardware offers the performance advantage of direct hardware execution and the flexibility of software-like programming [5]. Figure 15.2 represents the trade-off between flexibility and performance for different implementation approaches.

15.2.1 FPGAs and Reconfigurable Logic

Reconfiguration in today's reconfigurable computers is provided through FPGAs [6]. An FPGA can be viewed as programmable logic blocks embedded in programmable interconnects as shown in Figure 15.3. FPGAs are composed of three fundamental components: logic blocks, I/O blocks, and programmable interconnects. The logic block is the basic building block in the FPGA. In Xilinx, which is currently the largest FPGA vendor, this logic block is called a Configurable Logic Block (CLB). Routing resources enable efficient communication among CLBs. The CLB usually consists of lookup tables (LUTs), carry logic, flip-flops, and programmable multiplexers, as shown in Figure 15.4. The device can be programmed using a hardware description language such as VHDL, or using schematic capture software. There are also many C-to-gates compilers from research groups and from commercial vendors. A circuit is

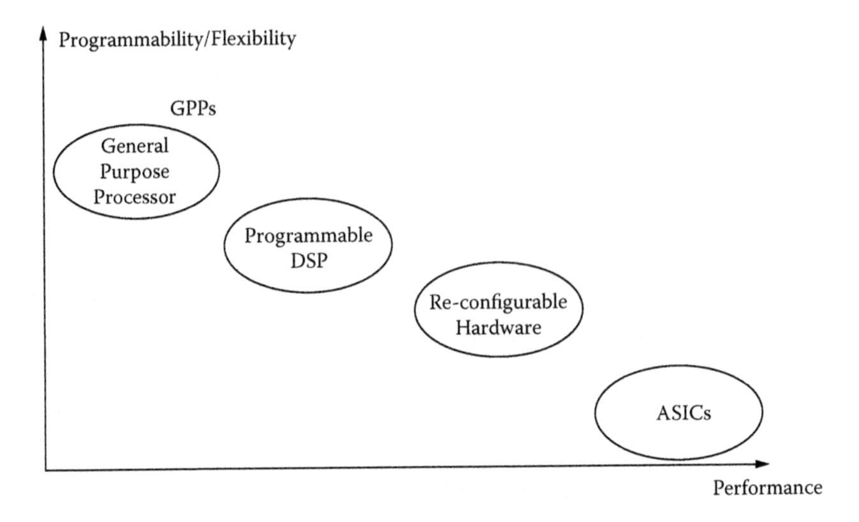

Figure 15.2 Trade-off between flexibility and performance [5].

implemented in an FPGA by programming each logic block to realize a portion of the logic required by the circuit, and each of the I/O blocks to act as either an input pad or an output pad, as required by the circuit. The programmable routing is configured to make all the necessary connections among the logic blocks and between logic blocks and I/O blocks. The programming technology determines the method of storing the configuration information.

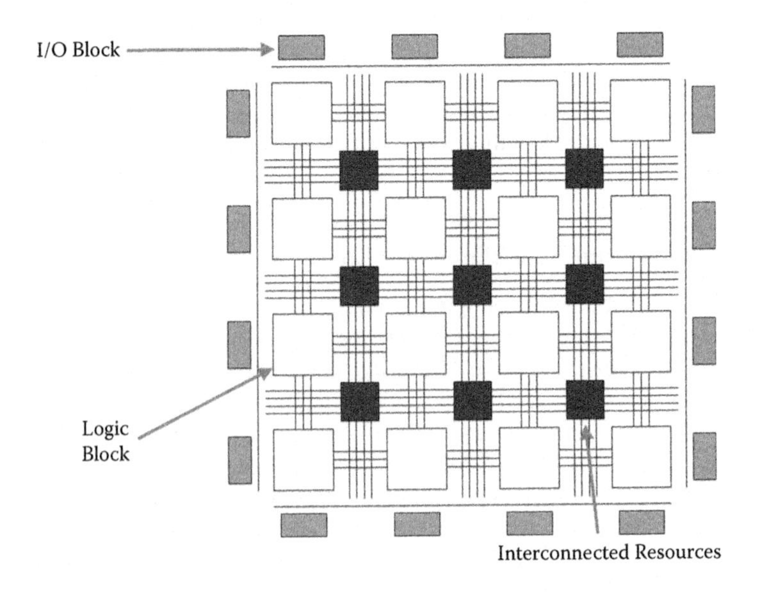

Figure 15.3 FPGA structure.

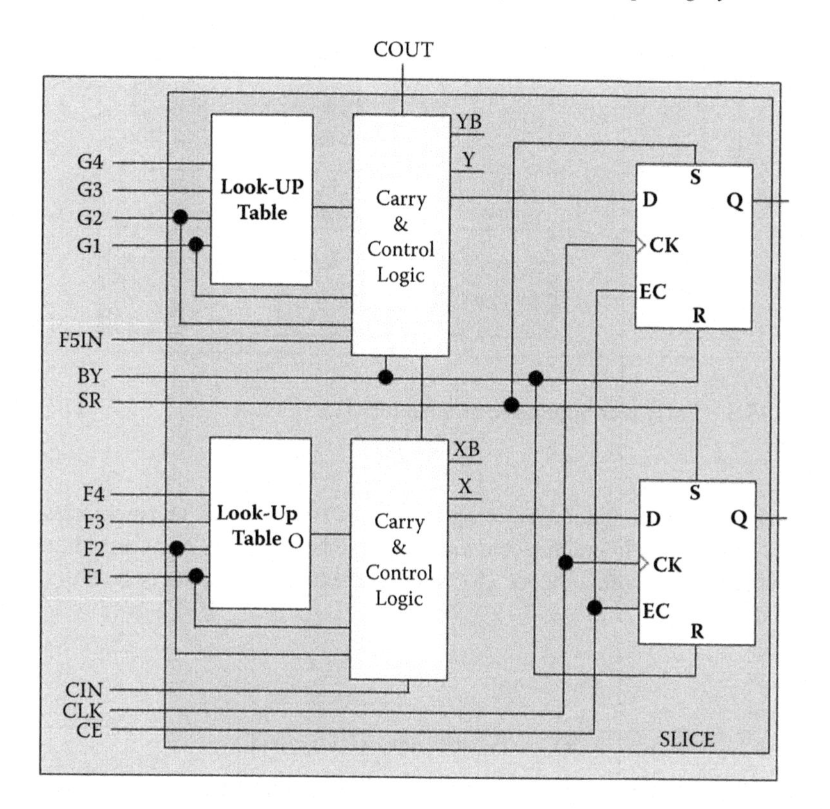

Figure 15.4 CLB structure.

15.2.2 Reconfigurable Computers

Reconfigurable computers [4, 7, 8, 9] are composed of one or more general-purpose processors and one or more reconfigurable chips, such as Field Programmable Gate Arrays (FPGAs) closely integrated with each other. The processor performs the operations that cannot be done efficiently in the reconfigurable logic, such as data-dependent control (e.g., loops and branches) and possibly memory accesses, while the computational cores are mapped to the reconfigurable hardware. Reconfigurable computing systems can be based on commodity PC boards (such as Pentium boards), where the reconfigurable processor sub-system is typically a commercial, off-the-shelf (COTS) FPGA accelerator board, such as WildStar II [10]. The reconfigurable board acts as a co-processor to the PC or workstation processor. Usually, they are interfaced to a computer via a PCI bus. RCs have recently evolved from accelerator boards to stand-alone general-purpose RCs and parallel reconfigurable supercomputers. Examples of such supercomputers are the Cray-XD1, SRC-6, and the SGI-Altix with FPGA bricks. These systems can leverage the synergism between conventional processors and FPGAs to provide low-level hardware functionality at the same level of programmability as general-purpose computers. A typical reconfigurable computer node

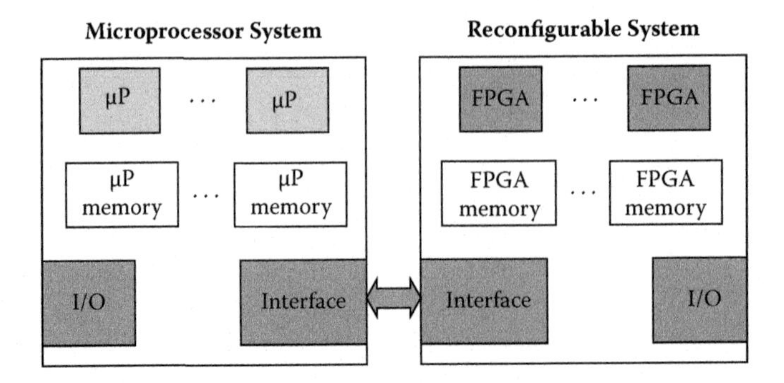

Figure 15.5 Early reconfigurable architecture [7].

can be thought of as a module of microprocessors, FPGA devices, and memories. Such modules can then be interconnected via some fixed (or perhaps even reconfigurable) interconnection networks. Figure 15.5 shows an example of an early reconfigurable architecture.

15.3 The Promise of Reconfigurable Computing for Remote Sensing

Many of the motivations and goals of reconfigurable computing are consistent with the needs of image processing and remote sensing applications.

Logic functionality of FPGAs with DSP resource and embedded processors can be customized to perform exactly the desired operation. This makes the FPGA a very good candidate for image and signal processing applications.

Most space systems have a need for reconfigurability for different reasons, such as the need for physical upgrade or repair of the unmanned spacecraft devices. This problem can be resolved by using RC. RC technology allows new hardware circuits to be uploaded via a radio link. This also allows us to change the system functionality according to changing mission requirements. Most reconfigurable devices and systems contain SRAM-programmable memory to allow logic and interconnect reconfigurations in the field.

During operation, the systems are physically remote from their operators, and all control of the spacecraft and new FPGA configurations can be transmitted over a wireless radio link. For example, the Mars Rover mission used a Xilinx FPGA that had not been completely designed at the time of launch. The FPGA configuration was uploaded to the spacecraft two months after the launch [11].

Reconfigurable computers have been widely used for accelerating low-level image processing algorithms. These algorithms are typically applied close to the raw sensor data and are characterized by large data volume. The fine-grained parallelism found in the FPGA devices is well-matched to the high sample rates and distributed computation often required of signal processing applications in areas such as image, audio, and speech processing.

15.4 Radiation-Hardened FPGAs

Space-based systems must operate in an environment in which radiation effects have an adverse impact on integrated circuit operation. Ionizing radiation can cause soft-errors in the static cells used to hold the configuration data. This will affect the circuit functionality and can cause system failure. This requires special FPGAs that provide on-chip configuration error-detection and/or correction circuitry.

Radiation-hardened FPGAs are in great demand for military and space applications to reduce cost and cycle time. Actel Corp. has been producing radiation-tolerant anti-fuse FPGAs for several years for high-reliability space-flight systems. Actel FPGAs have been onboard more than 100 launches. Xilinx FPGAs have been used in more than 50 missions [11].

15.5 Case Studies of Remote Sensing Applications

15.5.1 Wavelet-Based Dimension Reduction of Hyperspectral Imagery

Hyperspectral imagery, by definition, provides valuable remote sensing observations at hundreds of frequency bands. Conventional image classification (interpretation) methods may not be used without dimension reduction preprocessing. Dimension reduction is the transformation that brings data from a high order spectral dimension to a low order spectral dimension. Dimension reduction has become a significant step for onboard preprocessing of hyperspectral imagery, e.g., image interpretation/classification. In remote sensing, one of the most widely used dimension reduction techniques is the Principal Component Analysis (PCA). PCA, by definition, computes orthogonal projections, which results in time-consuming computations, inefficient use of the memory hierarchy, and, finally, large interprocessor communication overhead. For these reasons, the novel Automatic Wavelet Dimension Reduction technique has been proven to yield better or comparable classification accuracy, while achieving substantial computational savings [12]. However, the large hyperspectral data volumes remain to present a challenge for traditional processing techniques even with the wavelet-based method. Therefore, there is always a pressing need for new, efficient, and powerful processing capabilities for the implementation of dimension reduction algorithms within the domain of hyperspectral imagery processing.

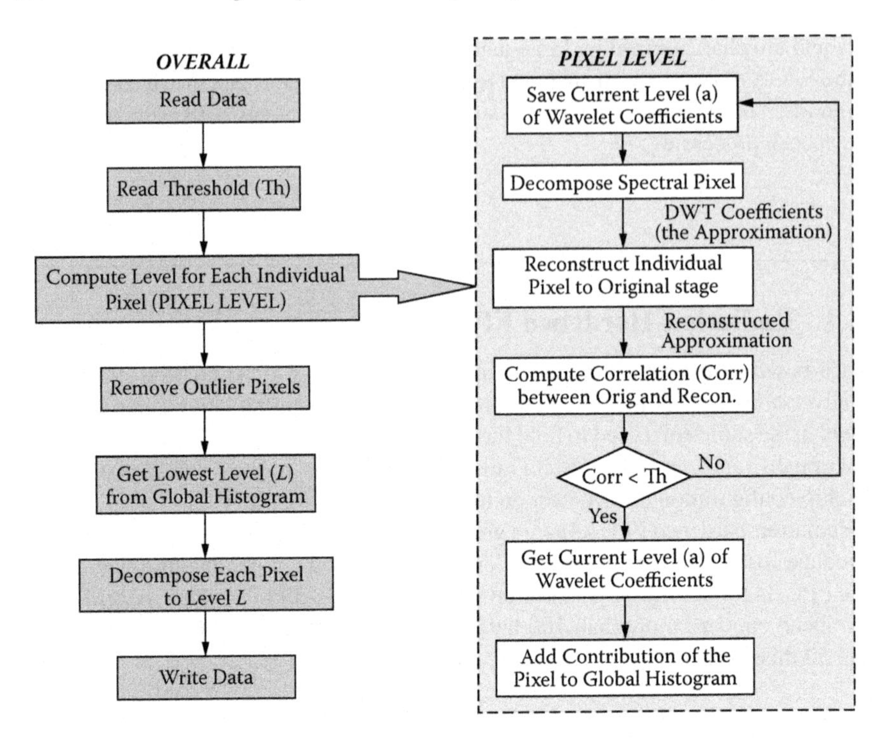

Figure 15.6 Automatic wavelet spectral dimension reduction algorithm.

The general description of the automatic wavelet dimension reduction algorithm is shown in Figure 15.6.

The correlation function between the original spectral signature (x), i.e., the original image, and its reconstructed approximation (y), which results from applying the discrete wavelet transform (DWT) and inverse discrete wavelet transform (IDWT), is defined by the following expression, in which N represents the original spectral dimension of the hyperspectral image:

$$\rho(x, y) = \frac{\sum_{i=1}^{N} x_i y_i - \frac{1}{N}\sum_{i=1}^{N} x_i \sum_{i=1}^{N} y_i}{\sqrt{\left(\sum_{i=1}^{N} x_i^2 - \frac{1}{N}\left(\sum_{i=1}^{N}\right)^2\right)\left(\sum_{i=1}^{N} y_i^2 - \frac{1}{N}\left(\sum_{i=1}^{N} y_i\right)^2\right)}} \tag{15.1}$$

Correlation is applied as a quantitative indicator, which measures the similarity between the original spectral signature and the reconstructed spectral approximation. The automatic wavelet spectral reduction algorithm is developed using this correlation measure and a user-specified threshold (Th).

Figure 15.7 shows the top hierarchical level of an architecture implementation [13]. The algorithm parallelism has been equally distributed along the pipelined architecture, which, in contrast to the sequential implementations on traditional computers, will expectedly yield significant speedup gains in performance.

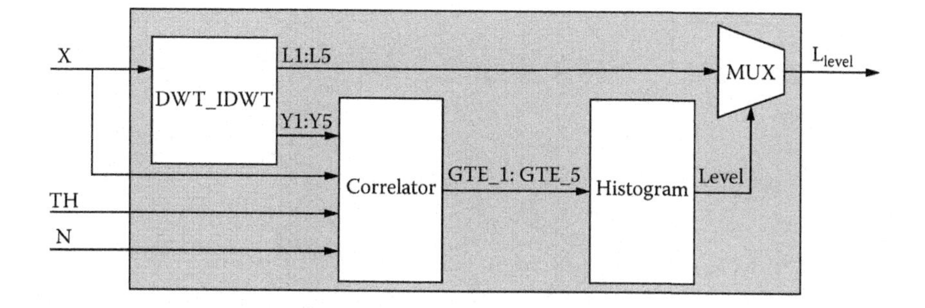

Figure 15.7 Top hierarchical architecture of the automatic wavelet dimension reduction algorithm.

As can be seen in Figure 15.8, the DWT_IDWT module, which performs both the Discrete Wavelet Transform (DWT) and the Inverse Discrete Wavelet Transform (IDWT) functions, produces the decomposition spectrum, i.e., $L1 - L5$, as well as the reconstruction spectrum, i.e., $Y1 - Y5$, respectively. The filtering (L, L') and down/up sampling operations are performed internally with full precision fixed-point signed (two's complement) data types. Truncation was used for quantization and saturated arithmetic was used for overflow handling. The data were externally interfaced into and out of this component using an 8-bit precision unsigned data type.

Figure 15.9 shows the correlator module; refer to equation 15.1. The division and the square-root operations have been avoided when evaluating the function. The threshold

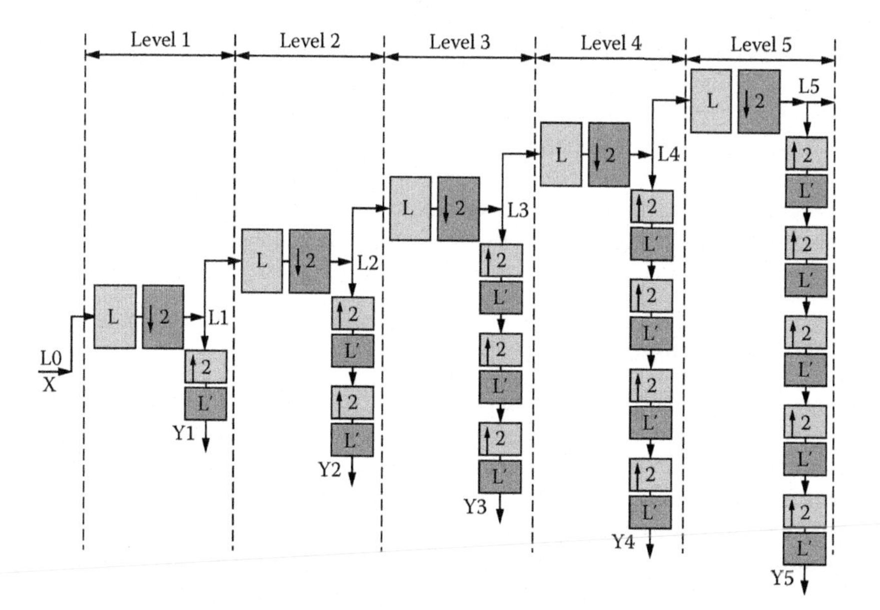

Figure 15.8 DWT_IDWT pipeline implementation.

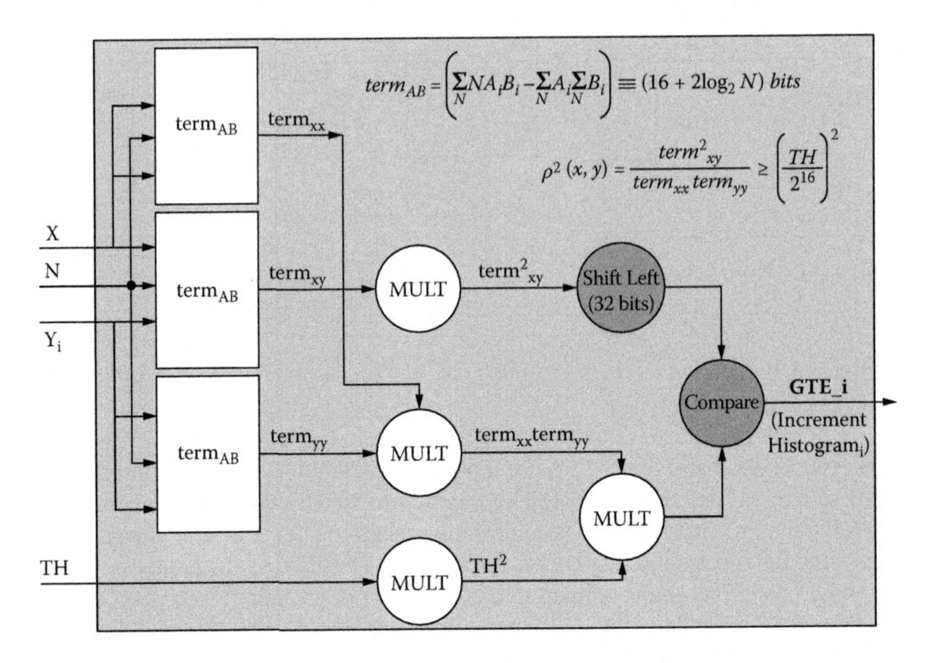

Figure 15.9 Correlator module.

(TH) has been interpreted with a 16-bit precision unsigned data type. The dimension of the hyperspectral image (N) has been interpreted with an 8-bit precision unsigned type, thus accommodating for the maximum possible number of hyperspectral bands, typically 220–240 bands.

Figure 15.10 shows the speedup of the SRC-6 implementation compared to a 1.8GHz Intel Xeon processor and it is crucial to note in Figure 15.10 that, because of the sequential manner by which the algorithm executes on traditional microprocessors, the execution time is proportional to the number of decomposition levels. In contrast to this, the corresponding time on SRC-6 is constant due to the fact that the algorithm has been fully pipelined.

15.5.2 Cloud Detection

The presence of cloud contamination can hinder the use of satellite data, and this requires a cloud detection process to mask out cloudy pixels from further processing. The Landsat 7 ETM+ (Enhanced Thematic Mapper) ACCA (Automatic Cloud Cover Assessment) algorithm is a compromise between the simplicity of earlier Landsat algorithms, e.g., ACCA for Landsat 4 and 5, and the complexity of later approaches such as the MODIS (Moderate Resolution Imaging Spectroradiometer) cloud mask [14].

The theory of Landsat 7 ETM+ ACCA algorithm is based on the observation that clouds are highly reflective and cold. The high reflectivity can be detected in the visible, near-, and mid-IR bands. The thermal properties of clouds can be detected in

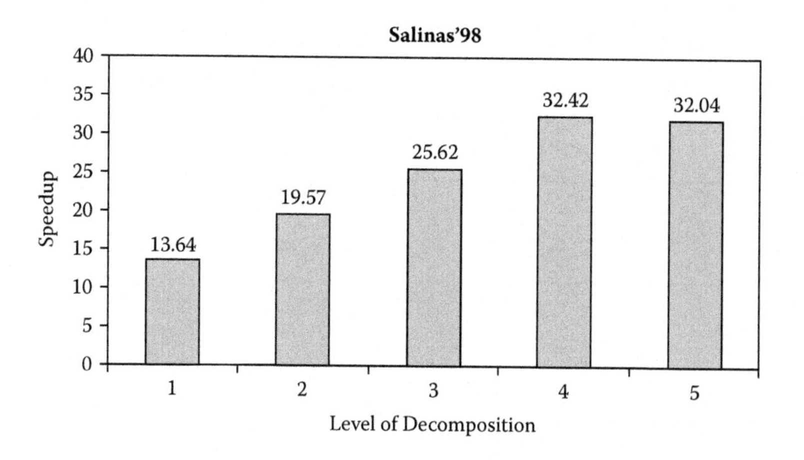

Figure 15.10 Speedup of wavelet-based hyperspectral dimension reduction algorithm.

the thermal IR band. The Landsat 7 ETM+ ACCA algorithm recognizes clouds by analyzing the scene twice.

The goal of Pass-One is to develop a reliable cloud signature for use in Pass-Two where the remaining clouds are identified. Omission errors, however, are expected. These errors create algorithm failure and must be minimized. Three categories result from Pass-One: clouds, non-clouds, and an ambiguous group that is revisited in Pass-Two. Williams et al. [15, 16] have used band mapping techniques to implement Landsat-based algorithms on MODIS data. The generalized and modified classification rules for Pass-One are shown in Figure 15.11.

Pass-Two resolves the detection ambiguity that resulted from Pass-One. Thermal properties of clouds identified during Pass-One are characterized and used to identify remaining cloud pixels. Band 6 statistical moments (mean, standard deviation, distribution skewness, kurtosis) are computed and new adaptive thresholds are determined accordingly. The 95th percentile, i.e., the smallest number that is greater than 95% of the numbers in the given set of pixels, becomes the new thermal threshold for Pass-Two.

After the two ACCA passes, a filter is applied to the cloud mask to fill in cloud holes. This filtering operation works by examining each non-cloud pixel in the mask. If 5 out of the 8 neighbors are clouds, then the pixel is reclassified as cloud. Cloud cover results when both Pass-One and Pass-Two are compared. Extreme differences are indicative of cloud signature corruption. When this occurs, Pass-Two results are ignored and all results are taken from Pass-One.

15.5.2.1 ACCA Hardware Architecture

The ACCA algorithm has been implemented targeting both conventional microprocessor (P) platforms and reconfigurable computing (RC) platforms. The μP

Classification	Rule
Snow	$\left(NSDI = \frac{B_2 - B_5}{B_2 + B_5} > 0.7\right)$ *AND* $(B_4 > 0.1)^A$
Desert	$\frac{B_4}{B_5} < 0.83^B$
NotCloud	$(B_3 < 0.08)$ *OR* $(B_6 > 300)$ *OR* (*Snow*)
Ambiguous	$\left(((1-B_5)\,B_6 > 225)\ OR\ \left(\frac{B_4}{B_3} > 2\right) OR \left(\frac{B_4}{B_2} > 2\right) OR\ (Desert)\right)\ AND\ (\sim NotCloud)$
ColdCloud	$((1-B_5)\,B_6 \geq 210)$ *AND* (~ *Ambiguous*) *AND* (~ *NotCloud*)
WarmCloud	$((1-B_5)\,B_6 < 210)$ *AND* (~ *Ambiguous*) *AND* (~ *NotCloud*)

Notes:

[A]The Band 4 brightness test, in the snow test, was added after observations that the NDSI (Normalized Difference Snow Index) algorithm applied to MODIS data incorrectly labeled many cloud pixels as snow.

[B]The desert detection threshold was lowered to 0.83, from the original ACCA value of 1.0, after it was observed that many cloud pixels were incorrectly classified as desert. The value of 0.83 was determined experimentally.

Figure 15.11 Generalized classification rules for Pass-One.

implementation has been performed using the C++ and MATLAB programs. The RC implementations have been performed using two designs, namely, full-precision fixed-point arithmetic and floating-point arithmetic.

Figure 15.12 shows the main functional/architectural units of the ACCA algorithm. As previously described, the ACCA algorithm handles the cloud population in each scene uniquely by examining the image data twice after a normalization

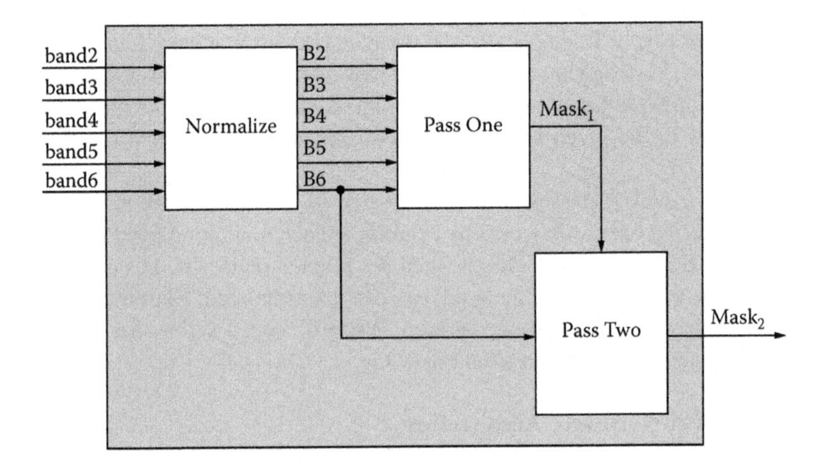

Figure 15.12 Top-level architecture of the ACCA algorithm.

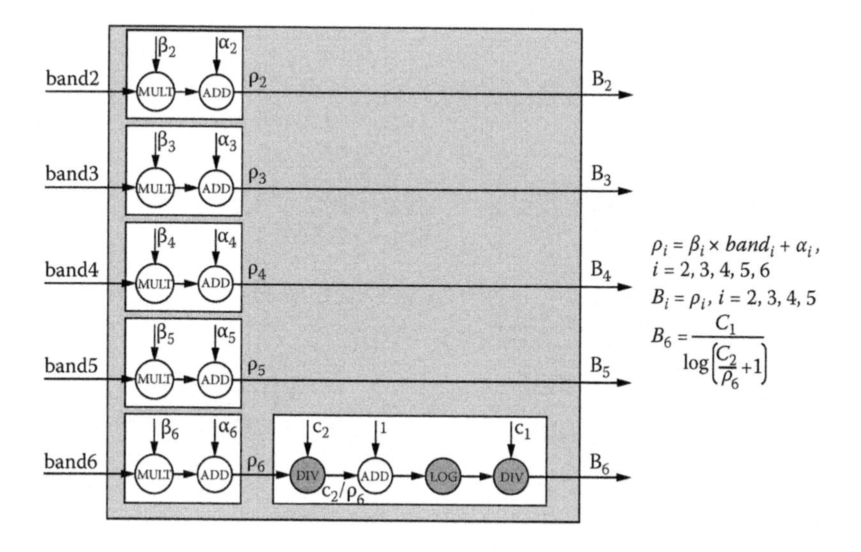

Figure 15.13 ACCA normalization module architecture: exact normalization operations.

step is performed on the raw data to compensate for temporal data characteristics. The first pass captures clouds using eight different filters. The goal of Pass-One is to develop a reliable cloud signature for Pass-Two. Pass-Two resolves the detection ambiguity that resulted from Pass-One where thermal properties of clouds identified during Pass-One are characterized and used to identify remaining cloud pixels.

15.5.2.2 Normalization module

ETM+ bands 2–5 are reflectance bands, while band 6 is a thermal band. The reflectance bands are normalized to correct for illumination (solar zenith) angle, yielding an estimated reflectance value. The thermal band is calibrated to an equivalent blackbody Brightness Temperature (BT). This normalization for the reflectance bands is a linear operation while it is non-linear for the thermal band; see Figure 15.13. In the onboard processing system, these operations are performed by the calibration stage [15]. Due to the high cost in terms of hardware resources required, a piecewise-linear approximation is used to implement the non-linear normalization function for the thermal band; see Figure 15.14.

15.5.2.3 Pass-One Module

The first pass of the ACCA algorithm is a cascading set of eight threshold-based filters. These filters are designed to classify each pixel into one of four classes: ColdCloud, WarmCloud, NotCloud, and Ambiguous. Pixels labeled Ambiguous are reprocessed in the second pass as previously discussed. Many of the tests in Pass-One are threshold

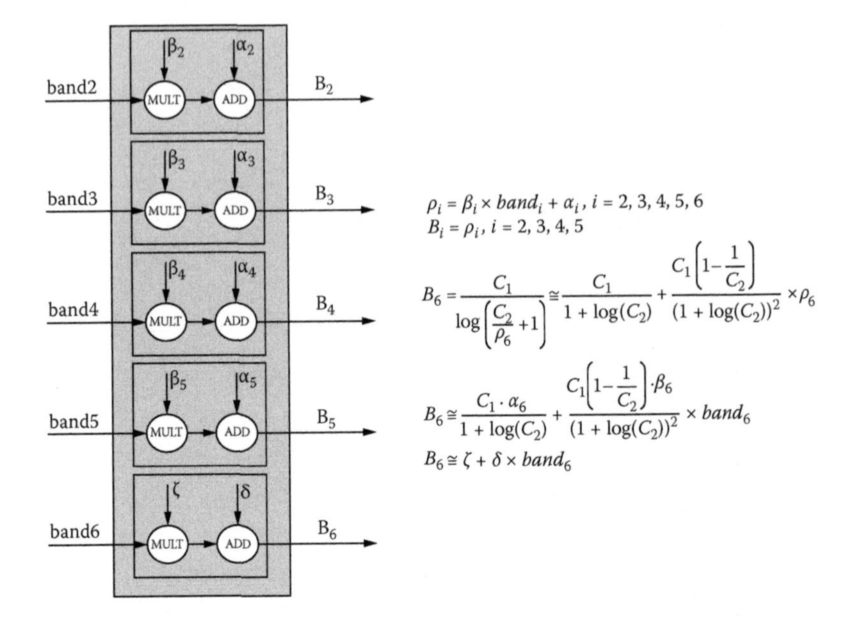

$$\rho_i = \beta_i \times band_i + \alpha_i,\ i = 2, 3, 4, 5, 6$$
$$B_i = \rho_i,\ i = 2, 3, 4, 5$$
$$B_6 = \frac{C_1}{\log\left(\frac{C_2}{\rho_6}+1\right)} \cong \frac{C_1}{1+\log(C_2)} + \frac{C_1\left(1-\frac{1}{C_2}\right)}{(1+\log(C_2))^2} \times \rho_6$$
$$B_6 \cong \frac{C_1 \cdot \alpha_6}{1+\log(C_2)} + \frac{C_1\left(1-\frac{1}{C_2}\right)\cdot \beta_6}{(1+\log(C_2))^2} \times band_6$$
$$B_6 \cong \zeta + \delta \times band_6$$

Figure 15.14 ACCA normalization module architecture: approximated normalization operations.

tests of ratio values, such as the snow test. It was more efficient, in terms of the required resources, to multiply one value by the threshold and compare it with the other value, instead of performing the division and then comparing it against the threshold. Figure 15.15 shows the equivalent hardware architecture of Pass-One.

The constraints to the design were the processing speed, as measured by throughput, and the hardware resources required for the design. The first constraint is approached through full-pipelining and superscaling of the design. The second constraint was approached through approximating the non-linear normalization step as mentioned earlier. Moreover, because many of the tests in Pass-One are threshold tests of ratio values such as the snow test (see Table 15.1), it was more efficient, in terms of the required resources, to multiply one value by the threshold and compare it with the other value, instead of performing the division and then comparing it against the threshold; see Figure 15.15.

15.5.2.4 Detection Accuracy

The criterion that defines the detection accuracy is based on the absolute error between the detected cloud mask and a reference mask produced by the software, C++/MATLAB, version of the ACCA algorithm. In addition, the goal of achieving high detection accuracy has been approached by minimizing the quantization errors

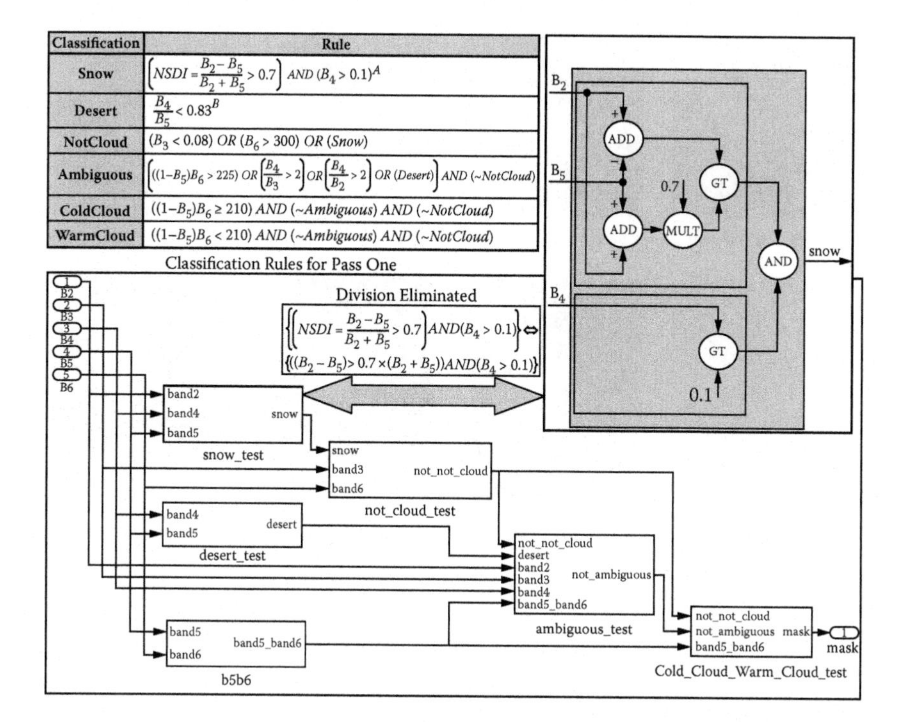

Figure 15.15 ACCA Pass-One architecture.

through full-precision fixed-point as well as floating-point arithmetic hardware implementations. Furthermore, saturated arithmetic has also been utilized in order to avoid overflow errors.

15.5.2.5 Experimental Results

The ACCA algorithm adapted for Landsat 7 ETM+ data has been implemented in both C++ and MATLAB, and Pass-One has been implemented and synthesized for the Xilinx XC2V6000 FPGA on SRC-6.

Figure 15.16 shows the image bands for a view taken for the city of Boston by Landsat 7 ETM+. Figure 15.16 also shows the reference mask produced by the software, C++/MATLAB, version of the ACCA algorithm as well as both hardware masks, i.e., fixed-point and floating-point. The results were obtained from a 2.8GHz Intel Xeon processor and from SRC-6. As shown in Figure 15.17, the linearization of the normalization step of the algorithm has introduced an error equal to 0.1028%. The hardware floating-point implementation has shown identical behavior to the software version of the algorithm. Figure 15.17 also shows that the hardware full-precision (23-bit) fixed-point version has improved the error due to quantization effects from 0.2676% to 0.1028%, which made it identical to the software/reference version.

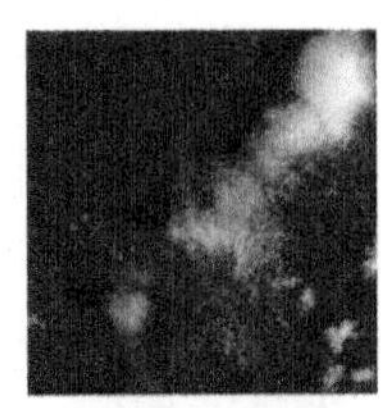
Band2 (Green Band)

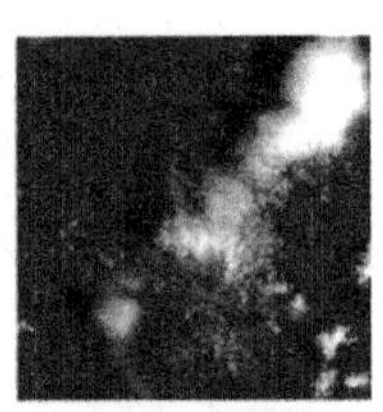
Band3 (Green Band)

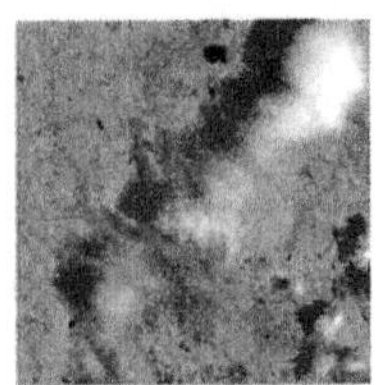
Band4 (Near -IR Band)

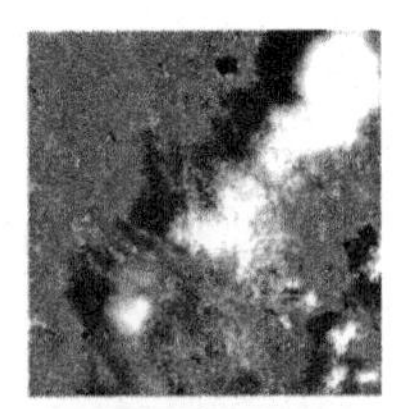
Band5 (Mid -IR Band)

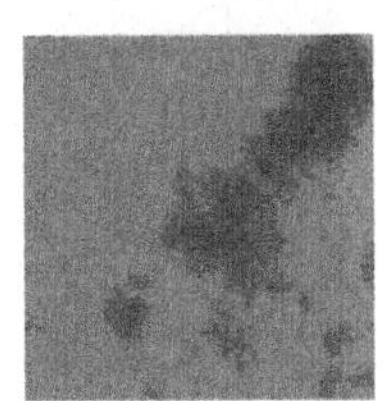
Band6 (Thermal IR Band)

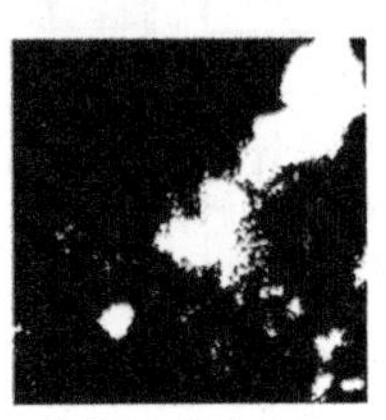
Software/Reference Mask

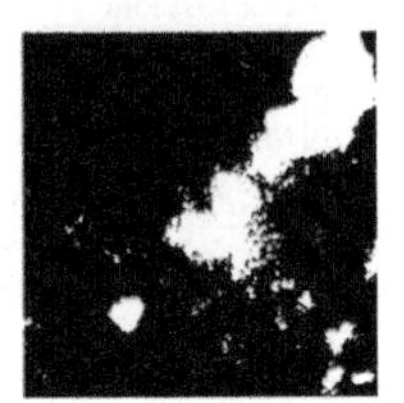
Hardware Floating-Point Mask (Approximate Normalization)

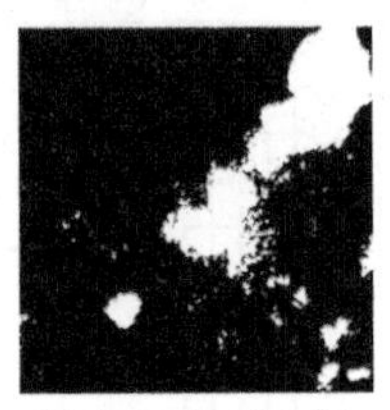
Hardware Fixed-Point Mask (Approximate Normalization)

Figure 15.16 Detection accuracy (based on the absolute error): image bands and cloud masks (software/reference mask, hardware masks).

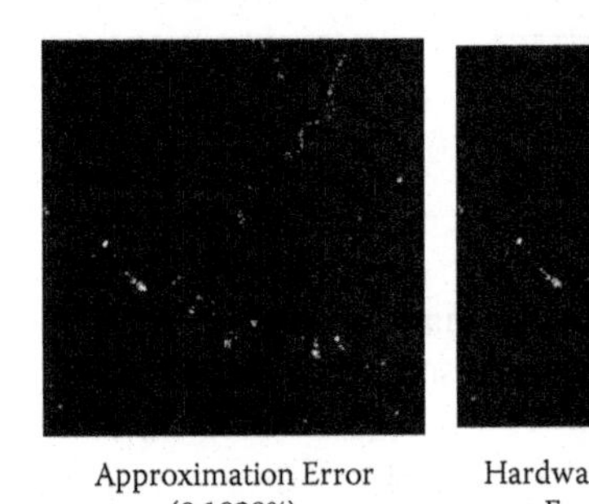
Approximation Error (0.1028%)

Hardware Floating-Point Error (0.1028%)

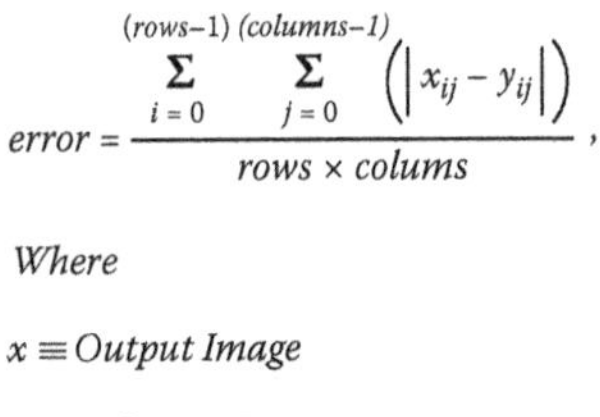

$$error = \frac{\sum_{i=0}^{(rows-1)} \sum_{j=0}^{(columns-1)} \left(\left| x_{ij} - y_{ij} \right| \right)}{rows \times colums},$$
Where
x ≡ Output Image
y ≡ Reference Image

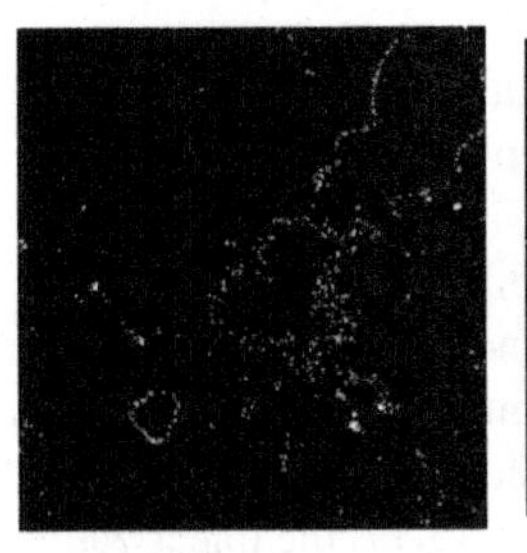
Hardware Fixed-Point (12-bit) Error (0.2676%)

Hardware Fixed-Point (23-bit) Error (0.1028%)

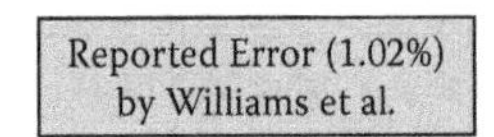

Figure 15.17 Detection accuracy (based on the absolute error): approximate normalization and quantization errors.

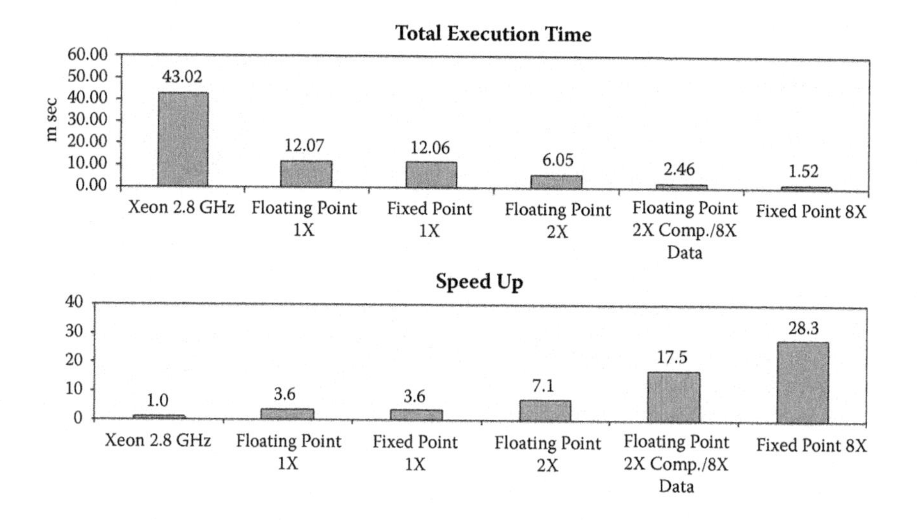

Figure 15.18 ACCA hardware-to-software performance.

The design was developed in VHDL, synthesized, placed and routed, and occupied approximately 7% of the available logic resources on the FPGA chip, Xilinx Virtex II-6000. This enabled the instantiation of eight concurrent processing engines of the design in the same chip, which increased the performance to eightfold of what was expected.

The maximum operational clock speed of the design is 100MHz, which resulted in 4000 Megapixels/sec (5 inputs × 8 engines × 100MHz) as the data input/consumption rate. Furthermore, the data output/production rate was 800 Megapixels/sec (1 output × 8 engines × 100MHz). The hardware implementations provided a higher performance (28 times faster) compared to the 2.8GHz Xeon implementation; see Figure 15.18. The superiority of RCs over traditional platforms for cloud detection is demonstrated through the performance plots shown in Figure 15.18.

15.6 Summary and Observations

Reconfigurable computing technology using SRAM-based FPGAs seems very promising technology to be used to implement image processing applications for remote sensing. Reconfigurable computing technology offers advantages such as high performance, low cost, and low power. In terms of flexibility, RC offers the ability to upgrade or repair spacecraft devices and/or change the system functionality at any time. FPGA chips are subject to some expected problems in space due to the high level of ionizing radiation to which these devices are exposed. This problem can be resolved by using radiation-hardened FPGAs.

References

[1] B. Neil and A. Dawood, Reconfigurable Computers in Space: Problems, Solutions and Future Directions, *MAPLD99: Military and Aerospace Applications of Programmable Logic Devices*, Laurel, Maryland, September, 1999.

[2] S. Montenegro, P. Behra, I. Rodionov, A. Rodionov and E. Fedounin, Hyperspectral Monitoring Data Processing, *4th IAA Symposium on Saml Satellites for Earth Observation*, 2003.

[3] M. A. Fischman, A. C. Berkun, F. T. Cheng, W. W. Chun, E. Im and R. Andraka, Design and Demonstration of an Advanced On-Board Processor for the Second-Generation Precipitation Radar, *IEEE Aerospace Conference*, vol. 2, pp. 1067–1075, 2003.

[4] K. Compton and S. Hauck, Reconfigurable Computing: A Survey of Systems and Software, *ACM Computing Surveys*, vol. 34, pp. 171–210, 2002.

[5] R. Tessier and W. Burleson, Reconfigurable Computing for Digital Signal Processing: A Survey, *Journal of VLSI Signal Processing Systems*, vol. 28, pp. 7–27, 2001.

[6] S. Brown and J. Rose, Architecture of FPGAs and CPLDs: A Tutorial, *IEEE Design and Test of Computers*, vol. 13, pp. 42–57, 1996.

[7] T. El-Ghazawi, K. Gaj, D. Buell and A. George, *Reconfigurable Supercomputing*, SuperComputing Tutorials (SC2005), Available online: http://hpcl.seas.gwu.edu/docs/sc2005_part1.pdf.

[8] A. DeHon and J. Wawrzynek, Reconfigurable Computing: What, Why, and Implications for Design Automation, *IEEE/ACM Design Automation Conference*, 1999.

[9] S. Hauck, The Roles of FPGAs in Reprogrammable Systems, *Proceedings of the IEEE*, vol. 86, pp. 615–638, 1998.

[10] *WildStarTM Reference Manual revision 3.0*, Annapolis Micro Systems Inc., 2000.

[11] J. T. Johnson, Rad Hard FPGAs, Available online: http://esl.eng.ohio-state.edu/ rstheory/iip/RadHardFPGA.doc.

[12] S. Kaewpijit, J. Le Moigne and T. El-Ghazawi, Automatic Reduction of Hyperspectral Imagery Using Wavelet Spectral Analysis, *IEEE Transactions on Geoscience and Remote Sensing*, vol. 41, pp. 863–871, 2003.

[13] E. El-Araby, T. El-Ghazawi, J. Le Moigne and K. Gaj, Wavelet Spectral Dimension Reduction of Hyperspectral Imagery on a Reconfigurable Computer, *Field Programmable Technology (FPT 2004)*, Brisbane, Australia, 2004.

[14] E. El-Araby, M. Taher, T. El-Ghazawi and J. Le Moigne, Prototyping Automatic Cloud Cover Assessment (ACCA) Algorithm for Remote Sensing On-Board Processing on a Reconfigurable Computer, *IEEE International Conference On Field-Programmable Technology (FPT 2005)*, 2005.

[15] J. A. Williams, A. S. Dawood and S. J. Visser, FPGA-based Cloud Detection for Real-Time Onboard Remote Sensing, *Proceedings of IEEE International Conference on Field-Programmable Technology (FPT 2002)*, pp. 110–116, 2002.

[16] J. A. Williams, A. S. Dawood and S. J. Visser, Real-Time Wildfire and Volcanic Plume Detection from Spaceborne Platforms with Reconfigurable Logic, *11th Australasian Remote Sensing and Photogrammetry Conference*, Brisbane, Australia, 2002.

[17] D. W. Scott, The Curse of Dimensionality and Dimension Reduction, in *Multivariate Density Estimation: Theory, Practice, and Visualization*, Chapter 7, John Wiley and Sons, pp. 195–217, 1992.

[18] M. Taher, E. El-Araby, T. El-Ghazawi and K. Gaj, Image Processing Library for Reconfigurable Computers, *ACM/SIGDA Thirteenth International Symposium on Field Programmable Gate Arrays (FPGA 2005)*, Monterey, California, 2005.

Chapter 16

FPGA Design for Real-Time Implementation of Constrained Energy Minimization for Hyperspectral Target Detection

Jianwei Wang,
University of Maryland, Baltimore County
Chein-I Chang,
University of Maryland, Baltimore County

Contents

The constrained energy minimization (CEM) has been widely used for hyperspectral detection and classification. The feasibility of implementing the CEM as a real-time processing algorithm in systolic arrays has also been demonstrated. The main challenge of realizing the CEM in hardware architecture is the computation of the inverse of the data correlation matrix performed in the CEM, which requires a complete set of data samples. In order to cope with this problem, the data correlation matrix must be calculated in a causal manner that only needs data samples up to the sample at the time it is processed. This chapter presents a Field Programmable Gate Arrays (FPGA) design of such a causal CEM. The main feature of the proposed FPGA design is to use the Coordinate Rotation Digital Computer (CORDIC) algorithm, which can convert a Givens rotation of a vector to a set of shift-add operations. As a result, the CORDIC algorithm can be easily implemented in hardware architectures, and therefore in FPGA. Since the computation of the inverse of the data correlation matrix involves a series of Givens rotations, the utility of the CORDIC algorithm allows the causal CEM to perform real-time processing in FPGA. In this chapter, an FPGA implementation of the causal CEM will be studied and its detailed architecture will be also described.

16.1 Introduction

The importance of real-time processing has been recently realized and recognized in many applications. In some applications, e.g., on-board spacecraft data processing system, it is very useful to have high levels of processing throughput. Specially, as the data rate generated by spacecraft instruments is increasing, onboard science data processing has been largely absent from remote sensing missions. Many advantages can result from real-time processing. One is the detection of moving targets. This is critical and crucial in battlefields when moving targets such as tanks or missile launching vehicles pose real threats to ground troops. Real-time data processing provides timely intelligence information that can help to reduce causality. Another is onboard data processing. For space-borne satellites, real-time data processing can significantly reduce mass storage of data volume. A third advantage is chip design. It can be implemented in parallel and reduce computation load. Furthermore, it can also reduce payload in aircrafts and satellites. Over the past years, many subpixel detection and mixed pixel algorithms have been developed and shown to be very versatile. However, their applicability to real-time processing problems is generally restricted by the very complex computational workloads.

In this chapter, we explore the feasibility of the Field Programmable Gate Arrays (FPGA) design for real-time implementation of a hyperspectral detection and classification algorithm, called constrained energy minimization (CEM) [1], which has shown promise in hyperspectral data exploitation. The issue of real-time processing for CEM was also studied in [2], where its systolic array implementation was developed. In recent years, rapid advances in VLSI technology have had a large impact on modern digital signal processing. Over the past thirty years, we have witnessed that the number of transistors per chip has doubled about once a year. Therefore, VLSI design of complex algorithms becomes more and more feasible. The major difficulty with implementing these algorithms in real time is the computation of the inverse of a matrix. Systolic arrays provide a possibility, but they require a series of Givens rotations to decompose a matrix into triangular matrices that can be implemented in real time. Unfortunately, such Givens rotations cannot be realized in hardware. In order to resolve this issue, the Givens rotations must be performed by operations such as adds, ORs, XORs, and shifts that can be realized in hardware architectures. In order to do so, we make use of the Coordinate Rotation Digital Computer (CORDIC) algorithm developed by Volder [3], which allows us to convert a Givens rotation to a series of shifts-adds operations. Using systolic arrays architecture in conjunction with the CORDIC algorithm, we can implement the computation of a matrix inverse in a set of shifts-adds operations. As a result, it makes the FPGA design of CEM possible. This chapter presents the detailed FPGA design layout for CEM.

16.2 Constrained Energy Minimization (CEM)

Assume that a remotely sensed image is a collection of image pixels denoted by $\{\mathbf{r}_1, \mathbf{r}_2, \cdots, \mathbf{r}_N\}$, where $\mathbf{r}_i = (r_{i1}, r_{i2}, \cdots, r_{iL})^T$ for $1 \leq i \leq N$ is an L-dimensional pixel vector, N is the total number of pixels in the image, and L is the total number of spectral channels. Suppose that $\mathbf{d} = (d_1, \cdots, d_L)^T$ is the desired target signature of interest in the image. The goal is to design a finite impulse response (FIR) linear filter specified by L filter coefficients $\{\mathbf{w}_1, \mathbf{w}_2, \cdots, \mathbf{w}_L\}$, denoted by an L-dimensional vector $w = (w_1, \cdots, w_L)^T$ that can be used to detect the signature $\mathbf{d}$ without knowing the image background. If we assume that y_i is the output of the designed FIR filter resulting from the input $\mathbf{r}_i$, then y_i can be expressed by

$$y_i = \sum_{l=1}^{L} w_l r_{il} = \mathbf{r}_i^T \mathbf{w} = \mathbf{w}^T \mathbf{r}_i \tag{16.1}$$

In order to detect the desired target signature d using the filter output y_i, the FIR filter must be constrained by the following equation:

$$\mathbf{d}^T \mathbf{w} = \sum_{l=1}^{L} d_l w_l = 1 \tag{16.2}$$

so that the $\mathbf{d}$ can pass through the filter while the output energies resulting from other signatures will be minimized. This problem is a well-known linearly constrained adaptive beamforming problem, called minimum variance distortionless response (MVDR), which can be cast as follows:

$$\min_{w}\{\mathbf{w}^T \mathbf{R}_{L\times L} \mathbf{w}\} \text{ subject to the constraint: } \mathbf{d}^T \mathbf{w} = 1 \tag{16.3}$$

where $\mathbf{R}_{L\times L} = (1/N)\left[\sum_{i=1}^{N} \mathbf{r}_i \mathbf{r}_i^T\right]$ is the autocorrelation sample matrix of the image and

$$\frac{1}{N}\left[\sum_{i=1}^{N} y_i^2\right] = \frac{1}{N}\left[\sum_{i=1}^{N} (\mathbf{r}_i^T \mathbf{w})^T (\mathbf{r}_i^T \mathbf{w})\right] = \mathbf{w}^T \left(\frac{1}{N}\left[\sum_{i=1}^{N} \mathbf{r}_i \mathbf{r}_i^T\right]\right) \mathbf{w} = \mathbf{w}^T \mathbf{R}_{L\times L} \mathbf{w} \tag{16.4}$$

The solution to Eq. 16.4 can be obtained by [1]:

$$\mathbf{w}^* = \frac{\mathbf{R}_{L\times L}^{-1} \mathbf{d}}{\mathbf{d}^T \mathbf{R}_{L\times L}^{-1} \mathbf{d}} \tag{16.5}$$

16.3 Real-Time Implementation of CEM

One of the significant advantages of the CEM is that the correlation matrix $\mathbf{R}_{L\times L}$ in the optimal weights specified by Eq. 16.5 can be decomposed into a product of a unitary matrix $\mathbf{Q}$ and an upper triangular matrix $\mathbf{R}$ by either the Givens rotations or the Householder transform. Such a decomposition is commonly referred to as QR-decomposition. Another advantage is that the CEM can be implemented in real time where the correlation matrix can be carried out either line-by-line or pixel-by-pixel from left to right and top to bottom. For illustrative purpose, we assume that the correlation matrix is performed line-by-line and at each line t a data matrix $\mathbf{X}_t = [\mathbf{r}_{t1}, \mathbf{r}_{t2}, \cdots, \mathbf{r}_{tN}]$ is formed up to this particular line. In this case, the $\mathbf{R}_{L\times L}$ in Eq. 16.3 is replaced by the data autocorrelation matrix of line t in the image, denoted by $\sum_t$:

$$\sum_t = \frac{1}{N}\left[\sum_{i=1}^{N} \mathbf{r}_{ti}\mathbf{r}_{ti}^T\right] = \frac{1}{N}\left[\mathbf{X}_t\mathbf{X}_t^T\right] \tag{16.6}$$

With QR-decomposition, $\mathbf{X}_t$ can be expressed by

$$\mathbf{X}_r = \mathbf{Q}_t\mathbf{R}_t \tag{16.7}$$

Here, $\mathbf{Q}_t$ is a unitary matrix with $\mathbf{Q}_t^{-1} = \mathbf{Q}_t^T$ and $\mathbf{R}_t =$ and $\mathbf{R} = \left[\begin{matrix}\mathbf{R}_t^{upper} \\ \mathbf{0}\end{matrix}\right]$ is not necessarily of full rank, where $\mathbf{0}$ is a zero vector and

$$\mathbf{R}_t^{upper} = \begin{bmatrix} * & * & \cdots & * \\ 0 & * & \cdots & * \\ \vdots & \ddots & \ddots & \vdots \\ 0 & \cdots & 0 & * \end{bmatrix} \tag{16.8}$$

is an upper triangular matrix, and $*$ in $\mathbf{R}_t^{upper}$ is a nonzero element. From Eq. 16.6, the inverse of $\sum_t$ can be computed as

$$\sum_t = N(\mathbf{X}\mathbf{X}^T) = N\left\{(\mathbf{R}_t^{upper})^{-1}[(\mathbf{R}_t^{upper})^T]^{-1}\right\} \tag{16.9}$$

where the unitary matrix $\mathbf{Q}_t$ is canceled out in Eq. 16.6 because $\mathbf{Q}_t^{-1} = \mathbf{Q}_t^T$. Substituting in Eq. 16.9 for $\mathbf{R}_{L\times L}^{-1}$ yields

$$w^* = \left\{(\mathbf{R}_t^{upper})^{-1}[(\mathbf{R}_t^{upper})^T]^{-1}\right\} \cdot \mathbf{d}(\mathbf{d}^T\left\{(\mathbf{R}_t^{upper})^{-1}[(\mathbf{R}_t^{upper})^T]^{-1}\right\}\mathbf{d})^{-1} \tag{16.10}$$

Since $\mathbf{R}_t^{upper}$ is an upper triangular matrix, so is $(\mathbf{R}_t^{upper})^{-1}$. Therefore, Eq. 16.10 does not require computation of $\mathbf{R}_{L\times L}^{-1}$. As a result, it can be implemented in real-time processing. In this chapter, a Givens rotation is used to perform QR-decomposition.

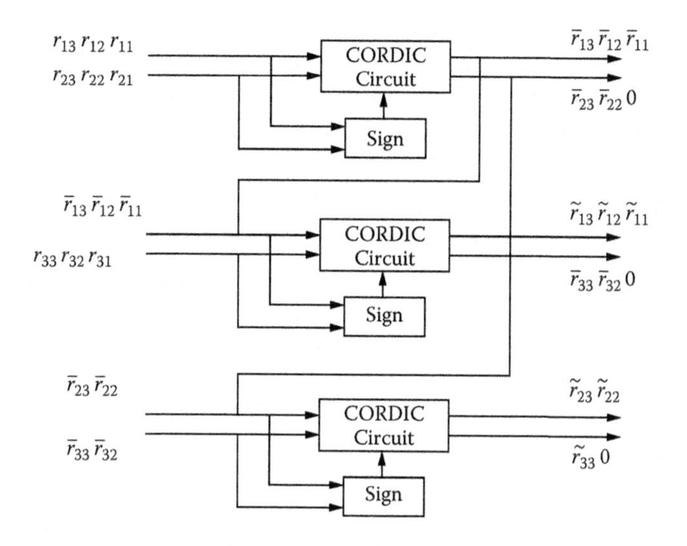

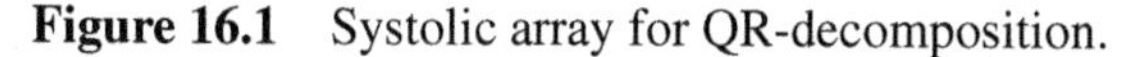

Figure 16.1 Systolic array for QR-decomposition.

16.3.1 Method 1: CEM Implementation

In Method 1, we decompose Eq. 16.5 into several components and each component is implemented by a separated hardware module:

$$w^* = \frac{\mathbf{R}_{L\times L}^{-1}\mathbf{d}}{\mathbf{d}^T\mathbf{R}_{L\times L}^{-1}\mathbf{d}} = \frac{(\mathbf{X}\mathbf{X}^T)^{-1}\mathbf{d}}{\mathbf{d}^T(\mathbf{X}\mathbf{X}^T)^{-1}\mathbf{d}} = \frac{(\mathbf{R}_t^{upper})^{-1}[(\mathbf{R}_t^{upper})^T]^{-1}\mathbf{d}}{\mathbf{d}^T(\mathbf{R}_t^{upper})^{-1}[(\mathbf{R}_t^{upper})^T]^{-1}]\mathbf{d}} \tag{16.11}$$

To implement Eq. 16.11, five modules are required:

- Array of CORDIC circuits shown in Figure 16.1, where the pixel stream is fed into the module and the upper triangular matrix $\mathbf{R}_t^{upper}$ is updated in realtime.
- Apply backsubstitution to obtain the inverse of $\mathbf{R}_t^{upper}$, called $inv\mathbf{R}$.
- Apply distributed arithmetic in order to calculate $c = [(\mathbf{R}_t^{upper})^T]^{-1}\mathbf{d} = inv\mathbf{R}^T * \mathbf{d}$.
- Compute $\mathbf{w} = inv\mathbf{R}^T * \mathbf{c}$.
- The filter output energy can be obtained by applying an FIR filter to the current input pixel streams.

The detailed implementation for each of the five modules is described as follows:

- *Generation of an upper triangular matrix.* A set of CORDIC circuits is applied to perform a Givens rotation, as shown in Figure 16.1. For demonstrative purpose, let us assume $L = 3$. At first, two pixel streams (row 1 and row 2) are fed into the CORDIC circuit, and as a result, the first zero, which will occupy the r_{21} position, is introduced by a Givens rotation. The first pair (r_{11}, r_{12}) is the leading pair, it decides the angle need to be rotated, and the other pairs are followers that will be rotated by the same angle accordingly. Then, the second zero, which will occupy the r_{31} position, is introduced by the second CORDIC circuit that operates on rows 1 and 2. The third zero, which will occupy the r_{32} position, is introduced by a CORDIC circuit that operates on rows 2 and 3. Finally, the output becomes an upper triangular matrix.

- *Backsubstitution*. Backsubstitution is applied to obtain the inverse of an upper triangular matrix $\mathbf{R}_t^{upper}$, $(\mathbf{R}_t^{upper})^{-1}$. For an illustrative purpose, we assume that the upper triangular matrix is $\mathbf{R}_t^{upper} = \begin{bmatrix} \hat{r}_{11} & \hat{r}_{12} & \hat{r}_{13} \\ 0 & \hat{r}_{22} & \hat{r}_{13} \\ 0 & 0 & \hat{r}_{13} \end{bmatrix}$. Its inverse is also an upper triangular matrix. If we let

$$(\mathbf{R}_t^{upper})^{-1} = \begin{bmatrix} \hat{r}_{11} & \hat{r}_{12} & \hat{r}_{13} \\ 0 & \hat{r}_{22} & \hat{r}_{13} \\ 0 & 0 & \hat{r}_{13} \end{bmatrix} = [\mathbf{a}_1 \mathbf{a}_2 \mathbf{a}_3] \tag{16.12}$$

then

$$\mathbf{R}_t^{upper} [\mathbf{a}_1 \mathbf{a}_2 \mathbf{a}_3] = \mathbf{I} = \begin{bmatrix} 1 & 0 & 0 \\ 0 & 1 & 0 \\ 0 & 0 & 1 \end{bmatrix} = [\mathbf{b}_1 \mathbf{b}_2 \mathbf{b}_3] \tag{16.13}$$

and it can be calculated via

$$\mathbf{R}_t^{upper} \mathbf{a}_k = \mathbf{b}_k \tag{16.14}$$

where the upper triangular matrix $\mathbf{R}_t^{upper}$ and the vector $\mathbf{b}_k$ are known and the vector $\mathbf{a}_k$ needs to be computed. Using Eq. 16.14, the backsubstitution can be described by the following recursive equation:

$$a_{kj} = \frac{b_{kj} - \sum_{j=i+1}^{n} r_{ij} a_{kj}}{r_{ii}} \tag{16.15}$$

with internal cells performing the summation given by

$$\sum_{j=i+1}^{n} r_{ij} a_{kj} \tag{16.16}$$

and boundary cells completing the calculation. The architecture of the backsubstitution array is shown in Figure 16.2, and the implementation of the boundary cell and internal cell is shown in Figure 16.3.

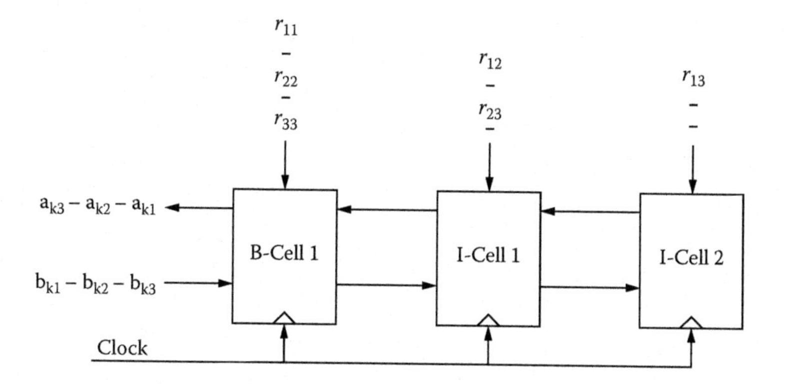

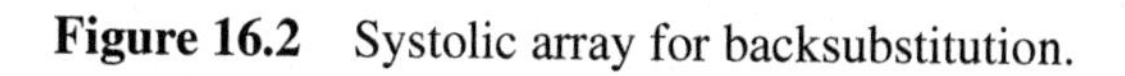
Figure 16.2 Systolic array for backsubstitution.

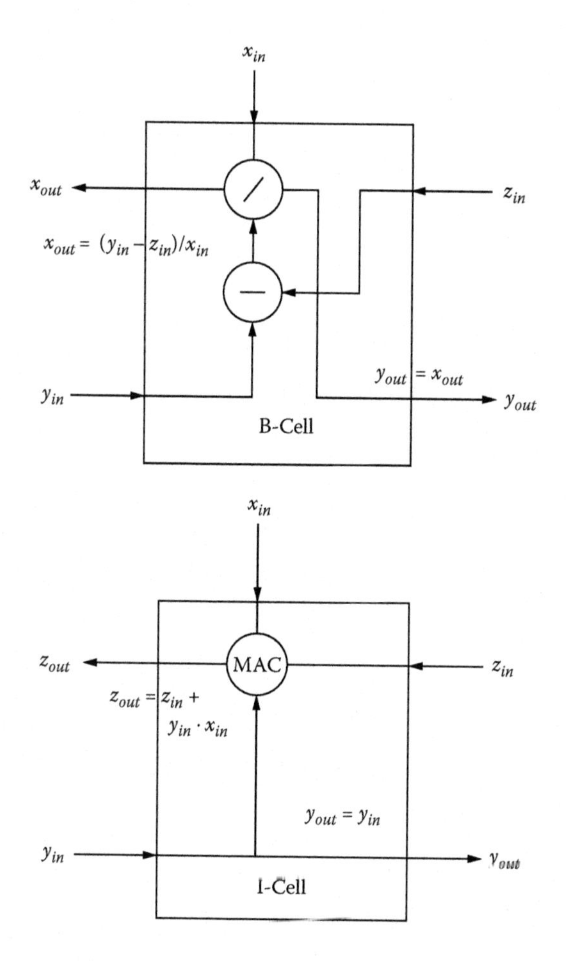

Figure 16.3 Boundary cell (left) and internal cell (right).

- *Distributed arithmetic architecture*. In this module, e.g., let $\mathbf{c} = (c_1, \cdots, c_L)^T$ and $inv\mathbf{R}^T = \begin{bmatrix} v_{11} & 0 & 0 & \cdots & 0 \\ v_{12} & v_{22} & 0 & \cdots & 0 \\ v_{13} & v_{23} & v_{33} & \cdots & 0 \\ \vdots & \vdots & \vdots & \ddots & \vdots \\ v_{1L} & v_{2L} & v_{3L} & \cdots & v_{LL} \end{bmatrix}$; then $\mathbf{c} = inv\mathbf{R}^T * \mathbf{d}$ can be represented as the inner product of two vectors as follows:

$$c_t = \sum_{l=1}^{L} v_{li} d_l \tag{16.17}$$

Because the desired signature d_k is known a priori, the term $v_{ki} d_k$ is simply a multiplication with a constant. In this case, a distributed arithmetic (DA) widely used in FPGA technology can be used to implement this module.

Assume that a B-bit system is used. The variable v_{li} can be represented by

$$v_{li} = \sum_{b=0}^{B-1} v(b) 2^b \text{ with } v(b) \in [0, 1] \tag{16.18}$$

where $v(b)$ denotes the bth bit of v_{li}. Therefore, the inner product in 16.17 can be represented as

$$c_l = \sum_{l=1}^{L} d_l \sum_{b=0}^{B-1} (v(b) 2^b) \tag{16.19}$$

Redistributing the order of summation results in

$$c_l = \sum_{b=0}^{B-1} 2^b \sum_{l=1}^{L} (d_l v_b(l)) = \sum_{b=0}^{B-1} 2^b \sum_{l=1}^{L} f(d_l, v_b(l)) \tag{16.20}$$

Implementation of the function $f(d_l, v_b(l))$ in 16.20 can be realized by a look-up table (LUT). That is, an LUT is preprogrammed to accept an L-bit input vector $\mathbf{v}_b = (v_b(0), v_b(1), \cdots, v_b(L-1))^T$ and output $f(d_l, v_b(l))$. The individual mapping $f(d_l, v_b(l))$ is weighted by an appropriate power-of-two factor and accumulated. The accumulation can be efficiently implemented using a shift-adder as shown in Figure 16.4. After L look-up cycles, the inner product c_i is computed.

- *Weight generation*. The inverse of $\mathbf{R}_t^{upper}$, $inv\mathbf{R}$, is an upper triangular matrix and it can be represented as

$$inv\mathbf{R}_t^{upper} = \begin{bmatrix} v_{11} & v_{12} & \cdots & v_{1L} \\ v_{21} & v_{22} & \cdots & v_{2L} \\ \vdots & \vdots & \ddots & \vdots \\ v_{L1} & v_{L2} & \cdots & v_{LL} \end{bmatrix} \tag{16.21}$$

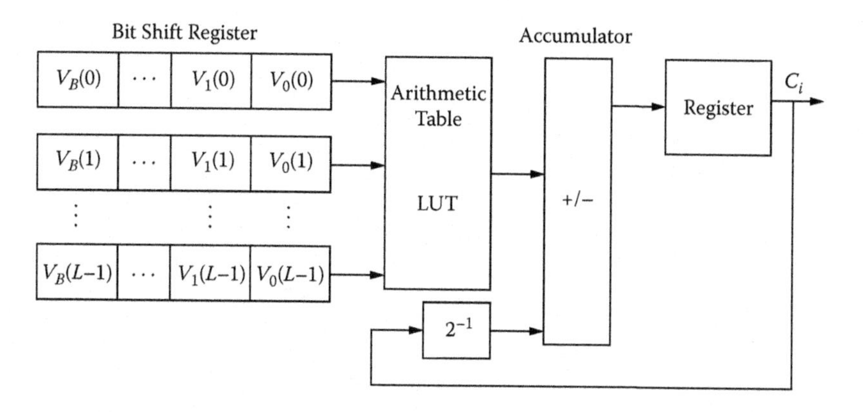

Figure 16.4 Shift-adder DA architecture.

where $v_{ij} = 0$ if $i > j$. The purpose of this module is to compute $\mathbf{w} = invR_t^{upper} * \mathbf{c}$.

Suppose $\mathbf{w} = (w_1, w_2, \cdots, w_L)^T$ and $\mathbf{c} = (c_1, c_2, \cdots, c_L)^T$. The kth element of $\mathbf{w}$ can be obtained by

$$w_l = \sum_{m=1}^{L} c_m v_{lm} \tag{16.22}$$

with its computation circuit given in Figure 16.5.

- *FIR filter*. Figure 16.6 delineates an FIR to produce estimated abundance fractions.

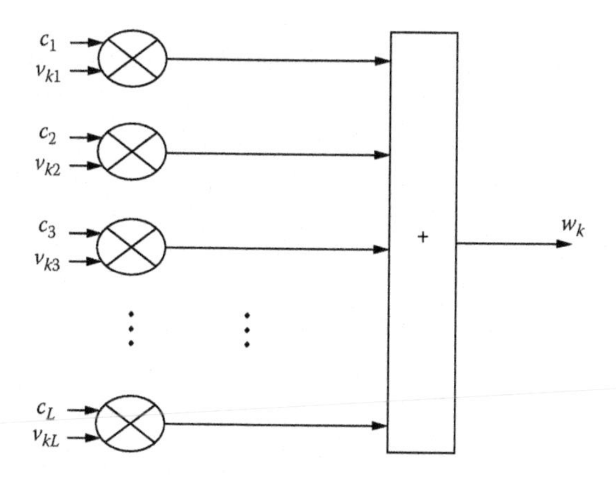

Figure 16.5 Computation of c_k.

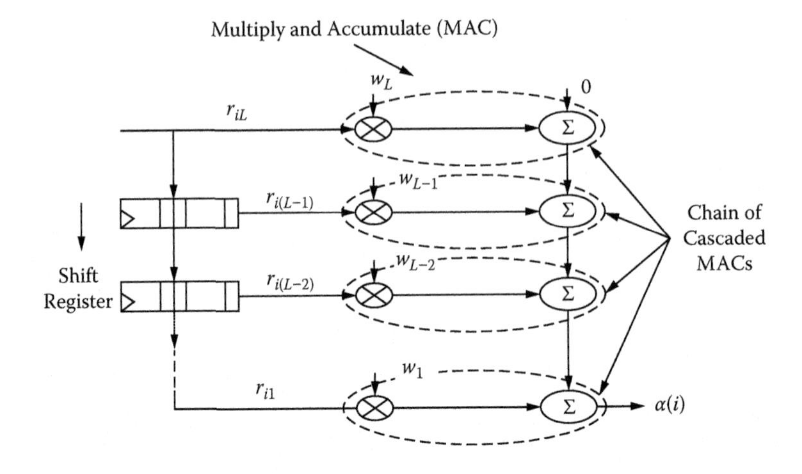

Figure 16.6 FIR filter for abundance estimation.

16.3.2 Method 2: CEM Implementation

In Method 2, the correlation matrix is calculated before QR-decomposition, as shown in Figure 16.6. Since $\mathbf{d}\mathbf{R}_{L\times L}^{-1}\mathbf{d}$ in Eq. 16.5 is a constant, $\mathbf{R}_{L\times L}^{-1}\mathbf{d}$ represents the relative strength of the detection power. Let $w = \mathbf{R}_{L\times L}^{-1}\mathbf{d}$, i.e., $\mathbf{R}_{L\times L}\mathbf{w} = \mathbf{d}$, where $\mathbf{R}_{L\times L}^{-1}\mathbf{d}$ can be implemented by a CORDIC module and followed by a backsubstitution module. As a result, four modules are required to implement a modified CEM detector:

- Auto-correlator.
- Apply CORDIC circuits to triangularize $[\mathbf{R}|\mathbf{d}]$.
- Apply backsubstitution to obtain $\mathbf{w}$.
- The filter output energy $\delta_{\text{CEM}}(\mathbf{r}) = \mathbf{w}^T\mathbf{r}$ can be obtained by applying an FIR filter to the current input pixel streams.
- Detailed implementation the for each of are four modules described as follows:
- *Auto-correlator.* This module generates the correlation matrix $\mathbf{R}_{L\times L}(i)$ with $\mathbf{R}_{L\times L}(i) = \mathbf{R}_{L\times L}(i-1) + \mathbf{r}_i\mathbf{r}_i^T$ and the update process is pixel-by-pixel. In other words, the correlation matrix is updated every time a new pixel arrives. For illustrative purposes, let us assume $L = 3$; then the new correlation matrix can be obtained by

$$\begin{bmatrix} R_{11} & R_{12} & R_{13} \\ R_{21} & R_{22} & R_{23} \\ R_{31} & R_{32} & R_{33} \end{bmatrix}(i) = \begin{bmatrix} R_{11} & R_{12} & R_{13} \\ R_{21} & R_{22} & R_{23} \\ R_{31} & R_{32} & R_{33} \end{bmatrix}(i-1) + \begin{bmatrix} r_{i1} \\ r_{i2} \\ r_{i3} \end{bmatrix}[r_{i1}\; r_{i2}\; r_{i3}] \tag{16.23}$$

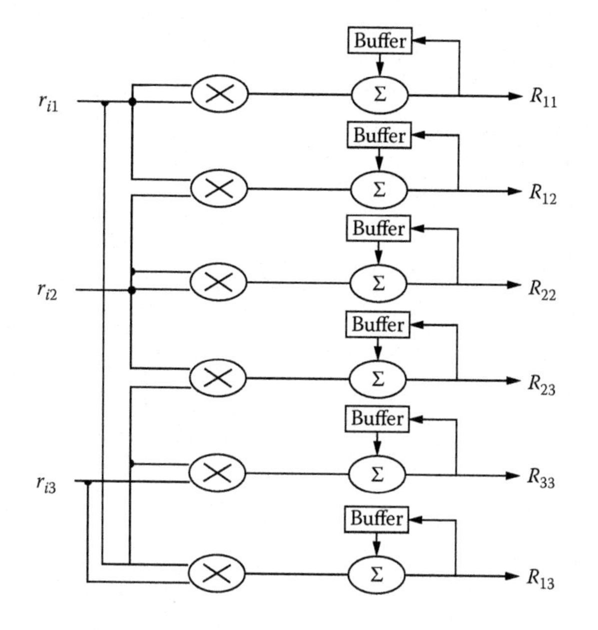

Figure 16.7 Block diagram of the auto-correlator.

The implementation of Eq. 16.23 is shown in Figure 16.7.

- *QR-decomposition of correlation matrix by CORDIC circuit.* The purpose of this module is to triangularize the matrix $[\mathbf{R}|\mathbf{d}]$, i.e., to convert the matrix $\begin{bmatrix} R_{11} & R_{12} & R_{13}|d_1 \\ R_{21} & R_{22} & R_{23}|d_2 \\ R_{31} & R_{32} & R_{33}|d_3 \end{bmatrix}$ to an upper triangular matrix given by $\begin{bmatrix} \hat{R}_{11} & \hat{R}_{12} & \hat{R}_{13}|\hat{d}_1 \\ 0 & \hat{R}_{22} & \hat{R}_{23}|\hat{d}_2 \\ 0 & 0 & \hat{R}_{33}|\hat{d}_3 \end{bmatrix}$ by a Givens rotation. Similar to Method 1, we implement the module by a set of CORDIC circuits. The difference is that we fed the pixel stream directly to the circuit for Method 1, while in this case we feed both the elements in the correlation matrix and the signatures of the desired target; for example, the first data stream feeding into the CORDIC circuit is R_{11} first, followed by R_{12}, and then R_{13}, the last one is d_1, as depicted in Figure 16.8.

  As soon as the first CORDIC in the chain has been fed by all the data it needs for row 1, it is free to accept and process data for the new row. The second CORDIC can begin its work on row 2 of the triangular matrix as soon as the first CORDIC has finished its work on the second element pair. The third CORDIC can begin its work as soon as the second CORDIC has finished its work, and so on.

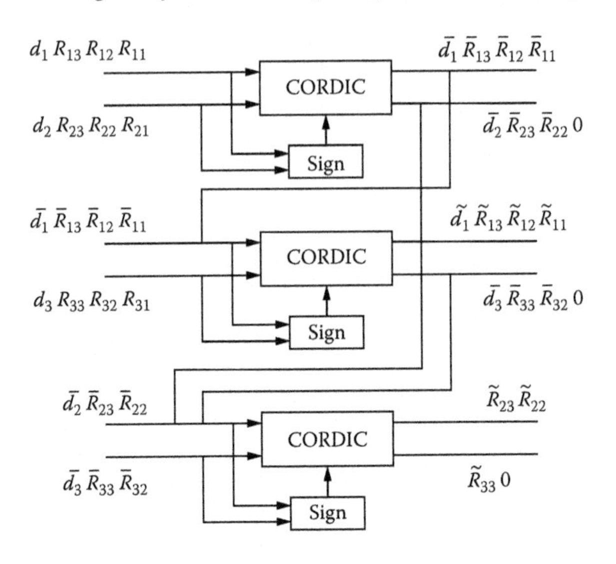

Figure 16.8 QR-decomposition by CORDIC circuit.

- *Backsubstitution*. Backsubstitution is applied to obtain the CEM filter coefficients $\mathbf{w}$ with $\mathbf{w} = \mathbf{R}_{L\times L}\mathbf{d} = \hat{\mathbf{R}}_{L\times L}^{-1}\hat{\mathbf{d}}$, where $\hat{\mathbf{R}}_{L\times L}$ is an upper triangular matrix. For an illustrative purpose, we assume that $\hat{\mathbf{R}}_{L\times L} = \begin{bmatrix} \hat{r}_{11} & \hat{r}_{12} & \hat{r}_{13} \\ 0 & \hat{r}_{22} & \hat{r}_{23} \\ 0 & 0 & \hat{r}_{33} \end{bmatrix}$. If we let the weight be $\mathbf{w} = (w_1, w_2, w_3)^T$, it can be calculated via

$$\hat{\mathbf{R}}_{L\times L}\mathbf{w} = \mathbf{d} \tag{16.24}$$

where the upper triangular matrix $\hat{\mathbf{R}}_{L\times L}$ and the vector $\mathbf{d}$ are known and the vector $\mathbf{w}$ needs to be computed. Using Eq. 16.24, the backsubstitution can be described by the following recursive equation:

$$w_i = \frac{(\hat{d}_i - \sum_{j=i+1}^{l} \hat{r}_{ij} - w_k}{r_{ij}} \tag{16.25}$$

with internal cells performing the summation given by

$$\sum_{j=i+1}^{n} r_{lj}a_{lj} \tag{16.26}$$

and boundary cells completing the calculation. The architecture of the backsubstitution array is shown in Figure 16.9, and the boundary cell and internal cell are depicted in Figure 16.10.

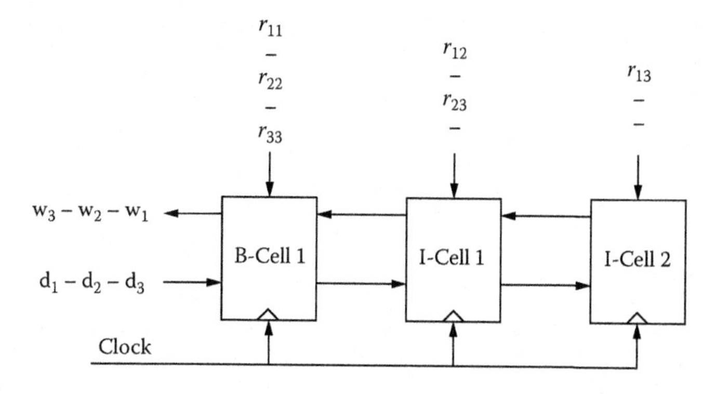

Figure 16.9 Systolic array for backsubstitution.

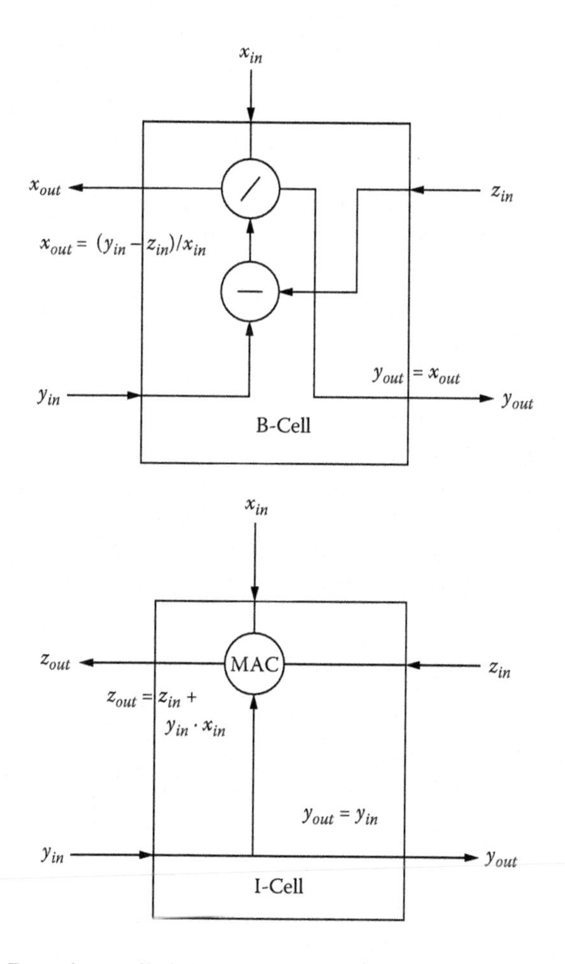

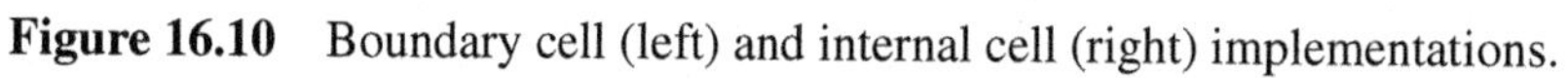

Figure 16.10 Boundary cell (left) and internal cell (right) implementations.

- *FIR filter*. In analogy with Method 1, the FIR to produce estimated abundance fractions can be also delineated by Figure 16.6.

16.4 Simulation Results

In this section, we present simulation results using the architecture used for Method 1. Here we use a Givens rotation to simulate our design. The code is written in C and uses the floating-point operation. As shown in Figure 16.11, the output of the CORDIC circuit is updated in real time, i.e., the upper triangular matrix receives more information as a new line is fed to the circuit.

Figure 16.11(a)–(h) are the results after the first 25, 50, 75, 100, 175, and 200 lines of pixels streams are fed into the CORDIC circuits. In the mean-time, the output of the weight-generated circuit also progressively updates its weight, as depicted in Figure 16.12. Additionally, the estimates of the FIR filter coefficients become more accurate and approximately approach the desired FIR coefficients as more pixels are fed to the circuit. As demonstrated in Figure 16.12, the weights generated after 100 lines are very close to the ones generated by using the complete set of pixel streams.

Finally, Figure 16.13 shows real-time detection results from the desired FIR filter, where Figure 16.13(a–h) are the detection results of the first 25 lines, the first 50 lines, and so on, until all the 200 lines are completed.

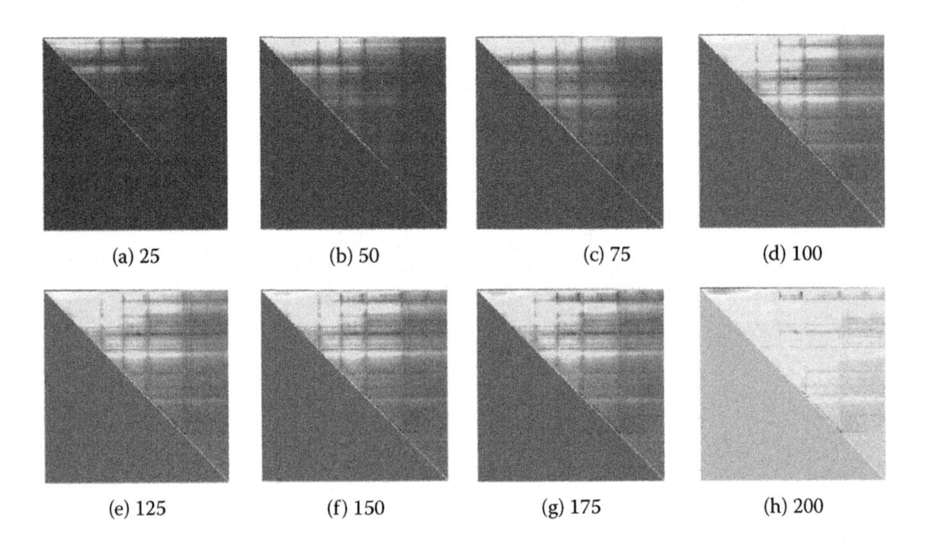

Figure 16.11 Real-time updated triangular matrix via CORDIC circuit.

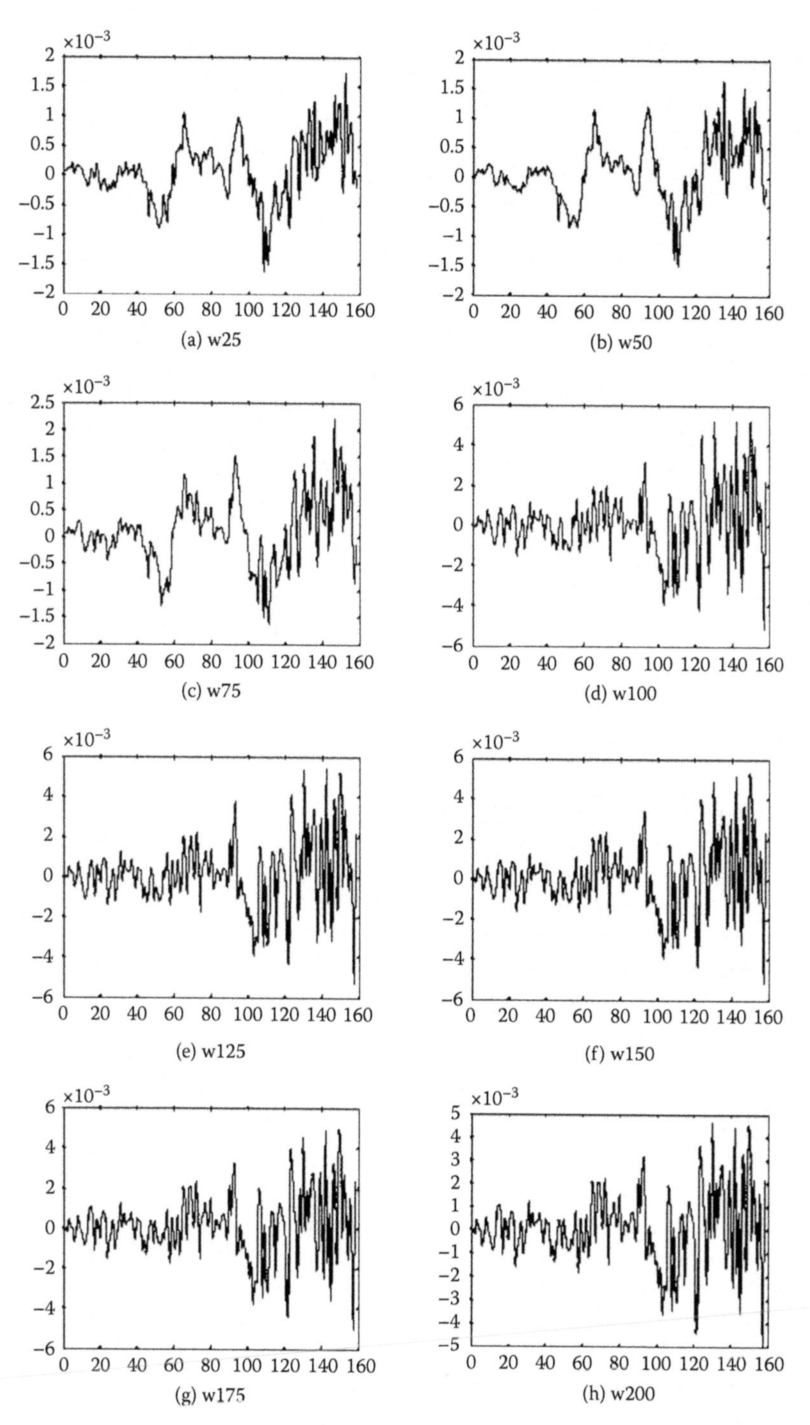

Figure 16.12 Real-time updated weights.

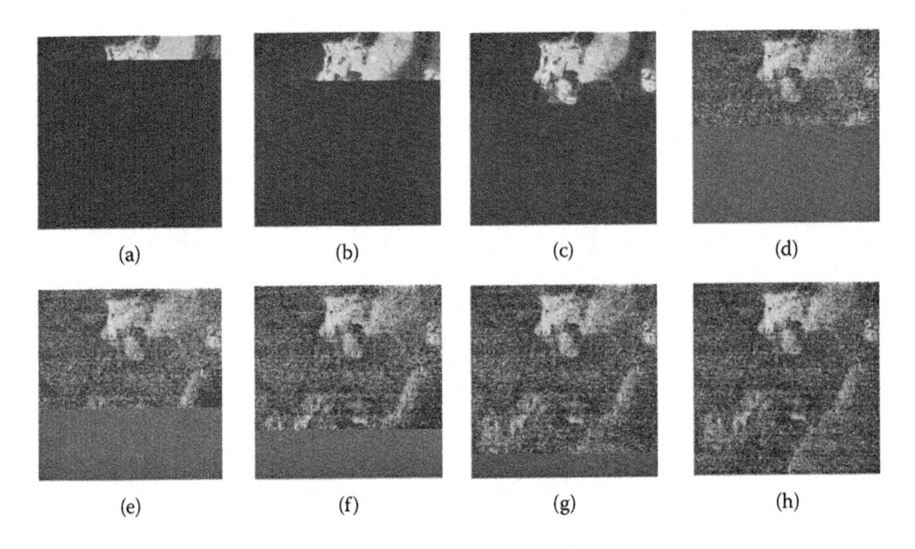

Figure 16.13 Real-time detection results.

16.5 Conclusions

In this chapter, the real-time implementation of the CEM is studied. Unlike the orthogonal subspace projection (OSP), the CEM involves the computation of the inverse of the sample correlation matrix instead of the computation of the pseudo-inverse of a matrix as did the OSP. In this case, the CORDIC algorithm is readily applicable. Depending upon how the input stream is computed, two methods are suggested in this chapter. Method 1 computes the input stream from image pixel vectors directly, while Method 2 computes the sample correlation matrix $\mathbf{R}$. As a result, five modules are proposed. The first module is to design an array of CORDIC circuits where the pixel stream is fed into the module and the upper triangular matrix $\mathbf{R}_t^{upper}$ is updated in realtime. This is followed by the second module, which applies backsubstitution to compute the inverse of $\mathbf{R}_t^{upper}$, $inv\mathbf{R}$. Then Module 3 uses a distributed arithmetic to calculate $\mathbf{c} = [(\mathbf{R}_t^{upper})^T]^{-1}\mathbf{d} = inv\mathbf{R}^T * \mathbf{d}$, where the $\mathbf{d}$ is the desired target signature. Next, Module 4 is developed to obtain the desired filter vector $\mathbf{w}$ by finding $\mathbf{w} = inv\mathbf{R} * \mathbf{c}$. Finally, Module 5 produces the results by applying an FIR filter to the current input pixel streams. Method 2 takes an alternative approach by first computing the auto-correlation matrix $\mathbf{R}$. Four modules are proposed for this method. Module 1 is the design of an auto-correlator that calculates the sample correlation matrix $\mathbf{R}$. It is then followed by Module 2, which uses the CORDIC circuits to triangularize $[\mathbf{R} \mid \mathbf{d}]$. Next, Module 3 applies the backsubstitution to obtain the desired filter vector $\mathbf{w}$. Finally, Module 4 produces the filter output energy $\delta_{CEM}(\mathbf{r}) = \mathbf{w}^T\mathbf{r}$ for target detection by applying an FIR filter to the current input pixel streams. Figure 16.14

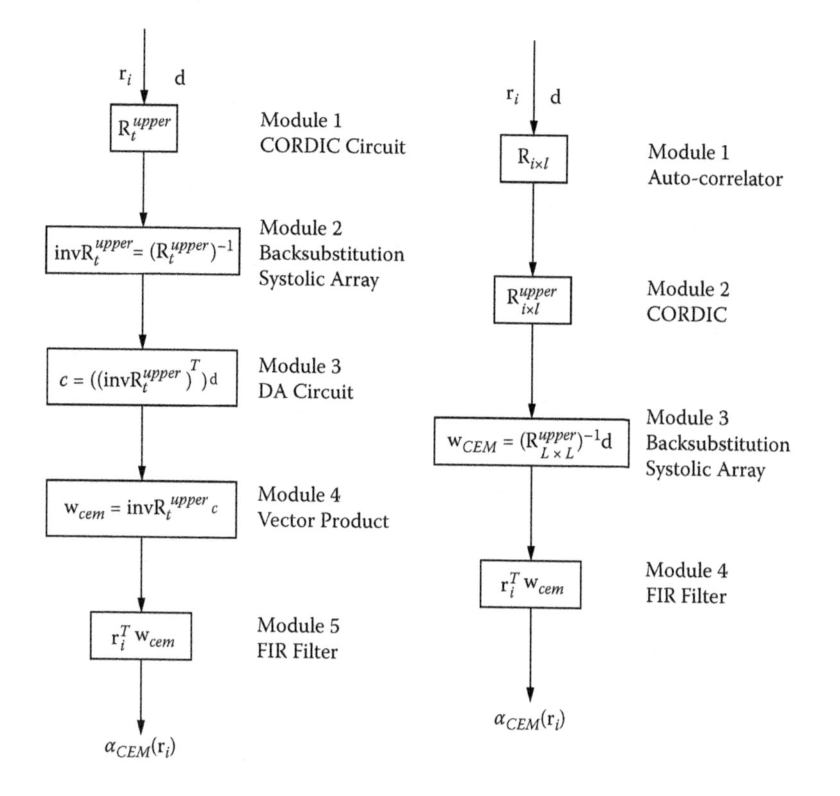

Figure 16.14 Block diagrams of Methods 1 (left) and 2 (right) to be used for FPGA designs of CEM.

depicts block diagrams of Methods 1 and 2 to be used for FPGA designs of the CEM, respectively.

References

[1] J.C. Harsanyi, *Detection and Classification of Subpixel Spectral Signatures in Hyperspectral Image Sequences*, Ph.D. dissertation, Department of Electrical Engineering, University of Maryland, Baltimore County, Baltimore, MD. 1993.

[2] C.-I Chang, H. Ren and S.S. Chiang, Real-time processing algorithms for target detection and classification in hyperspectral imagery, *IEEE Transactions on Geoscience and Remote Sensing*, vol. 39, no. 4, pp. 760–768, April 2001.

[3] J.E. Volder, The CORDIC trigonometric computing technique, *IEEE Transactions on Electronic Computing*, pp. 330–334, 1959.

Chapter 17

Real-Time Online Processing of Hyperspectral Imagery for Target Detection and Discrimination

Qian Du,
Missisipi State University

Contents

Hyperspectral imaging is a new technology in remote sensing. It acquires hundreds of images in very narrow spectral bands (normally 10nm wide) for the same area on the Earth. Because of higher spectral resolutions and the resultant contiguous spectral signatures, hyperspectral image data are capable of providing more accurate identification of surface materials than multispectral data, and are particularly useful in national defense related applications. The major challenge of hyperspectral imaging is how to take full advantage of the plenty spectral information while efficiently handling the data with vast volume.

In some cases, such as national disaster assessment, law enforcement activities, and military applications, real-time data processing is inevitable to quickly process data and provide the information for immediate response. In this chapter, we present a real-time online processing technique using hyperspectral imagery for the purpose of target detection and discrimination. This technique is developed for our proposed algorithm, called the constrained linear discriminant analysis (CLDA) approach. However, it is applicable to quite a few target detection algorithms employing matched filters. The implementation scheme is also developed for different remote sensing data formats, such as band interleaved by pixel (BIP), band interleaved by line (BIL), and band sequential (BSQ).

17.1 Introduction

We have developed the constrained linear discriminant analysis (CLDA) algorithm for hyperspectral image classification [1, 2]. In CLDA, the original high-dimensional data are projected onto a low-dimensional space as done by Fisher's LDA, but different classes are forced to be along different directions in this low-dimensional space. Thus all classes are expected to be better separated and the classification is achieved simultaneously with the CLDA transform. The transformation matrix in CLDA maximizes the ratio of interclass distance to intraclass distance while satisfying the constraint that the means of different classes are aligned with different directions, which can be constructed by using an orthogonal subspace projection (OSP) method [3] coupled with a data whitening process. The experimental results in [1],[2] demonstrated that the CLDA algorithm could provide more accurate classification results than other popular methods in hyperspectral image processing, such as the OSP classifier [3] and the constrained energy minimization (CEM) operator [4]. It is particularly useful to detect and discriminate small man-made targets with similar spectral signatures.

Assume that there are c classes and the k-th class contains N_k patterns. Let $N = N_1 + N_2 + \cdots N_c$ be the number of pixels. The j-th pattern in the k-th class, denoted by $\mathbf{x}_j^k = [x_{1j}^k, x_{2j}^k, \cdots, x_{Lj}^k]^T$, is an L-dimensional pixel vector (L is the number of spectral bands, i.e., data dimensionality). Let $\mu_k = \frac{1}{N_k}\sum_{j=1}^{N_k} x_j^k$ be the mean of the k-th class. Define $J(F)$ to be the ratio of the interclass distance to the intraclass distance after a linear transformation F, which is given by

$$J(F) = \frac{\frac{2}{c(c-1)}\sum_{i=1}^{c-1}\sum_{j=i+1}^{c}\|F(\mu_i) - F(\mu_j)\|^2}{\frac{1}{CN}\sum_{k=1}^{c}[\sum_{j=1}^{N_k}\|F(\mathbf{x}_j^{N_k}) - F(\mu_k)\|^2]} \tag{17.1}$$

and

$$F(\mathbf{x}) = (\mathbf{W}_{L\times c})^T; \mathbf{x} = [\mathbf{w}_1, \mathbf{w}_2, \cdots, \mathbf{w}_c]^T\mathbf{x} \tag{17.2}$$

The optimal linear transformation $F^*$is the one that maximizes $J(F)$ subject to $\mathbf{t}_k = F(\mu_k)$ for all k, where $\mathbf{t}_k = (0\cdots 01\cdots 0)^T$ is a $c \times 1$ unit column vector with one in the k-th component and zeros elsewhere. F^* can be determined by

$$\mathbf{w}_i^* = \hat{\mu}_i{}^T \mathbf{P}_{\hat{\mathbf{U}}_i}^{\perp} \tag{17.3}$$

where

$$\mathbf{P}_{\hat{\mathbf{U}}_i}^{\perp} = \mathbf{I} - \hat{\mathbf{U}}_i(\hat{\mathbf{U}}_i^T\hat{\mathbf{U}}_i)^{-1}\hat{\mathbf{U}}_i^T \tag{17.4}$$

with $\hat{\mathbf{U}}_i = [\hat{\mu}_1 \cdots \hat{\mu}_j \cdots \hat{\mu}_c]_{j \neq i}$ and $\mathbf{I}$ the identity matrix. The 'hat' operator specifies the whitened data, i.e., $\hat{\mathbf{x}} = \mathbf{P}_w^T\mathbf{x}$, where $\mathbf{P}_w$ is the data whitening operator.

Let $\mathbf{S}$ denote the entire class signature matrix, i.e., c class means. It was proved in [2] that the CLDA-based classifier using Eqs. (17.4)–(17.5) can be equivalently expressed as

$$\mathbf{P}_k^T = [0 \cdots 010 \cdots 0]\left(\mathbf{S}^T {\sum}^{-1} \mathbf{S}\right)^{-1} \mathbf{S}^T {\sum}^{-1} \tag{17.5}$$

for classifying the k-th class in $\mathbf{S}$, where $\sum$ is the sample covariance matrix.

17.2 Real-Time Implementation

In our research, we assume that an image is acquired from left to right and from top to bottom. Three real-time processing fashions will be discussed to fit the three remote sensing data formats: pixel-by-pixel processing for BIP formats, line-by-line processing for BIL formats, and band-by-band processing for BSQ formants. In the pixel-by-pixel fashion, a pixel vector is processed right after it is received and the analysis result is generated within an acceptable delay; in the line-by-line fashion, a line of pixel vectors is processed after the entire line is received; in the band-by-band fashion, a band is processed after it is received.

In order to implement the CLDA algorithm in real time, Eq. (17.6) is used. The major advantage of using Eq. (17.6) instead of Eqs. (17.4) and (17.5) is the simplicity of real-time implementation since the data whitening process is avoided. So the key becomes the adaptation of $\sum^{-1}$, the inverse sample covariance matrix. In other words, $\sum^{-1}$ at time t can be quickly calculated by updating the previous $\sum^{-1}$ at $t-1$ using the data received at time t, without recalculating the $\sum$ and $\sum^{-1}$ completely. As a result, the intermediate data analysis result (e.g., target detection) is available in support of decision-making even when the entire data set is not received; and when the entire data set is received, the final data analysis result is completed (within a reasonable delay).

17.2.1 BIP Format

This format is easy to handle because a pixel vector of size $L \times 1$ is received continuously. It fits well a spectral-analysis based algorithm, such as CLDA.

Let the sample correlation matrix $\mathbf{R}$ be defined as $\mathbf{R} = \frac{1}{N}\sum_{i=1}^{N} \mathbf{x}_i \cdot \mathbf{x}_i^T$, which can be related to $\sum$ and sample mean μ by

$$\sum = \mathbf{R} - \mu \cdot \mu^T \tag{17.6}$$

Using the data matrix $\mathbf{X}$, Eq. (17.7) can be written as $N \cdot \sum = \mathbf{X} \cdot \mathbf{X}^T - N \cdot \mu \cdot \mu^T$. If $\tilde{\sum}$ denotes $N \cdot \sum$, $\tilde{\mathbf{R}}$ denotes $N \cdot R$, and $\tilde{\mu}$ denotes $N \cdot \tilde{\mu}$, then

$$\tilde{\sum} = \tilde{\mathbf{R}} - \frac{1}{N_t} \cdot \tilde{\mu} \cdot \tilde{\mu}^T \tag{17.7}$$

Suppose that at time t we receive the pixel vector $\mathbf{x}_t$. The data matrix $\mathbf{X}_t$ including all the pixels received up to time t is $\mathbf{X}_t = [\mathbf{x}_1, \mathbf{x}_2, \cdots, \mathbf{x}_t]$ with N_t pixel vectors. The sample mean, sample correlation, and covariance matrices at time t are denoted as μ_t, $\mathbf{R}_t$, and $\sum_t$, respectively. Then Eq. (17.8) becomes

$$\tilde{\sum}_t = \tilde{\mathbf{R}}_t - \frac{1}{N_t} \cdot \tilde{\mu}_t \cdot \tilde{\mu}_t^T \tag{17.8}$$

The following Woodbury's formula can be used to update $\tilde{\sum}_t^{-1}$:

$$(\mathbf{A} + \mathbf{BCD})^{-1} = \mathbf{A}^{-1} - \mathbf{A}^{-1}\mathbf{B}(\mathbf{C}^{-1} + \mathbf{DA}^{-1}\mathbf{B})^{-1}\mathbf{DA}^{-1} \tag{17.9}$$

where $\mathbf{A}$ and $\mathbf{C}$ are two positive-definite matrices, and the sizes of matrices $\mathbf{A}$, $\mathbf{B}$, $\mathbf{C}$, and $\mathbf{D}$ allow the operation $(\mathbf{A} + \mathbf{BCD})$. It should be noted that Eq. (17.10) is for the most general case. Actually, $\mathbf{A}$, $\mathbf{B}$, $\mathbf{C}$, and $\mathbf{D}$ can be reduced to vector or scalar as long as Eq. (17.10) is applicable. Comparing Eq. (17.9) with Eq. (17.10), $\mathbf{A} = \tilde{\mathbf{R}}_t$, $B = \tilde{\mu}_t$, $\mathbf{C} = -\frac{1}{N_t}$, $D = \tilde{\mu}_t^T$, $\tilde{\sum}_t^{-1}$ can be calculated using the variables at time $(t-1)$ as

$$\tilde{\sum}_t^{-1} = \tilde{\mathbf{R}}_t^{-1} + \tilde{\mathbf{R}}_t^{-1}\tilde{\mathbf{u}}_t(N_t - \tilde{\mathbf{u}}_t^T\tilde{\mathbf{R}}_t^{-1}\tilde{\mathbf{u}}_t)^{-1}\tilde{\mathbf{u}}_t^T\tilde{\mathbf{R}}_t^{-1} \tag{17.10}$$

The $\tilde{\mu}_t$ can be updated by

$$\tilde{\mu}_t = \tilde{\mu}_{t-1} + \mathbf{x}_t \tag{17.11}$$

Since $\tilde{\mathbf{R}}_t$ and $\tilde{\mathbf{R}}_{t-1}$ can be related as

$$\tilde{\mathbf{R}}_t = \tilde{\mathbf{R}}_{t-1} + \mathbf{x}_t \cdot \mathbf{x}_t^T \tag{17.12}$$

$\tilde{\mathbf{R}}_t^{-1}$ in Eq. (17.12) can be updated by using the Woodbury's formula again:

$$\tilde{\mathbf{R}}_t^{-1} = \tilde{\mathbf{R}}_{t-1}^{-1} - \tilde{\mathbf{R}}_{t-1}^{-1}\mathbf{x}_t(1 + \mathbf{x}_t^T\tilde{\mathbf{R}}_{t-1}^{-1}\mathbf{x}_t)^{-1}\mathbf{x}_t^T\tilde{\mathbf{R}}_{t-1}^{-1} \tag{17.13}$$

Note that $(1 + \mathbf{x}_t^T\tilde{\mathbf{R}}_{t-1}^{-1}\mathbf{x}_t)$ in Eq. (17.14) and $(N_t - \tilde{\mathbf{u}}_t^T\tilde{\mathbf{R}}_t^{-1}\tilde{\mathbf{u}}_t)$ in Eq. (17.11) are scalars. This means no matrix inversion is involved in each adaptation.

In summary, the real-time CLDA algorithm includes the following steps:

- Use Eq. (17.14) to update the inverse sample correlation matrix $\tilde{\mathbf{R}}_t^{-1}$ at time t.
- Use Eq. (17.12) to update the sample mean μ_{t+1} at time $t+1$.
- Use Eq. (17.11) to update the inverse sample covariance matrix $\tilde{\sum}_{t+1}^{-1}$ at time $t+1$.
- Use Eq. (17.6) to generate the CLDA result.

17.2.2 BIL Format

If the data are in BIL format, we can simply wait for all the pixels in a line to be received. Let M be the total number of pixels in each line. M pixel vectors can be constructed by sorting the received data. Assume the data processing is carried out line-by-line from left to right and top to bottom in an image, the line received at time t forms a data matrix $\mathbf{Y}_t = [\mathbf{x}_{t1}\mathbf{x}_{t2}\cdots\mathbf{x}_{tM}]$. Assume that the number of lines received up to time t is K_t, then Eq. (17.10) remains almost the same as

$$\tilde{\sum}_t^{-1} = \tilde{\mathbf{R}}_{t-1}^{-1} - \tilde{\mathbf{R}}_{t-1}^{-1}\tilde{\mathbf{u}}_t(K_t M - \tilde{\mathbf{u}}_t^T\tilde{\mathbf{R}}_t^{-1}\tilde{\mathbf{u}}_t)^{-1}\tilde{\mathbf{u}}_t^T\tilde{\mathbf{R}}_t^{-1} \tag{17.14}$$

Eq. (17.11) becomes

$$\tilde{\mu}_t = \tilde{\mu}_{t-1} + \sum_{i=1}^{M}\mathbf{x}_{ti} \tag{17.15}$$

and Eq. (17.12) becomes

$$\tilde{\mathbf{R}}_t^{-1} = \tilde{\mathbf{R}}_{t-1}^{-1} - \tilde{\mathbf{R}}_{t-1}^{-1}\mathbf{Y}_t(\mathbf{I}_{M\times M} + \mathbf{Y}_t^T\tilde{\mathbf{R}}_{t-1}^{-1}\mathbf{Y}_t)^{-1}\mathbf{Y}_t^T\tilde{\mathbf{R}}_{t-1}^{-1} \tag{17.16}$$

where $\mathbf{I}_{M\times M}$ is an $M \times M$ identity matrix. Note that $\left(\mathbf{I}_{M\times M} + \mathbf{Y}_t^T\tilde{\mathbf{R}}_{t-1}^{-1}\mathbf{Y}_t\right)$ in Eq. (17.16) is a matrix. This means the matrix inversion is involved in each adaptation.

17.2.3 BSQ Format

If the data format is BSQ, the sample covariance matrix $\sum$ and its inverse $\sum^{-1}$ have to be updated in a different way, because no single completed pixel vector is available until all of the data are received.

Let $\sum_1$ denote the covariance matrix when Band 1 is received, which actually is a scalar, calculated by the average of pixel squared values in Band 1. Then $\sum_1$ can be related to $\sum_2$ as $\sum_2 = \begin{bmatrix} \sum_1 & \sum_{12} \\ \sum_{21} & \sum_{22} \end{bmatrix}$, where $\sum_{22}$ is the average of pixel squared values in Band 2, $\sum_{12} = \sum_{21}$ is the average of the products of corresponding pixel values in Band 1 and 2. Therefore, $\sum_t$ can be related to $\sum_{t-1}$ as

$$\sum_t = \begin{bmatrix} \sum_{t-1} & \sum_{t-1,t} \\ \sum_{t-1,t}^T & \sum_{t,t} \end{bmatrix} \tag{17.17}$$

where $\sum_{t,t}$ is the average of pixel squared values in Band t and $\sum_{t-1,t} = [\sum_{1,t}, \cdots, \sum_{j,t} \cdots, \sum_{t-1,t}]^T$ is a $(t-1) \times 1$ vector with $\sum_{j,t}$ being the average of the products of corresponding pixel values in Band j and t. Equation (17.17) shows that the dimension of $\sum$ is increased as more bands are received.

When $\sum_{t-1}^{-1}$ is available, it is more cost-effective to calculate $\sum_t^{-1}$ by modifying $\sum_{t-1}^{-1}$ with $\sum_{t,t}$ and $\sum_{t-1,t}$. The following partitioned matrix inversion formula can be used for $\sum^{-1}$ adaptation.

Let a matrix $\mathbf{A}$ be partitioned as $A = [\begin{matrix} \mathbf{A}_{11} & \mathbf{A}_{12} \\ \mathbf{A}_{21} & \mathbf{A}_{22} \end{matrix}]$. Then its inverse matrix $\mathbf{A}^{-1}$ can be calculated as

$$\begin{bmatrix} \left(\mathbf{A}_{11} - \mathbf{A12}\mathbf{A}_{22}^{-1}\mathbf{A}_{21}\right)^{-1} & -\left(\mathbf{A}_{11} - \mathbf{A12}\mathbf{A}_{22}^{-1}\mathbf{A}_{21}\right)^{-1}\mathbf{A}_{12}\mathbf{A}_{22}^{-1} \\ -\left(\mathbf{A}_{22} - \mathbf{A21}\mathbf{A}_{11}^{-1}\mathbf{A}_{12}\right)^{-1}\mathbf{A}_{21}\mathbf{A}_{11}^{-1} & \left(\mathbf{A}_{22} - \mathbf{A21}\mathbf{A}_{22}^{-1}\mathbf{A}_{12}\right)^{-1} \end{bmatrix} \tag{17.18}$$

Let $\mathbf{A}_{11} = \sum_{t-1}$, $\mathbf{A}_{22} = \sum_{t,t}$, $\mathbf{A}_{12} = \sum_{t-1,t}$, and $\mathbf{A}_{21} = \sum_{t-1,t}^T$. All these elements can be generated by simple matrix multiplication. Actually, in this case, no operation of matrix inversion is used when reaching the final $\sum^{-1}$.

The intermediate result still can be generated by applying the $\sum_t^{-1}$ to the first t bands. This means the spectral features in these t bands are used for target detection and discrimination. This may help to find targets at early processing stages.

17.3 Computer Simulation

The HYDICE image scene shown in Figure 17.1 was collected in Maryland in 1995 from a flight altitude of 10,000 feet with approximately 1.5m spatial resolution in 0.4–2.5 μm spectral region. The atmospheric water bands with low signal-to-noise ratio were removed, reducing the data dimensionality from 210 to 169. The image scene has 128 lines and the number of pixels in each line M is 64, so the total number of pixel vectors is $128 \times 64 = 4096$. This scene includes 15 panels arranged in a 15×3 matrix. Each element in this matrix is denoted by p_{ij} with rows indexed by $i = 1, \cdots, 5$ and columns indexed by $i = a, b, c$. The three panels in the same row p_{ia}, p_{ib}, p_{ic} were made from the same material of size 3m $\times$ 3m, 2m $\times$ 2m, 1m $\times$ 1m, respectively, which could be considered as one class, p_i. As shown in Figure 17.1(c), these ten classes have very similar spectral signatures. In the computer simulation, we simulated the three cases when data were received pixel-by-pixel, line-by-line,

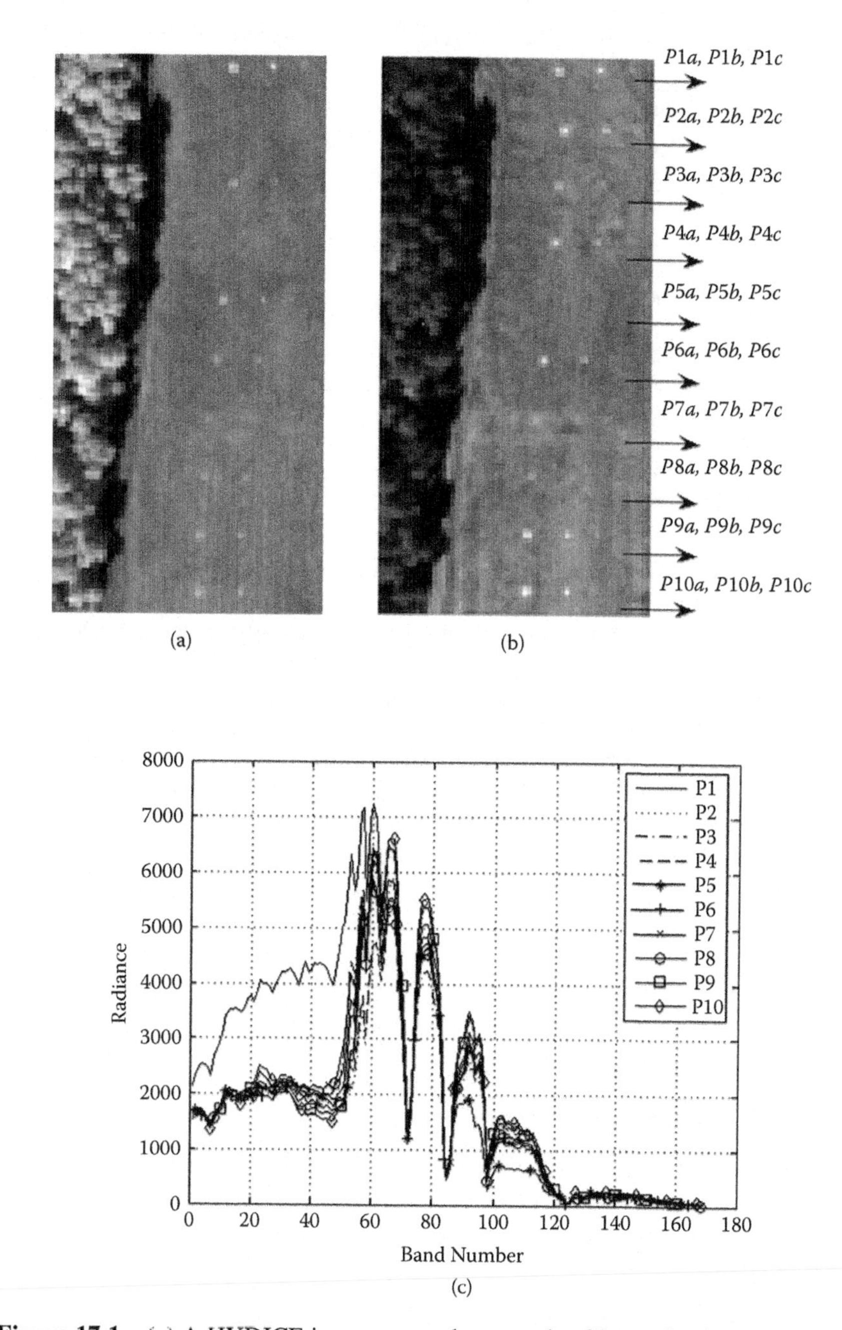

Figure 17.1 (a) A HYDICE image scene that contains 30 panels. (b) Spatial locations of 30 panels provided by ground truth. (c) Spectra from P1 to P10.

TABLE 17.1 Classification Accuracy N_D Using the CLDA Algorithm (in Al Cases, The Number of False Alarm Pixels $N_F = 0$).

Panel #	Pure Pixels	Offline Proc.	Online Proc. (BIP)	Online Proc. (BIL)	Online Proc. (BSQ)
P1	3	2	2	2	2
P2	3	2	2	2	2
P3	4	3	3	3	3
P4	3	2	2	2	2
P5	6	5	6	6	6
P6	3	2	2	2	2
P7	4	3	3	3	3
P8	4	3	3	3	3
P9	4	3	3	3	3
P10	4	3	3	3	3
Total	38	28	29	29	29

and band-by-band. Then the CLDA results were compared with the result from the off-line processing.

In order to compare with the pixel-level ground truth, the generated gray-scale classification maps were normalized into [0,1] dynamic range and converted into binary images using a threshold 0.5. The numbers of correctly classified pure panel pixels N_D in the different cases were counted and listed in Table 17.1. Here the number of false alarm pixels is $N_F = 0$ in all the cases, which means the ten panel classes were well separated. As shown in Table 17.1, all three cases of online processing can correctly classify 29 out of 38 panel pixels, while the offline CLDA algorithm can correctly classify 28 out of 38 panel pixels. We can see that these performances are comparable.

17.4 Practical Considerations

17.4.1 Algorithm Simplification Using $\mathbf{R}^{-1}$

According to Section 17.2, $\mathbf{R}^{-1}$ update only includes one step, while $\sum^{-1}$ update has three steps. The number of multiplications saved by using $\mathbf{R}^{-1}$ is $5 \times L^2$ for each update. Obviously, using $\mathbf{R}^{-1}$ instead of $\sum^{-1}$ can also reduce the number of modules in the chip. Then Eq. (17.6) will be changed to

$$\mathbf{P}_k^T = \left[0 \cdots 010 \cdots 0\right] \left(\mathbf{S}^T \mathbf{R}^{-1} \mathbf{S}\right)^{-1} \mathbf{S}^T \mathbf{R}^{-1} \tag{17.19}$$

for classifying the k-th class in $\mathbf{S}$. From the image processing point of view, the functions of $\mathbf{R}^{-1}$ and $\sum^{-1}$ in the operator are both for suppressing the undesired background pixels before applying the match filter $\mathbf{S}^T$. Based on our experience on different hyperspectral/multispectral image scenes, using $\mathbf{R}^{-1}$ generates very close results to using $\sum^{-1}$. Detailed performance comparisons can be found in [5].

17.4.2 Algorithm Implementation with Matrix Inversion

The major difficulty in hardware implementation is the expensiveness of a matrix inversion module, in particular, when the dimension of $\mathbf{R}$ or $\sum$ (i.e., the number of bands L) is large. A possible way to tackle this problem is to partition a large matrix into four smaller matrices and derive the original inverse matrix by using the partitioned matrix inversion formula in Eq. (17.19).

17.4.3 Unsupervised Processing

The CLDA is a supervised approach, i.e., the class spectral signatures need to be known a priori. But in practice, this information may be difficult or even impossible to obtain, in particular, when dealing with remote sensing images. This is due to the facts that: 1) any atmospheric, background, and environmental factors may have an impact on the spectral signature of the same material, which makes the in-field spectral signature of a material or object not be well correlated to the one defined in a spectral library; 2) a hyperspectral sensor may extract many unknown signal sources because of its very high spectral resolution, whose spectral signatures are difficult to be pre-determined; and 3) an airborne or spaceborne hyperspectral sensor can take images from anywhere, whose prior background information may be unknown and difficult to obtain.

The target and background signatures in $\mathbf{S}$ can be generated from the image scene directly in an unsupervised fashion [6]. In this section, we present an unsupervised class signature generation algorithm based on constrained least squares linear unmixing error and quadratic programming. After the class signatures in $\mathbf{S}$ are determined, Eq. (17.6) or Eq. (17.21) can be applied directly.

Because of the relatively rough spatial resolution, it is generally assumed that the reflectance of a pixel in a remotely sensed image is the linear mixture of reflectances of all the materials in the area covered by this pixel. According to the linear mixture model, a pixel vector $\mathbf{x}$ can be represented as

$$\mathbf{x} = \mathbf{S}\alpha + \mathbf{n} \tag{17.20}$$

where $\mathbf{S} = \left[\mathbf{s}_1, \mathbf{s}_2, \cdots, \mathbf{s}_p\right]$ is an $L \times p$ signature matrix with p linearly independent endmembers (including desired targets, undesired targets, and background objects) and $\mathbf{s}_i$ is the i-th endmember signature; $\alpha = (\alpha_1 \alpha_2 \cdots \alpha_p)^T$ is a $p \times 1$ abundance fraction vector, where the i-th element α_i represents the abundance fraction of s_i present in that pixel; $\mathbf{n}$ is an $L \times 1$ vector that can be interpreted as a noise term or model error. Abundances of all the endmembers in a pixel are related as

$$\sum_{i=1}^{p} \alpha_i = 1, 0 \leq \alpha_i \leq 1, \text{ for any } i \tag{17.21}$$

which are referred to as sum-to-one and non-negativity constraints.

Now our task is to estimate α with Eq. (17.22) being satisfied for a pixel. It should be noted that $\mathbf{S}$ is the same for all the pixels in the image scene, while α varies from pixel to pixel. Therefore, when $\mathbf{S}$ is known, there are p unknown variables to be estimated with L equations and $L >> p$. This means the problem is overdetermined, and no solution exists. However, we can formulate a least squares problem to estimate the optimal $\hat{\alpha}$ such that the estimation error defined as below is minimized:

$$e = \|\mathbf{x} - \mathbf{S}\hat{\alpha}\|^2 = \mathbf{x}^T\mathbf{x} - 2\hat{\alpha}^T\mathbf{M}^T\mathbf{x} + \hat{\alpha}^T\mathbf{M}^T\mathbf{M}\hat{\alpha} \tag{17.22}$$

When the constraints in Eq. (17.22) are to be relaxed simultaneously, there is no closed form solution. Fortunately, if $\mathbf{S}$ is known, this constrained optimization problem defined by Eqs. (17.22) and (17.23) can be formulated into a typical quadratic programming problem:

$$\text{Minimize } f(\alpha) = \mathbf{r}^T\mathbf{r} - 2\mathbf{r}^T\mathbf{M}\alpha + \alpha^T\mathbf{M}^T\mathbf{M}\alpha \tag{17.23}$$

subject to $\alpha_1 + \alpha_2 + \cdots + \alpha_p = 1$ and $0 \leq \alpha_i \leq 1$, for $1 \leq p$. Quadratic programming (QP) refers to an optimization problem with a quadratic objective function and linear constraints (including equality and inequality constraints). It can be solved using nonlinear optimization techniques. But we prefer to use linear optimization based techniques in our research since they are simpler and faster [7].

When $\mathbf{S}$ is unknown, endmembers can be generated using the algorithm based on linear unmixing error [8] and quadratic programming. Initially, a pixel vector is selected as an initial signature denoted by $\mathbf{s}_0$. Then it is assumed that all other pixel vectors in the image scene are made up of $\mathbf{s}_0$ with 100 percent abundance. This assumption certainly creates estimation errors. The pixel vector that has the largest least squares error (LSE) between itself and $\mathbf{s}_0$ is selected as a first endmember signature denoted by $\mathbf{s}_1$. Because the LSE between $\mathbf{s}_0$ and $\mathbf{s}_1$ is the largest, it can be expected that $\mathbf{s}_1$ is most distinct from $\mathbf{s}_0$. The signature matrix $\mathbf{S} = [\mathbf{s}_0\mathbf{s}_1]$ is then formed to estimate the abundance fractions for $\mathbf{s}_0$ and $\mathbf{s}_1$, denoted by $\hat{\alpha}_0(\mathbf{x})$ and $\hat{\alpha}_1(\mathbf{x})$ for pixel $\mathbf{x}$, respectively, by using the QP-based constrained linear unmixing technique in Section 17.3.1. Now the optimal constrained linear mixture of $\mathbf{s}_0$ and $\mathbf{s}_1$, $\hat{\alpha}_0(\mathbf{x})\mathbf{s}_0 + \hat{\alpha}_1(\mathbf{x})\mathbf{s}_1$, is used to approximate the $\mathbf{x}$. The LSE between r and its estimated linear mixture $\hat{\alpha}_0(\mathbf{x})\mathbf{s}_0 + \hat{\alpha}_1(\mathbf{x})\mathbf{s}_1$ is calculated for all pixel vectors. Once again, a pixel vector that yields the largest LSE between itself and its estimated linear mixture will be selected to be a second endmember signature $\mathbf{s}_2$. As expected, the pixel that yields the largest LSE is the most dissimilar to $\mathbf{s}_0$ and $\mathbf{s}_1$, and most likely to be an endmember pixel yet to be found. The same procedure with $\mathbf{S} = [\mathbf{s}_0\mathbf{s}_1\mathbf{s}_2]$ is repeated until the resulting LSE is below a prescribed error threshold η.

17.5 Application to Other Techniques

The real-time implementation concept of the CLDA algorithm can be applied to several other target detection techniques. They employ the matched filter and require the computation of $\mathbf{R}^{-1}$ or $\sum^{-1}$. The difference from the CLDA algorithm is that they can only detect the targets, but the CLDA algorithm can detect targets and discriminate different targets from each other.

- RX algorithm [9]: The well-known RX algorithm is an anomaly detector, which does not require any target spectral information. The original formula is $\mathbf{w}_{RX} = \mathbf{x}^T \sum^{-1} \mathbf{x}$, which was simplified as $\tilde{\mathbf{w}}_{RX} = \mathbf{x}^T \mathbf{R}^{-1} \mathbf{x}$ [10].
- Constrained energy minimization (CEM) [4]: The CEM detector can be written as $\mathbf{w}_{CEM} = \frac{\mathbf{R}^{-1}\mathbf{d}}{\mathbf{d}^T\mathbf{R}^{-1}\mathbf{d}}$, where $\mathbf{d}$ is the desired target spectral signature. To detect if $\mathbf{d}$ is contained in a pixel $\mathbf{x}$, we can simply apply $\mathbf{w}_{CEM}^T\mathbf{x}$, i.e., $\frac{\mathbf{d}^T\mathbf{R}^{-1}\mathbf{x}}{\mathbf{d}^T\mathbf{R}^{-1}\mathbf{d}}$.
- Kelly's generalized likelihood ratio test (KGLRT) [11]: This generalized likelihood ratio test is given by $\frac{(\mathbf{d}^T \sum^{-1} \mathbf{x})^2}{(\mathbf{d}^T \sum^{-1} \mathbf{d})(1+\mathbf{x}^T \sum^{-1} \mathbf{x}/N)}$, where N is the number of samples used in the estimation of $\sum$.
- Adaptive matched filter (AMF) [12]: When the number of samples N is a very large value, the KGLRT is reduced to a simple format: $\frac{(\mathbf{d}^T \sum^{-1} \mathbf{x})^2}{\mathbf{d}^T \sum^{-1} \mathbf{d}}$. We can see that it is close to the CEM except that the numerator has a power of two.
- Adaptive coherence estimator (ACE) [13]: The estimator can be written as $\frac{(\mathbf{d}^T \sum^{-1} \mathbf{x})^2}{(\mathbf{d}^T \sum^{-1} \mathbf{d})(\mathbf{x}^T \sum^{-1} \mathbf{x})}$. It is similar to AMF except that a term similar to the RX algorithm is included in the denominator.

Some quantitative performance comparisons between these algorithms can be found in [14].

17.6 Summary

In this chapter, we discussed the constrained linear discriminant analysis (CLDA) algorithm and its real-time implementation. This is to meet the need in practical applications of remote sensing image analysis when the immediate data analysis result is desired for real-time or near-real-time decision-making. The strategy is developed for each data format, i.e., BIP, BIL, and BSQ. The basic concept is to real-time update the inverse covariance matrix $\sum^{-1}$ or inverse correlation matrix $\mathbf{R}^{-1}$ in the CLDA algorithm as the data; (i.e., a pixel vector, or a line of pixel vectors, or a spectral band) coming in, then the intermediate target detection and discrimination result are generated for quick response, and the final product is available right after (or with a reasonable delay) when the entire data set is received. Several practical implementation issues are discussed. The computer simulation shows the online results are similar to the offline results. But its performance when onboard actual platforms needs further investigation.

Although the real-time implementation scheme is originally developed for the CLDA algorithm, it is applicable to any detection algorithm involving $\sum^{-1}$ or $\mathbf{R}^{-1}$ computation, such as RX, CEM, KGLRT, AMF, and ACE algorithms.

As a final note, we believe the developed real-time implementation scheme is more suitable to airborne platforms, where the atmospheric correction is not critical for relatively small monitoring fields. Due to its complex nature, onboard atmospheric correction is almost impossible. After the real-time data calibration is completed onboard, the developed algorithm can be used to generate the intermediate and quick final products onboard.

Acknowledgment

The author would like to thank Professor Chein-I Chang at the University of Maryland Baltimore County for providing the data used in the experiment.

References

[1] Q. Du and C.-I Chang. Linear constrained distance-based discriminant analysis for hyperspectral image classification, *Pattern Recognition*, vol. 34, pp. 361–373, 2001.

[2] Q. Du and H. Ren. Real-time constrained linear discriminant analysis to target detection and classification in hyperspectral imagery, *Pattern Recognition*, vol. 36, pp. 1–12, 2003.

[3] J.C. Harsanyi and C.-I Chang. Hyperspectral image classification and dimensionality reduction: an orthogonal subspace projection, *IEEE Transactions on Geoscience and Remote Sensing*, vol. 32, pp. 779–785, 1994.

[4] W.H. Farrand and J.C. Harsanyi. Mapping the distribution of mine tailing in the coeur d'Alene river valley, Idaho through the use of constrained energy minimization technique, *Remote Sensing of Environment*, vol. 59, pp. 64–76, 1997.

[5] Q. Du and R. Nekovei. Implementation of real-time constrained linear discriminant analysis to remote sensing image classification, *Pattern Recognition*, vol. 38, pp. 459–471, 2005.

[6] Q. Du. Unsupervised real-time constrained linear discriminant analysis to hyperspectral image classification, *Pattern Recognition*, in press.

[7] P. Venkataraman. *Applied optimization with MATLAB programming*, Wiley-Interscience, 2002.

[8] D. Heinz and C.-I Chang. Fully constrained least squares linear mixture analysis for material quantification in hyperspectral imagery, *IEEE Transactions on Geoscience and Remote Sensing*, vol. 39, pp. 529–545, 2001.

[9] I. S. Reed and X. Yu. Adaptive multiple-band CFAR detection of an optical pattern with unknown spectral distribution, *IEEE Trans. on Acoustic, Speech and Signal Processing*, vol. 38, pp. 1760–1770, 1990.

[10] C.-I Chang and D. Heinz. Subpixel spectral detection for remotely sensed images, *IEEE Transactions on Geoscience and Remote Sensing*, vol. 38, pp. 1144–1159, 2000.

[11] E. J. Kelly. An adaptive detection algorithm, *IEEE Transactions on Aerospace and Electronic Systems*, vol. 22, pp. 115–127, 1986.

[12] F. C. Robey, D. R. Fuhrmann, E. J. Kelly, and R. Nitzberg. A CFAR adaptive matched filter detector, *IEEE Transactions on Aerospace and Electronic Systems*, vol. 28, pp. 208–216, 1992.

[13] S. Kraut and L. L. Sharf. The CFAR adaptive subspace detector is a scale-invariant GLRT, *IEEE Transactions on Signal Processing*, vol. 47, pp. 2538–2541, 1999.

[14] Q. Du. On the performance of target detection algorithms for hyperspectral imagery analysis, *Proceedings of SPIE*, Vol. 5995, pp. 599505-1–599505-8, 2005.

Chapter 18

Real-Time Onboard Hyperspectral Image Processing Using Programmable Graphics Hardware

Javier Setoain,
Complutense University of Madrid, Spain

Manuel Prieto,
Complutense University of Madrid, Spain

Christian Tenllado,
Complutense University of Madrid, Spain

Francisco Tirado,
Complutense University of Madrid, Spain

Contents

This chapter focuses on mapping hyperspectral imaging algorithms to graphics processing units (GPU). The performance and parallel processing capabilities of these units, coupled with their compact size and relative low cost, make them appealing for onboard data processing. We begin by giving a short review of GPU architectures. We then outline a methodology for mapping image processing algorithms to these architectures, and illustrate the key code transformation and algorithm trade-offs involved in this process. To make this methodology precise, we conclude with an example in which we map a hyperspectral endmember extraction algorithm to a modern GPU.

18.1 Introduction

Domain-specific systems built on custom designed processors have been extensively used during the last decade in order to meet the computational demands of image and multimedia processing. However, the difficulties that arise in adapting specific designs to the rapid evolution of applications have hastened their decline in favor of other architectures. Programmability is now a key requirement for versatile platform designs to follow new generations of applications and standards.

At the other extreme of the design spectrum we find general-purpose architectures. The increasing importance of media applications in desktop computing has promoted the extension of their cores with multimedia enhancements, such as SIMD instruction sets (the Intel's MMX/SSE of the Pentium family and IBM-Motorola's AltiVec are well-know examples). Unfortunately, the cost of delivering instructions to the ALUs poses a serious bottleneck in these architectures and makes them still unsuited to meet more stringent (real-time) multimedia demands.

Graphics processing units (GPUs) seem to have taken the best from both worlds. Initially designed as expensive application-specific units with control and communication structures that enable the effective use of many ALUs and hide latencies in the memory accesses, they have evolved into highly parallel multipipelined processors with enough flexibility to allow a (limited) programming model. Their numbers are impressive. Today's fastest GPU can deliver a peak performance in the order of 360 Gflops, more than seven times the performance of the fastest x86 dual-core processor (around 50 Gflops) [11]. Moreover, they evolve faster than more-specialized platforms, such as field programmable gate arrays (FPGAs) [23], since the high-volume game market fuels their development.

Obviously, GPUs are optimized for the demands of 3D scene rendering, which makes software development of other applications a complicated task. In fact, their astonishing performance has captured the attention of many researchers in different areas, who are using GPUs to speed up their own applications [1]. Most of the research activity in general-purpose computing on GPUs (GPGPU) works towards finding efficient methodologies and techniques to map algorithms to these architectures. Generally speaking, it involves developing new implementation strategies following a stream programming model, in which the available data parallelism is explicitly uncovered, so that it can be exploited by the hardware. This adaptation presents numerous implementation challenges, and GPGPU developers must be proficient not only in the target application domain but also in parallel computing and 3D graphics programming.

The new hyperspectral image analysis techniques, which naturally integrate both the spatial and spectral information, are excellent candidates to benefit from these kinds of platforms. These algorithms, which treat a hyperspectral image as an image cube made up of spatially arranged pixel vectors [18, 22, 12] (see Figure 18.1), exhibit regular data access patterns and inherent data parallelism across both pixel vectors (coarse-grained pixel-level parallelism) and spectral information (fine-grained spectral-level parallelism). As a result, they map nicely to massively parallel systems made up of commodity CPUs (e.g., Beowulf clusters) [20]. Unfortunately, these systems are generally expensive and difficult to adapt to onboard remote sensing data processing scenarios, in which low-weight integrated components are essential to reduce mission payload. Conversely, the compact size and relative low cost are what make modern GPUs appealing to onboard data processing.

The rest of this chapter is organized as follows. Section 18.2 begins with an overview of the traditional rendering pipeline and eventually goes over the structure of modern

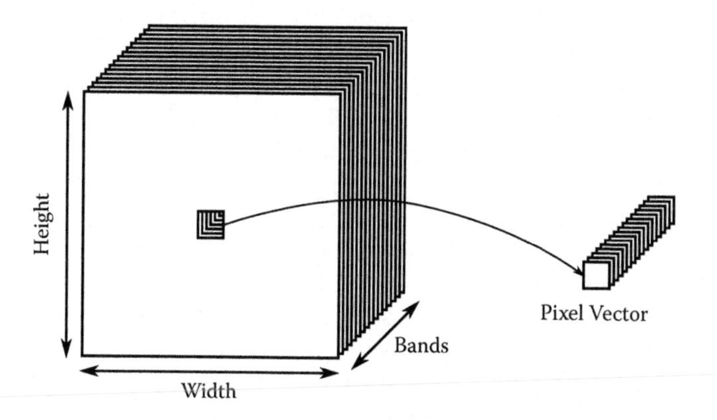

Figure 18.1 A hyperspectral image as a cube made up of spatially arranged pixel vectors.

GPUs in detail. Section 18.3, in turn, covers the GPU programming model. First, it introduces an abstract stream programming model that simplifies the mapping of image processing applications to the GPU. Then it focuses on describing the essential code transformations and algorithm trade-offs involved in this mapping process. After this comprehensive introduction, Section 18.4 describes the Automatic Morphological Endmember Extraction (AMEE) algorithm and its mapping to a modern GPU. Section 18.5 evaluates the proposed GPU-based implementation from the viewpoint of both endmember extraction accuracy (compared to other standard approaches) and parallel performance. Section 18.6 concludes with some remarks and provides hints at plausible future research.

18.2 Architecture of Modern GPUs

This section provides background on the architecture of modern GPUs. For this introduction, it is useful to begin with a description of the traditional rendering pipeline [8, 16], in order to understand the basic graphics operations that have to be performed. Subsection 18.2.1 starts on the top of this pipeline, where data are fed from the CPU to the GPU, and work their way down through multiple processing stages until a pixel is finally drawn on the screen. It then shows how this logical pipeline translates into the actual hardware of a modern GPU and describes some specific details of the different graphics cards manufactured by the two major GPU makers, NVIDIA and ATI/AMD. Finally, Subsection 18.2.2 outlines recent trends in GPU design.

18.2.1 The Graphics Pipeline

Figure 18.2 shows a rough description of the traditional 3D rendering pipeline. It consists of several stages, but the bulk of the work is performed by four of them: *vertex-processing (vertex shading)*, *geometry*, *rasterization*, and *fragment-processing (fragment shading)*. The rendering process begins with the CPU sending a stream of vertex from a 3D polygonal mesh and a *virtual camera* viewpoint to the GPU, using some graphics API commands. The final output is a 2D array of pixels to be displayed on the screen.

In the vertex stage the 3D coordinates of each vertex from the input mesh are transformed (projected) onto a 2D screen position, also applying lighting to determine their colors. Once transformed, vertices are grouped into rendering primitives, such as triangles, and scan-converted by the rasterizer into a stream of pixel fragments. These fragments are discrete portions of the triangle surface that correspond to the pixels of the rendered image. The vertex attributes, such as texture coordinates, are then interpolated across the primitive surface storing the interpolated values at each fragment. In the fragment stage, the color of each fragment is computed. This computation usually depends on the interpolated attributes and the information retrieved from the

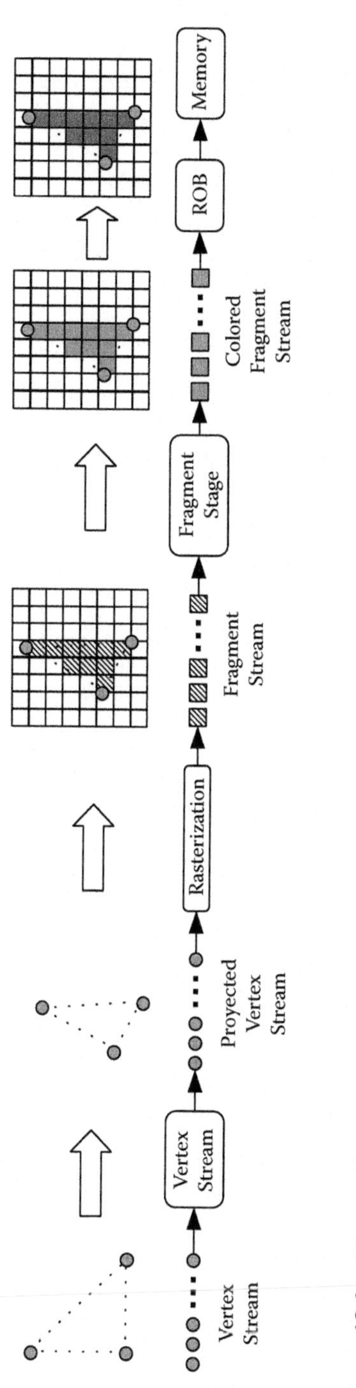

Figure 18.2 3D graphics pipeline.

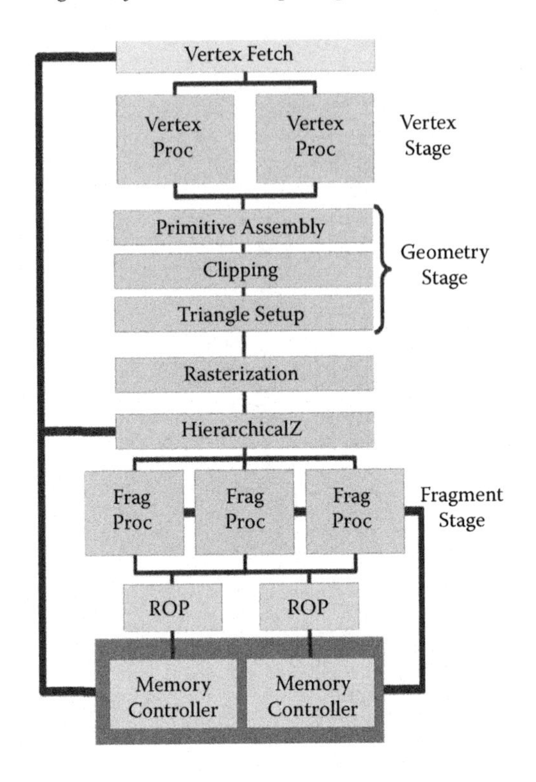

Figure 18.3 Fourth generation of GPUs block diagram. These GPUs incorporate fully programmable vertexes and fragment processors.

graphics card memory by texture lookups.[1] The colored fragments are sent to the ROP stage,[2] where Z-buffer checking ensures only visible fragments are processed further. Those partially transparent fragments are blended with the existing frame buffer pixel. Finally, if enabled, fragments are antialiazed to produce the ultimate colors.

Figure 18.3 shows the actual pipeline of a modern GPU. A detailed description of this hardware is out of the scope of this book. Basically, major pipeline stages corresponds 1-to-1 with the logical pipeline. We focus instead on two key features of this hardware: programmability and parallelism.

- *Programmability*. Until only a few years ago, commercial GPUs were implemented using a hard-wired (fixed-function) rendering pipeline. However, most GPUs today include fully programmable vertex and fragment stages.[3] The programs they execute are usually called vertex and fragment programs

[1]This process is usually called *texture mapping*.

[2]ROP denotes raster operations (NVIDIA's terminology).

[3]The vertex stage was the first one to be programmable. Since 2002, the fragment stage is also programmable.

(or shaders), respectively, and can be written using C-like high-level languages such as Cg [6]. This feature is what allows for the implementation of non-graphics applications on the GPUs.

- *Parallelism*. The actual hardware of a modern GPU integrates hundreds of physical pipeline stages per major processing stage to increase the throughput as well as the GPU's clock frequency [2]. Furthermore, replicated stages take advantage of the inherent data parallelism of the rendering process. For instance, the vertex and fragment processing stages include several replicated units known as *vertex* and *fragment processors*, respectively.[4] Basically, the GPU launches a thread per incoming vertex (or per group of fragments), which is dispatched to an idle processor. The vertex and fragment processors, in turn, exploit multithreading to hide memory accesses, i.e., they support multiple in-flight threads, and can execute independent shader instructions in parallel as well. For instance, fragment processors often include vector units that operate on 4-element vectors (Red/Gree/Blue/Alpha channels) in an SIMD fashion.

Industry observers have identified different generations of GPUs. The description above corresponds to the fourth generation[5] [7]. For the sake of completeness, we conclude this subsection reproducing in Figure 18.4 the block diagram of two representative examples of that generation: NVIDIA's G70 and ATI's Radeon R500 families. Obviously, there are some differences in their specific implementations, both in the overall structure and in the internals of some particular stages. For instance, in the G70 family the interpolation units are the first stage in the pipeline of each fragment processor, while in the R500 family they are arranged in a completely separate hardware block, outside the fragment processors. A similar thing happens with the texture access units. In the G70 family they are located inside each fragment processor, coupled to one of their vector units [16, 2]. This reduces the fragment processors performance in case of a texture access, because the associated vector unit remains blocked until the texture data are fetched from memory. To avoid this problem, the R500 family places all the texture access units together in a separate block.

18.2.2 State-of-the-art GPUs: An Overview

The recently released NVIDIA G80 families have introduced important new features, which anticipate future GPU design trends. Figure 18.5 shows the structure of the GeForce 8800 GTX, which is the most powerful G80 implementation introduced so far. Two features stand out over previous generations:

- *Unified Pipeline*. The G80's pipeline only includes one kind of programmable unit, which is able to execute three different kinds of shaders: vertex, geometry,

[4]The number of fragment processors usually exceeds the number of vertex processors, which follows from the general assumption that there are frequently more pixels to be shaded than vertexes to be projected

[5]The fourth generation of GPUs dates from 2002 and begins with NVIDIA's GeForce FX series and ATI's Radeon 9700 [7].

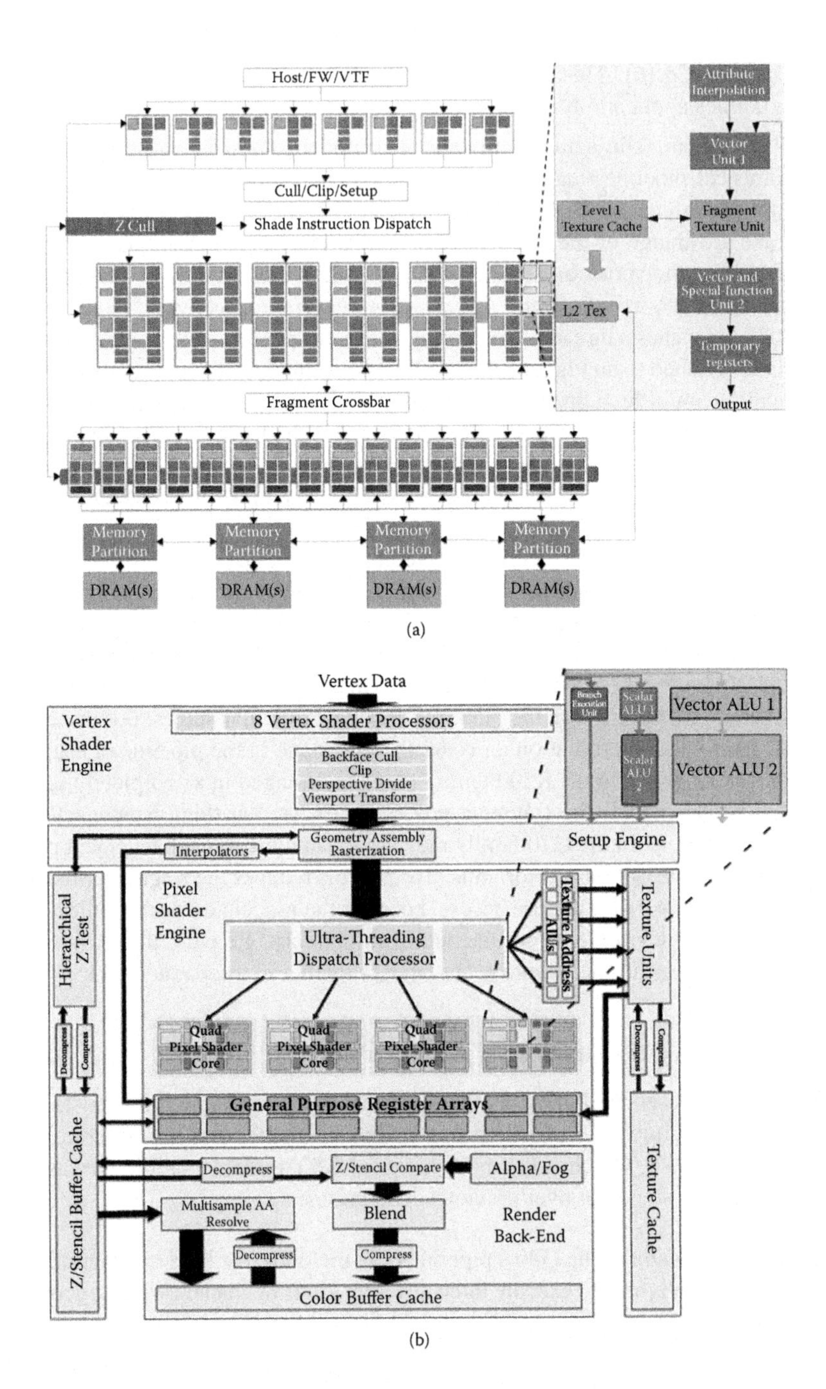

Figure 18.4 NVIDIA G70 (a) and ATI-RADEON R520 (b) block diagrams.

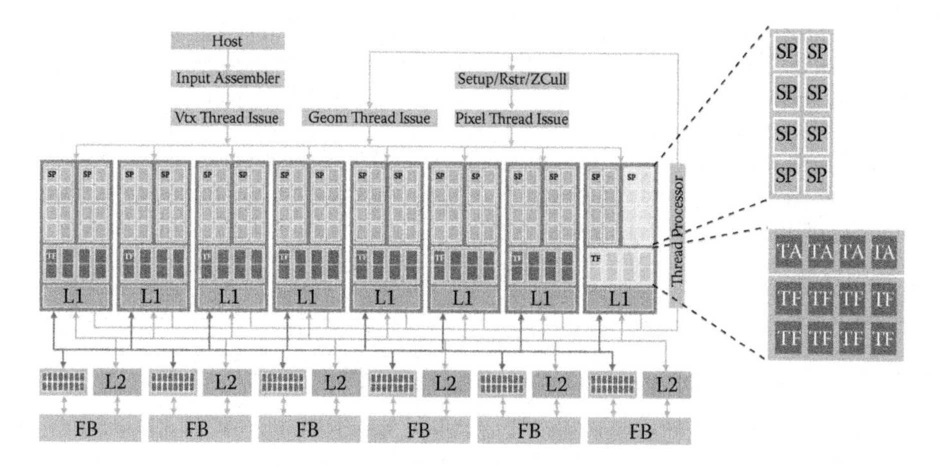

Figure 18.5 Block diagram of the NVIDIA's Geforce 8800 GTX.

and fragment. This design reduces the number of pipeline stages and changes the sequential flow to be more looping oriented. Inputs are fed to the top of the unified *shader core*, and outputs are written to registers and then fed back into the top of the *shader core* for the next operation. This unified architecture promises to improve the performance for those programs dominated by only one type of shader, which would otherwise be limited by the number of specific processors available [2].

- *Scalar Processors*. Another important change introduced in the NVIDIA's G80 family over previous generations is the scalar nature of the programmable units. In previous architectures both the vertex and fragment processors had SIMD (vector) functional units, which were able to operate in parallel on the different components of a vertex/fragment (e.g., the RGBA channels in a fragment). However, modern shaders tend to use a mix of vector and scalar instructions. Scalar computations are difficult to compile and schedule efficiently on a vector pipeline. For this reason, NVIDIA's G80 engineers decided to incorporate only scalar units, called Stream Processors (SPs), in NVIDIA parlance [2]. The GeForce 8800 GTX includes 128 of these SPs driven by a high-speed clock,[6] which can be dynamically assigned to any specific shader operation. Overall, thousands of independent threads can be in flight in any given instant.

SIMD instructions can be implemented across groupings of SPs in close proximity. Figure 18.5 highlights one of these groups with the associated Texture Filtering (TF), Texture Addressing (TA), and Cache units. Using dedicated units for texture access (TA) avoids the blocking problem of previous NVIDIA generations mentioned above.

[6]The SPs are driven by a high-speed clock (1.35 GHz), separated from the core clock (575 MHz) that drives the rest of the chip.

In summary, GPU makers will continue the battle for dominance in the consumer gaming industry, producing a competitive environment with rapid innovation cycles. New features will constantly be added to next-generation GPUs, which will keep delivering outstanding performance-per-dollar and performance-per-square millimeter. Hyperspectral imaging algorithms fit relatively well with the programming environment the GPU offers, and can benefit from this competition. The following section focuses on this programming environment.

18.3 General Purpose Computing on GPUs

For non-graphics applications, the GPU can be better thought of as a *stream coprocessor* that performs computations through the use of *streams* and *kernels*. A stream is just an ordered collection of elements requiring similar processing, whereas kernels are data-parallel functions that process input streams and produce new output streams. For relatively simple algorithms this programming model may be easy to use, but for more complex algorithms, organizing an application into streams and kernels could prove difficult and require significant coding efforts. A kernel is a data-parallel function, i.e., its outcome must not depend on the order in which output elements are produced, which forces programmers to explicitly expose data parallelism to the hardware.

This section illustrates how to map an algorithm to the GPU using this model. As an illustrative example we have chosen the 2D Discrete Wavelet Transform (2D-DWT), which has been used in the context of hyperspectral image processing for principal component analysis [9], image fusion [15, 24], and registration [17] (among others). Despite its simplicity, the comparison between the GPU-based implementations of the popular *Lifting* (LS) and *Filter-Bank* (FBS) schemes of the DWT allows us to illustrate some of the algorithmic trade-offs that have to be considered. This section begins with the basic transformations that convert loop nests into an abstract stream programming model. Eventually it goes over the actual mapping to the GPU using standard 3D graphics API and describes the structure of the main program that orchestrates kernel execution. Finally, it introduces a compact C++ GPU framework that simplifies this mapping process, hiding the complexity of 3D graphics APIs.

18.3.1 Stream Programming Model

Our stream programming model focuses on data-parallel kernels that operate on arrays using gather operations, i.e., operations that read from random locations in an input array. Storing the input and output arrays as textures, this kind of kernel can be easily mapped to the GPU using fragment programs.[7] The following subsections illustrates how to identify this kind of kernel and map them efficiently to the GPU.

[7]Scatter operations write into random locations of a destination array. They are also common in certain applications, but fragment programs only support gather operations.

18.3.1.1 Kernel Recognition

The first step in the modeling process consists in identifying a set of potential kernels. Specifically, we want to partition the target algorithm into a set of code blocks tagged as kernel and non-kernel blocks. A kernel block is a parallel loop nest, i.e., a loop nest that carries no data dependences, that can be modeled as Listing 1.

Listing 1 Kernel block. D_{OUT} and D_{IN} denote output and input arrays, respectively. IDX denotes index matrices for indirect array accesses

for all (i,j) **do**
$D_{OUT_1}[i, j] = F(i, j, D_{IN_1}(IDX_{11}(i, j)), ...);$
$D_{OUT_2}[i, j] = F'(i, j, D_{IN_1}(IDX'_{11}(i, j)), ...);$
...
end for

The computations performed inside these loop nests define the kernels of our stream model. The output streams are defined by the set of elements of the output arrays D_{OUT} that are written in the loop nest. Stream elements are arranged according to their respective induction variables i and j. The input streams are defined by the set of array elements read in the loop. Index arrays (IDX) allow for indirect access to the input arrays D_{IN} and eventually translate into texture coordinates. A non-kernel block is whatever other construct that cannot be modeled as Listing 1, which accounts for non-parallel loops and other sequential parts of the application such as control flow statements, including the control flow of kernel blocks. These non-kernel blocks will eventually be part of the main program that orchestrates and chains the kernel blocks to satisfy data dependences.

Certain loop transformations could be useful for uncovering parallelism and enhancing kernel extraction. One of these is *loop distribution* (also know as *loop fission*), which can be used for splitting a loop nest that does not match listing 1 into smaller loop nests that do match that pattern.

The horizontal lifting algorithm helps us to illustrate this transformation. The conventional implementation of LS shown in Listing 2 contains loop-carried flow dependences and cannot be run in parallel. However, we can safely transform the loop nest in Listing 2 into Listing 3 since it preserves all the data dependences of the original code.[8] Notice that the transformed nested loops are free of loop-carried data dependences and match our definition of a kernel block.

In general, this transformation can also be useful to restructure existing loop nests in order to separate potentially parallel code (kernel blocks) from code that must be sequentialized (non-kernel blocks). Nevertheless, it must be applied judiciously since loop distribution results into finer granularity, which may deteriorate temporal locality and increase the overhead caused by kernel setup[9] and synchronization:

[8]Loop distribution is a safe transformation when all the data dependences point forward [14].

[9]Every kernel launch incurs a certain fixed CPU time to set up and issue the kernel on the GPU.

Distribution converts loop-independent and forward-carried dependences into dependences between kernels, which forces kernel synchronization and reduces kernel level parallelism. In fact, *loop fusion*, which performs the opposite operation, i.e., it merges multiple loops into a single loop, may be beneficial when it creates a larger kernel that still fits Listing 1.

Returning to our example, we are able to identify six kernels in the transformed code, one for each loop nest. All of them read input data from two arrays and produce one or two output streams (the first and the sixth loops produce two output streams, whereas the others produce only one). As mentioned above, the loop-independent and forward-carried dependences of the original LS loopnest convert into dependences between these kernels, which forces synchronization between them to avoid race conditions.

Obviously, more complex loop nests might require additional transformations to uncover parallelism, such as loop interchange, scalar expansion, array renaming, etc. [14]. Nevertheless, uncovering data parallelism is not enough to get an efficient GPU mapping. The following subsection illustrates another transformation that deals with specific GPU limitations.

Listing 2 Original horizontal LS loopnest. Specific boundary processing is not shown.

```
for i = 0 to N − 1 do
   for j = 0 to (N − 1)/2 do
      App[i,j] = A[i,2*j];
      Det[i,j] = A[i,2*j+1];
   end for
end for
{left boundary processing...}
for i = 0 to N − 1 do
   for j = 0 to (N − 6)/2 − 1 do
      Det[i,j+2] = Det[i,j+2] + alpha*(App[i,j+2] + App[i,j+3]);
      App[i,j+2] = App[i,j+2] + beta *(Det[i,j+1] + Det[i,j+2]);
      Det[i,j+1] = Det[i,j+1] + gamma*(App[i,j+1] + App[i,j+2]);
      App[i,j+1] = App[i,j+1] + delta*(Det[i,j] + Det[i,j+1]);
      Det[i,j] = Det[i,j]/phi;
      App[i,j] = App[i,j]*phi;
   end for
end for
{left boundary processing...}
```

18.3.1.2 Platform-Dependent Transformations

Once we have uncovered enough data parallelism and extracted the initial kernels and streams, we have to perform some additional transformations to efficiently map the stream model to the target GPU.

One of those transformations is *branch removal*. Although some GPUs tolerate branches, they normally reduce performance, hence eliminating conditional sentences from the kernel loops previously detected that would be useful. In some cases, removing the branch from the kernel loop body transfers the flow control to the main program, which will select between kernels based on a condition.

Listing 3 Transformed horizontal LS loopnests. The original loop has been distributed to increase kernel extraction. Specific boundary processing is not shown.

```
for i = 0 to N − 1 do
  for j = 0 to (N − 1)/2 do
    App[i,j] = A[i,2*j];
    Det[i,j] = A[i,2*j+1];
  end for
end for
{left boundary processing...}
for i = 0 to N − 1 do
  for j = 0 to (N − 6)/2 − 1 do
    Det[i,j+2] = Det[i,j+2] + alpha*(App[i,j+2] + App[i,j+3]);
  end for
end for
for i = 0 to N − 1 do
  for j = 0 to (N − 6)/2 − 1 do
    App[i,j+2] = App[i,j+2] + beta *(Det[i,j+1] + Det[i,j+2]);
  end for
end for
for i = 0 to N − 1 do
  for j = 0 to (N − 6)/2 − 1 do
    Det[i,j+1] = Det[i,j+1] + gamma*(App[i,j+1] + App[i,j+2]);
  end for
end for
for i = 0 to N − 1 do
  for j = 0 to (N − 6)/2 − 1 do
    App[i,j+1] = App[i,j+1] + delta*(Det[i,j] + Det[i,j+1]);
  end for
end for
for i = 0 to N − 1 do
  for j = 0 to (N − 6)/2 − 1 do
    Det[i,j] = Det[i,j]/phi;
    App[i,j] = App[i,j]*phi;
  end for
end for
{right boundary processing...}
```

Listing 4, which sketches the FBS scheme of the DWT, illustrates a common example, where branch removal provides significant performance gains. The second loop (the j loop) matches Listing 1, but its body includes two branches associated with the non-parallel inner loops (the k loops). These inner loops perform a reduction whose outcomes are finally written on the output arrays. In this example, the inner loop bounds are known at compile time. Therefore, they can be fully unrolled (actually this is what NVIDIA's Cg compiler generally does). However, removing loop branches through *unrolling* is not always possible since there is a limit on the number of instructions per kernel.

Listing 4 Original horizontal FBS loopnest. Specific boundary processing is not shown.

```
{left boundary processing...}
for i = 0 to N − 1 do
  for j = 2 to (N − 6)/2 do
    aux = 0;
    for k = 0 to LENGTH_H do
      aux = aux + h[k]*A[i,2 ∗ j − LENGTH_H/2 + k];
    end for
    App[i,j] = aux;
    aux = 0;
    for k = 0 to LENGTH_G do
      aux = aux + g[k]*A[i,2 ∗ j − LENGTH_G/2 + k];
    end for
    Det[i,j] = aux;
  end for
end for
{right boundary processing...}
```

Loop distribution can also be required to meet GPU *render target* (number of shader outputs) constraints. Some GPUs do not permit several render targets, i.e., output streams, in a *fragment shader*, or have a limit on the number of targets. For instance, if we run LS on a GPU that only allows one render target, the first and sixth loops in Listing 3 have to be distributed into two kernels, each one writing to a different array. Notice that in this case, unlike the previous distribution that converts Listing 2 into Listing 3, the new kernels can run in parallel, since both loopnests are free of dependences.

Finally, GPU *memory constraints* have to be considered. Obviously, we need to restrict the size of the kernel loop nests so that the amount of elements accessed in these loops fits into this memory. This is usually achieved by *tiling* or *strip-mining* the kernel loop nests. For instance, if the input array in the FBS algorithm is too large, we should tile the loops in Listing 4. Listing 5 shows a transformed FBS code after applying loop tiling and distributing the loops in order to meet render target constraints. The external loops (ti, tj) have been fused to improve temporal locality, i.e., the two filter loops have been tiled in a way that both kernels read from the same data in every iteration of the external loops. This way, we reduce memory transfers between the GPU and the main memory, since data have to be transferred to the GPU only once.

Listing 5 Transformed horizontal FBS loopnest. The original loopnest has been tiled and distributed to meet memory and render target constraints (assuming only one render target is possible). Specific boundary processing is not shown.

```
{left boundary processing...}
for ti = 0 to (N − 1)/TI do
  for tj = 2 to ((N − 6)/2)/TJ do
    for i = ti * TI to (ti + 1) * TI − 1 do
      for j = tj * TJ to (tj + 1) * TJ − 1 do
        aux = 0;
        for k = 0 to LENGTH_H do
          aux = aux + h[k]*A[i,2 * j − LENGTH_H/2 + k];
        end for
        App[i,j] = aux;
      end for
    end for
    for i = ti * TI to (ti + 1) * TI − 1 do
      for j = tj * TJ to (tj + 1) * TJ − 1 do
        aux = 0;
        for k = 0 to LENGTH_G do
          aux = aux + g[k]*A[i,2 * j − LENGTH_G/2 + k];
        end for
        Det[i,j] = aux;
      end for:kernel-block
    end for
  end for
end for
{right boundary processing...}
```

Loop tiling is also useful to optimize cache locality. GPU texture caches are heavily optimized for graphics rendering. Therefore, given that the reference patterns of GPGPU applications usually differ from those for rendering, GPGPU applications can lack cache performance. We do know that these caches are organized to capture 2D locality [10], but we do not know their exact specifications today, as they are not released by GPU makers. This lack of information complicates the practical application of tiling since the structure of the target memory hierarchy is the principal factor in determining the tile size. Therefore, some sort of memory model or empirical tests will be needed to make this transformation useful.

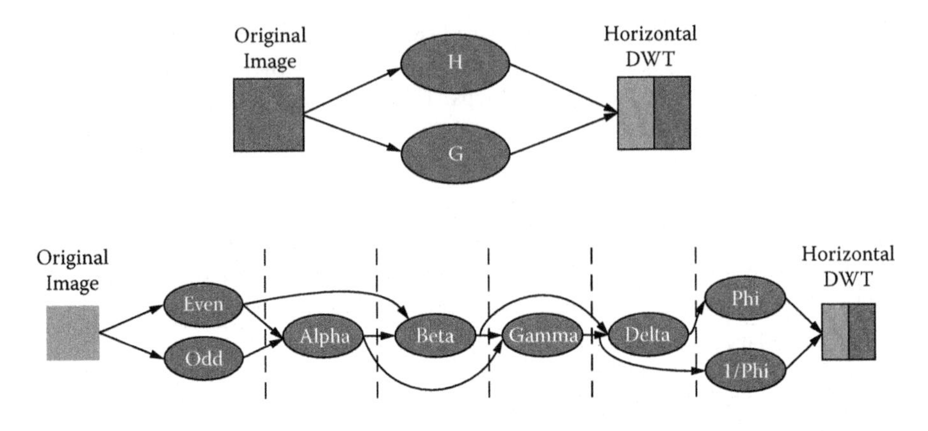

Figure 18.6 Stream graphs of the GPU-based (a) filter-bank (FBS) and (b) lifting (LS) implementations.

18.3.1.3 The 2D-DWT in the Stream Programming Model

Figures 18.6(a) and 18.6(b) graphically illustrate the implementation of the two DWT algorithms in the stream programming model. These stream graphs have been extracted from the sequential code applying the previous transformations. For the FBS we only need two kernels, one for each filter. Furthermore, these kernels can be run in parallel (without synchronization) as both write on different parts of the output arrays and do not depend on the results of each other. On the other hand, the dependences between LS steps translate into a higher number of kernels, which results in finer grain parallelism (each LS step is performed by a different kernel) and explicit synchronization barriers between them to avoid race conditions.

These algorithms also allow us to highlight the parallelism versus complexity trade-off that developers usually face. Theoretically, LS requires less arithmetic operations than its FBS counterpart, down to one half depending on the type and length of the wavelet filter [4]. In fact, LS is usually the most efficient strategy in general-purpose microprocessors [13]. However, its FBS fits better the programming environment the GPU offers. In practice, performance models or empirical tests are needed to evaluate these kinds of trade-offs.

18.3.2 Stream Management and Kernel Invocation

As mentioned above, kernel programs can be written in high-level C-like languages such as Cg [7]. However, we must still use a 3D graphics API, such as OpenGL, to organize data into streams, transfer those data streams to and from the GPU as 2D textures, upload kernels, and perform the sequence of kernel calls dictated by the application flow. In order to illustrate these concepts, Figure 18.8 shows some of the OpenGL commands and the respective Cg code that performs one lifting step (the ALPHA step). The following subsections describe this example code in detail.

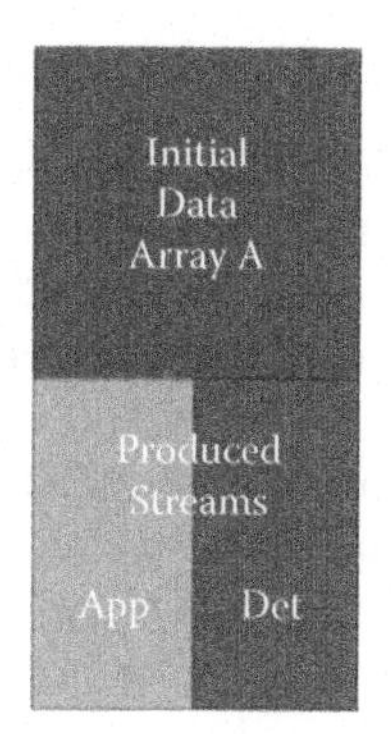

Figure 18.7 2D texture layout.

18.3.2.1 Mapping Streams to 2D Textures

In our programming model, the stream management is performed by allocating a single 2D texture, large enough to pack all the input and output data streams (not shown in Figure 18.8). Given that textures are made up of 4-channel elements, known as *texels*, different data-to-texel mappings are possible. The most profitable one depends on the operations being performed in the the kernel loops, since this mapping determines the following key performance factors:

- SIMD parallelism. As mentioned above, *fragment processors* usually have vector units that process the four elements of a *texel* in a SIMD fashion.
- Locality. Texel mapping influences the memory access behavior of the kernels since fetching one *texel* only requires one memory access.
- Automatic texture addressing. Texture mapping may also determine how texture coordinates (addresses) are computed. If the number of texture addresses needed by a kernel does not exceed the number of available hardware interpolators, memory address calculations can be accelerated by hardware.

For the DWT, a 2D layout is an efficient mapping, i.e., packing two elements from two consecutive rows of the original array into each *texel*. This layout permits all the memory (texture) address calculations to be performed by the hardware interpolators. Nevertheless, for the sake of simplicity we will consider a simpler texel mapping, in which each *texel* contains only one array element, in the rest of this section.

Apart from the texel mapping, we should also define the size and aspect ratio of the allocated 2D texture as well as the actual allocation of input and output arrays within this texture. For example, Figure 18.7 illustrates these decisions for our DWT implementations. We use a 2D texture twice as large as the original array. The initial data (array A in Listing 3) are allocated on the top half of this texture, whereas the bottom half will eventually contain the produced streams (the App and Det in Listing 3) .

18.3.2.2 Orchestrating Memory Transfers and Kernel Calls

With data streams mapped onto 2D textures, our programming model uses the GPU fragment processors to execute kernels (fragment shaders) over the stream elements. In an initialization phase, the main program uploads these *fragment shaders* into the graphics hardware. Later on, they are invoked on demand according to the application flow.[10]

To invoke a kernel, the size of the output stream must be defined. This definition is done by drawing a rectangle that covers the region of pixels matching the output stream. The *glVertex2f* OpenGL commands define the vertex coordinates of this rectangle, i.e., they delimit the output area, which is equivalent to specifying the kernel loop bounds. The vertex processor and the rasterizer transform the rectangle to a stream of fragments, which are then processed by the active fragment program.

Among the attributes of the generated fragment, we find hardware interpolated 2D texture coordinates, which are used as indexes to fetch the input data associated to that fragment.[11] To delimit those input areas, the *glMultiTexCoord2fvARB* OpenGL commands assign texture coordinates at each vertex of the quad. In our example, we have three equal-sized input areas, which partially overlap with each other, since we must fetch three different elements (Det[i][j], App[i][j] and App[i][j+1]) per output value.

In the example, both the input and output areas have the same size and aspect ratio, but they can be different. For instance, the FBS version takes advantage of the linear interpolation to perform downsampling by defining input areas twice as wide as the output one.

As mention above, there is a limit on the number of texture coordinates (per fragment) that can be hardware interpolated, which depends on the target GPU. As long as the number of input elements that we must fetch per output value is lower than this limit (as in the example), memory addresses are computed by hardware. Otherwise, texture coordinates must be computed explicitly on the fragment program.

Finally, synchronization between consumers and producers is performed using the OpenGL *glFinish()* command. This function does not return until the effects of all previously invoked GL commands are completed and it can be seen as a synchronization barrier. In the example, we need barriers between every lifting step.

18.3.3 GPGPU Framework

As shown in the previous section, in order to exploit the GPU following a stream programming model, we have to deal with the many abstraction layers that the system introduces to ease the access to graphics hardware in graphics applications. As we can observe in Figure 18.8, these layers do nothing but cloud the resources we want to use. Therefore, it is useful for us to build some abstraction layers that bring together our programming model and the graphics hardware, so we can move away

---

[10] In our example, *Active_fp(Alpha_fp)* enables the *Alpha_fp* fragment program. Kernels always operate on the active texture, which is selected by*Active_texture*.

[11] This operation is also known as texture lookup in graphics terminology.

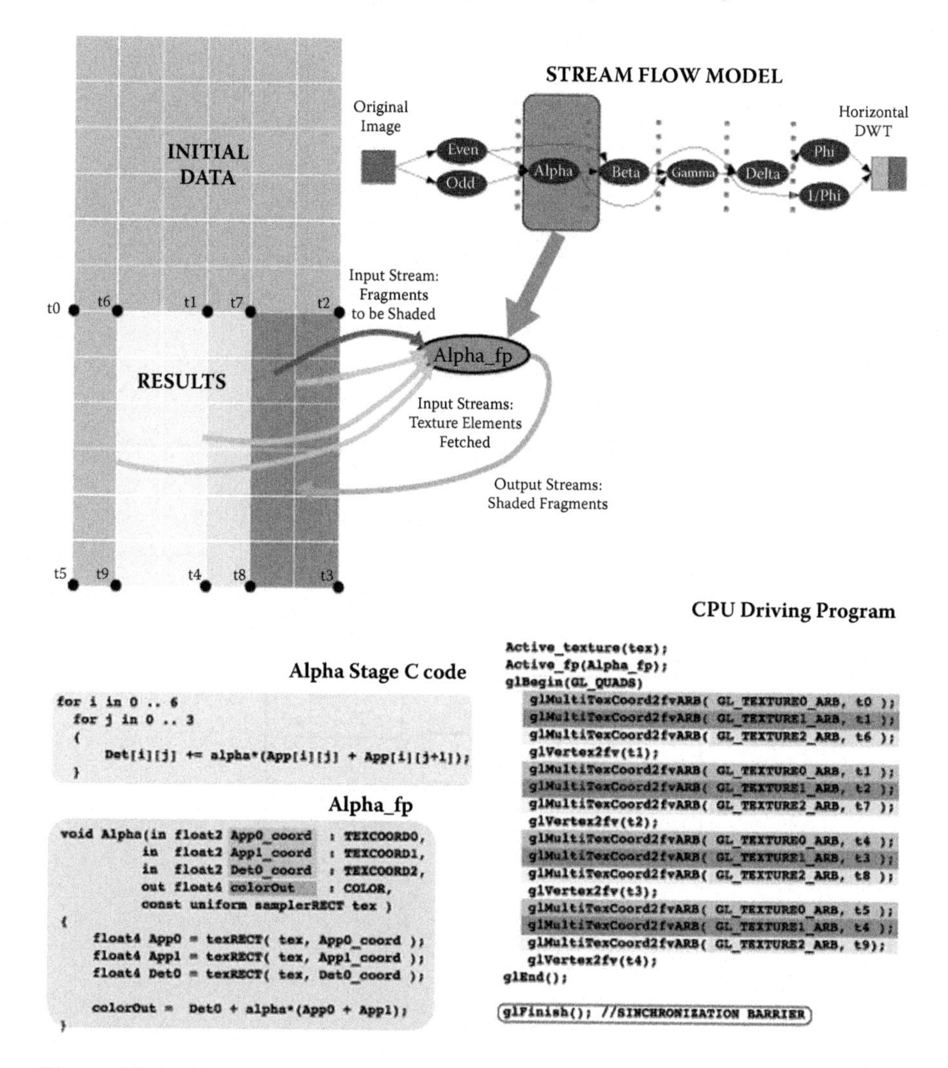

Figure 18.8 Mapping one lifting step onto the GPU.

the graphics API, worthless – even harmful – in our case. In this section, we present the API of the framework we have been using in our research to clarify how we can easily utilize a graphics card to implement the stream flow models developed for our target algorithms.

18.3.3.1 The Operating System and the Graphics Hardware

In an operating system, we find that access to the graphics hardware implies going through a series of graphics libraries and extensions to the windowing system. First of all, we have to install a driver for the graphics card, which exports an API to the windowing system. Then, the windowing system exports an extension for initializing

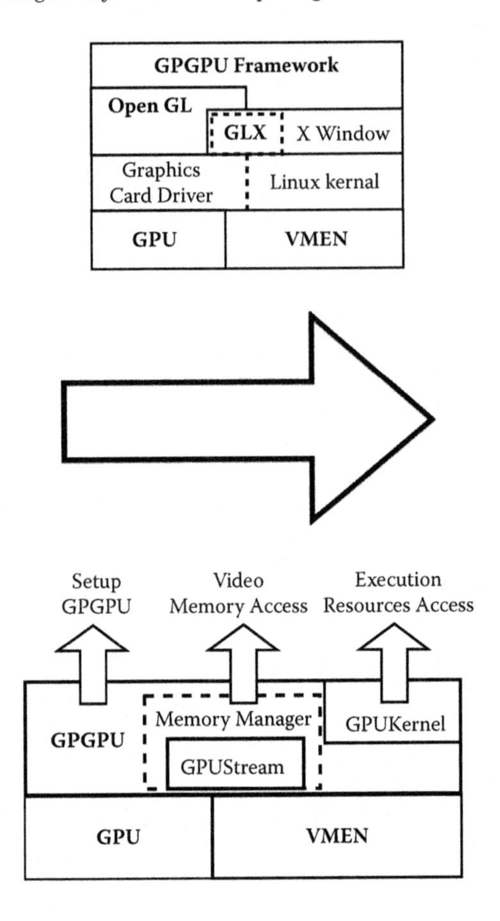

Figure 18.9 Implementation of the GPGPU Framework.

our graphics card, so it can be used through the common graphics libraries – like OpenGL or DirectX – that provide higher level primitives for 3D graphics applications.

In GNU/Linux (see Figure 18.9(a)), the driver is supplied by the graphics card's manufacturer, the windowing system, is the X Window System, and the graphics library is OpenGL, which can be initialized through the GLX extension. Our GPGPU framework hides the graphics-related complexity, i.e., the X Window System, the GLX, and the OpenGL library.

Figure 18.9(b) illustrates the software architecture of the GPGPU framework we implement. It consists of three classes: GPUStreams, GPUKernels, and a GPGPU static class for GPU and video memory managing.

The execution resources are handled through the GPUKernel class, which represents our execution kernels. We can control the GPGPU mode through the GPGPU class and transfer data to and from the video memory using the GPUStream class. This

TABLE 18.1 GPGPU Class Methods

GPGPU		
method	*input params*	*output params*
initialize	width, height, chunks	*(void)*
getGPUStream	width, height	GPUStream*
freeGPUStream	GPUStream*	*(void)*
waitForCompletion	*(void)*	*(void)*

way, we provide a stream model friendly API that allows us to directly implement the stream flow models of our target applications, avoiding most of the graphics issues.

18.3.3.2 The GPGPU Framework

The static class GPGPU allows us to initialize and finalize the GPGPU mode and wait for the GPU to complete any kernel execution or data transference in progress. In addition, it incorporates a memory manager to allocate and free video memory. Two possible implementations of the memory manager are possible: a multi-texture or a single-texture model. In the former, the memory manager allocates the different streams as different textures that provide a noticeable amount of streams (up to sixteen) that can be managed at a time in our kernels. In addition, all these textures can be independently accessed using the results from the eight hardware interpolators, i.e., a single coordinate references one element on each stream as the texture coordinates are shared among them.

On the other hand, the single-texture model allocates all the streams as different regions of a single 2D texture. This model limits the amount of memory that can be managed.[12] Furthermore, each hardware-interpolated texture coordinate can only be used to access one element in one stream, i.e., one element in the texture. However, it is always possible to compute extra coordinates in the *fragment shader*. Despite of all these limitations, a single-texture model has a definitive advantage: a unified address space. This allows us to identify a stream by its location in the unified address space and store this location as data in other streams, i.e., we can use pointers to streams. On the contrary, this is not possible in a multi-texture model since we cannot store a texture identifier as data. This limitation in the multi-texture model makes it difficult to dereferencing streams based on the output of previous kernels (it may be very inefficient or even impossible).

Because of the benefits of the single address space, we decided to implement the memory manager following the single-texture model, using a first-fit policy to allocate the streams on the texture. The interface of the GPGPU class is shown in Table 18.1.

[12]The maximum size for a texture that OpenGL allows us to allocate is 4096×4096 texels, so we are limited to 256MB of video memory ($4096 \times 4096 \times 4$(RGBA)$\times 4$(floating point elements)).

TABLE 18.2 GPUStream Class Methods

GPUStream		
method	***input params***	***output params***
writeAll	float*	*(void)*
write	x_0, y_0, width, height, float*	*(void)*
readAll	*(void)*	float*
read	x_0, y_0, width, height	float*
setValue	float	*(void)*
getSubStream	x_{off}, y_{off}, width, height	GPUStream*
getX/getY	*(void)*	int
getWidth/getHeight	*(void)*	int

Once the GPGPU mode has been set up, we have access to the video memory through *GPUStream* objects, whose methods are shown in Table 18.2. We can transfer data between the main memory and the GPUStreams by using the *read[All]* and *write[All]* methods. *getSubStream* allow, us to obtain a displaced reference to a stream: We can specify an off-set from the original position of the stream in memory (x_{off}, y_{off}), and the width and height of the new stream. This way, we can use different regions of the same stream as different streams, as the example in Figure 18.10 illustrates.

Our kernels, written in Cg [7], are managed in the application through the *GPUKernel* objects, whose methods are shown in Table 18.3. We can use these objects to run kernels on different streams, which are passed as parameters. Apart from streams, we can also pass constant arguments through the *set[Local}Named]Param* method. The *launch* method is a non-blocking version of *run*.

These three classes (*GPGPU*, *GPUKernel*, and *GPUStream*) abstract all the basic functionality required to map applications in the stream programming model to the GPU. For instance, Listing 6 shows the C++ code for the implementation of the algorithm in Figure 18.10, while Listings 7 and 8 show the Cg codes of the kernels used in this example.

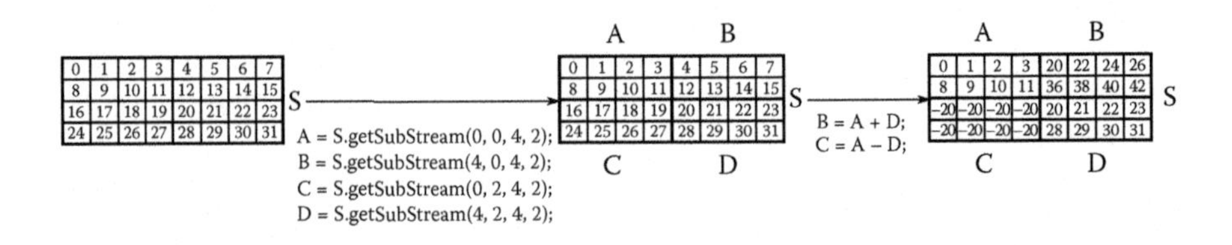

Figure 18.10 We allocate a stream S of dimension 8×4 and initialize its content to a sequence of numbers (from 0 to 31). Then, we ask four substreams dividing the original stream into four quadrants (A, B, C, and D). Finally, we add quadrants A and D and store the result in B, and we substract D from A and store the result in C.

TABLE 18.3 GPUKernel Class Methods

GPUKernel		
method	***input params***	***output params***
setNamedParam	char* *name, StreamElement*	*(void)*
setLocalParam	int *pos, StreamElement*	*(void)*
launch	GPUStream*, ...	GPUStream*
run	GPUStream*, ...	GPUStream*

Listing 6 C++ Main program for the example in Figure 18.10.

```
#include "GPGPU.h"

int main( )
{
        // Allocate enough video memory
        GPGPU::Initialize( 128, 128, 1 );

        // Allocate the main stream
        GPUStream* S = GPGPU::getGPUStream( 8, 4 );

        float data[32];
        for( int i = 0; i < 32; i++ ) data[i] = i;

        // Write the initial data to the stream
        S->writeAll( data );

        // Create 4 streams as references to four quadrants of S
        GPUStream* A = S->getSubStream( 0, 0, 4, 2 );
        GPUStream* B = S->getSubStream( 4, 0, 4, 2 );
        GPUStream* C = S->getSubStream( 0, 2, 4, 2 );
        GPUStream* D = S->getSubStream( 4, 2, 4, 2 );

        // Load kernels for addition and substraction
        GPUKernel* add( "add.cg" );
        GPUKernel* sub( "sub.cg" );

        // Run them in parallel
        add->launch( A, D, B ); // Asynchronous
        sub->run( A, D, C ); // Synchronous

        GPGPU::Finalize( );
}
```

Listing 7 Cg code for an addition kernel, which takes two streams and adds them.

```
void add(      in float2 s1_coord : TEXCOORD0,
               in float2 s2_coord : TEXCOORD1,
               out float s1_plus_s2 : COLOR,
               const uniform samplerRECT mem )
{
        // We dereference the corresponding stream elements
        float s1 = texRECT( mem, s1_coord );
        float s2 = texRECT( mem, s2_coord );
        // We add them and return the result
        s1_plus_s2 = s1 + s2;
}
```

Listing 8 Cg code for a substraction kernel, which takes two streams and substract them.

```
void sub(      in float2 s1_coord : TEXCOORD0,
               in float2 s2_coord : TEXCOORD1,
               out float s1_minus_s2 : COLOR,
               const uniform samplerRECT mem )
{
        // We dereference the corresponding stream elements
        float s1 = texRECT( mem, s1_coord );
        float s2 = texRECT( mem, s2_coord );
        // We substract s2 from s1 and return the result
        s1_minus_s2 = s1 - s2;
}
```

18.4 Automatic Morphological Endmember Extraction on GPUs

This section develops a GPU-based implementation of the Automatic Morphological Endmember Extraction (AMEE) algorithm following the design guidelines outlined above. First, we provide a high-level overview of the algorithm, and then we discuss the specific aspects about its implementation on a GPU.

18.4.1 AMEE

Let us denote by f a hyperspectral data set defined on an N-dimensional (N-D) space, where N is the number of channels or spectral bands. The main idea of the AMEE algorithm is to impose an ordering relation in terms of spectral purity in the set of pixel vectors lying within a spatial search window or structuring element around each image pixel vector [21]. To do so, we first define a cumulative distance between one particular pixel $\boldsymbol{f}(x, y)$, i.e., an N-D vector at discrete spatial coordinates (x, y), and all the pixel vectors in the spatial neighborhood given by B (B-neighborhood) as follows [18]:

$$D_B(\boldsymbol{f}(x, y)) = \sum_{(i,j)\in Z^2(B)} Dist(\boldsymbol{f}(x, y), \boldsymbol{f}(i, j)) \tag{18.1}$$

where (i, j) are the spatial coordinates in the B-neighborhood discrete domain, represented by $Z^2(B)$, and *Dist* is a pointwise distance measure between two N-D vectors. The choice of *Dist* is a key topic in the resulting ordering relation. The AMEE algorithm makes use of the spectral angle mapper (SAM), a standard measure in hyperspectral analysis [3]. For illustrative purposes, let us assume that $\boldsymbol{s}_i = (s_{i1}, s_{i2}, \ldots, s_{iN})^T$ and $\boldsymbol{s}_j = (s_{j1}, s_{j2}, \ldots, s_{jN})^T$ are two N-D signatures. Here, the term 'spectral signature' does not necessarily imply 'pixel vector' and hence spatial coordinates are omitted from the two signatures above, although the following argumentation would be the same if pixel vectors were considered. The SAM between $\boldsymbol{s}_i$ and $\boldsymbol{s}_j$ is given by

$$\text{SAM}(\boldsymbol{s}_i, \boldsymbol{s}_j) = cos^{-1}(\boldsymbol{s}_i \cdot \boldsymbol{s}_j / \|\boldsymbol{s}_i\| \cdot \|\boldsymbol{s}_j\|) \tag{18.2}$$

It should be noted that SAM is invariant in the multiplication of input vectors by constants and, consequently, is invariant to unknown multiplicative scalings that may arise due to differences in illumination and sensor observation angles.

In contrast, SID is based on the concept of divergence and measures the discrepancy of probabilistic behaviors between two spectral signatures. If we assume that $\boldsymbol{s}_i$ and $\boldsymbol{s}_j$ are nonnegative entries, then two probabilistic measures can be respectively defined as follows:

$$M[s_{ik}] = p_k = s_{ik} / \sum_{l=1}^{N} s_{il} \quad M[s_{jk}] = q_k = s_{jk} / \sum_{l=1}^{N} s_{jl} \tag{18.3}$$

Using the above definitions, the self-information provided by $\boldsymbol{s}_j$ for band l is given by $I_l(\boldsymbol{s}_j) = -log\ q_l$. We can further define the entropy of $\boldsymbol{s}_j$ with respect to $\boldsymbol{s}_i$ by

$$\begin{aligned} D(\boldsymbol{s}_i \| \boldsymbol{s}_j) &= \sum_{l=1}^{N} p_l \ D_l(\boldsymbol{s}_i \| \boldsymbol{s}_j) \\ &= \sum_{l=1}^{N} p_l (I_l(\boldsymbol{s}_j) - I_l(\boldsymbol{s}_i)) = \sum_{l=1}^{N} p_l \ log(p_l / q_l) \end{aligned} \tag{18.4}$$

By means of equation (18.4), SID is defined as follows:

$$\text{SID}(\boldsymbol{s}_i, \boldsymbol{s}_j) = D(\boldsymbol{s}_i \| \boldsymbol{s}_j) + D(\boldsymbol{s}_j \| \boldsymbol{s}_i) \tag{18.5}$$

With the above definitions in mind, we provide below a step-by-step description of the AMEE algorithm that corresponds to the implementation used in [19]. The inputs to the algorithm are a hyperspectral data cube $\boldsymbol{f}$, a structuring element B with size of $t \times t$ pixels, a maximum number of algorithm iterations I_{max}, and a maximum number of endmembers to be extracted p. The output is a set of endmembers $\{\boldsymbol{e}_i\}_{i=1}^{q}$, with $q \leq p$.

1. Set $i = 1$ and initialize a morphological eccentricity index score $\text{MEI}(x, y) = 0$ for each pixel.
2. Move B through all the pixels of $\boldsymbol{f}$, defining a local spatial search area around each pixel $\boldsymbol{f}(x, y)$, and calculate the maximum and minimum pixel vectors at each B-neighborhood using morphological erosion and dilation [21], respectively, as follows:

$$(\boldsymbol{f} \ominus B)(x, y) = argmin_{(i,j) \in Z^2(B)}\{D_B[\boldsymbol{f}(x+i, y+j)]\} \quad (18.6)$$

$$(\boldsymbol{f} \oplus B)(x, y) = argmax_{(i,j) \in Z^2(B)}\{D_B[\boldsymbol{f}(x+i, y+j)]\} \quad (18.7)$$

3. Update the MEI at each spatial location (x, y) using $MEI(x, y) = Dist[(\boldsymbol{f} \ominus B)(x, y), (\boldsymbol{f} \oplus B)(x, y)]$.
4. Set $i = i + 1$. If $i = I_{max}$, then go to step 5. Otherwise, set $\boldsymbol{f} = \boldsymbol{f} \oplus B$ and go to step 2.
5. Select the set of p pixel vectors with higher associated *MEI* scores (called endmember candidates) and form a unique spectral set of $\{\boldsymbol{e}_i\}_{i=1}^{q}$ pixels, with $q \leq p$, by calculating the *Dist* for all pixel vector pairs.

18.4.2 GPU-Based AMEE Implementation

This subsection describes how to implement in parallel the first four steps of the AMEE algorithm (finding the MEI score map) using a GPU architecture. It should be noted that these steps account for most of the execution time involved in the endmember extraction process.

The first issue that needs to be addressed is how to map a hyperspectral image onto the memory of the GPU. Since the size of hyperspectral images usually exceeds the capacity of such memory, we split them into multiple spatial partitions made up of entire pixel vectors (called *spatial regions* or SRs), i.e., each SR incorporates all the spectral information on a localized spatial region and is composed of spatially adjacent pixel vectors. As shown in Figure 18.11, SRs are further divided into 4-band tiles (called SR tiles), which are stored in different *GPUStreams* (Subsection 18.3.3). Using stream elements with four floating-point components each allows us to map four consecutive spectral bands onto the same *GPUStream* so we can make full use of the vector operations available in the GPU. Apart from the SR tiles, we also allocate additional *GPUStreams* to hold intermediate information such as inner products, norms, pointwise distances, and cumulative distances.

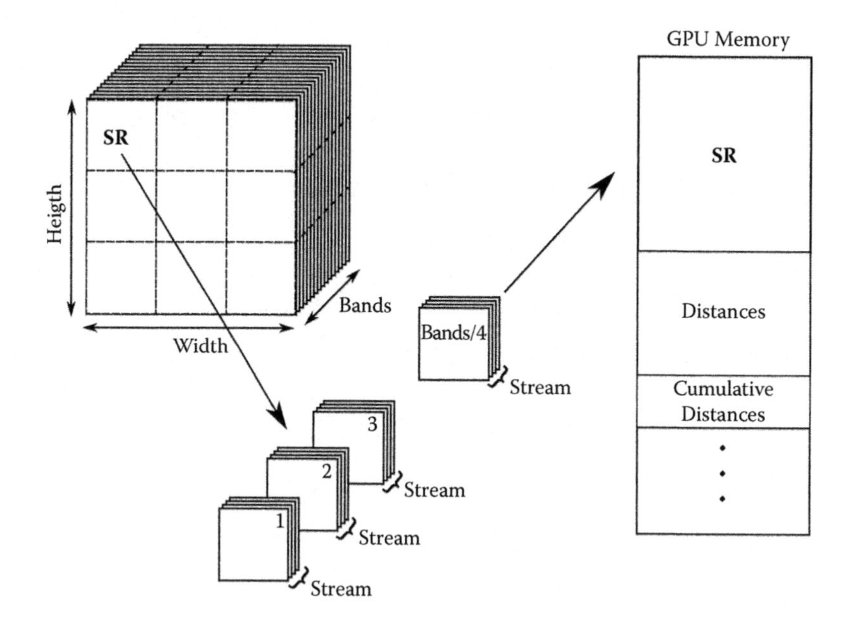

Figure 18.11 Mapping of a hyperspectral image onto the GPU memory.

Figure 18.12 shows a flowchart describing our GPU-based implementation of the AMEE algorithm using SAM as pointwise distance. The *stream uploading* stage performs the data partitioning and mapping operations described above, i.e., dividing the image into SRs and writing them as a set of SR tiles (*GPUStreams*). The remaining stages perform the actual computation and comprise the following stages:

1. *Inner products and norms.* The SR tiles stored in our *GPUStreams* are considered as input streams to this stage, which obtain all the inner products and norms necessary to compute the required point-wise distances. Keeping in mind that the size of the structuring element is $t \times t$ pixels, it will be necessary to compute $\frac{t^2(t^2+1)}{2}$ distances per pixel. However, taking advantage of the redundancy between adjacent structuring elements, it is possible to reduce this amount to $\lceil \frac{(2t-1)^2}{2} \rceil$. As shown in Figure 18.13, for $t = 3$ we only need to compute $12 + 1$ inner products per pixel: one product of the vector with itself (to find the norm) and twelve with the pixel vectors within its *region of influence* (RI).[13] Since *GPUStreams* are actually SR tiles, the implementation

[13]The region of influence (RI) of a pixel includes 4-connected neighbors to the pixel – the southwest (SW), south (S), southeast (SE), and east (E) neighbors – as well as their respective W, SW, S, SE, and E neighbors within the structuring element. It is worth noting that other alternative definitions for the RI are possible by adopting different connectivity criteria in the selection of neighbors, as far as the chosen RI contains a minimum set of neighbors that cover all the instances.

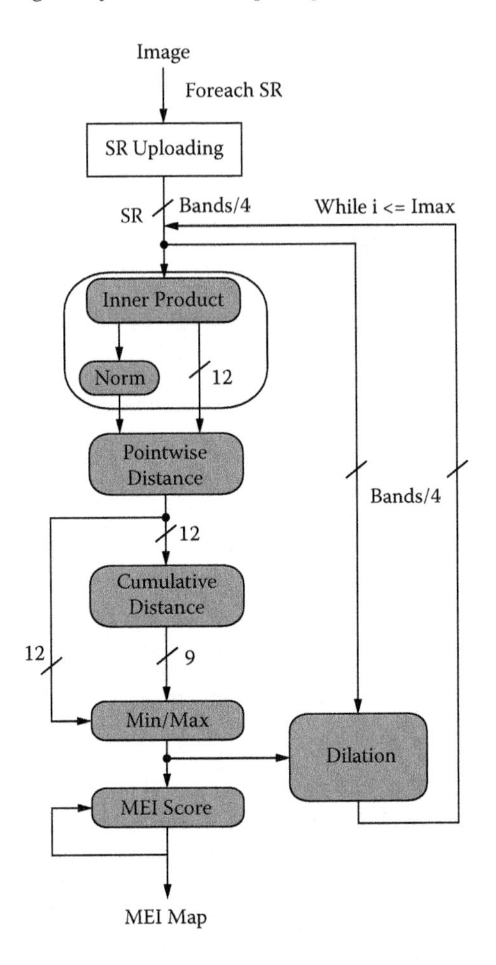

Figure 18.12 Flowchart of the proposed stream-based GPU implementation of the AMEE algorithm using SAM as pointwise distance.

of this stage is based on two *GPUKernels*, denoted as *multiply and add* (MAD) and *4-to-1* in Figure 18.13. The former is a multi-pass *GPUKernel* that implements an element-wise multiply and add operation iterating over each SR tile, thus producing four partial inner products stored in the four components of our *GPUStreams* (we can see an example of this *GPUKernel* in Figure 18.14). The latter is a single-pass *GPUKernel* that computes the final inner products, performing the sum reduction of these four-element vectors. A third single-pass *GPUKernel* produces the norm of every pixel vector.

2. *Pointwise distance*. For each pixel vector, this stage computes the SAM with all the neighbor pixels within its RI. It is based on a single-pass *GPUKernel* that computes the SAM between two pixel vectors using the inner products and norms produced by the previous stage.

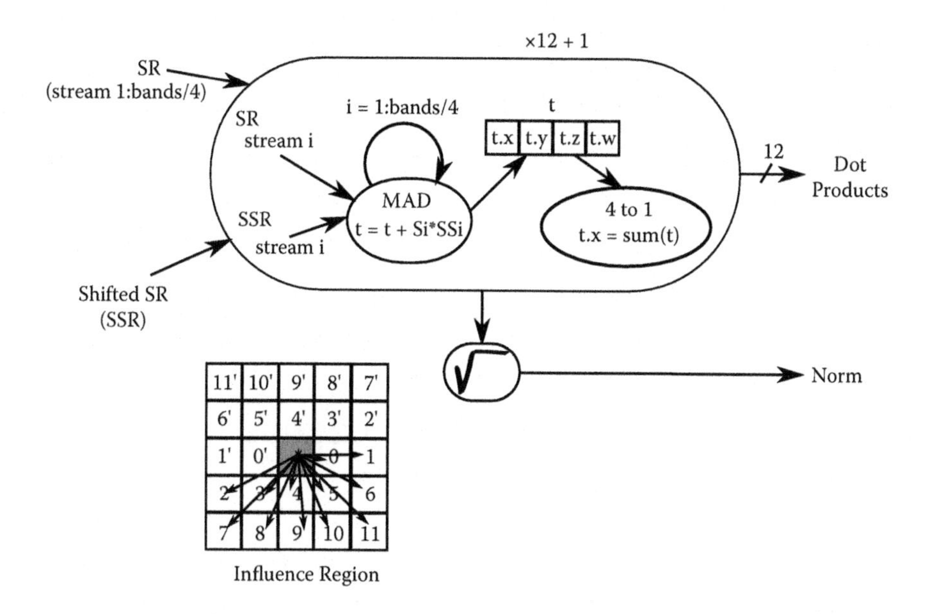

Figure 18.13 Kernels involved in the computation of the inner products/norms and definition of a region of influence (RI) for a given pixel defined by an SE with $t = 3$.

3. *Cumulative distance.* For each pixel vector, this stage produces t^2 cumulative distance streams for each of the t^2 neighbors defined by the structuring element. It is based on a single-pass *GPUKernel* that accumulates up to eight pointwise distances.[14]
4. *Maximum/minimum finding.* Erosion and dilation are finalized at this stage through a *GPUKernel* that applies minimum and maximum reductions. This *GPUKernel* uses as inputs the cumulative distances generated in the previous stage, and produces a *GPUStream* containing, for each pixel vector, the relative coordinates of the neighboring pixels with maximum and minimum cumulative distances.
5. *Dilation.* If $i < I_{max}$, this stage propagates the purest pixels in the current iteration to produce the SR tiles for the next iteration. It is based on a *GPUKernel* that takes as input the *GPUStream* containing the maximum and the *GPUStreams* containing the SR tiles, and produces a new set of *GPUStreams* that contain the SR tiles for the next iteration.
6. *MEI score update.* Finally, this stage updates the MEI scores using the maximum/minimum and point-wise distance *GPUStreams*. We can take advantage of the unified addressing space provided by our *GPGPU framework*, and use

[14]The number of texture interpolations that can be performed in a fragment program by hardware is limited to eight in our target GPUs

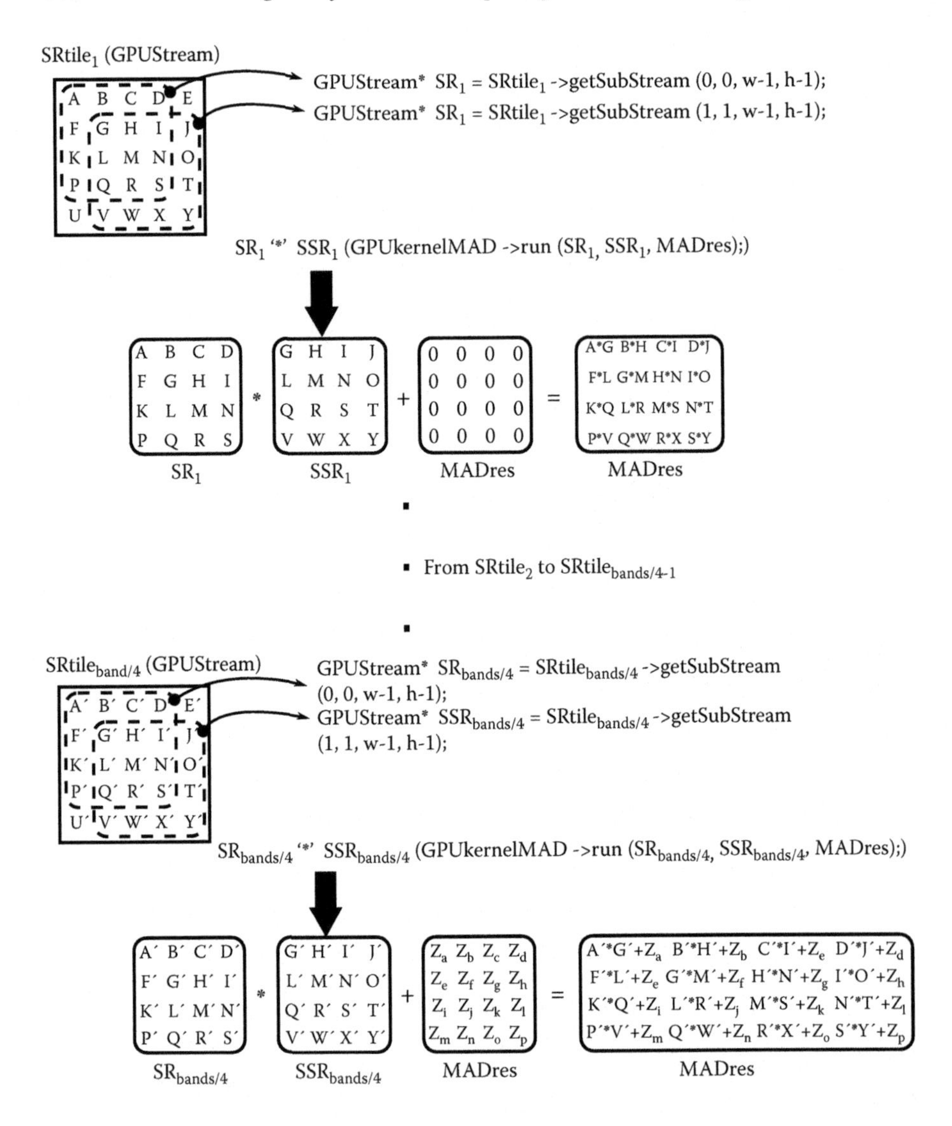

Figure 18.14 Computation of the partial inner products for distance 5: each pixel-vector with its south-east nearest neighbor. Notice that the elements in the *GPU-Streams* are four-element vectors, i.e., A, B, C ... contains four floating-point values each, and vector operations are element-wise.

the maximum and minimum values to index a reference table containing the corresponding distances, so we don't need to compute again the distance between the most similar and the most distinct pixel vector; we just have to reference the adecuate point-wise distance previously computed.

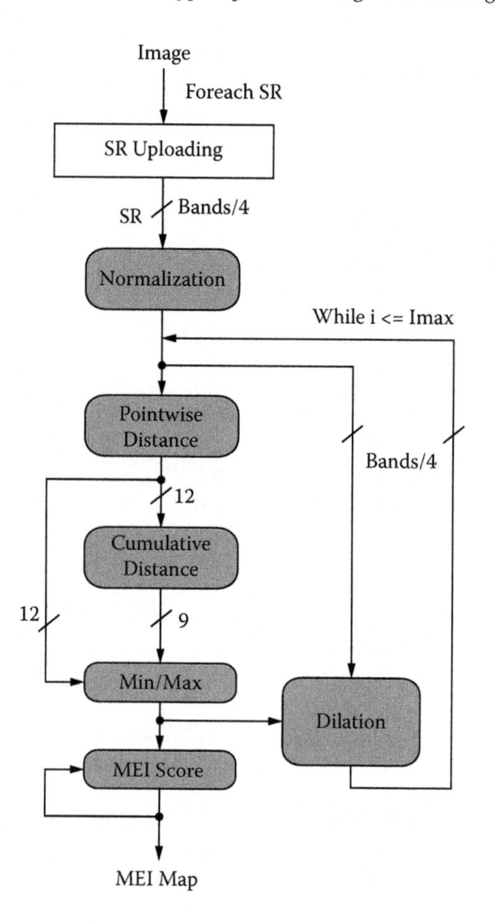

Figure 18.15 Flowchart of the proposed stream-based GPU implementation of the AMEE algorithm using SID as pointwise distance.

As shown in Figure 18.15, the stream-based implementation of the AMEE algorithm using SID as pointwise distance is similar. Basically, there is a pre-normalization stage, but the computation of the pointwise distance does not require intermediate inner products.

18.5 Experimental Results

18.5.1 GPU Architectures

The proposed endmember extraction algorithm has been implemented on a state-of-the-art GPU, as well as on an older (3-year-old) system in order to account for

TABLE 18.4 Experimental GPU Features

	F × 5950 Ultra	**7800 GTX**
Year	2003	2005
Architecture	NV38	G70
Bus	AGP × 8	PCI Express
Video Memory	256 MB	256 MB
Core Clock	475 MHz	430 MHz
Memory Clock	950 MHz	1.2 GHz GDDR3
Memory Interface	256-bit	256-bit
Memory bandwidth	30.4 GB/s	38.4 GB/s
#Pixel shader processors	4	24
Texture fill rate	3800 MTexels/s	10320 MTexels/s

TABLE 18.5 Experimental CPU Features

	Pentium 4 (Northwood C)	**Prescott (6 × 2)**
Year	2003	2005
FSB	800 MHz, 6.4 GB/s	800 MHz, 6.4 GB/s
ICache L1	12 KB	12 KB
DCache L1	8 KB	16 KB
L2 Cache	512 KB	2 M
Memory	1 GB	2 GB
Clock	2.8 GHz	3.4 GHz

potential improvements that might be achievable on future generations of GPUs. The generations considered correspond to the NV30 and G70 families (see Table 18.4), and the programs were coded using Cg [7]. For perspective, we have also reported performance results on contemporary Intel CPUs (see Table 18.5). The CPU implementations were developed using the Intel C/C++ compiler and optimized via compilation flags to exploit data locality and avoid redundant computations of common point-wise distances between adjacent structuring elements.

18.5.2 Hyperspectral Data

The hyperspectral data set used in the experiments is the well-known AVIRIS Cuprite scene, available online[15] (in reflectance units). It was collected by an AVIRIS flight over the Cuprite mining district in Nevada. Figure 18.16 shows a subscene of the full flightline, which is centered at a region with high mineral diversity. The full scene comprises a relatively large area, with 677 × 1939 pixels, spatial resolution of 20 meters, 204 narrow spectral bands between 0.4 and 2.5 μm, and nominal spectral resolution of 10 nm (for a total size of 512 MB). It should be noted that a total of 19 bands were removed prior to the analysis due to water absorption and low SNR in those bands. The site is well understood mineralogically and has several

[15]http://aviris.jpl.nasa.gov/html/aviris.freedata.html.

Figure 18.16 Subscene of the full AVIRIS hyperspectral data cube collected over the Cuprite mining district in Nevada.

exposed minerals of interest. The reflectance spectra of ten U.S. Geological Survey (USGS) ground mineral spectra: *alunite*, *buddingtonite*, *calcite*, *chlorite*, *kaolinite*, *jarosite*, *montmorillonite*, *muscovite*, *nontronite*, and *pyrophilite* (all available from http://speclab.cr.usgs.gov) were used as ground-truth spectra to illustrate endmember extraction accuracy. Figure 18.17 plots the spectra of the ten above-mentioned minerals of interest. Finally, in order to study the scalability of our CPU- and GPU-based implementations, we tested them on different image sizes, where the largest one corresponds to a 512 MB subscene while the others correspond to cropped portions of the same subscene.

18.5.3 Performance Evaluation

Before empirically investigating the performance of the proposed GPU-based implementation, we first briefly discuss endmember extraction accuracy of the proposed morphological method in comparison with other available approaches. Table 18.6 tabulates the SAM spectral similarity scores obtained after comparing USGS library spectra with the corresponding endmembers extracted by standard endmember extraction algorithms, including the PPI, N-FINDR, VCA and IEA (the smaller the scores across the ten minerals considered in Table 18.6, the better the results). On the other hand, Table 18.7 displays the spectral similarity scores achieved by the endmembers extracted by the proposed AMEE algorithm, using different numbers of iterations, ranging from $I_{max} = 1$ to $I_{max} = 5$, and a constant SE with $t = 3$. The number of endmembers to be extracted in all cases was set to 16 after calculating the intrinsic dimensionality of the data [3]. The value obtained relates very well

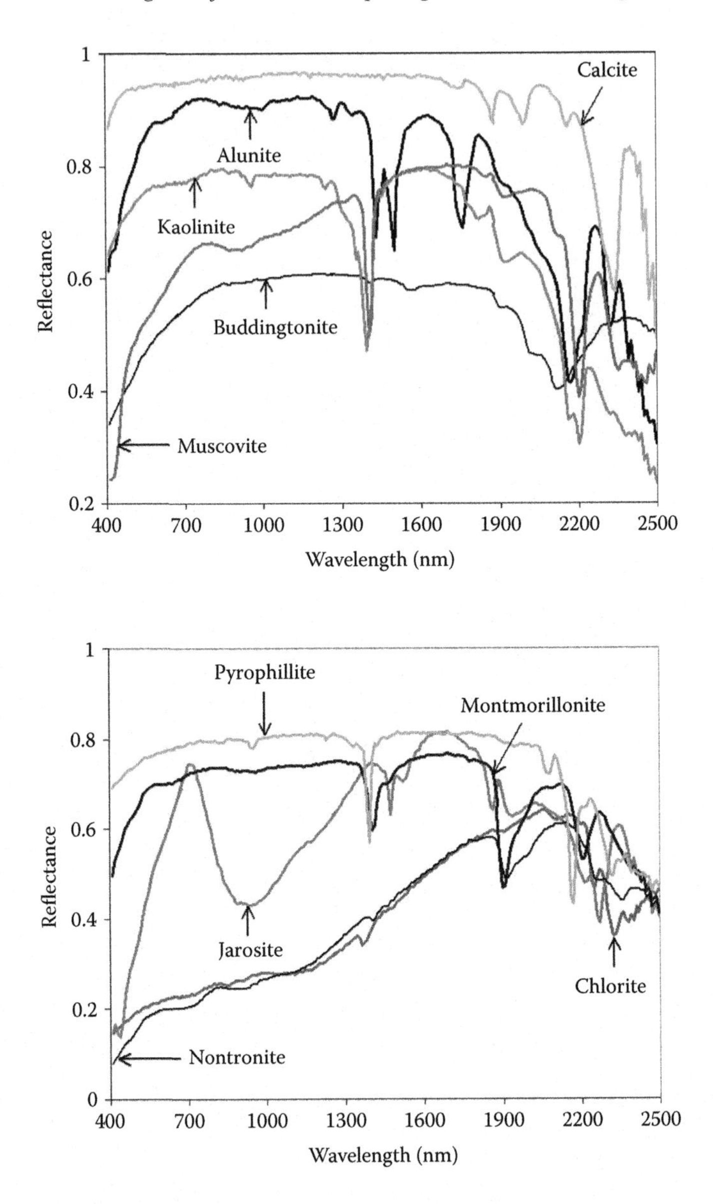

Figure 18.17 Ground USGS spectra for ten minerals of interest in the AVIRIS Cuprite scene.

to the ground-truth information available for the scene. The two above-mentioned tables reveal that the spatial/spectral AMEE algorithm was able to improve the results obtained by other methods which rely on using the spectral information alone, in particular, as the number of algorithm iterations was increased to account for richer spatial/spectral information.

TABLE 18.6 SAM-Based Spectral Similarity Scores Among USGS Mineral Spectra and Endmembers Produced by Different Algorithms

	PPI	N-FINDR	VCA	IEA
Alunite	0.084	0.081	0.084	0.084
Buddingtonite	0.106	0.084	0.112	0.094
Calcite	0.105	0.105	0.093	0.110
Chlorite	0.125	0.136	0.096	0.096
Kaolinite	0.136	0.152	0.134	0.134
Jarosite	0.112	0.102	0.112	0.108
Montmorillonite	0.106	0.089	0.120	0.096
Muscovite	0.108	0.094	0.105	0.106
Nontronite	0.102	0.099	0.099	0.099
Pyrophilite	0.094	0.090	0.112	0.090

TABLE 18.7 SAM-Based Spectral Similarity Scores Among USGS Mineral Spectra and Endmembers Produced by the AMEE Algorithm (Implemented Using Both SAM And SID, and Considering Different Numbers of Algorithm Iterations)

	AMEE (using SAM)			AMEE (using SID)		
	$I_{max} = 1$	$I_{max} = 3$	$I_{max} = 5$	$I_{max} = 1$	$I_{max} = 3$	$I_{max} = 5$
Alunite	0.084	0.081	0.079	0.081	0.081	0.079
Buddingtonite	0.112	0.086	0.081	0.103	0.084	0.082
Calcite	0.106	0.102	0.093	0.101	0.095	0.090
Chlorite	0.122	0.110	0.096	0.112	0.106	0.084
Kaolinite	0.136	0.136	0.106	0.136	0.136	0.102
Jarosite	0.115	0.103	0.094	0.108	0.103	0.094
Montmorillonite	0.108	0.105	0.101	0.102	0.099	0.092
Muscovite	0.109	0.099	0.092	0.109	0.095	0.078
Nontronite	0.101	0.095	0.090	0.101	0.092	0.085
Pyrophilite	0.098	0.092	0.079	0.095	0.086	0.071

TABLE 18.8 Execution Time (In Milliseconds) for the CPU Implementations

	AMEE-1 (SAM)		AMEE-2 (SID)	
Size (MB)	Pentium 4	Prescott	Pentium 4	Prescott
16	6588.76	4133.57	22369.8	16667.2
32	13200.3	8259.66	45928	33826.7
64	26405.6	16526.7	92566.6	68185
128	52991.8	33274.9	187760	137412
256	106287	66733.7	377530	277331
512	212738	133436	756982	557923

TABLE 18.9 Execution Time (In Milliseconds) for the GPU Implementations

	AMEE-1 (SAM)		AMEE-2 (SID)	
Size (MB)	**FX5950 U**	**7800 GLX**	**FX5950 U**	**7800 GLX**
16	1923.63	457.37	898.36	513
32	3909.91	905.93	1817.52	1034.42
64	7873.9	1781.58	3714.86	2035.01
128	15963.1	3573.3	7364.12	4144.82
256	31854.5	7311.05	14877.2	8299.07
512	63983.9	14616	29794.8	16692.2

Although the use of SAM or SID does not seem to play a very relevant role, the endmember spectra produced using SID show slightly better similarity to USGS reference spectra.

Table 18.9 shows the execution times of our CPU- and GPU-based AMEE implementations for different image sizes. First, it is worth noting that the GPU version is able to process the full 677 × 1939-pixel data cube extremely fast: For $I_{max} = 5$, the SAM (SID) version requires 14 (16) seconds, in spite of the overheads involved in data transfer between the main memory and the GPU. This confirms our intuition that GPUs are indeed suitable for spatial/spectral processing of hyperspectral data sets. Figures 18.18 and 18.19 further demonstrate that the complexity of the implementation scales linearly with the problem size, i.e., doubling the image size simply doubles the execution time.

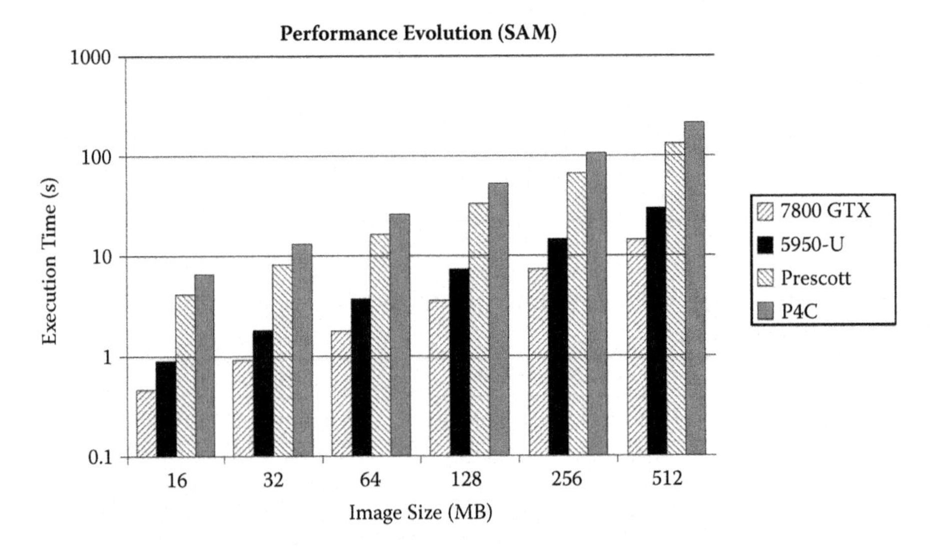

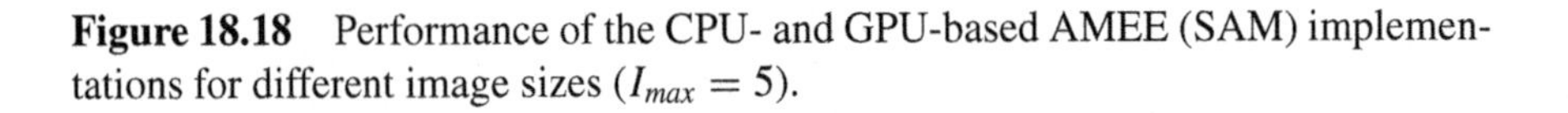
Figure 18.18 Performance of the CPU- and GPU-based AMEE (SAM) implementations for different image sizes ($I_{max} = 5$).

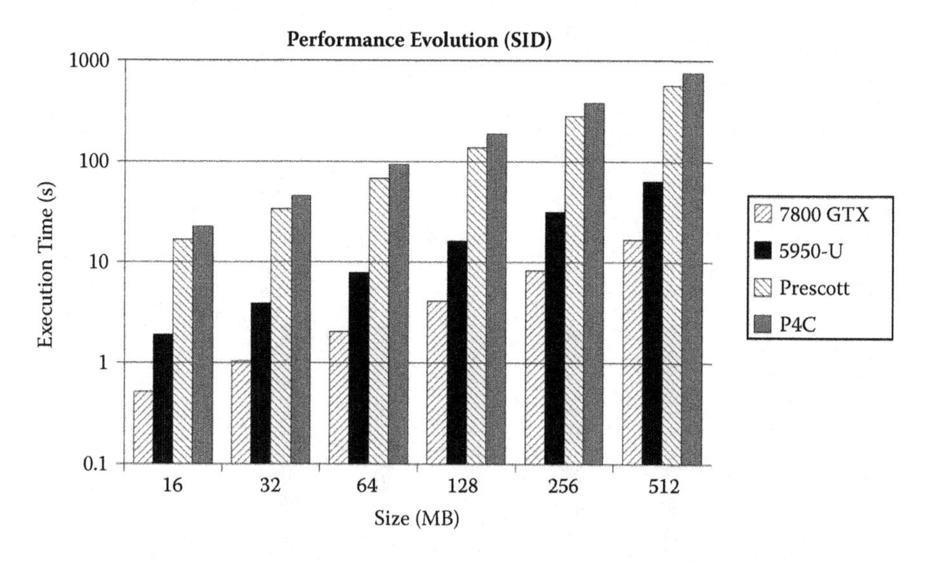

Figure 18.19 Performance of the CPU- and GPU-based AMEE (SID) implementations for different image sizes ($I_{max} = 5$).

As shown in Figure 18.20, the speedups achieved by the GPU implementation over their CPU counterparts are outstanding. As expected, these speedups grow with the number of iterations since the respective $transfer\_overhead$ to $GPU\_computation$ ratios improve. For the SAM version, they are up to 10. For the SID version, which

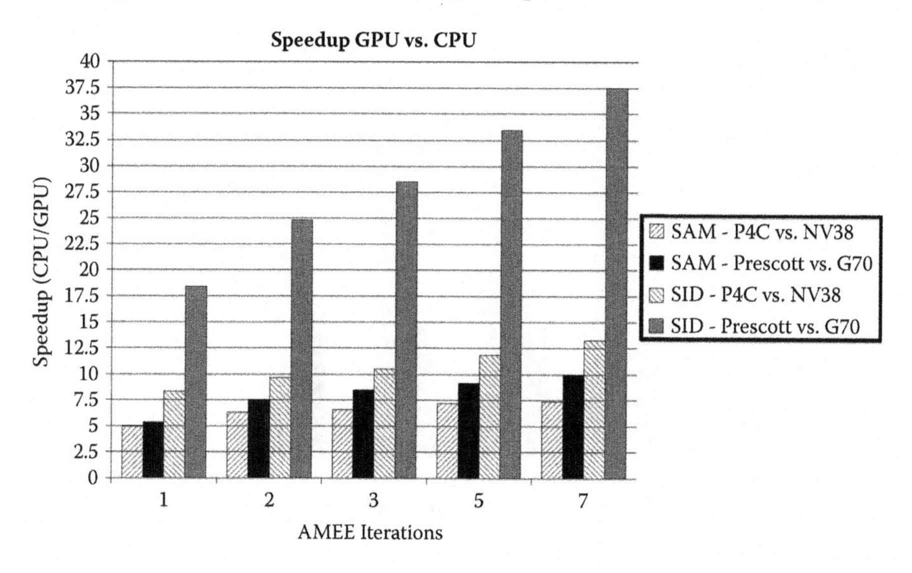

Figure 18.20 Speedups of the GPU-based AMEE implementations for different numbers of iterations.

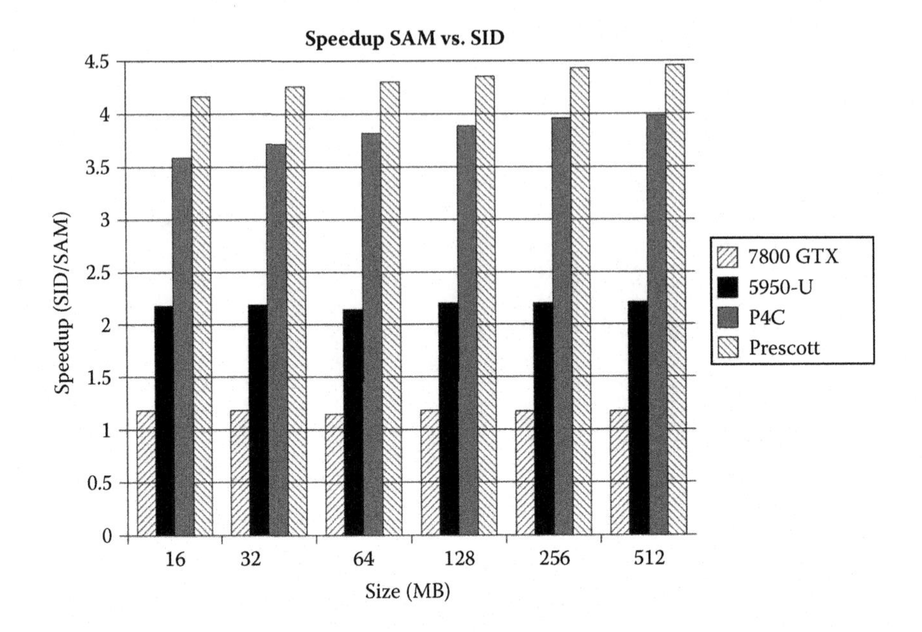

Figure 18.21 Speedup comparison between the two different implementations of AMEE (SID and SAM) in the different execution platforms ($I_{max} = 5$).

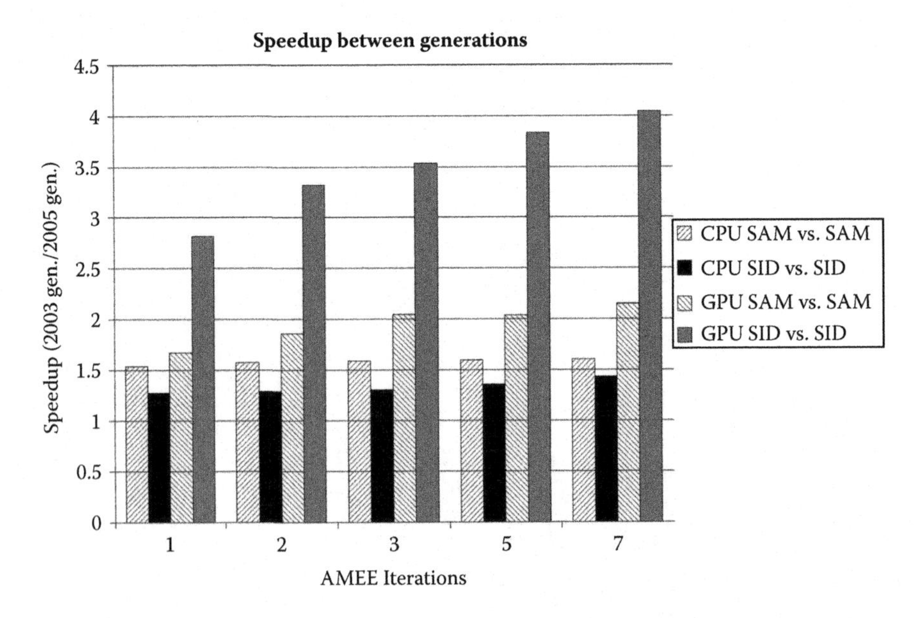

Figure 18.22 Speedup comparison between the two generations of CPUs, P4 Northwood (2003) and Prescott (2005), and the two generations of GPUs, 5950 Ultra (2003) and 7800 GTX (2005).

has a higher arithmetic intensity, it grows to an impressive $17 - 35\times$ factor. As a consequence of this behavior, the execution times of both versions become quite similar on the latest GPU (see Figure 18.21), despite SAM being around four times faster on the CPU. Although these figures are already remarkable, we should also highlight that multi-GPU systems or even clusters of GPUs [5] may significantly increase the reported performance gains.

Finally, we must also note the remarkable relative evolution of GPUs when compared to that of CPUs (see Figure 18.22). The performance gain caused by the evolution of Intel CPUs is significant by below 60% in both versions. However, the speedup factor observed as a result of the evolution of GPUs is around $2\times$ for SAM and up to $4\times$ for SID. This comes as no surprise, since the latest generation has multiplied by six the number of fragment processors as well as increased the onboard memory bandwidth.

18.6 Conclusions

In this chapter, we have explored the viability of using GPUs for efficiently implementing spatial/spectral endmember extraction algorithms. This approach represents a cost-effective alternative to other high-performance systems, such as Beowulf-type clusters, which are expensive and difficult to adapt to onboard processing scenarios. The outstanding speedups reported in experiments, together with the low cost and impressive evolution of GPUs, anticipate a significant impact of these hardware devices in the remote sensing community. In future developments, we will explore additional partitioning strategies to balance the workload between the CPU and the GPU. Further research will also include experiments with multi-GPU systems and clusters of GPUs, with the ultimate goal of adapting these commodity components to onboard hyperspectral data processing.

18.7 Acknowledgment

This work has been supported in part by the Spanish government through the research contract CYCIT-TIN 2005/5619.

References

[1] General Purpose Computations on GPU's. http://www.gpgpu.org.

[2] NVIDIA GeForce 8800 GPU Architecture Overview. Technical Brief TB-02787-001 v0.9, November 2006.

[3] C.-I. Chang. *Hyperspectral Imaging: Techniques for Spectral Detection and Classification*. New York, Kluwer, 2003.

[4] I. Daubechies and W. Sweldens. Factoring Wavelet Transforms into Lifting Steps. *J. Fourier Anal. Appl.*, vol. 4, pp. 245–267, 1998.

[5] Z. Fan, F. Qin, A. Kaufman, and S. Yoakum-Stover. GPU Cluster for High Performance Computing. *SC'04: Proceedings of the 2004 ACM/IEEE Conference on Supercomputing*, pp. 47, 2004.

[6] R. Fernando and M. J. Kilgard. *The Cg Tutorial: The Definitive Guide to Programmable Real-Time Graphics*. Addison-Wesley: Boston, 2003.

[7] R. Fernando and M. J. Kilgard. *The Cg Tutorial: The Definitive Guide to Programmable Real-Time Graphics*. Addison-Wesley Longman Publishing Co., Inc., Boston, MA, USA, 2003.

[8] J. Foley, A. van Dam, S. K. Feiner, and J. F. Hughes. *Computer Graphics. Principles and Practice*. 2nd Edition, Addison-Wesley, 1996.

[9] M. Gupta and N. Jacobson. Wavelet Principal Component Analysis and Its Application to Hyperspectral Images. *International Conference on Image Processing*, October 2006.

[10] Z. S. Hakura and A. Gupta. The Design and Analysis of a Cache Architecture for Texture Mapping. *ISCA '97: Proceedings of the 24th Annual International Symposium on Computer Architecture*, pp. 108–120, New York, NY, USA, 1997.

[11] T. R. Halfhill. Number Crunching with GPUs. In-Stat Microprocessor Report, (10/2/06-01), October 2006.

[12] L. Jimenez, J. Rivera-Medina, E. Rodriguez-Diaz, E. Arzuaga-Cruz, and M. Ramirez-Velez. Integration of Spatial and Spectral Information by Means of Unsupervised Extraction and Classification of Homogeneous Objects Applied to Multispectral and Hyperspectral Data. *IEEE Transactions on Geoscience and Remote Sensing*, vol. 43, pp. 844–851, 2005.

[13] A Comparison of Hardware Implementations of the Biorthogonal 9/7 DWT: Convolution versus Lifting. *IEEE Transactions on Circuits and Systems II: Express Briefs*, pp. 256–260, 2005.

[14] K. Kennedy and J. R. Allen. *Optimizing Compilers for Modern Architectures: A Dependence-Based Approach*. Morgan Kaufmann Publishers Inc., San Francisco, CA, 2002.

[15] H. Li, B. S. Manjunath, and S. K. Mitra. Multisensor Image Fusion Using the Wavelet Transform. *Graph. Models Image Process.*, vol. 57, pp. 235–245, 1995.

[16] J. Montrym and H. Moreton. The GeForce 6800. *IEEE Micro Magazine*, vol. 25, pp. 41–51, 2005.

[17] H. Okumura, M. Suezaki, H. Sueyasu, and K. Arai. Automated Corresponding Point Candidate Selection for Image Registration Using Wavelet Transformation Neural Network with Rotation Invariant Inputs and Context Information about Neighboring Candidates. *Proceedings of the SPIE*, vol 4885, pp. 401–410, 2003.

[18] A. Plaza, P. Martinez, R. Perez, and J. Plaza. Spatial/Spectral Endmember Extraction by Multidimensional Morphological Operations. *IEEE Transactions on Geoscience Remote Sensing*, vol. 40, pp. 2025–2041, 2002.

[19] A. Plaza, D. Valencia, J. Plaza, and C.-I. Chang. Parallel Implementation of Endmember Extraction Algorithms from Hyperspectral Imagery. *IEEE Geoscience and Remote Sensing Letters*, vol. 3, pp. 334–338, 2006.

[20] A. Plaza, D. Valencia, J. Plaza, and P. Martinez. Commodity Cluster- Based Parallel Processing of Hyperspectral Imagery. *Journal of Parallel and Distributed Computing*, vol. 66, pp. 345–358, 2006.

[21] P. Soille. *Morphological Image Analysis: Principles and Applications*, 2nd ed. Springer, Berlin, 2003.

[22] J. Tilton. Analysis of Hierarchically Related Image Segmentations. *Proceedings of the IEEE Workshop on Advances in Techniques for Analysis of Remotely Sensed Data*, pp. 27–28, 2003.

[23] D. Valencia and A. Plaza. FPGA-Based Compression of Hyperspectral Imagery Using Spectral Unmixing and the Pixel Purity Index Algorithm. *Lect. Notes Comput. Sci.*, vol. 3991, pp. 888–891, 2006.

[24] Z. Zhang and R. Blum. Multisensor Image Fusion Using a Region-Based Wavelet Transform Approach. *DARPA IUW*, pp. 1447–1451, 1997.

Index

D

E

F

H

T

U

V

W

X

Y